Orthopedic Biomaterials

Bingyun Li • Thomas Webster

Editors

Orthopedic Biomaterials

Progress in Biology, Manufacturing, and Industry Perspectives

 Springer

Editors
Bingyun Li
Department of Orthopedics
School of Medicine
West Virginia University
Morgantown, WV, USA

Thomas Webster
Department of Chemical Engineering
Northeastern University
Boston, MA, USA

ISBN 978-3-030-07798-3   ISBN 978-3-319-89542-0   (eBook)
https://doi.org/10.1007/978-3-319-89542-0

This Springer imprint is published by the registered company Springer Nature Switzerland AG
The registered company address is: Gewerbestrasse 11, 6330 Cham, Switzerland

Preface

It is clear that advances in orthopedic biomaterials need strong input from clinicians as well as those in industry and academia. Neither can operate in a vacuum to provide solutions that will help bone health. This book provides just that. We have chapters from clinicians, industry and academia to discuss Orthopedic Biomaterials: Progress in Biology, Manufacturing, and Industry Perspectives.

Importantly, while many biomaterial textbooks have omitted the merging of these key constituencies, they have also neglected to emphasize the role that manufacturing plays in biomaterial properties—and this could not be any more important today when we think of soft biomaterials, nanofabrication, new polymers, self-assembled materials, and biologics to name a few. Think of all of the clinical input and academic research that goes into generating the next generation of biomaterials only to be lost due to poor manufacturing processes? Or, the development of an elegant nanofabrication process to modify bone screws that can not be implemented clinically due to cost or manufacturing constraints? Or, a new bone tissue engineering material that can be manufactured, but not implanted since clinicians do not have the necessary tools to do so? Or, what about, biologics that can not be purified during manufacturing without altering their attractive bone growth properties?

All of these issues are so intertwined, yet, unfortunately, many are only realized at the end when researchers try to commercialize technologies. And then follows frustration, wasted resources, and lost technologies. This book puts such issues front and center and in doing so, encourages us all to think about them before one single experiment is conducted. Only then, will we satisfy our growing needs for improved orthopedic biomaterials.

So, please enjoy this book and think of your own approaches that need to integrate biology and manufacturing from an industry perspective before you conduct your first experiment! We will all be better for it and, yes, then the train does not need to stop!

Morgantown, WV, USA   Bingyun Li
Boston, MA, USA   Thomas J. Webster

Contents

Part I
Design, Manufacturing, Assessment, and Applications

Nanotechnology for Orthopedic Applications: From Manufacturing Processes to Clinical Applications

Dan Hickey and Thomas Webster

Keywords Bone · Spine · Nanomedicine · Growth · Inflammation · Infection · Nanotextured · Nano-topography · Fibroblasts

1 Introduction

This chapter covers the integration of artificial materials into natural tissues of the human body, particularly bone, and what can be achieved through a couple of key nano-manufacturing techniques (such as shot peening and electrophoretic deposition). To achieve proper mechanical anchorage and integration, orthopedic implanted materials should resemble the tissues they are replacing as much as possible. Thus, provided here is an overview of the structure and function of bone tissues, as well as a review of the concepts and methods used by other researchers attempting to regenerate orthopedic tissues, with a focus on nanotechnology.

2 The Extracellular Matrix (ECM)

The extracellular matrix (ECM) is the principle extracellular component of all tissues and organs. It serves as a shelter to house cells, and it relays important chemical and mechanical signals between cells and their environment. The ECM is a structural material which controls the spatial organization of tissues across all length scales, connecting nano-scale features to the larger-scale organization that controls cell positioning and motility. The ECM also mediates cellular attachment, proliferation, and differentiation through a multitude of proteins, growth factors, and cytokines.

D. Hickey · T. Webster (✉)
Department of Chemical Engineering, Northeastern University, Boston, MA, USA
e-mail: th.webster@neu.edu

© Springer International Publishing AG, part of Springer Nature 2018
B. Li, T. Webster (eds.), *Orthopedic Biomaterials*,
https://doi.org/10.1007/978-3-319-89542-0_1

The ECM is a constantly evolving system which reacts to chemical stimuli as well as external stresses. For example, Magnusson *et al.* showed that acute exercise in humans was followed by an increase in both the synthesis and degradation of collagen in tendons [1]. Thus, the ECM changes with time, and its composition and organization can vary significantly with location within the body (e.g., bone ECM vs. heart ECM) [2, 3]. Therefore, from a tissue engineering standpoint, it is important to fabricate tissue scaffolds with the capacity to develop the appropriate signaling pathways that will allow cells within the scaffold to shape the synthesis of new tissue.

2.1 ECM Composition

The ECM is composed primarily of glycoproteins, such as collagen, as well as proteoglycans, elastins, and glycosaminoglycans such as hyaluronic acid. These structural proteins exist as nanofibers which range in diameter from 50 to 500 nm [4]. The randomly oriented ECM nanofibers of cartilage, formed by chondrocytes, are displayed in Fig. 1 [5]. In order to accommodate cell adhesion and migration, as well as the ingrowth of vessels and other biomolecules, the ECM is quite porous. However, mature tissues in the body undergo decreased remodeling and do not typically exhibit the extensive porosity that is required for a tissue scaffold to generate neotissue [6].

2.2 The ECM as a Molecular Reservoir

In addition to its structural functionality, the ECM also acts as a reservoir for growth factors, cytokines, matrix-degrading enzymes, and their inhibitors. The ECM protects these molecules from degradation, and it also presents them more efficiently to

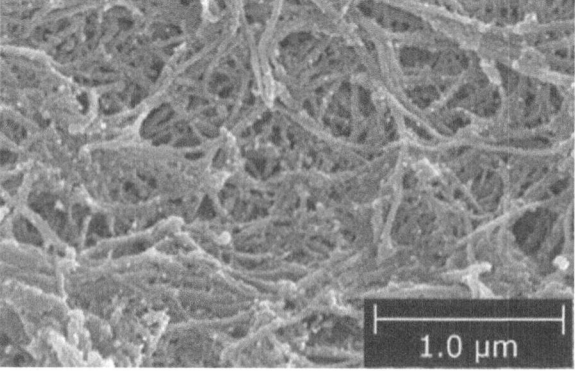

Fig. 1 Scanning electron micrograph exhibiting the nanofibrous morphology of the self-generated native environment of chondrocytes, containing a structural matrix of proteins linked to polysaccharides consisting of glycosaminoglycan units [5]

their receptors, affecting their synthesis [7]. In a reciprocal relationship, these growth factors and cytokines can signal cells to alter the production of ECM molecules, their receptors, and/or their inhibitors [8, 9]. The localization of growth factors and cytokines to particular tissues contributes to the variations in ECM composition and organization observed for different locations in the body [10].

2.3 Cell-ECM Interactions

Cells and their ECM are in constant communication with each other. Morphological and functional relationships are established between basal membranes, ECM, pericellular matrix, and cytoskeleton through cell surface receptors, including integrins, immunoglobulins, and selectins. The majority of these receptors, such as stretch-sensitive ion channels and G-protein coupled receptors respond to chemical cues, whereas only the integrins and cadherins appear to be capable of responding to mechanical stimuli [11]. Figure 2 shows how integrin molecules span the cellular membrane to relay signals between the cytoplasm and the ECM. The smallest

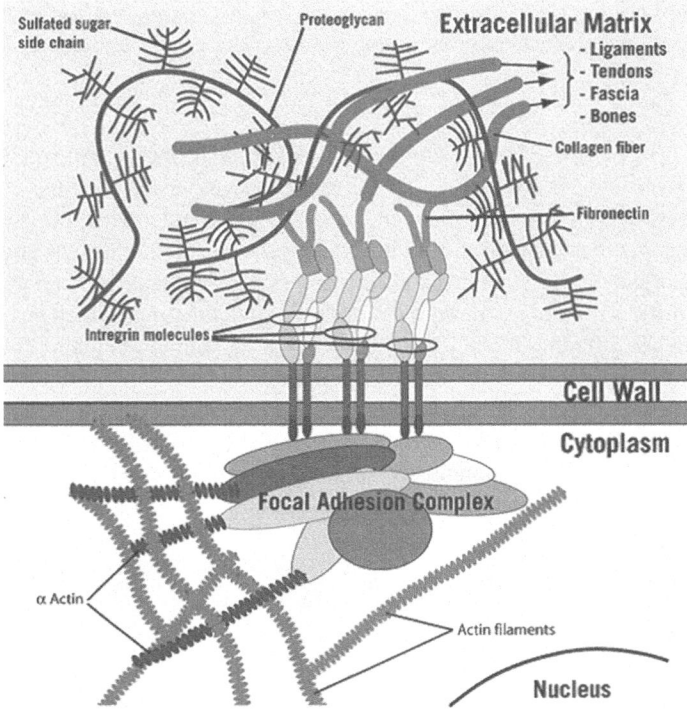

Fig. 2 Interaction of a cell with the ECM. Integrin receptors span the cellular membrane to transmit signals between extracellular proteins and the actin filaments of the cytoplasm [15]

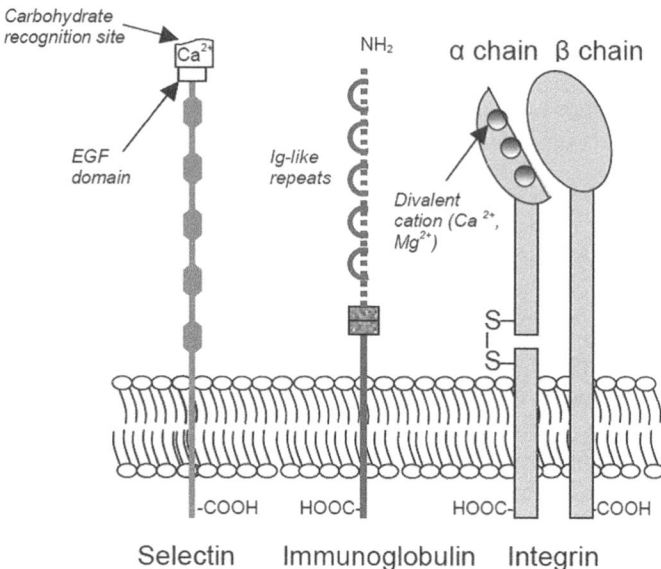

Fig. 3 Interactions of cells with the ECM is mediated through ligand binding to cell surface receptors such as selectins, immunoglobulins, and integrins. Divalent cations (Ca^{2+} and Mg^{2+}) attach to α chain sites to activate conformational activation of integrins for ligand binding [10]

collagen fibers of the ECM attach to proteoglycans and fibronectin, which interact directly with integrin molecules. Within the cell cytoplasm, integrins connect to focal adhesion complexes which elicit contractile responses in actin filaments. The actin filaments then transmit forces throughout the cell to the nucleus and to other signaling organelles [12]. In this way, integrin mediated signaling from the ECM is known to induce changes in cell morphology and function, which may lead to growth and/or differentiation [13, 14]. This signaling pathway explains how our tissues respond to growth, work, play, and injuries [11].

Integrins were the first cell surface receptors to be discovered, and they have since been studied most extensively [15]. These receptors are heterodimeric transmembrane glycoproteins composed of α and β subunits. To properly attach ligands, integrins rely on a population of α chain sites with Ca^{2+} and Mg^{2+} cations, which stimulate the integrin into a conformation that is accepting of ligand binding (Fig. 3) [10]. So far, at least 18 α and 8 β subunits have been identified, and they pair with each other in a variety of combinations, giving rise to a large family of receptors that recognize very specific peptide sequences of ECM molecules. Some combinations bind only very specific peptide sequences, while others bind to several different epitopes, providing tissues with built-in plasticity and redundancy [16].

2.4 Bone

This research focuses on the integration of biomaterials with bone, and as such, an understanding of the structure and composition of bone is essential. Bone is a nano-composite material comprised of a porous organic matrix with a high degree of interstitial mineralization. This composite internal organization gives bone important physical properties which allow it to function properly within the skeletal structure – the hard mineral constituents provide bone with high compressive strength, and the pliable organic matrix allow bone to bend and resist fracture from lateral stresses. At the macroscale, bone can be divided into two primary types: cortical bone and cancellous bone. Examining a cross-section, long bones such as the femur exhibit a dense cortical shell surrounding a porous, cancellous interior.

2.4.1 Cortical Bone

Cortical (or compact) bone is more mature than cancellous bone, and it contains a highly ordered microstructure, which is easily distinguished from that of cancellous bone through histological evaluation [17]. In cortical bone, at the nanoscale, aligned triple helix collagen molecules are infiltrated and surrounded by mineral (bone apatite). Mature apatite crystals are plate-shaped (not needle-shaped) with average dimensions of $50 \times 25 \times 3$ nm, and a composition very similar to that of hydroxyapatite (HA) [18]. The longest dimension of bone mineral corresponds to the axis along which collagen molecules align. Collagen molecules, separated in a regular fashion by apatite crystals, bundle into collagen fibrils, which bundle further to form collagen fibers approximately 150–250 nm in diameter. Continuing up the hierarchical structure of cortical bone, these mineralized collagen fibers form into planar arrangements called lamellae (3–7 μm wide), which often wrap in concentric circles to form an osteon, or Haversian system. Osteons (cylinders roughly 200 μm in diameter) group together running roughly parallel to the long axis of the bone. The hierarchical architecture of cortical bone described above is illustrated in Fig. 4. Some differing forms of cortical bone do not form osteons; however, the basic mineralized collagen fiber structure remains constant.

2.4.2 Cancellous Bone

Cancellous (or trabecular) bone is generally more metabolically active than cortical bone. It is remodeled more often, and is therefore 'younger' on average than cortical bone [17]. Cancellous bone is often called 'spongy' bone because it is highly porous, allowing bone cells to migrate and proliferate more freely. Because most tissue scaffolds are made to degrade over time and be replaced by native tissue, it is most often the structure of cancellous bone that must be created to achieve proper bone regeneration. Therefore, tissue scaffolds for bone tissue grafts do not need to

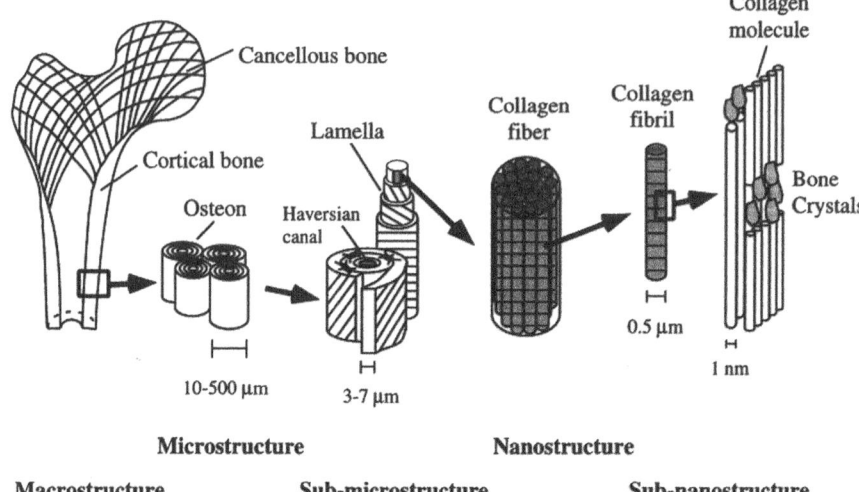

Fig. 4 Hierarchical architecture of cortical bone. At the nanoscale, bone apatite crystals fit into interstitial spaces within individual collagen molecules which bundle into the larger micron-scale structures that give cortical bone its characteristic strength [17]

possess the strength of cortical bone, which is very high (the Young's modulus is around 17 GPa). Instead, bone tissue scaffolds can be made to have an elastic modulus matching that of cancellous bone, which ranges between 0.1 and 4.5 GPa [19].

3 Tissue Engineering

The field of tissue engineering emerged from the clinical setting in the late 1980s as a new approach to repair or replace damaged or degrading tissues [20]. Building upon the basic principles of *in vitro* cell culturing techniques, Robert Langer, Joseph Vacanti, and others showed that a biocompatible "scaffold" could be seeded with cells and implanted into the affected region to enhance natural tissue regeneration (Fig. 5) [21–23]. The scaffold would then be expected to degrade into biocompatible side products as it is replaced by native tissue. The first successful application of this technology was seen for skin grafts, but today more complicated tissues, and even whole organs, are being fabricated to treat a variety of ailments in human patients [21, 23].

3.1 Nanotechnology for Tissue Engineering

The progression of scaffolds for tissue engineering has been aided by a number of technological advancements, including imaging techniques that allow researchers to observe smaller systems, and more sensitive biological assays that are capable of

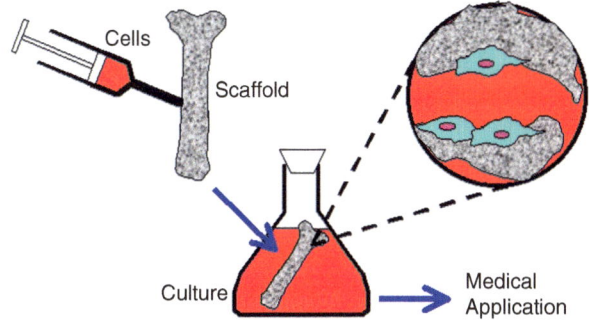

Fig. 5 Diagram of the basic tissue engineering process. A biocompatible scaffold is seeded with cells and cultured in appropriate media until it is ready to be used in the body [24]

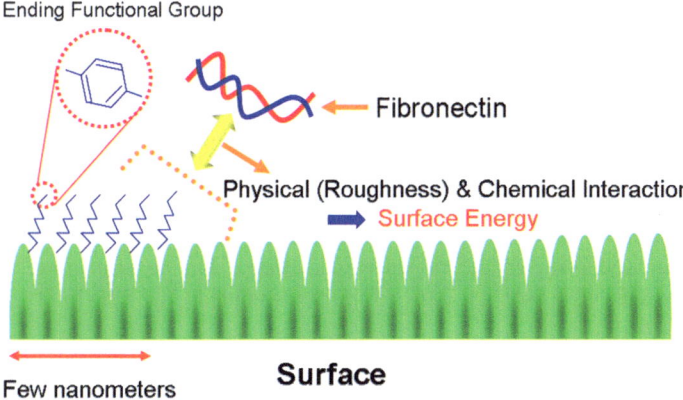

Fig. 6 Nanoscale surface roughness enhances surface energy and resembles structures found in the body, and can lead to the increased adsorption of cell surface proteins [25]

elucidating biological processes at a protein-based size scale. Additionally, the development of new material fabrication techniques has provided researchers with the ability to control material properties on an atom-by-atom basis. In general, these advancements can be catalogued under the umbrella of nanotechnologies.

Broadly defined, nanotechnology is "the use of materials whose components exhibit significantly changed properties by gaining control of structures at the atomic, molecular, and supramolecular levels" [25]. For instance, enhanced surface energy on nanorough surfaces can lead to increased and selective adsorption of proteins important to cell adhesion and functions, such as fibronectin (Fig. 6) [26]. Applied to tissue engineering, nanotechnology has aided in the development of tissue scaffolds which more closely resemble the native extracellular matrix (ECM)—the structural support system for all cells in the body. Tissue engineers aim to mimic the ECM in order to provide cells with an environment which closely resembles the tissue they will be implanted into. Therefore, the body serves as a model for the fabrication of artificial tissue scaffolds, and because events and interactions within the body occur in the nanoscale, the most fruitful control over the interactions of engineered scaffolds within the body can be achieved by application of nanotechnology.

3.2   Control of Cell Functions Using Nanotechnology

The introduction of nanofeatures into the cellular environment has been shown to improve cell functions. For instance, attachment proteins such as fibronectin, vitronectin, and laminin were found to absorb to nanofibrous tissue scaffolds 2.6–3.9 times more than solid-walled scaffolds [27]. Additionally, the adsorption of these proteins was found to be selective, indicating that variations in the nanostructure of a substrate are important for regulating cellular functions, as well as the signaling that causes formation of different tissues at different locations within the body.

Although three-dimensional scaffolds are required for *in vivo* applications, studies employing two-dimensional (2D) topographies offer a great deal of information on controlling cell fate and morphology. A number of 2D nanopatterned geometries aimed at mimicking the length scales of native ECM have been tested, including nanogrooves, nanoposts, and nanopit arrays [28]. These substrate topographies have been shown to influence a variety of cellular processes, including changes in shape, differentiation, and adhesion [28]. For example, epithelial cells elongate and align along patterns of grooves and ridges with feature dimensions as small as 70 nm, whereas on smooth substrates, cells are mostly rounded (Fig. 7) [29].

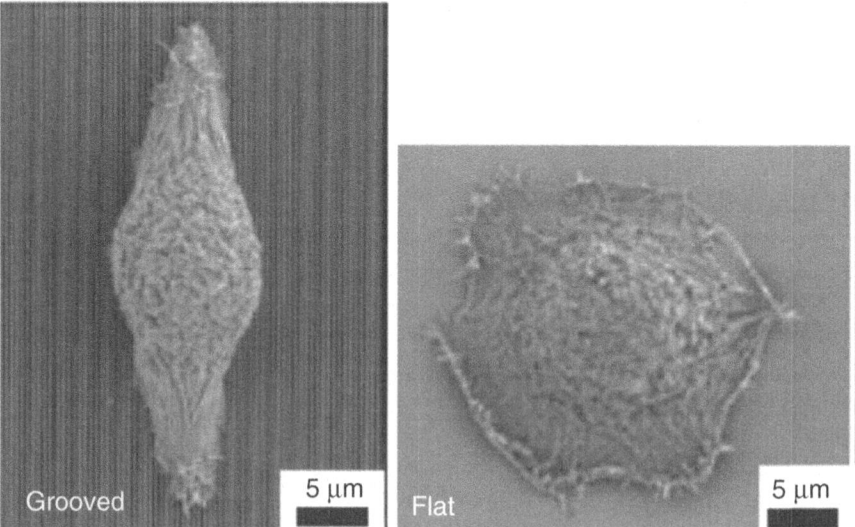

Fig. 7   Left: Epithelial cells respond to nanopatterning by alignment and elongation along the grating axis. Right: On smooth substrates, cells are mostly rounded [29]

3.3 Cell Sensitivity to Nanofeatures

It is interesting to consider exactly what size topographical features that cells can sense and respond to. The small diameter of ECM protein fibers are thought to have a significant impact on the function of cells, which are 1–2 orders of magnitude larger than typical ECM nanofibers. Work by Wojciak-Stothard et al. suggested that P388D1 macrophage-like cells react to surface feature dimensions down to 44 nm [30]. However, other cell types do not appear to be as sensitive, as epitena, epithelia, fibroblasts, and endothelia did not react to groove depths less than 70 nm [30]. Taken as a whole, the utilization of such nanotopographies, possibly in combination with chemical cues, could potentially lead to tissue scaffolds with very precise control over cell morphologies and functions.

3.4 Important Features of Scaffolds for Tissue Engineering

To provide transplanted cells with the optimum environment for tissue regeneration, it is desirable to construct scaffolds that mimic the native ECM in terms of structure and chemistry, with one exception: tissue repair via tissue engineering processes is expected to progress more quickly than natural tissue development. Therefore, scaffolds should exhibit large degrees of porosity and inter-pore connectivity to allow for accelerated mass transfer and neovascularization. Increased porosity also encourages cells to migrate and reorganize their extracellular environment in accordance with the hierarchical architecture of the local tissue [31].

Other aspects of scaffolds for tissue engineering are designed to resemble as much as possible the tissue that they are meant to regenerate or replace. These properties include nanofiber length and diameter, mechanical stiffness and flexibility, and nanotopography, as well as the biochemical factors that influence proper cell-cell and cell-scaffold interactions. Because ECM properties vary widely from tissue to tissue, care must be taken in selecting the proper materials and fabrication technique for each particular application.

3.5 Materials for Scaffold Construction

Scaffolds for tissue engineering can be constructed from synthetic or natural materials, such as collagen or hyaluronic acid. Metals and ceramics have been considered in the past (primarily for bone tissue engineering), but the stiffness of these materials typically leads to a mismatch in mechanical properties between the implant and the surrounding native tissues, leading to large interfacial stress concentrations and eventual loosening and failure [32]. Today, synthetic and natural polymers are the primary materials of construction, and for bone tissue scaffolds, success has been achieved by mineralizing polymers with nanoparticles of metal or ceramic, which closely resembles the mineralized collagen matrix of natural bone [33].

Poly (α-hydroxy) esters (such as poly lactic acid, poly glycolic acid, and combinations thereof) have found the most widespread application for tissue engineering due to their biocompatibility and biodegradability. These polymers are readily available and FDA approved for clinical use. One drawback is that they do not possess the functionalities found in natural ECM fibers that enhance cellular functions and signaling [34]. Therefore, natural polymers such as collagen and hyaluronic acid have also been widely investigated as materials for tissue engineering. Yet, even these natural polymers have some disadvantages, including immunogenicity and variations in mechanical properties, degradation, and reproducibility [35]. Ultimately, research has focused on improving the cell-specific properties of these materials by functionalizing them with bioactive molecules and through the application of various fabrication techniques.

4 Unmet Clinical Need

Despite the above advances, improved biomaterials for the regeneration of bone are still needed to treat the growing population of people with damaged and degrading bone [36–38]. This need is highlighted by the fact that there are more than 4,000,000 operations involving bone grafting or bone substitutes performed annually around the world [38]. These procedures are being used to repair bone fractures, alleviate spinal issues, replace joints, and repair bone tumor resections [38], and the total associated medical burden costs tens of billions of dollars each year [39].

Complications can arise following bone grafting procedures, often necessitating secondary procedures and incurring more cost. Delayed healing or non-union occurs in 20% of all high-impact fractures [40], and 35% in the case of spinal fusions [41]. Additionally, bacterial infection of surgical wounds poses a serious risk to patients and hospitals. Approximately 30% of implanted bone fracture-fixation devices acquire bacterial infections [42], and the dangers associated with bacterial infections are becoming much more challenging with the rise of antibiotic resistance [43–45].

Autografts and allografts currently provide the best tissue-to-implant healing. However, autograft usage is limited by donor-site morbidity and supply, and allografts can transmit disease or illicit an immunogenic response by the host to the foreign tissue [46–49]. Ultimately, new solutions are needed to meet the demand for orthopedic biomaterials that can improve tissue integration and resist bacterial infection. But to develop biomaterials with the necessary chemical and physical properties, it is necessary to more fully understand how changing the features of the material can modulate their interactions with their environment, and nanotechnology may hold the answer.

4.1 Substrate Properties for Osseointegration

Proper mechanical inter-locking into bone is essential to the success of implants. The surface characteristics of the implant play a pivotal role in directing cell activity and ultimately the integration of the implant within surrounding tissues. Implants with smooth surfaces at the nanoscale can become encased in a fibrous membrane, thus damaging osseointegration and leading to implant failure. Alternatively, implants with increased nanoscale roughness have demonstrated significantly improved osteoblast adhesion, and on some nanotopographies have even reduced fibroblast adherence compared to osteoblast cells [50–54].

Surface micro−/nano-scale topography, grain structure, stiffness, and chemistry can modulate cellular functions at the cell-substrate interface [55–60]. These substrate properties also significantly affect surface energy, a parameter that has been shown to play a large role in directing protein adsorption and cellular activity [26]. Several studies have demonstrated that increased surface energy increased the adherence of osteoblasts and significantly improved the integration of the material with bone [61–64].

4.2 Substrate Properties to Resist Bacterial Infection

If bacteria find their way onto an implant, they can attach almost immediately and subsequently proliferate rapidly [65, 66]. If the bacteria progress to form a biofilm, the infection becomes highly resistant to antibiotics and virtually impossible to eradicate, necessitating surgical resection [67–69]. Thus, it is desirable to create biomaterial surfaces that intrinsically resist the attachment of bacteria (while still promoting cell adhesion). Such a material would then allow the immune system more time to locate and destroy infection-causing bacteria before they can colonize and form a biofilm.

Like cells, bacterial adhesion and biofilm formation is also directly affected by changing substrate properties [70, 71]. Increasing surface roughness has been reported to both deter and attract bacterial adhesion [72–76]. However, the scale/feature size of the surface roughness is a critical parameter that must be considered. In general, decreasing the substrate feature size below the size of the bacteria (and into the nanoscale) results in decreased bacterial retention [76]. It has been shown that bacterial colonization depends heavily on both surface roughness and surface energy and from a thermodynamics perspective, the tendency for bacteria to attach to a surface can be determined by calculating the free energy of adhesion, which is a function of the substrate-bacteria, substrate-liquid, and liquid-bacteria interfacial free energies. In the next two sections, we provide two processes which can easily create nanoscale surface features on everyday orthopedic implants to decrease bacteria growth (and promote bone growth).

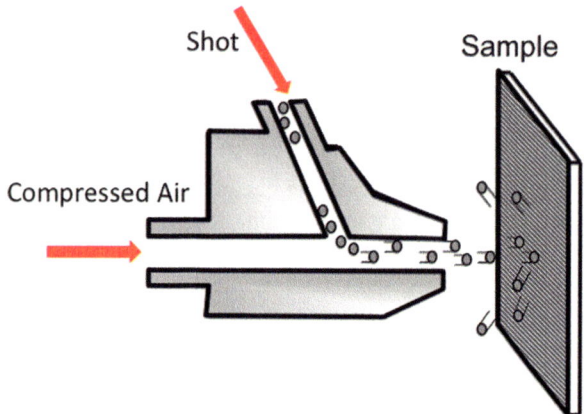

Fig. 8 Schematic of the shot peening apparatus

4.2.1   Shot Peened 316 L Stainless Steel

Compared to the biodegradable polymer scaffolds that are gaining great interest for tissue engineering, permanent metal implants provide a much simpler and more consistent platform on which to study the effects of surface modifications on cell/bacteria-substrate interactions. Accordingly, a variety of metallic surface modification approaches have been tested. Among the applied techniques, severe plastic deformation (SPD) procedures are metal forming processes that result in exceptional grain refinement by imposing high strain rates at relatively low temperatures without considerably altering the overall dimensions of the material. Severe shot peening (SSP) can be described as one of the most effective and, at the same time, the least demanding SPD techniques.

SSP is based on impacting the material surface with high energy shots as demonstrated in Fig. 8. The effect of the treatment is to introduce numerous defects, dislocations and grain boundaries onto the surface layer of the material, subsequently transforming the coarse grained structure into a nanostructured one. The characteristics of the affected surface layer in terms of grain size, thickness, surface roughness, and work hardening as well as induced residual stresses can be tailored by the proper choice of peening parameters, including the Almen intensity (a measure of shot stream kinematic energy during the shot peening process) and surface coverage (the ratio of the area covered by plastic indentation to the whole surface area). Studies performed on SSP in recent years have indicated its ability to significantly improve the mechanical properties of treated materials in terms of hardness, fatigue strength, corrosion, wear, scratch resistance and so on, contributing to enhanced functionality and service characteristics of the material [77, 78].

In this work, 316 L stainless steel (which is the most widely used stainless steel for orthopedic, cardiovascular and craniofacial applications due to its good corrosion resistance and formability) was used to assess the interaction between the shot-peened surfaces and primary human osteoblasts (bone forming cells), as well as a

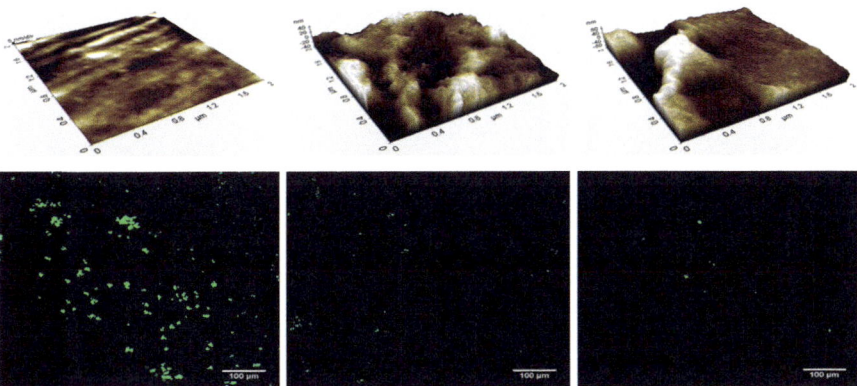

Fig. 9 Changes in nanoscale surface features for stainless steel severely shot-peened compared to untreated samples. Top: AFM images. Bottom: *Staph. aureus* growth after 1 day stained with a green fluorescent dye. Results showed that the presence of nanoscale features from severely shot-peened and conventional shot peening stainless steel were imperative for reducing bacteria adhesion and increasing bone growth [79]. Left column conventional (or no treatment), middle (conventional shot peening) and right (severe shot peening)

range of bacteria, including the gram-positive strains *Staphylococcus aureus* and *Staphylococcus epidermidis,* and gram-negative *Pseudomonas aeruginosa* and ampicillin-resistant *Escherichia coli* (Fig. 9). To examine the effects of grain size refinement alone (without the confounding changes in roughness), a group of samples were surface ground after the SSP treatments to remove differences in surface roughness between sample groups (while the samples retained their differences in underlying grain structure). The mechanical and physical properties of the substrates were fully characterized and related to the response of osteoblasts and bacteria seeded onto different sample groups. Results demonstrated considerable promise to enhance the mechanical and cytocompatibility properties of 316 L stainless steel using SSP treatments alone without resorting to the use of pharmaceutical agents [79]. In fact, when polishing the samples, no differences in bacteria growth were observed while significantly decreased bacterial growth was observed on the SSP compared to conventional shot peened and no treated samples. Bone cell growth was enhanced on all treated compared to untreated samples, even those was there polished compared to untreated samples. This study also provided significant insight into how surface roughness and surface energy affect cell and bacteria functions.

4.2.2 Electrophoretic Deposition

Another technique that can be used to impart antibacterial properties to materials is electrophoretic deposition (EPD). EPD can offer the best characteristics to obtain a coating of highly exposed nanoparticles (that would retain their nanostructure after coating) of great versatility for the greatest bacteria-killing activity. EPD is a

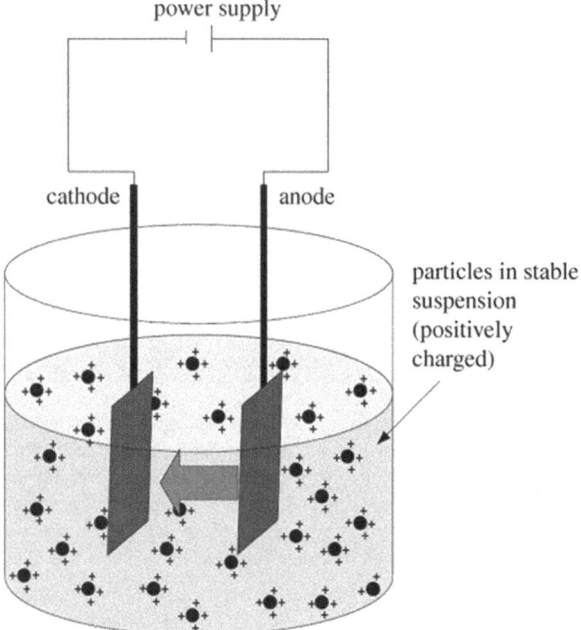

Fig. 10 Electrophoretic deposition (EPD) cell showing positively charged particles in suspension migrating towards the negative electrode [80]

colloidal processing technique employing a electrophoresis mechanism to deposit a thick or thin film of charged particles in an ordered manner onto a substrate. This process is typically performed within a two-electrode cell, as depicted schematically in Fig. 10 [80].

The coating thickness achieved using EPD can vary widely from around 100 nm to more than 100 μm, and depends directly on the applied voltage, processing time, and the concentration of particles in the EPD solution. However, regarding the particle concentration in the EPD solution, there is an upper limit at which particle flocculation caused by migrations near the depositing electrode may cause sub-optimal coating. Studies have shown that by coating titanium with nanoscale hydroxyapatite one can decrease bacteria attachment and growth and promote bone growth without releasing any growth factors (Fig. 11) [81].

5   Conclusions

It is now abundantly clear that nanotechnology can improve bone growth and, even more recently, inhibit bacteria growth. In this chapter, the fundamental reasons why nanotechnology can control cell functions is presented and several concrete manufacturing process are presented which can implement nanoscale features on today's

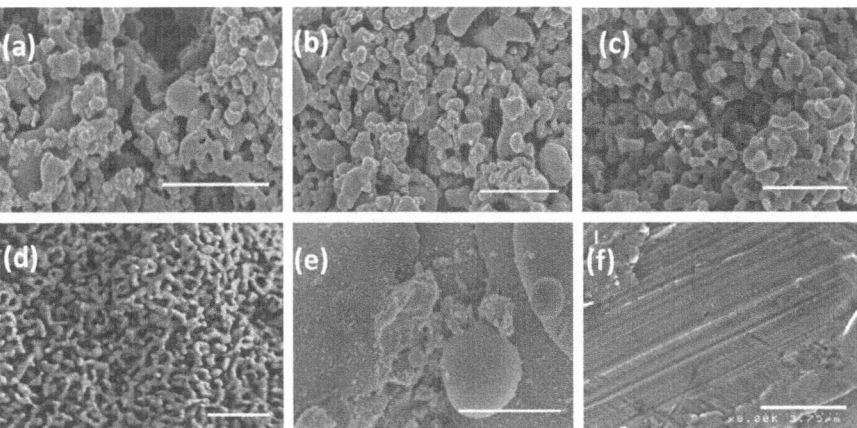

Fig. 11 Electrophoretic coated titanium with nanoparticles of hydroxyapatite that can reduce bacteria attachment and growth and promote bone growth without releasing drugs. SEM images of the Ti surfaces coated with (**a**) 170 nm, (**b**) 150 nm, (**c**) 130 nm, (**d**)110 nm hydroxyapatite powders, (**e**) plasma sprayed micron sized hydroxyapatite onto Ti and (**f**) plain Ti. Scale bar: 1 μm

implants. The ability of nanoscale surface features to change surface energy to in turn control initial protein adsorption events that dictate cell responses will be continually integrated into medicine to improve human health.

References

1. Magnusson SP, Langberg H, Kjaer M. The pathogenesis of tendinopathy: balancing the response to loading. Nat Rev Rheumatol. 2010;6(5):262–8.
2. Sechler JL, Corbett SA, Wenk MB, Schwarzbauer JE. Modulation of cell-extracellular matrix interactions. Ann N Y Acad Sci. 1998;857:143–54.
3. Sanes JR, Engvall E, Butkowski R, Hunter DD. Molecular heterogeneity of basal laminae: isoforms of laminin and collagen IV at the neuromuscular junction and elsewhere. J Cell Biol. 1990;111(4):1685–99.
4. Barnes CP, Sell SA, Boland ED, Simpson DG, Bowlin GL. Nanofiber technology: designing the next generation of tissue engineering scaffolds. Adv Drug Deliv Rev. 2007;59:1413–33.
5. Slavik GJ, Ragetly G, Ganesh N, Griffon DJ, Cunningham B. A replica molding technique for producing fibrous chitosan scaffolds for caritlage engineering. J Mater Chem. 2007;17:4095–101.
6. Ma P. Biomimetic materials for tissue engineering. Adv Drug Deliv Rev. 2008;60(2):184–98.
7. Flaumenhaft R, Rifkin DB. The extracellular regulation of growth factor action. Mol Biol Cell. 1992;3(10):1057–65.
8. Streuli CH, Schmidhauser C, Kobrin M, Bissell MJ, Derynck R. Extracellular matrix regulates expression of the TGF-beta 1 gene. J Cell Biol. 1993;120(1):253–60.
9. Schuppan D, Schmid M, Somasundaram R, Ackermann R, Ruehl M, Nakamura T, Riecken E. Collagens in the liver extracellular matrix bind hepatocyte growth factor. Gastroenterology. 1998;114(1):139–52.

10. Benoit DSW, Anseth KS. Nanostructured scaffolds for tissue engineering. In: Peppas NA, editor. Nanotechnology in therapeutics: current technology and applications. London: Taylor & Francis; 2007. p. 205–38.
11. Ravin T. Tensegrity to tendinosis. J Prolother. 2011;3(4):826–35.
12. Giancotti FG. Integrin signaling: specificity and control of cell survival and cell cycle progression. Curr Opin Cell Biol. 1997;9(5):691–700.
13. Hemler ME. Integrin associated proteins. Curr Opin Cell Biol. 1998;10(5):578–85.
14. Ruoslahti E, Yamaguchi Y, Hildebrand A, Border WA. Extracellular matrix/growth factor interactions. Cold Spring Harb Symp Quant Biol. 1992;57:309–15.
15. Hynes RO. Integrins: a family of cell surface receptors. Cell. 1987;48(4):549–54.
16. Hynes RO. Integrins: versatility, modulation, and signaling in cell adhesion. Cell. 1992;69(1):11–25.
17. Rho JY, Juhn-Spearing L, Zioupos P. Mechanical properties and the hierarchical structure of bone. Med Eng Phys. 1998;20(2):92–102.
18. Weiner S, Traub W. Bone structure: from angstroms to microns. FASEB J. 1992;6(3):879–85.
19. Rho JY, Ashman RB, Turner CH. Young's modulus of trabecular and cortical bone material: ultrasonic and microtensile measurements. J Biomech. 1993;26(2):111–9.
20. Viola J, Lal B, Grad O. The emergence of tissue engineering as a research field. Arlington, VA: The National Science Foundation; 2003.
21. Berthiaume F, Maguire TJ, Yarmush ML. Tissue engineering and regenerative medicine: history, progress, and challenges. Annu Rev Chem Biomol Eng. 2011;2:403–30.
22. Langer R, Vacanti JP. Tissue engineering. Science. 1993;260(5110):920–6.
23. Nerem R. Regenerative medicine: the emergence of an industry. J R Soc Interface. 2010;7:S771–5.
24. "Goldstein Research Group Homepage." [Online]. https://secure.hosting.vt.edu/www.tissue.che.vt.edu/home_frame.htm.
25. Webster TJ. From nanotechnology to picotechnology: what is on the horizon? Nanotek-2013, OMICS International, Nanotek Expo, 2013.
26. Zhang L, Webster TJ. Nanotechnology and nanomaterials: promises for improved tissue regeneration. Nano Today. 2009;4(1):66–80.
27. Woo KM, Chen VJ, Ma PX. Nano-fibrous scaffolding architecture selectively enhances protein adsorption contributing to cell attachment. J Biomed Mater Res A. 2003;67(2):531–7.
28. Bettinger CJ, Langer R, Borenstein JT. Engineering substrate topography at the micro- and nanoscale to control cell function. Angew Chem Int Ed Engl. 2009;48(30):5406–15.
29. Teixeira AI, Abrams GA, Bertics PJ, Murphy CJ, Nealey PF. Epithelial contact guidance on well-defined micro- and nanostructured substrates. J Cell Sci. 2003;116:1881–92.
30. Wójciak-Stothard B, Curtis AS, Monaghan W, McGrath M, Sommer I, Wilkinson CD. Role of the cytoskeleton in the reaction of fibroblasts to multiple grooved substrata. Cell Motil Cytoskeleton. 1995;31(2):147–58.
31. Dvir T, Timko BP, Kohane DS, Langer R. Nanotechnological strategies for engineering complex tissues. Nat Nanotechnol. 2011;6:13–22.
32. Weiner S, Wagner HD. THE MATERIAL BONE: structure-mechanical function relations. Annu Rev Mater Sci. 1998;28(1):271–98.
33. Song J, Malathong V, Bertozzi CR. Mineralization of synthetic polymer scaffolds: a bottom-up approach for the development of artificial bone. J Am Chem Soc. 2005;127(10):3366–72.
34. Willmann G. Coating of implants with hydroxyapatite—material connections between bone and metal. Adv Eng Mater. 1999;1(2):95–105.
35. Danie Kingsley J, Ranjan S, Dasgupta N, Saha P. Nanotechnology for tissue engineering: need, techniques and applications. J Pharm Res. 2013;7(2):200–4.
36. Navarro M, Michiardi A, Castano O, Planell JA. Biomaterials in orthopaedics. J R Soc Interface. 2008;5(27):1137–58.
37. Laurencin CT, Ambrosio AM, Borden MD, Cooper JA Jr. Tissue engineering: orthopedic applications. Annu Rev Biomed Eng. 1999;1:19–46.
38. Brydone A, Meek D, Maclaine S. Bone grafting, orthopaedic biomaterials, and the clinical need for bone engineering. Proc Inst Mech Eng H. 2010;224(12):1329–43.

39. Reid RL. Hernia through an iliac bone-graft donor site. A case report. J Bone Joint Surg Am. 1968;50(4):757–60.
40. Dickson G, Buchanan F, Marsh D, Harkin-Jones E, Little U, McCaigue M. Orthopaedic tissue engineering and bone regeneration. Technol Health Care. 2007;15(1):57–67.
41. Ludwig SC, Kowalski JM, Boden SD. Osteoinductive bone graft substitutes. Eur Spine J. 2000;9(Suppl 1):S119–25.
42. Trampuz A, Zimmerli W. Diagnosis and treatment of infections associated with fracture-fixation devices. Injury. 2006;37(Suppl 2):S59–66.
43. Spellberg B, Bartlett JG, Gilbert DN. The future of antibiotics and resistance. N Engl J Med. 2013;368(4):299–302.
44. Laxminarayan R, Duse A, Wattal C, Zaidi AKM, Wertheim HFL, Sumpradit N, Vlieghe E, Hara GL, Gould IM, Goossens H, Greko C, So AD, Bigdeli M, Tomson G, Woodhouse W, Ombaka E, Peralta AQ, Qamar FN, Mir F, Kariuki S, Bhutta ZA, Coates A, Bergstrom R, Wright GD, Brown ED, Cars O. Antibiotic resistance—the need for global solutions. Lancet Infect Dis. 2013;13(12):1057–98.
45. Neu HC. The crisis in antibiotic resistance. Science. 1992;257(5073):1064–73.
46. Laurencin C, Khan Y, El-Amin SF. Bone graft substitutes. Expert Rev Med Devices. 2006;3(1):49–57.
47. Desai BM. Osteobiologics. Am J Orthop (Belle Mead NJ). 2007;36(4 Suppl):8–11.
48. De Long WG, Einhorn TA, Koval K, McKee M, Smith W, Sanders R, Watson T. Bone grafts and bone graft substitutes in orthopaedic trauma surgery. A critical analysis. J Bone Joint Surg Am. 2007;89(3):649–58.
49. Toolan BC. Current concepts review: orthobiologics. Foot Ankle Int. 2006;27(7):561–6.
50. Zhao G, Zinger O, Schwartz Z, Wieland M, Landolt D, Boyan BD. Osteoblast-like cells are sensitive to submicron-scale surface structure. Clin Oral Implants Res. 2006;17(3):258–64.
51. McManus AJ, Doremus RH, Siegel RW, Bizios R. Evaluation of cytocompatibility and bending modulus of nanoceramic/polymer composites. J Biomed Mater Res A. 2005;72(1):98–106.
52. Webster TJ, Ergun C, Doremus RH, Siegel RW, Bizios R. Specific proteins mediate enhanced osteoblast adhesion on nanophase ceramics. J Biomed Mater Res. 2000;51(3):475–83.
53. Price RL, Gutwein LG, Kaledin L, Tepper F, Webster TJ. Osteoblast function on nanophase alumina materials: influence of chemistry, phase, and topography. J Biomed Mater Res A. 2003;67(4):1284–93.
54. Shalabi MM, Gortemaker A, Van't Hof MA, Jansen JA, Creugers NHJ. Implant surface roughness and bone healing: a systematic review. J Dent Res. 2006;85(6):496–500.
55. Dolatshahi-Pirouz A, Nikkhah M, Kolind K, Dokmeci MR, Khademhosseini A. Micro- and nanoengineering approaches to control stem cell-biomaterial interactions. J Funct Biomater. 2011;2(4):88–106.
56. Dolatshahi-Pirouz A, Jensen T, Kraft DC, Foss M, Kingshott P, Hansen JL, Larsen AN, Chevallier J, Besenbacher F. Fibronectin adsorption, cell adhesion, and proliferation on nanostructured tantalum surfaces. ACS Nano. 2010;4(5):2874–82.
57. Bagherifard S, Ghelichi R, Khademhosseini A, Guagliano M. Cell response to Nanocrystallized metallic substrates obtained through severe plastic deformation. ACS Appl Mater Interfaces. 2014;6(11):7963–85.
58. Nikkhah M, Edalat F, Manoucheri S, Khademhosseini A. Engineering microscale topographies to control the cell-substrate interface. Biomaterials. 2012;33(21):5230–46.
59. Webster TJ, Ergun C, Doremus RH, Siegel RW, Bizios R. Enhanced functions of osteoblasts on nanophase ceramics. Biomaterials. 2000;21(17):1803–10.
60. Guvendiren M, Burdick JA. Stiffening hydrogels to probe short- and long-term cellular responses to dynamic mechanics. Nat Commun. 2012;3:792.
61. Hallab NJ, Bundy KJ, O'Connor K, Moses RL, Jacobs JJ. Evaluation of metallic and polymeric biomaterial surface energy and surface roughness characteristics for directed cell adhesion. Tissue Eng. 2001;7(1):55–71.
62. Kieswetter K, Schwartz Z, Dean DD, Boyan BD. The role of implant surface characteristics in the healing of bone. Crit Rev Oral Biol Med. 1996;7(4):329–45.

63. Zhao G, Raines AL, Wieland M, Schwartz Z, Boyan BD. Requirement for both micron- and submicron scale structure for synergistic responses of osteoblasts to substrate surface energy and topography. Biomaterials. 2007;28(18):2821–9.
64. Zhao G, Schwartz Z, Wieland M, Rupp F, Geis-Gerstorfer J, Cochran DL, Boyan BD. High surface energy enhances cell response to titanium substrate microstructure. J Biomed Mater Res Part A. 2005;74A(1):49–58.
65. Ribeiro M, Monteiro FJ, Ferraz MP. Infection of orthopedic implants with emphasis on bacterial adhesion process and techniques used in studying bacterial-material interactions. Biomatter. 2012;2(4):176–94.
66. Donlan RM. Biofilms: microbial life on surfaces. Emerg Infect Dis. 2002;8(9):881–90.
67. Nazhat SN, Young AM, Pratten J. Sterility and infection. In: Biomedical materials. Boston, MA: Springer US; 2009. p. 239–60.
68. Davies D. Understanding biofilm resistance to antibacterial agents. Nat Rev Drug Discov. 2003;2(2):114–22.
69. Gristina AG, Naylor P, Myrvik Q. Infections from biomaterials and implants: a race for the surface. Med Prog Technol. 14(3–4):205–24.
70. Puckett SD, Taylor E, Raimondo T, Webster TJ. The relationship between the nanostructure of titanium surfaces and bacterial attachment. Biomaterials. 2010;31(4):706–13.
71. Epstein AK, Hochbaum AI, Kim P, Aizenberg J. Control of bacterial biofilm growth on surfaces by nanostructural mechanics and geometry. Nanotechnology. 2011;22(49):494007.
72. Ivanova EP, Truong VK, Wang JY, Berndt CC, Jones RT, Yusuf II, Peake I, Schmidt HW, Fluke C, Barnes D, Crawford RJ. Impact of nanoscale roughness of titanium thin film surfaces on bacterial retention. Langmuir. 2010;26(3):1973–82.
73. Kerr A, Cowling MJ. The effects of surface topography on the accumulation of biofouling. Philos Mag. 2003;83(24):2779–95.
74. Whitehead KA, Verran J. The effect of surface topography on the retention of microorganisms. Food Bioprod Process. 2006;84(4):253–9.
75. Graham M, Cady N. Nano and Microscale topographies for the prevention of bacterial surface fouling. Coatings. 2014;4(1):37–59.
76. Helbig R, Günther D, Friedrichs J, Rößler F, Lasagni A, Werner C. The impact of structure dimensions on initial bacterial adhesion. Biomater Sci. 2016;4(7):1074–8.
77. Bagherifard S, Guagliano M. Fatigue behavior of a low-alloy steel with nanostructured surface obtained by severe shot peening. Eng Fract Mech. 2012;81:56–68.
78. Bagherifard S, Fernandez-Pariente I, Ghelichi R, Guagliano M. Fatigue behavior of notched steel specimens with nanocrystallized surface obtained by severe shot peening. Mater Des. 2013;45:497–503.
79. Bagherifard S, Hickey DJ, de Luca AC, Malheiro VN, Markaki AE, Guagliano M, Webster TJ. The influence of nanostructured features on bacterial adhesion and bone cell functions on severely shot peened 316L stainless steel. Biomaterials. 2015;73:185–97.
80. Besra L, Liu M. A review on fundamentals and applications of electrophoretic deposition (EPD). Prog Mater Sci. 2007;52(1):1–61.
81. Mathew D, Bhardwaj G, Wang Q, Webster TJ. Decreased Staphylococcus aureus and increased osteoblast density on nanostructured electrophoretic-deposited hydroxyapatite on titanium without the use of pharmaceuticals. Int J Nanomed. 2014;9:1775–81.

Additive Manufacturing of Orthopedic Implants

Maryam Tilton, Gregory S. Lewis, and Guha P. Manogharan

Keywords Additive manufacturing · Orthopedics · Patient-specific design · Biomaterial · Biocompatibility · Microarchitecture · Large bone defects · Arthroplasty · Electron beam melting · Laser-powder bed fusion · 3D printing · Custom implants · Biomechanics · Finite element analysis · Reverse engineering · Reconstruction

1 Introduction

Additive Manufacturing (AM) is the process of selectively joining materials to fabricate objects in a layer-by-layer approach using digital part information, i.e. 3D CAD models. This definition highlights the fundamental difference between AM process and traditional manufacturing methods such as subtractive processes (e.g. machining), forming processes (e.g. forging) and bulk solidification processes (e.g. casting). AM is often also called 3D printing, additive processes, freeform fabrication and layered manufacturing. When compared to traditional processes, AM offers unique advantages to economically produce low volume batches (one to a few) of highly complex products. Since AM does not require design and/or material dependent tooling (e.g. jigs and fixtures), AM is an ideal candidate for the next generation design and manufacturing of orthopedic implants. Although "customization" of product specifications implants has been around long before the introduction of AM technology to the medical field, the lack of tooling requirement for each design in AM makes it economically viable for patient-specific orthopedic implant production. Finally, design freedom that can be easily achieved through AM technology enables introduction of porous structures for bone ingrowth and biological implant

M. Tilton · G. P. Manogharan (✉)
Department of Mechanical and Nuclear Engineering, College of Engineering, Pennsylvania State University, University Park, PA, USA
e-mail: gum53@psu.edu

G. S. Lewis
Department of Orthopedics and Rehabilitation, College of Medicine, Pennsylvania State University, Hershey, PA, USA

© Springer International Publishing AG, part of Springer Nature 2018
B. Li, T. Webster (eds.), *Orthopedic Biomaterials*,
https://doi.org/10.1007/978-3-319-89542-0_2

fixation. The motivation for this chapter is to understand the current state of orthopedic applications of AM which have been shown to economically produce highly customized and highly complex design features in low volumes.

First, this chapter details the basic AM workflow to manufacture orthopedic implants and principles of various metal AM methods. Subsequently, a broad review of biocompatible materials that can be currently processed via AM and resulting mechanical properties are outlined. Beyond feasibility of biomaterials for AM processing, other criteria such as biocompatibility, strength and prevention of stress shielding are important for the success of an implant. It is evident that chemical composition is not the only factor that differentiates the microstructure and mechanical properties of materials; manufacturing processes can be as important as chemical composition. The inherent layer-by-layer nature of AM processes affects the resulting mechanical properties of the implant (e.g. anisotropy, residual thermal stress), which could be explored in a beneficial manner through careful design-process planning for manufacturing. This chapter also outlines a detailed methodology for patient specific design processes for AM (dfAM) and ability of AM to seamlessly generate porous structures for osseointegration. Findings from reported studies on clinical applications of AM in orthopedic implants, surgical guides and other surgeries are presented in this chapter. Finally, a summary of commercial AM solutions, existing gaps in ongoing research and future direction in AM of orthopedic implants are provided.

2   Additive Manufacturing Techniques

As noted earlier, the basis of AM lies in fabricating a part through a layer-wise approach by selectively joining materials. The CAD file of the required part is sliced across each layer along the direction of fabrication as shown in Fig. 1. Upon generating process files for the machine, the file is transferred to the machine and built layer-by-layer from the bottom to top layer. The part is then removed from the AM machine for post-processing such as removal of sacrificial supports for any overhanging edges. After conducting the required post-processing operations, the part is ready for application. These steps are illustrated in Figs. 1 and 2.

Currently, a wide range of materials can be processed using AM: polymers, metals, and ceramics. The techniques for combining those materials in layers vary from pneumatically extruding suspended collagen for scaffold fabrication, using a laser for selectively solidifying photo-curable polymers to using electron-beam for selectively melting super-alloys. Several approaches have been proposed to categorize all AM methods based on the nature of raw material, aggregation geometry during deposition, and energy source. A functional classification of major AM methods such as Binder-Jetting, Powder Bed Fusion Methods (PBF) using Electron Beam Melting (EBM), Laser methods (L-PBF), Directed Energy Deposition (DED) and Material Extrusion based on the framework is presented as in Fig. 3. This functional classification is based on the raw material (i.e. feedstock), pattern of material

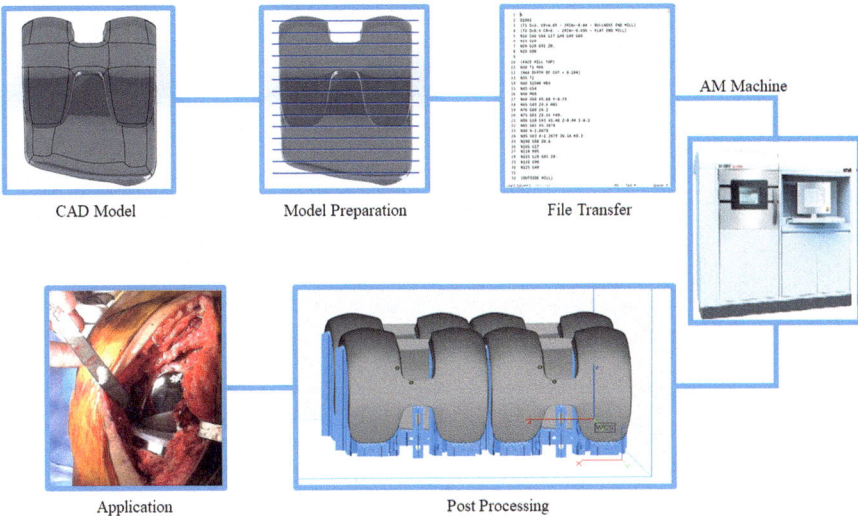

Fig. 1 Principle of AM for biomedical applications resulting in implantation. Picture demonstrating the application is reprinted with permission from the reference [1]. Copyright 2015 Springer International Publishing Switzerland

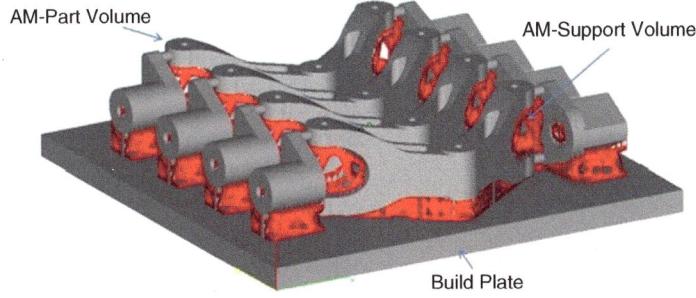

Fig. 2 Part volume and sacrificial support volume [2]

distribution, energy source, physical phenomena and support structures. It should be noted that this section provides a detailed overview of only the commonly used AM methods for fabrication of orthopedic implants and other AM methods [3] such as material-jetting and sheet lamination are not discussed in this chapter.

2.1 Binder Jetting

In the process of binder jetting, a layer of material (metal, polymers, sand or ceramics) is spread with selective deposition of binders as illustrated in Fig. 4.

	Store Material	Pattern Material	Pattern Energy	Create Primitive	Provide New Material	Support Material
Binder-Jetting	Single Phase Powder	No Material Patterning	3D Surface Tension	Fusion	Recoat by Spreading	Bed of Build Material
EBM - PBF						
L-PBF (Metal)						
L-PBF (Polymer)	Coated Powder					
L-DED	Single Phase Powder		1D Heat	Solidify Melt		5 Axis Deposition
E-DED	Single Phase Wire					
Material Extrusion	Single Phase Wire	1D Deposition			Direct Material Addition	2 Axis Deposition
	Single Phase Pellets					
	Multi-Phase Suspensions		3D Surface Tension	Fusion		

Fig. 3 Functional classification of thermal energy based AM methods

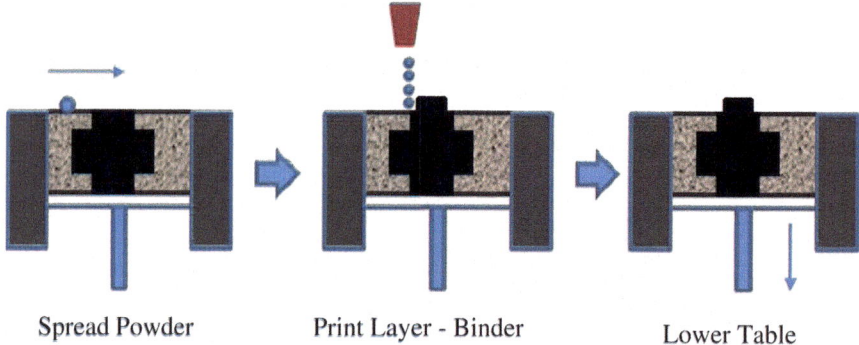

Spread Powder   Print Layer - Binder   Lower Table

Fig. 4 Principle of binder jetting

Based on the part information, i.e. design features in respective slices of the CAD file, binders are selectively applied using inkjet-printing technology. Subsequently, the next layer of material is spread across the platform and this process is continued until the final layer is finished. After dispensing the binder of the final layer, the bound metal or ceramic part (i.e. the so-called 'green part') is metallurgically bonded through sintering in a furnace to produce the final part, often through infiltration of other materials (e.g. bronze) to achieve full density. Whereas in the case of 3D sand printing, which is used in metal casting industry, the binder-jet molds are directly employed to produce metal castings [4]. Some of the unique characteristics of this AM technology are the ability to change binder chemistry (e.g. aqueous, organic, phenolic) based on the material being processed and absence of phase transformation in the AM machine which lends itself to process materials with high melting temperature [5]. It should be noted that this AM technology is being widely studied for its applications in producing porous bone scaffolds and bioactive surfaces [6, 7].

Fig. 5 Principle of L-DED

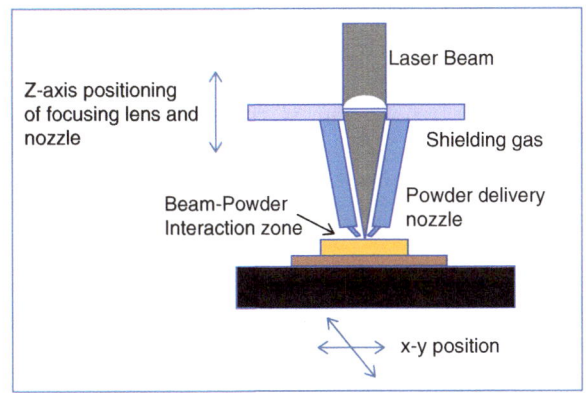

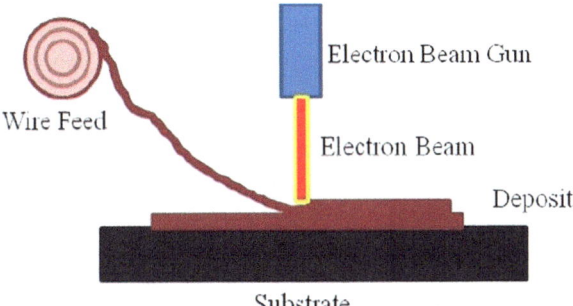

Fig. 6 Principle of E-DED

2.2 *Directed Energy Deposition (DED)*

In the case of DED AM technology, an energy source (laser or electron beam) is targeted at a synchronized flow of material (powder or wire) and deposits material based on information from each layer of the CAD file through precisely coordinated Computer Numeric Control (CNC) system between the energy source and material supply. An illustration of directed energy principle based on L-DED process using laser and directed powder supply is presented in Fig. 5. In some cases, the substrate/build plate and/or laser-powder supply is mounted on a 5-axis platform in order to create parts with overhanging edges without creating sacrificial supports.

Another example of the directed energy deposition process is the E-DED, also known as Electron Beam Freeform fabrication- EBF3 where the electron beam is the energy source and the material is supplied through a wire feeding system. Directed energy deposition process using electron beam and wire-fed material is shown in Fig. 6. Similar to the L-DED process, studies have been conducted to optimize the EBF process by varying the wire feed angle and the electron beam process parameters taking place in the vacuum environment [8]. It should be noted that electron beam based AM systems operate under controlled vacuum and laser based AM systems operate under shielding gas to avoid inter-layer oxidation. Some of the unique characteristics of DED technology include its major applications in repairing existing parts [9, 10], fabrication of Functionally Graded Material (FGM) components [11, 12] and integration within CNC machining for sequential hybrid processing, i.e. AM and machining/grinding [13, 14]. However, DED AM methods are limited in part design complexity due to absence of sacrificial support structures (see Fig. 2).

2.3 Powder Bed Fusion (PBF)

In PBF AM methods, an energy source (laser or electron beam) is targeted at a spread layer of powder material (metal) to specifically melt segments of material based on information from each layer of the CAD file. A unique aspect of such a metal based powder-bed fusion is the ability to vary the processing parameters (power, speed, energy distribution, etc.) to easily generate support structures as shown in Fig. 2 for complex overhanging surfaces with relatively automated process planning. Figure 7 outlines the powder bed-fusion principle in the case of the EBM process using an electron beam as the energy source with spread powder layers.

EBM is classified as a hot-bed AM process since the entire powder bed is maintained at an elevated temperature throughout the build process. When compared to other L-PBF (cold bed) processes, EBM functions in a vacuum as opposed to inert gas and accelerated electron beam employs higher power to completely melt the powder particles. It should also be noted that currently, there is only one EBM based AM process in the market (Arcam, Sweden). Due to the hot-bed processing conditions, EBM does not result in significant residual thermal stresses in the part when compared to L-PBF [15, 16]. However, the layer thickness in EBM (50–75 μm) results in inferior surface finish and part feature resolution when compared to L-PBF methods (25 μm) [17].

Another example of the powder bed-fusion process is the L-PBF where the laser beam is the energy source and the powder material (e.g. metal, coated polymers) is spread across the build platform as shown in Fig. 8. When compared to electromagnetic control in EBM, electromechanical control of laser scanning in L-PBF leads to lower scanning speed, i.e. cold-bed process where spread powders are directly melted without pre-heating. This results in high thermal residual stress, warping and the need for volume supports that require significant thermal and mechanical post-processing [18, 19].

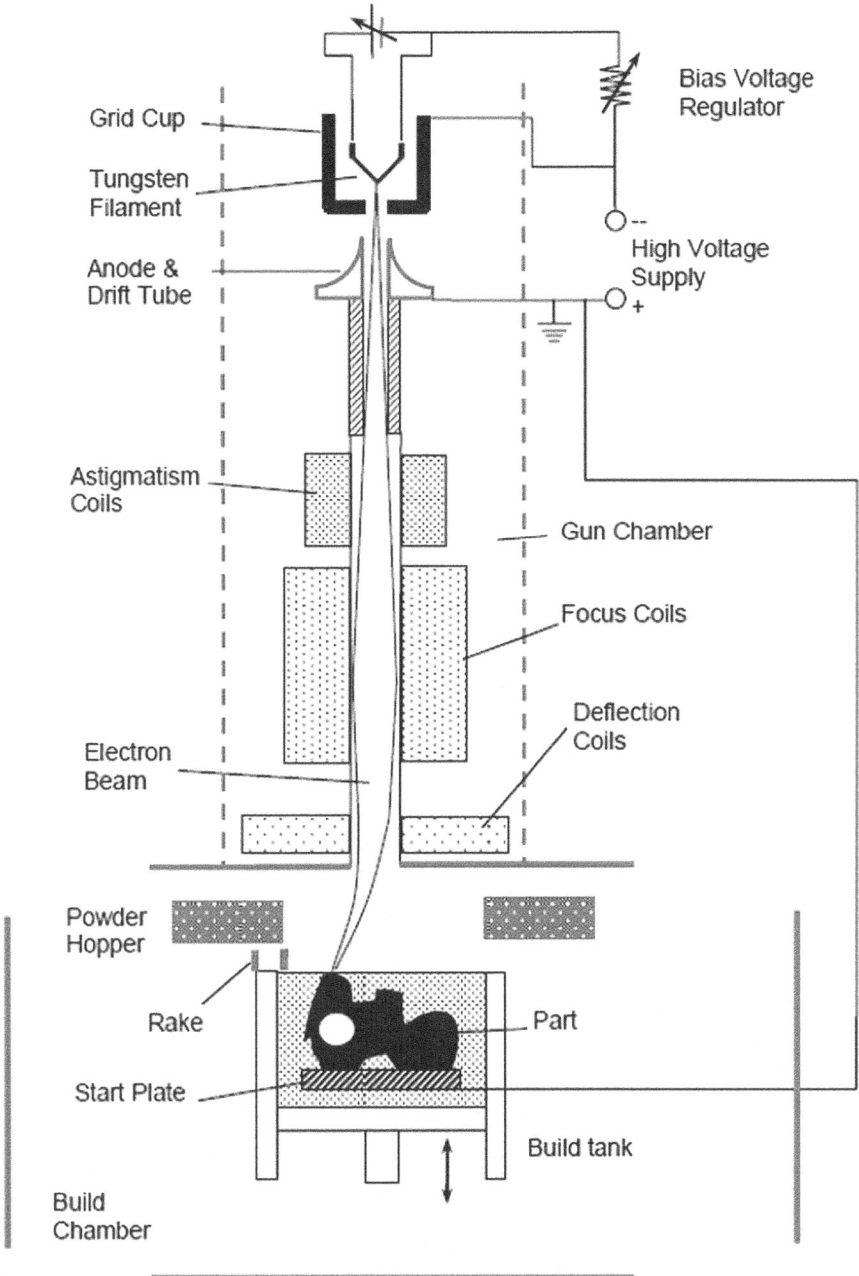

Fig. 7 Principle of EBM

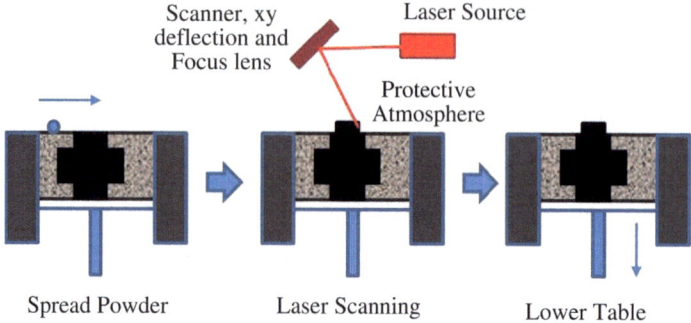

Fig. 8 Principle of L-PBF

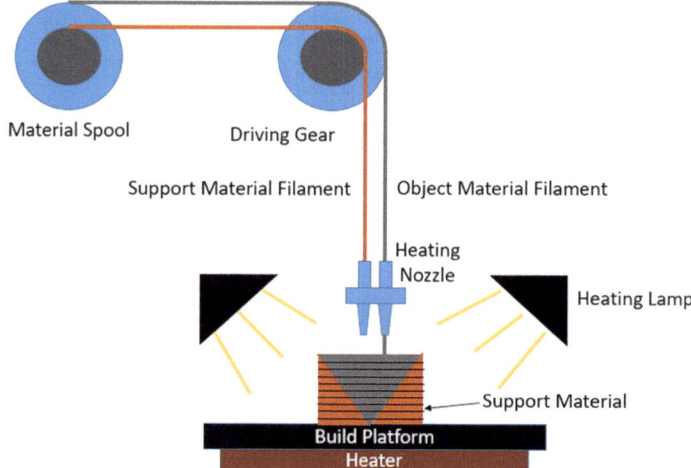

Fig. 9 Principle of filament-based extrusion

2.4   *Material Extrusion*

In material extrusion AM processes, the material (e.g. liquified polymer, colloidal suspension) is selectively dispensed through a nozzle or orifice using a sliced CAD part model as shown in Fig. 9. In polymer material extrusion AM, the filament material is fed to a nozzle using a driving gear. Then, the melted filament is deposited onto the build platform layer by layer to build the complex 3D structure often using a heated build platform and temperature controlled AM envelope. Temperature of the nozzle is sufficiently high to melt the filament material (e.g., 200–230 °C for ABS and 180–200 °C for PLA polymers). The ambient temperature is controlled using heating lamps surrounding the build platform to prevent part warpage and delamination [20]. Support structures for downward facing surfaces are created using a low temperature material with designs like infill structures for easier removal

Table 1 Mechanical properties of metallic biomaterials [29]

Material	E modulus [GPa]	Yield strength [MPa]	Ultimate tensile strength [MPa]
Stainless steel	190	221–1213	586–1351
Co-Cr alloys	210–253	448–1606	655–1896
Titanium	110	485	760
Ti-6Al-4 V	116	896–1034	965–1103
Cortical bone	15–30	30–70	70–150

[21]. When compared to other AM technologies, this technique is relatively inexpensive, easier to operate and maintain, and have found a range of biomedical applications using biological cells [22], ceramics [23] and polymers [24].

3   Additively Manufactured Biomaterials

In the orthopedic industry, processed biomaterials are categorized into three main groups: metallic biomaterials, polymers, and bio-ceramics. These biomaterials must satisfy all the clinical, manufacturing and economic requirements in order to be used for orthopedic implants. Additionally, to satisfy the clinical requirements, the manufactured implant should have mechanical properties compatible with bone and physiological loading conditions, corrosion resistant in the biological environment of the human body and be biocompatible. In this sense, the implanted material is considered as a biomaterial. A definition of biomaterial states that: "A biomaterial is a substance that has been engineered to take a form which, alone or as part of a complex system, is used to direct, by control of interactions with components of the living system, the course of any therapeutic or diagnostic procedure, in human or veterinary medicine" [25]. It should be noted that not all biomaterials are biocompatible. In other words, biocompatibility is not a material property, but it is viewed as the biological performance of the implant when it is placed in a biological environment. Biomaterials which are nontoxic and non-carcinogenic, and they do not elicit an adverse response including from the genetic and immune system of the patient, are known to be biocompatible. This section is dedicated to the common biomaterials used for AM orthopedic implants. Different mechanical properties obtained from various AM manufacturing techniques are compared along with discussions on criteria for selecting biomaterials.

3.1   Metallic Biomaterials

Metallic materials used in orthopedic implants are known as conventional biomaterials. They have always been used in load-bearing implants because of their high strength and good corrosion resistance. Currently, metallic biomaterials used in AM orthopedic implants are stainless steel, Co-Cr, Titanium, and their alloys. Properties including elastic modulus, yield and fatigue strength, toughness, hardness, and bio-compatibility are important for different surgical procedures including common joint replacements and fracture fixation. Unfortunately, the elastic modulus of all the existing metallic biomaterials is significantly higher than the cortical bone (Table 1). When the modulus of elasticity of an implant is higher than the host (i.e. bone), all the stress is induced to the implant and the neighboring bone experiences lower stress. This phenomenon is known as stress shielding. Stress shielding may lead to bone resorption and loosening of the implant [26, 27]. There are different approaches to control the mechanical properties of an orthopedic implant, including adjustment of alloying elements and structural design, and introducing composite materials are two of the conventional methods for this purpose. However, unprecedented design freedom in AM allows us to achieve this goal by altering the topology and controlling the macro-porosity of the implant [27, 28]. It is evident that mechanical properties of an implant depend on the number of cells in the lattice structure as well as its topology (shape) and size. Microarchitecture and optimization of AM orthopedic implants will be discussed in detail in the next section of this chapter.

3.1.1 Stainless Steel

Material properties including mechanical strength, corrosion resistance, biocompatibility and relatively lower cost make stainless steel a common material used to manufacture orthopedic implants especially in fracture fixation. Stainless Steel (SS) 316 has been widely accepted as the SS grade in place of SS 302 (the first generation of stainless steel used in medical implants). Like other metals, stainless steel possesses a higher elastic modulus when compared to cortical bone as shown in Table 1 which can result in stress shielding.

It is known that the existence of alloying elements including chromium, nickel and molybdenum enhance the corrosion resistance of SS 316. The presence of the carbon SS composition increases its mechanical strength and stiffness. The American Society of Testing Material (ASTM) recommends stainless steel type 316 L over the 316 for biomedical implants due to lower carbon composition in 316 L when compared to 316. Despite this advantage of SS316L over SS316, under high stress and low oxygen levels SS316L tends to be more reactive to corrosion [30]. SS316L has been used for permanent orthopedic implants, though some studies have cited stainless steel as an unsuitable biomaterial for permanent implants [31, 32]. Also, Ni and Cr released from SS316L implants are considered to be toxic and studies have reported skin related diseases such as dermatitis due to the toxicity of Ni [33].

Zhong et al. examined the microstructure of both L-PBF SS316L and EBM SS316L specimens in two separate additive manufacturing studies [34, 35]. Zhong et al. were able to achieve a relative density of 99.80% by the EBM process and over 99.90% through the L-PBF process and secondary post-processing, i.e. heat-treatment and Hot Isostatic Pressing (HIP). Hardness of EBM and SLM samples

Table 2 Mechanical properties of L-PBF SS316L, EBM SS316L, and forged SS316L

Material	Yield strength [MPa]	Ultimate tensile strength [MPa]	Elongation [%]
L-PBF SS316L [34]	487	594	49
EBM SS316L [35]	253	509	59
Forged SS316L [36]	1241	1344	1262
Wrought SS316L [36]	345	563	30

Table 3 Mechanical properties of CoCr alloys processed by SLM, EBM, and casting

Material	Yield strength [MPa]	Ultimate tensile strength [MPa]	Elongation [%]
EBM Co-Cr-Mo [37]	510	1450	36
L-PBF Co-Cr alloy [38]	580	1050	32
Cast Co-Cr alloy [38]	540	800	10

were reported at 165 [35] and 228 (at the cross-section) [34] respectively; while the micro-hardness of the HIP SS316L sample was 220. Result of mechanical testing at room temperature from Zhong's studies were compared with the mechanical properties of the forged SS316L in the Table 2. These studies showed that the mechanical performance of metal AM parts of the same alloy can greatly vary based on the manufacturing processes. L-PBF SS316L has the highest yield strength compared to EBM SS316L and cold-forged SS316L. This could be due to the higher cooling rate in the L-PBF process. A higher cooling rate leads to the existence of a nano-structure in L-PBF SS316L (formation of nano-cells) [34]. Also, as the grain size decreases and number of the grains increases in the structure, the yield strength of the specimen increases based on the grain boundary strengthening theory.

3.1.2 Co-Cr Alloys

Prior to the introduction of AM in metal biomedical applications, CoCr alloys have been extensively employed to manufacture implants (e.g. knee replacements) through casting. In recent years, many groups have investigated the clinical and mechanical behavior of AM-made CoCr alloys as orthopedic implants. Gaytan et al. studied the microstructure and mechanical properties of EBM Co-26Cr-6Mo-0.2C which showed better mechanical properties compared to the ASTM F75 standard (i.e. casting Co-28Cr-6Mo) [37]. Kim et al. compared the mechanical properties of L-PBF CoCr alloys with cast specimens as shown in Table 3. Although, the average yield strength is similar for all the listed manufacturing techniques, Ultimate Tensile Strength (UTS) and elongation % varied noticeably. Both L-PBF and EBM processed parts have shown superior properties to cast parts.

3.1.3   Titanium Alloys

Titanium alloys including Ti-6Al-4 V are well-established biomaterials in orthopedics because of their superior mechanical properties, higher strength to density ratio, biocompatibility and corrosion resistance. Murr et al. compared the microstructure and mechanical behavior of EBM processed Ti-6Al-4 V with SLM processed Ti-6Al-4 V. It was observed that the parts processed using L-PBF have predominant microstructures of α' martensitic with intermix of α'' phase, and Ti-6Al-4 V EBM parts showed a similar microstructure as wrought Ti-6Al-4 V, α phase with β along the phase boundaries [39]. The transition of the β phase to α' martensite is a result of varying solidification rates and cyclic thermal loading on prior layers. Due to the cold-bed processing conditions in L-PBF. Ti-6Al-4 V experiences higher temperature gradients which causes a more rapid cooling of the β phase and its transformation to α' martensite [40]. On the other hand, hot-bed processing conditions in Ti-6Al-4 V EBM occurs with a lower temperature gradient across the build layer and the entire build chamber is maintained at an elevated temperature throughout the EBM process (e.g. 650–750 °C in Ti-6Al-4 V). Such varying solidification conditions are attributed with the difference in microstructure and resulting mechanical behavior between L-PBF and EBM. However, it should be noted that the mechanical properties of Ti-6Al-4 V processed using both L-PBF and EBM were superior to wrought or cast Ti-6Al-4 V and the tensile strength of the EBM part increased by 50% over the wrought Ti-6Al-4 V.

Recent studies have shown that in the long term, Ti-6Al-4 V implants lead to the release of Al and V ions. Both Al and V ions are found to be strongly cytotoxic and can cause long term health problems such as Alzheimer disease [33]. This finding has led to interest in the research and development of alternative titanium alloys with nontoxic elements. In the process of alloying Titanium, an alloying element is either an α stabilizer, β stabilizer, or has an indistinct effect on the phase equilibrium [42]. It is empirically proven that in order to maintain the β phase at ambient temperature, the amount of β stabilizer required is about 10 wt% of molybdenum [42].

The Ti-15Zr-4Nb-4Ta alloy developed by Okazaki showed a lower toxicity compared to Ti6Al-4 V and a higher corrosion resistance [41]. Fukuda processed this alloy using EBM and found that the mechanical properties of specimens processed through EBM with the Ti-15Zr-4Nb-4Ta were very similar to the standard values given by JIS T7401-4 (tension strength of 860 MPa or higher, 0.2% proof stress of 790 MPa, and an elongation of 10% or higher) [40].

Sing et al. studied the microstructure and mechanical properties of a L-PBF titanium-tantalum (TiTa) alloy with 50 wt% of each element where Ta was used to stabilize the β phase as well as lowering the elastic modulus of Ti [42]. In the same study, microstructure analysis showed the existence of only the β phase which was attributed to Ta and might have caused more rapid solidification than in Ti-6Al-4 V. Mechanical properties of TiTa are compared with Ti-6Al-4 V and pure Ti in the Table 4. It can be inferred that the elastic modulus of TiTa is closer to cortical bone when compared to Ti-6Al-4 V which could help with reducing the risk of stress shielding in orthopedic implants.

Table 4 Mechanical properties of SLM processed TiTa, Ti-6Al-4 V, and commercially pure Ti [42]

Material	Young's modulus [GPa]	Ultimate tensile strength [MPa]	Yield strength [MPa]	Elongation [%]
TiTa	75	924	882	11
Ti-6Al-4 V	131	1165	1055	6
Ti	111	703	619	5
Cortical bone [29]	15–30	30–70	70–150	–

3.1.4 Tantalum

Porous Tantalum (Ta) is known for its good chemical resistance and mechanical properties [43]. Many publications have also reported Tantalum's biocompatibility and non-toxic behavior [44, 45]. However, when compared to SS and Ti-6Al-4 V, the higher cost, lower strength and higher density of Tantalum can be disadvantages for implants and grafts. Bulk Ta with an elastic modulus of 186 GPa may lead to stress shielding. In order to minimize the risk of stress shielding in Ta implants, incorporation of significant macro-porosity is highly recommended. Additionally, bulk Ta is very difficult to process using traditional manufacturing techniques (i.e. subtractive manufacturing). On the other hand, it is possible to make nearly 100% dense parts that satisfy both chemical and mechanical requirements for ISO 13782'Unalloyed Tantalum for Surgical Applications' by L-PBF [46].

Wauthle et al. used the L-PBF technique to manufacture a porous (80%) pure tantalum implant with fully interconnected pores. *In-vivo* animal studies on rat femurs were performed to evaluate the osteo-conductivity of L-PBF implants. Based on the observations from this study, L-PBF Ta shows better osteoconductivity, biocompatibility and normalized fatigue strength when compared to identical L-PBF porous Ti-6Al-4 V samples [43].

3.2 Other Biomaterials

3.2.1 PEEK

PEEK (PolyarylEtherEtherKeton) is a prominently used biomaterial in orthopedic industry, especially for AM orthopedic implants. Properties including desired mechanical properties, stability at high temperature and biocompatibility make PEEK ideal for biomedical applications. Unlike other common polymers, PEEK material undergoes four thermal transitions during thermal processing using material extrusion in AM heated: glass transition temperature (T_g), melt temperature (T_m), flow temperature (T_f), and re-crystallization transition temperature (T_c).

PEEK has a bulk elastic modulus of about 3.5 GPa which can be tailored to match the desired bone structure by varying its chemical or architectural composition.

Table 5 Mechanical properties of PEEK samples [20]

Sample	Yield strength [MPa]	Yield strain [MPa]	Ave. compressive strength [GPa]
38% porosity	29.34	0.044	–
0% porosity (solid)	102.38	0.056	1.82

Table 6 Mechanical properties of common bioceramics [49]

Material	E modulus [GPa]	Compressive strength [MPa]	Tensile strength [MPa]
Alumina	380	4500	350
Bioglass ceramics	22	500	56–83
Calcium phosphate	18–28	517	280–560

Like many other biomaterials, the addition of carbon to PEEK improves the stiffness of this biomaterial. PEEK is often mixed with barium sulfate to improve the radiolucent properties for post-operative imaging purposes.

Bulk implants made of PEEK biomaterials are shown to be bio-inert in both soft and hard tissue; although, they do not allow bone in-growth [47]. For this reason, coating, surface modification, and microarchitecture enhancement of PEEK implants are used to improve their bioactive and osseointegration properties.

In a study, Vaezi and Yang [20] investigated the effect of controlled temperature (in the Nozzle, build plate and printing environment) on the quality of 3D printed parts. In order to avoid warpage and delamination, the Nozzle temperature within the range of 400–430 °C, with the build plate and ambient temperature of 130 °C and 80 °C respectively, were suggested for an extrusion rate of 2.2 mg/s. Results of mechanical testing of extruded PEEK samples with 38% porosity is compared with solid PEEK-OPTIMA LT1 samples in Table 5.

3.2.2 Ceramics

This group of biomaterials is often used for coatings of orthopedic implants. Since the main objective of many orthopedic implants is load-bearing, brittleness of bulk ceramics is a concern for orthopedic implants. Ceramic materials like Alumina and Zirconia have shown high fatigue strength and excellent tribological properties. For these reasons, use of these ceramics for total joint replacement, especially femoral heads, have had some success. Ceramics can also add bioactive characteristics to the metallic implant. Although, metallic implants meet a majority of the mechanical, manufacturing, and economic requirements of orthopedic implants, they have lower osseo-conductivity with surrounding tissues when compared to metal oxides. This poor interfacial bonding between an implant and tissue may lead to aseptic loosening of the implant at the site and ultimately failure of the device [48]. Coating the implants with bioactive materials (i.e. bio-ceramics) alleviates this weakness of metallic implants. Mechanical properties of some of the common bio-ceramics are listed in the Table 6.

Currently, Calcium Phosphate (CaP) and hydroxyapatite (HA), $Ca_5(PO_4)_3OH$ which has a modulus of elasticity close to cortical bone, is the most common bioceramic used for coating implants. HA is one of the main components of natural bone which helps with functional load bearing. Although, several methods including solution-based coatings and vapor deposition techniques have been developed for surface coating purposes, plasma spraying is the only FDA approved technique [48]. In this technique, molten hydroxyapatite is sprayed on to the surface of an implant at high temperature (up to 12,000 °C) with high velocity. Unfortunately, difficulties in the control of process variables in plasma spraying and higher processing temperature can cause serious damage to the implant, including formation of an amorphous CaP phase in the film and alternation of the surface and even internal structure of an implant (especially those with complex microarchitecture) [48]. Several studies argue that solution-based methods are better alternatives due to these disadvantages in the plasma spraying technique [48]. However, all these methods have their own challenges and result in the need for expensive secondary coating processes to the manufacturing steps of an orthopedic implant. An alternative solution could involve integrating a metal-hydroxyapatite mix in the manufacturing phase by taking advantage of AM's layer-wise inherent property. In other words, AM could be theoretically used to create complex microarchitecture and have an ability to use different material systems with spatially controlled chemical compositions at every single layer and/or region of a single part. Unfortunately, insufficient studies have been published to evaluate this approach in orthopedic applications.

4 AM Design Considerations

In traditional manufacturing, expensive investment is required for tooling (e.g. jigs and fixtures) which limits the commercially offered implant design (i.e. shape and geometry). However, with the "freeform" layer-wise approach in AM, custom implants with an additional ability to control surface porosity can be fabricated in order to enhance osseointegration [50, 51]. AM allows us to design scaffolds with a microarchitecture closer to bone [52]. Additionally, the ability of AM provides a potential for the design of Functionally Graded Material (FGM) implants such as fracture fixation devices [53]. Clearly different implants have different functions, so their structure and design should be based on their specific functional requirements. The hierarchical structure of human bone is very complex and at the same time very efficient. As illustrated in Fig. 10, the hierarchical structure of bone consists of 4 main levels: macro-structure, micro-structure, nano-structure, and subnano-structure. Each of these levels has their own mechanical and biological function [28]. In the macroscale, the bone structure is classified as cortical bone (compact bone) or trabecular bone (cancellous). These two classes of bone have different microarchitecture and porosity. Therefore, an ability to precisely design and control the porosity may be important in an orthopedic implant to encourage vascularization and bone ingrowth and to mimic the properties of the site. Based on published

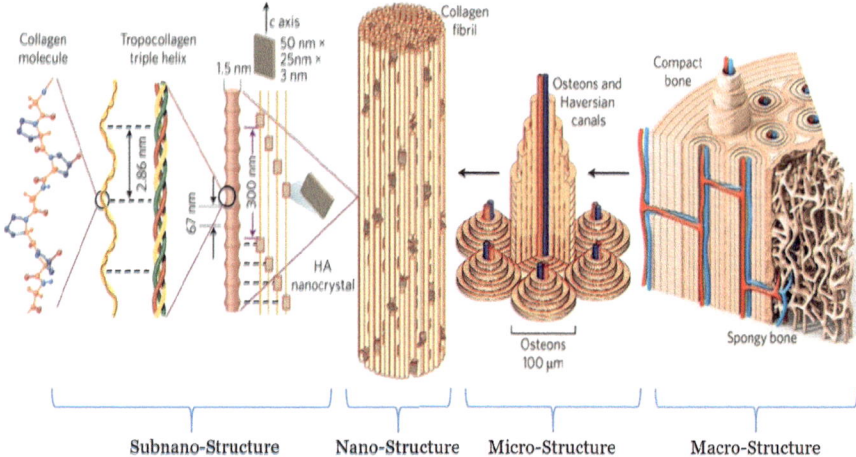

Fig. 10 Hierarchical structure of bone. Reprinted with permission from the reference [55]. Copyright 2014 Macmillan Publishers Limited

reports, the size of the cellular (or porous) structures in compact bone ranges from 10 to 500 μm [28] and varies between 50 and 90% in trabecular bone porosity [28]. Although, it may not be required to employ nano and sub-nano details of human bone into the orthopedic implant, the macro and micro design must be efficient and hierarchical. Concurrently, manufacturing processes should be able to create scaffolds or implants which can cover the range of both cortical and trabecular bone properties [54].

Conventional orthopedic implants manufactured through casting or forging do not have the structural characteristics of human bone. In some cases, they might resemble the overall shape of the bone but they lack the complex microarchitecture of human bone. In recent years, several groups have studied the porosity and cellular microarchitecture of bone and have employed these characteristics to create various types of patient-specific orthopedic implants using AM [28, 54].

Microarchitecture of an implant will significantly affect its mechanical properties, biodegradability, biological properties, and osseointegration characteristics. But, controlling all these parameters by changing the microarchitecture of an implant is very difficult using conventional manufacturing processes. However, advanced design methodologies, such as topology optimization techniques [56], can achieve multiple requirements of an ideal orthopedic implants simultaneously. Topology optimization of orthopedic implants is a fast growing area of interest by exploring design freedom in AM [28, 57, 58]. This section explains the overall design procedure of patient-specific orthopedic implants. Additionally, it provides a summary of the current state of design and topology optimization for AM orthopedic implants.

4.1 Patient-Specific Design Procedures

The first step in the design of a patient-specific orthopedic implant is collecting the anatomical data of the patient through medical imaging. Medical imaging techniques used to obtain the Digital Imaging and Communications in Medicine (DICOM) data varies for different anatomical and biological sites. Computed Tomography (CT scans) and Magnetic Resonance Imaging (MRI) are the most common imaging techniques used for 3D anatomy reconstruction purposes. CT provides superior imaging for bone. Images taken using MRI have better soft tissue contrast than CT, but MRI is more expensive and it lasts 30 min to an hour, while CT scan is a relatively faster process (5–20 min).

Resolution and noise present in the DICOM images determines the quality and accuracy of the 3D reconstructed model. Sometimes additional image processing, such as filtering (i.e. noise removal), is required to enhance the quality of data captured in the images [59]. 3D anatomy reconstruction out of DICOM images requires specific software applications. Some software offers joint segmentation and 3D reconstruction such as Mimics, developed by Materialise. There are other software programs like ITK-Snap which only offer a segmentation toolbox where the user needs to obtain the 3D model from segmented images. Depending on the resolution of the initial DICOM images and the algorithms used for segmentation and 3D reconstruction, the final anatomy model can be "noisy" [60]. In this case, smoothening and secondary filtering of the 3D surface model is required. Some of the commonly used software for this purpose are Magics and Geomagic Wrap. Usually, designing an implant for an anatomical model that consists of a lesser number of triangular mesh elements is easier and faster. Subsequently, the STL (stereolithography) file of the 3D reconstructed anatomy is now ready to be used for the design of a patient-specific implants using Computer Aided Designing (CAD) software. Preparation of the implant for manufacturing is explained earlier (Fig. 1). For design validation purposes, cadaver testing may be carried out and compared with Finite Element Analysis (FEA) results. Figure 11 illustrates the discussed steps in the design of an orthopedic implant using AM. It needs to be mentioned that the design procedures are not limited to these steps. In fact, it is a common practice to perform an FEA during and after the design of the implant is completed. Results of FEA (e.g. stress during mechanical loading) determine the need to redesign or proceeding with its fabrication using AM. Unfortunately, because of the difficulty in obtaining the loading conditions directly from the patient, it is a common practice to rely on the estimated boundary conditions from previously published reports. But, mechanical loads could widely vary based on patient age, sex, anthropometric measurements and activity levels.

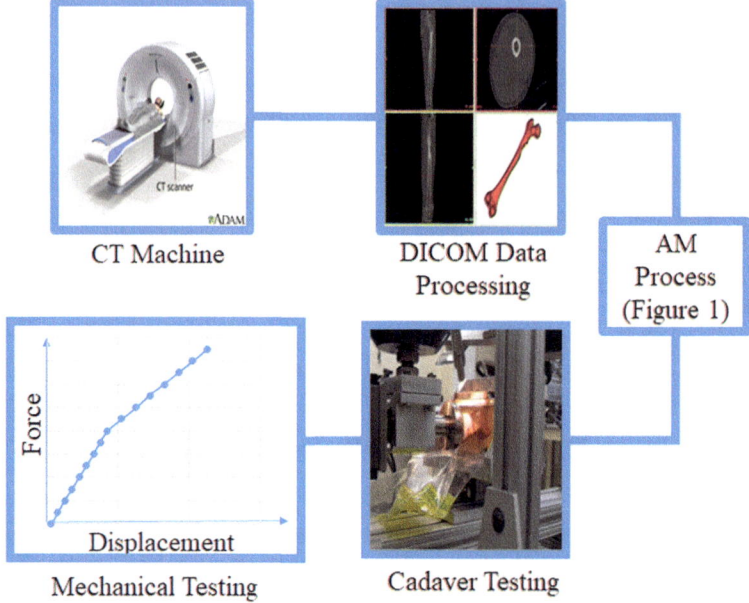

Fig. 11 Patient-specific orthopedic implant's design process flow

4.2 Porosity

According to Wolff's law, human bone remodels its internal structure based on applied load distribution [61, 62]. Therefore, in order to preserve the surrounding tissue (after implantation), the design of the implant should enable load distribution to facilitate an appropriate mechanical stimulation of the neighboring osteons during physiological loading [63]. Another important design criteria arises when the implant undergoes a large amount of stresses, the implant should not magnify the high stress peaks [63]. Porous structures in the implant can be incorporated through AM as an effective tool to design and manufacture an implant that satisfies criteria.

Various studies have been performed to investigate the relationship between the porosity and mechanical and biological behavior of AM orthopedic implants. During the design process of porous implants, there are three key parameters (pore configuration, ratio between beam thickness and beam length, and pore density) which can be varied for design optimization. Initial studies in this area showed that the elastic modulus and stiffness of an implant can be controlled by varying the number of pores [62]. By increasing the number of pores (or cellular structures), the total surface area in contact with neighboring tissue is also increased; this leads to better osseointegration properties for the implant [62]. Van Bael et al. conducted a study to analyze the effect of both pore shape and pore size using Ti-6Al-4 V scaffolds with three different unit cell shapes (triangular, hexagonal, and rectangular) and two different sizes (500 µm and 1000 µm) that were manufactured using SLM [54]. It was found that in larger pores, cells tend to attach to a single strut and in scaffolds with smaller pores, cells attach to several struts. Incorporating these observation on porous implant designs suggest that large pore sizes should be used on the outer surface of the implant to avoid pore occlusion and smaller pore size in the inner volume to decrease permeability. Also, it was noticed that pore occlusion can be delayed with triangular shaped pores [54]. Based on these findings, it can be concluded that in order to achieve efficient microarchitecture as bone, implants with a gradient porosity could be an alternative solution.

In summary, porous structures in AM orthopedic implants offer several advantages namely: (1) Finer control of stiffness of the implant to prevent stress shielding by mimicking the stiffness of the neighboring bone, (2) reduction in the weight of the implant and (3) improvement in osseointegration characteristics of the implant by increasing the contact surface area.

4.3   Clinical Applications

In orthopedic surgical procedures, surgeons often implant artificial components for structural purposes. In joint arthroplasty, these components serve to replace joints such as the hip, knee, shoulder and ankle with low friction bearing surfaces. These joints are often replaced due to debilitating osteoarthritis, but joint replacements can also be used for other reasons such as trauma. In fracture fixation, artificial implants typically serve to stabilize bone fragments. These fragments can be separated by simple fractures or complex, comminuted fractures that may result in many fragments or a bone void. Large bone voids can also result from surgical bone resections in patients with bone tumors or infected bone tissue. Patients with these large bone voids are not as common as those receiving joint replacements or fracture fixation, but they can be the most difficult to treat. The potential of AM implants in each of these procedures is discussed further below. Artificial implants are also used for orthopedic procedures such as joint arthrodesis and soft tissue reconstruction. However, the ability to capture and process 3D imaging data (e.g. osteophytes, poor quality bone that may be resected normally during surgery) in a short clinical timeline (e.g. trauma) needs to be improved for broader adoption of AM in patient-specific orthopedic implants.

4.4   Patient Variability

An important aspect in orthopedic patients that raises a potential need for AM is patient variability. First, anatomic variability exists naturally across healthy humans. Variability is observed at the macro-scale in the shape of bones [64] and soft tissues and connections among tissues [65]. For example, the glenoid version (angulation of the shoulder socket in the horizontal plane) in healthy individuals has a range of at least 20° [66]. It is obvious that bones have vastly different lengths and sizes across individuals of different height and weight. Variability is also evident at the micro-scale in morphology and mechanical properties. Young individuals, on average, have bones with stronger, denser trabecular tissue and thicker cortices than bones in older females.

Anatomic variability is magnified in orthopedic patients. The glenoid version, instead of having a range of approximately 20° in healthy individuals, ranges from 1.4° to 56.8° [67] in patients with osteoarthritis. The trabecular bone in osteopenic or osteoporotic patients has vastly degraded morphology, and in some patients, regions of trabecular bone have completely resorbed away. Many classification systems have been proposed for fracture patterns in trauma patients, but in more severe cases the fracture may not clearly fall within a category or the patient presents with complicating factors such as additional fractures or compromised soft tissue. Bone defects associated with cancer and infection can be found in many different locations.

Substantial variability also exists in patient factors other than anatomy. Patients with impaired bone healing, associated with genetic or environmental factors such as heavy smoking may have slower implant osseointegration or may subject implants to larger mechanical loads for a longer period of time. The mechanical loads with which an orthopedic implant serves under varies for different patient sizes and activity levels. Anticipated mechanical loads dictate the necessary material static and fatigue strength. Several major studies using instrumented joint replacements confirm that the lower limb joints are loaded with several times the bodyweight during gait, and even the shoulder can be loaded by greater than full bodyweight, due in part to contractions of antagonistic muscles [68]. In addition to joint replacements, these large forces are applicable in fracture fixation and other orthopedic implants. Athletes anxious to return to their sport might subject an implant to magnified mechanical loads, whereas a patient instructed to not bear weight might subject an implant to very little loads. Patients with lower limb injuries are often instructed to bear weight as tolerated, however this weight varies across patients due to both physical and psychological factors. A surgeon must also consider that patients may not be compliant with their postoperative rehabilitation orders, or may have accidents that subject implants to large single loading events.

4.5 Shoulder and Other Joint Replacements

Joint arthroplasty refers to replacing one or both sides of a joint with synthetic components. The most frequent reason is degenerated cartilage and subchondral bone associated with osteoarthritis. However, joint replacements can also be used to relieve symptoms of rheumatoid arthritis, post-traumatic arthritis, avascular necrosis and other conditions. Hip and knee arthritis and replacements are more common

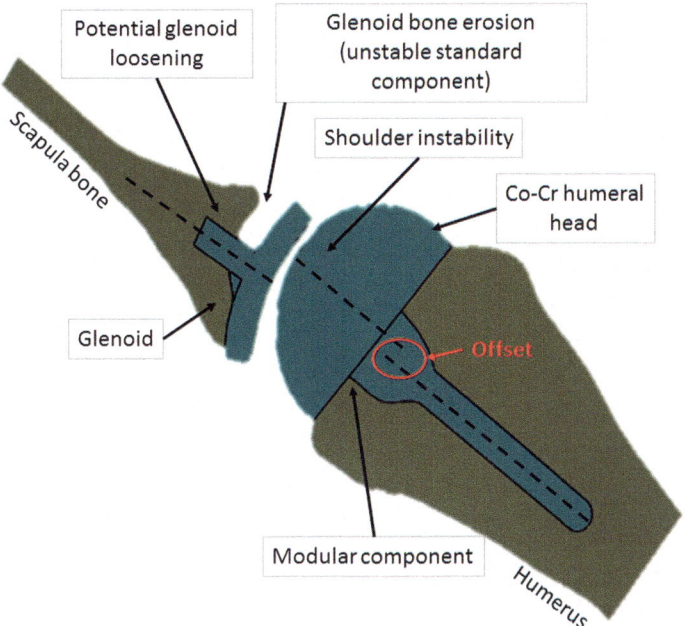

Fig. 12 Top view schematic of standard total shoulder replacement, along with some potential limitations

than arthritis of the shoulder, ankle, or other joints. It should be noted that the rate of arthroplasty for each joint is dramatically increasing [69]. Shoulder, or glenohumeral, joint replacement is focused in this section but many of the concepts also apply to other forms of joint replacement.

The most frequent complications following shoulder replacement are shoulder instability and glenoid (shoulder socket part of the scapula) component loosening. The most challenging patient may be the individual suffering from glenoid bone erosion combined with shoulder subluxation prior to surgery. Glenoid bone erosion associated with the arthritic process results in the highly variable glenoid version angle, along with other variable three-dimensional morphology characteristics as illustrated in Fig. 12 [70]. The volume of glenoid bone present for component fixation is drastically reduced and the perforation of the cortex during drilling occasionally occurs [71]. Most implant manufacturers provide several different sizes of glenoid components for the surgeon to choose from in their standard surgical tray. The surgeon must ream down the 'high side' of the already limited glenoid bone or utilize specialized bone grafting techniques to implant a standard glenoid component at an anatomically normal version angle [71]. Furthermore, it is difficult for the surgeon to know what proper alignment is because of limited intraoperative exposure of the glenoid and without any access to the scapula body. Because of these challenges, patients with severe glenoid deformity typically cannot receive an anatomic total shoulder replacement, and currently the surgeon must resort to less

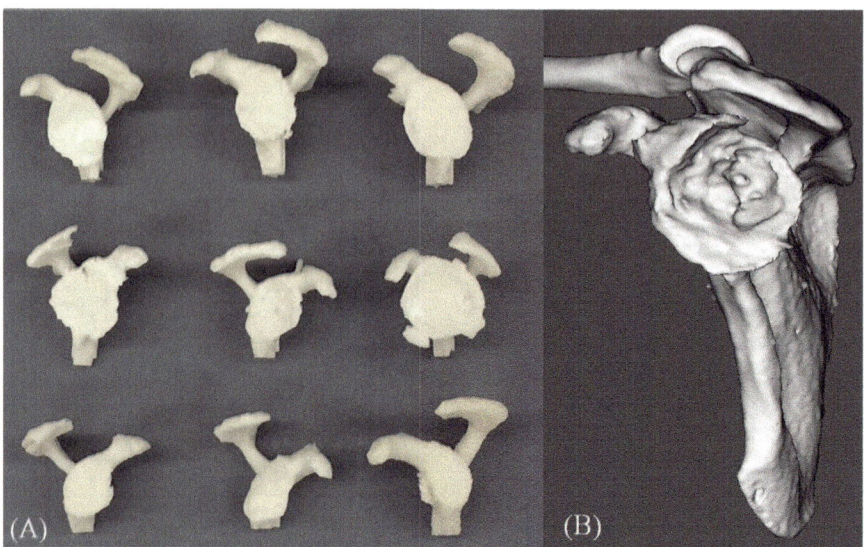

Fig. 13 (**a**) 3D printed models of 9 shoulder replacement patients demonstrating variable glenoid morphology. (**b**) Patient with severe posterior glenoid bone deformity

desirable alternatives such as replacement of only the humeral side of the joint (Fig. 13).

Additively manufacturing the glenoid component leads to enhanced flexibility in morphology of the medial 'back-side' of the component by improving the mate and fastening to glenoid bone. This may take the form of geometric features that interlock with locations of strong cortical bone, fixation pegs that are precisely centered within the glenoid vault bone and/or screws that are positioned for best purchase in dense bone [72]. An AM made glenoid with custom backing that geometrically mates to the bone could also address the challenges in the proper alignment of the components intraoperatively. Studies have demonstrated that augmented components (with standardized asymmetric backings, albeit not produced by AM) can help restore normal glenoid anatomical alignment and may improve component fixation [73]. In addition, AM offers new possibilities in the fabrication of porous metal backings that could improve bone ingrowth for long term un-cemented fixation component.

On the humeral side of the shoulder replacement, surgeons typically aim to restore an anatomically correct version angle and offset of the humeral head. The humeral head is a highly polished metal such as cobalt chrome and the stem may be a different metal such as titanium. Some popular implant systems include modularity in which the head and stem are separate components that are press fit together immediately prior to implantation. Currently, humeral heads are commercially available in discrete standard sizes. The humeral head angulation and positional offset relative to the stem can be adjusted on a patient-specific basis by the angle at which the head is fastened to the stem with an eccentric Morse taper connection.

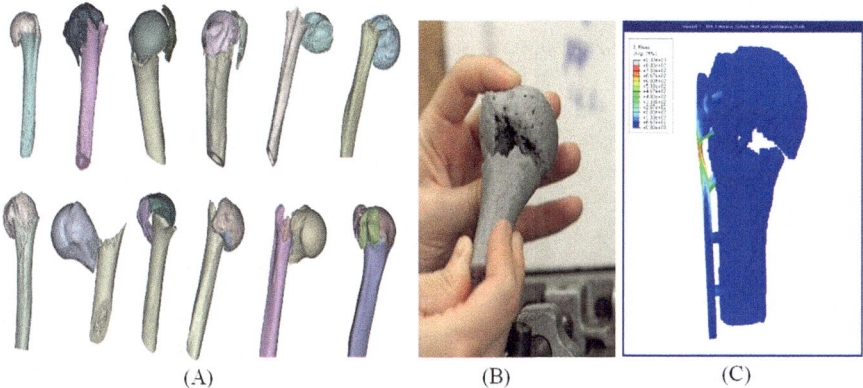

(A)   (B)   (C)

Fig. 14 Proximal humerus fracture fixation models derived from patient CT scans. (**a**) A series of geometric models of different patients demonstrate substantial variability in fracture pattern. (**b**) 3D printed models were created for these patients, and fractures were induced by a surgeon and poses digitized. (**c**) Finite element models have been generated which demonstrate large plate and screw stresses when stable reduction is not obtained

Finer control in the humeral head offset may help centralize joint contact pressures and stabilize the shoulder according to recent laboratory and clinical studies [74, 75]. It should be noted that standard sizes of components and component modularity are also available for other joint replacements.

AM of the humeral component would provide additional flexibility in head size, version, and offset, as well as stem length and diameter. An AM humeral component could reduce the need for modular connections that are prone to micromotions and fretting corrosion. As with the glenoid, new porous metal structures integrated into the humeral stem could improve bone ingrowth for uncemented component fixation.

4.6 Fracture Fixation

In order to provide effective patient-specific AM solutions in fracture fixation, many acute trauma cases will require compressing the AM implant design and fabrication timeline to realistically fit into a clinical workflow of hours or days and not weeks. Bone fractures are by nature highly variable due to a wide variety of injury mechanisms and the complexity of fracture mechanics in bone. Higher energy injuries tend to create more complex, comminuted fractures having more separate fragments [76]. Many classification systems have been proposed for categorizing fractures based on visual fracture pattern, but the inter-observer reliability demonstrated by these classification systems reflects the fact that no two fractures are the same. In addition, as mentioned before, patients present a variety of bone qualities, body types, activity levels, and healing capacity (Fig. 14).

Surgical fixation of fractures usually involves reducing the bone fragments to restore preoperative anatomy and fastening a stainless steel or titanium-based implant to the fragments with screws. The screws are often similar in material to the main implant to reduce galvanic corrosion. Two basic types of implants are most common. Plates are positioned outside the bone cortex and conversely nails are inserted within the cortex down the intramedullary canal of long bones. Straight and pre-contoured anatomic plates and nails are commercially available. Bicortical screws provide the strongest fixation. Unicortical or cancellous screws may rely on fixation within weaker trabecular bone, but are often necessary to avoid damage of surrounding soft tissues or worse induce penetration into a joint space. Screws may be non-locking or locking, the latter meaning that the screw head fastens to the plate with a rigid connection. Simple fractures that heal by primary bone healing can benefit from compression of the fracture site using one of several techniques. Conversely, comminuted fractures or simple fractures that experience interfragmentary motion heal by secondary bone healing through callus formation. In such fractures, mechanical loads experienced by the implant are larger until sufficient tissue stiffening occurs at the fracture site. Postoperative stresses and strains experienced in a fracture fixation construct depend on surgical related variables including implant size and material, and construct design (e.g. positioning of screws). These stresses and strains also depend on patient-related variables mentioned above.

Complications in fracture fixation post-surgery include delayed or poor fracture healing, fracture healing with residual bone deformity, implant or screw yield or fatigue failure and screw loosening. Primary bone healing requires very small strains at the fracture gap. Secondary bone healing is believed to benefit from a moderate level of axial strain (perpendicular to the fracture plane) but to be inhibited by shear strain. Screw loosening is generally associated with cyclic loading and progressive damage of surrounding bone due to high stresses and strains.

Although the aforementioned clinical workflow issues are a substantial obstacle, AM may hold significant potential for improving fracture fixation in a subset of patients. Implant manufacturers already provide a large array of options to surgeons including plates and nails of multiple lengths and sizes, holes for screw insertion at various positions, and screws with lengths varying by 1–2 mm. However, two areas of opportunity are focused on here: (1) patient-specific implant shape; and (2) patient-specific screw position and orientation.

Patient-specific intramedullary nails for long bones such as the femur could better match the anatomic curvature of the patient's intramedullary canal. The radius of the femur varies, with one study reporting a mean of 120 cm with standard deviation of 36 cm [77]. Standardizing a single standard radius for femur intramedullary nails has been challenging for implant manufacturers with different manufacturers offering different radii. Non-anatomically conforming nail radii can result in painful pressing of the nail on the inner cortex, injury during nail insertion, or angular deformity in the limb as the bone conforms to the nail through motion at the fracture site. The anatomic radius for the patient could be obtained from the non-injured contralateral limb (if applicable). Additionally, the nail diameter could be customized beyond standard options offered.

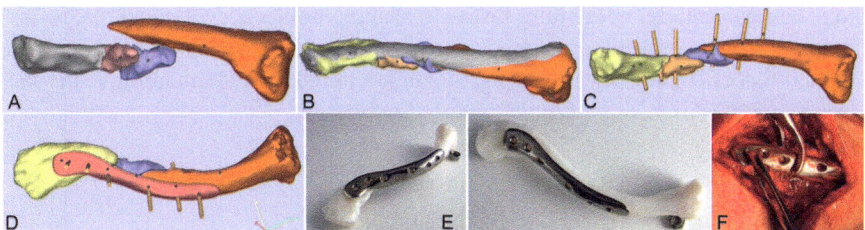

Fig. 15 Process flow from anatomy reconstruction to test fitting of the EBM plate in surgery. Reprinted with permission from the reference [83]. Copyright 2015 ASME

Additively manufactured plates could open new design possibilities for stabilizing difficult fractures. For example, in proximal humerus and proximal femur fractures in patients with poorer bone quality, a major clinical concern is varus collapse of the head of the bone associated with eccentric deforming joint forces and cancellous screw 'cut-out' (loosening) [78, 79]. Long term construct stability is difficult using only screw-based fixation of the plate. Novel AM plates may provide radically different approaches for bone fragment stabilization. Patient-specific plates could also conform better to anatomic bone shape and assist with obtaining proper reduction without requiring intraoperative plate contouring. Plate conformity with the outer cortex is important when using non-locking screws in which mechanical stability of the construct depends on plate-to-bone friction. Locking plates and screws, conversely, often have a 1–2 mm gap between the plate and bone.

Patient-specific screw position and orientation could also be beneficial to a subset of patients. A common goal for plate fixation, especially for osteopenic or osteoporotic patients, is to insert screws into denser, higher quality bone to mitigate the risk of screw loosening. In patients with preexisting joint replacements near a fracture, it is necessary for screws to avoid the prosthesis [80]. In more complex fractures it is not uncommon for a screw to inadvertently pass through the fracture site, which reduces mechanical stability and can impair healing. Variable angle locking screws are a recent technology that enable flexibility in screw orientation, but these systems have been shown to have decreased mechanical strength [81] and increased rates of clinical construct failure [82] relative to traditional systems.

As an example, clavicular fractures are often treated nonoperatively [83] but may benefit from internal fixation in a subset of patients. Commercial clavicle fixation devices are known to be difficult to use and fitted at the implantation site. This is because, clavicle bone has a complex geometry which involves large variations in curvature, torsion and inclination; also it is located close to essential organs [83]. In a study done by Cronskar et al., the feasibility of customization of clavicle fixation was investigated. In this study, titanium plates for three case studies were manufactured using EBM. All the plates were post-processed, cleaned and sterilized at the hospital; then they were fitted at the implantation site in place of commercial plates by the same surgeon. Process flow, from reconstruction of the patient's anatomy to design, validation and test fitting (in surgery), for one of the case studies, is illustrated in Fig. 15. It was confirmed that EBM fracture plates were conforming to the bone contour and did not require further reshaping unlike conventional plates. It is known that screw positioning and angles can be crucial factors in determining the clinical success of an operation. Surgical guide offered by AM can improve the success rate in this aspect [84, 85].

4.7 Large Bone Defects

There are various causes of large bone defects in patients, including trauma and resection associated with tumors or bone infection. Historically, management for severe tumors of the bone and soft tissue required amputation with wide margins. With advances in surgery, radiation therapy and chemotherapy, long-term survival for patients with osteosarcoma has increased, e.g. from 20 to 70% for osteosarcoma of the distal femur. Limb salvage surgeries have been developed which incorporate an implant along with reconstruction of soft tissue to attain biomechanical functioning of a limb. Limb salvage surgeries may prevent chronic pain, "phantom pain", and neuromas instead of severing neurovasculature during amputations. Unfortunately, amputations are still often needed especially in pediatric patients. Most implants are designed for the adult population, and are not scaled to pediatric patients. As a result, an 8 years old pediatric patient with a primary malignant tumor of the bone is recommended for amputation because of this lack of implant options. Amputations can cause multiple complications for patients, including wound care, chronic pain, neuromas, recurrent ulceration, poor residual limb padding and decreased function [86].

The emergence of AM now provides the ability to create a design that fits a patient's specific anatomy. For example, the implant can be married with patient's medullary canal to create a "press fit design". A patient specific implant can reduce strain which is associated with implant loosening, periprosthetic fractures and dislocations. Recently, AM has been incorporated into the preoperative and operative decisions for bone tumor resections [87]. AM-made titanium endo-prosthetics have been demonstrated to be effective in limb salvage surgeries [88]. AM made implants can also be applied to patient specific anatomy that is especially challenged by available implants, such as distal tibia, pelvis, talus, calcaneus, distal humerus and other reconstructive procedures (Fig. 16).

In 2015, H. Fan et al. published a report on three cases that underwent limb salvage surgeries to receive patient-specific prosthesis [89]. In one case, a 35-year-old patient was diagnosed by right scapular Elwing's sarcoma. The radiography results showed severe bone loss. With follow-up CT and coronal MRI imaging, the team could identify the origin of the tumor to be the scapula which was later extended into neighboring muscles and forming large soft tissue mass at that site [89]. The design of the scapular prosthesis was based on the 3D reconstructed anatomy of the contralateral site. After an adjuvant chemotherapy, cancerous scapular was removed from the patient, then a porous Ti-6Al-4 V prosthesis manufactured using EBM was implanted and secured at the site. Figure 17 demonstrates the explained process flow. Postoperative follow-ups for 21 months provided no evidence of loosening, length discrepancy nor failure [89].

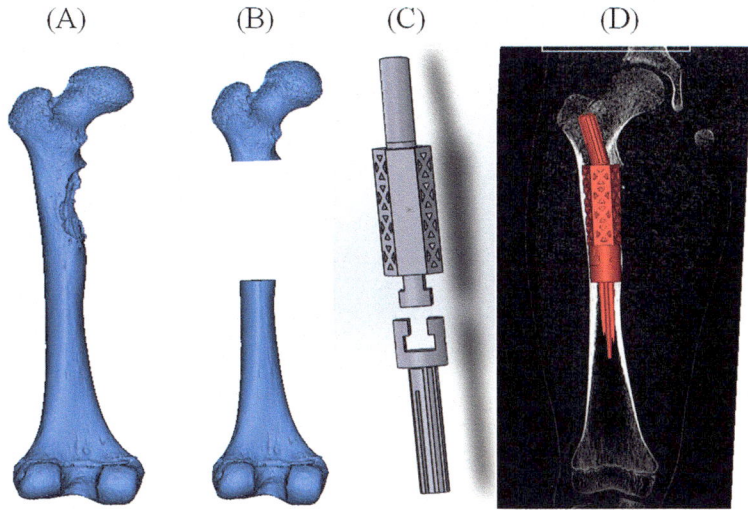

Fig. 16 Preliminary custom implant designs for a pediatric male previously seen for osteosarcoma of the proximal femur (**a**, **b**). The implant includes a modular locking mechanism and truss structure (**c**). The implant shape is designed to restore stability and anatomy while fitting the existing intramedullary canal (**d**), although additional adjustments are being made for sagittal plane alignment

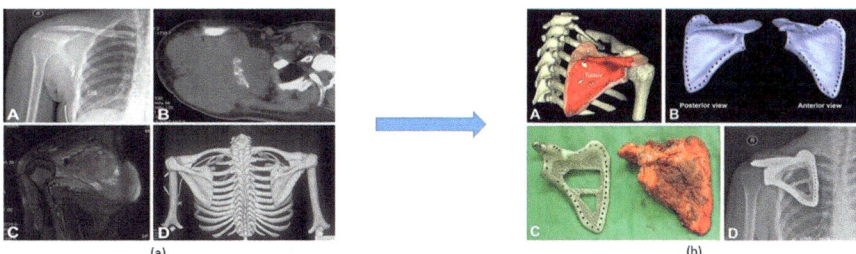

Fig. 17 Scapular replacement process flow [89]; (**a**) Patient's medical image data, (**b**) (*A*) Anatomy reconstruction, (*B*) Implant design, (*C*) Corresponding AM manufactured prosthesis, and (*D*) Implantation and postoperative imaging. Reprinted with permission from reference [89]. Copyright 2015 Fan et al.

4.8 Surgical Guides

Fabrication of patient-specific surgical guides using AM involves surgical planning in its process flow unlike the traditional method. The first maxillofacial osteotomy guide using AM was reported by Richard Bibb et al. [90]. In the reported case, distraction osteogenesis was performed, during which maxilla was gradually moved relative to the rest of the skull by mounting it on two precision locator devices. In this operation, two separate cuts were made across the maxilla under the nose and to remove some bone from the skull. The designed surgical guides for each cut were

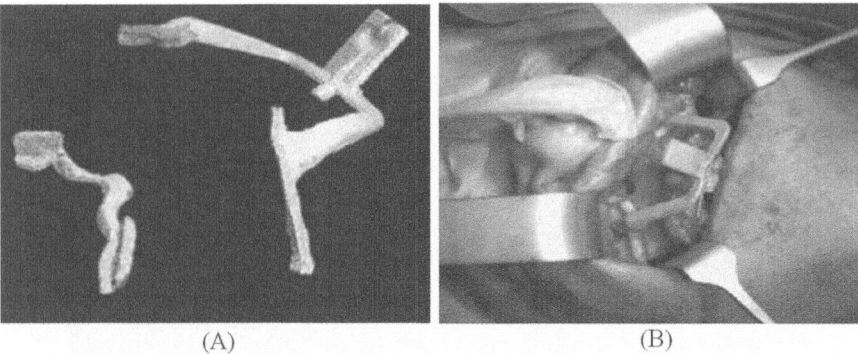

(A)   (B)

Fig. 18 (**a**) Two osteotomy cutting guides, (**b**) Use of surgical guide in operation. Reprinted with permission from reference [90]. Copyright 2009 Emerald Group Publishing Limited

modeled using FreeForm software and fabricated using L-PBF. The designed surgical guides helped the surgeon to have a better control during the surgery [72]. Orthopedic implant companies now provide 'patient specific instrumentation' (PSI) for procedures such as knee replacement. Despite potential improvements that could be achieved using custom surgical guides, additional investigation is required to streamline the surgical planning-design-AM fabrication of patient-specific guides (Fig. 18).

4.9 Additional Clinical Examples

In another case study reported by Jardini et al. [91], a 22-year-old patient with a large cranial defect received a patient-specific cranial implant. In this case, the restoration had to be done for missing area of approximately 106 cm^2; this large bone defect was the result of a decompressive craniectomy. After, obtaining the medical images from the patient, and processing the DICOM data, anatomy of the patient's skull was reconstructed and converted into an STL file. Subsequently, the reconstructed anatomy was used for both surgical planning and design of patient-specific implant. The design of the implant was based on the contralateral side of the skull. The team used L-PBF AM process to fabricate the implant from Ti-6Al-4 V alloy. In addition, reconstructed anatomy model was processed using binder-jetting AM process (Zprinter 510) for surgical planning and form fit testing. Figure 19 details the discrete steps from processing the medical images of the patient to fitting the customized cranial implant and final surgery for implantation into the patient's skull.

In 2016, Nanfang Xu and his team [92] published a report on the first customized AM vertebra implantation. Cervical replacement surgery involves complicated procedures. In this case report, a 12-year-old patient diagnosed with a C2 Ewing sarcoma underwent a pre-planned staged intralesional spondylectomy [92]. After

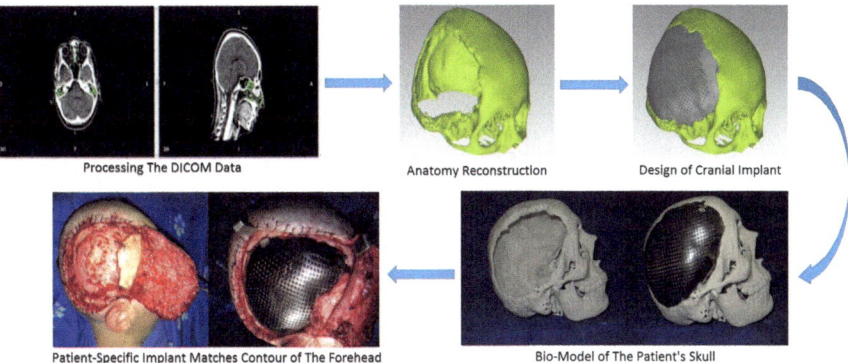

Fig. 19 Design procedure of patient-specific cranial implant. Reprinted with permission from [91]. Copyright 2017 Elsevier Inc

excision of C2 vertebra, the inferior and superior surfaces of C1 and C3 were decorticated, respectively. This allowed a proper fitting of the patient-specific self-stabilizing artificial vertebral body (Fig. 20). The porous component promoted osseointegration and was manufactured using an EBM technique from a titanium alloy. Using a porous AM clavicle implant eliminated the need for bone grafting and ventral cages. However, performing a similar operation with conventional implants requires both bone grafting and ventral cages to augment instrumentation by dorsal approach. Another advantage of AM clavicle implant over a conventional implant was the improved local stability which is the result of a customized design.

5   Summary

This chapter provided a review of AM technologies, materials, design processes and survey of reported clinical applications of AM in orthopedics. Despite recent advancements in AM of orthopedic implants, there are several challenges that need to be addressed for its widespread adoption in orthopedics. In the realm of patient-specific implants, these challenges include a need to semi-automate the data collection and processing of patient's data (DICOM data post processing, anatomic reconstruction and rendering), optimize custom design and AM process planning, and concurrent planning of design-AM-implantation due to surgical and clinical implications. However, most of these issues can be resolved to some extent through better communication between surgeons, physicians and engineers. Additionally, a more reliable validation of the designed implant could be achieved by employing advanced computational modeling approaches. Post-operative follow up from previous case studies is a valuable asset for further development of AM orthopedic implants.

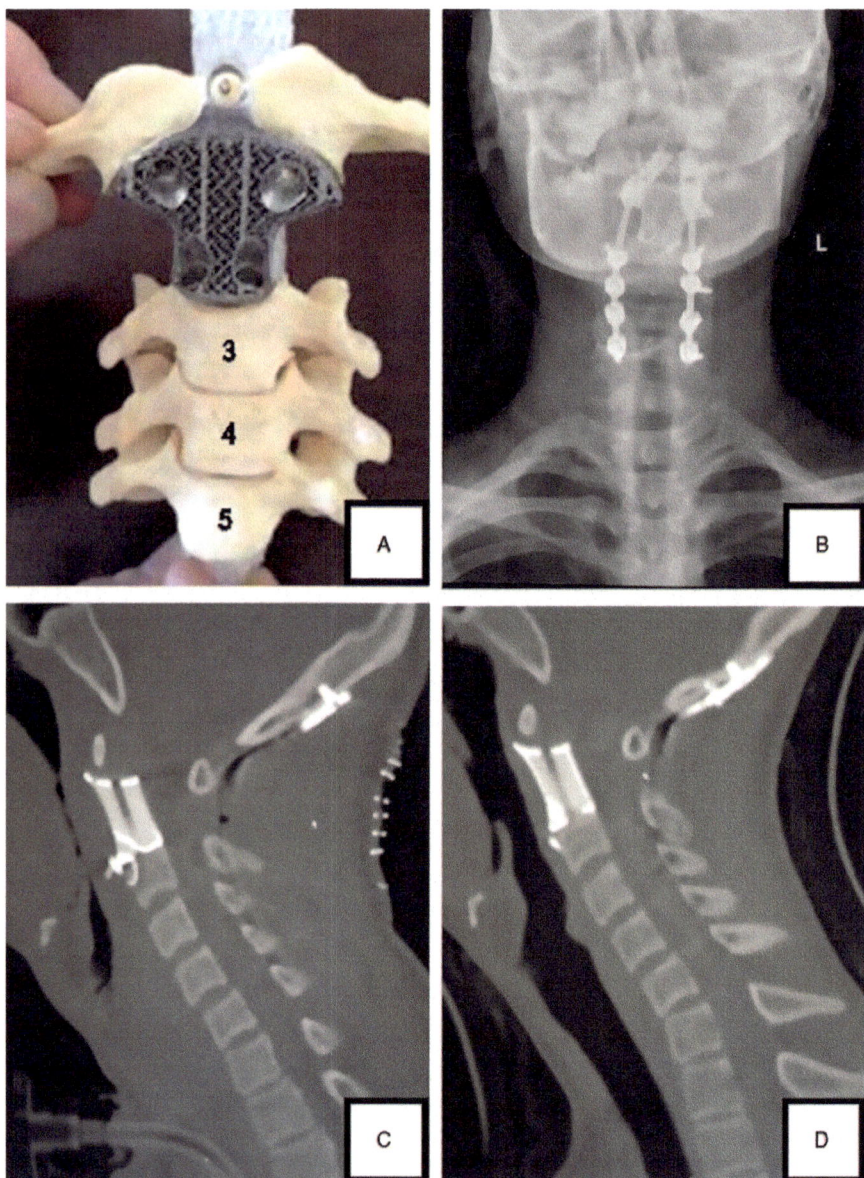

Fig. 20 (a) Fitting test of the AM fabricated vertebral component; (b) Postoperative X-ray; (c) Sagittal reconstruction; (d) 1-year-followup X-ray showing implant osseointegration. Reprinted with permission from reference [92]. Copyright 2015 Wolters Kluwer Health Inc

Achieving enhanced biological compatibility for orthopedic implants is still a big challenge. However, the layer-wise approach in AM allows for implementation of a gradient lattice structure and FGM in design of implants which may lead to better osteoconductive and osseointegration properties.

Acknowledgment We acknowledge Dr. April D. Armstrong, Professor and Assistant Director from the Penn State Hershey Bone and Joint Institute. We also thank Evan Roush for his contribution in preparing the anatomical models and processing the DICOM data. We also acknowledge Conner Zale for his contribution in discussing large bone defects.

References

1. Bennett DWF. Partial custom knee replacement in Sarasota, Florida. 2014. [Online]. http://www.bennettorthosportsmed.com/custom-knee-replacement/.
2. Manogharan G. Hybrid manufacturing: analysis of integrating additive and subtractive methods. Raleigh, NC: North Carolina State University; 2014.
3. Gibson I, Rosen D, Stucker B. Additive manufacturing technologies. New York: Springer; 2010.
4. Almaghariz ES, Conner BP, Lenner L, Gullapalli R, Manogharan GP, Lamoncha B, Fang M. Quantifying the role of part design complexity in using 3d sand printing for molds and cores. Int J Met. 2016;10:240.
5. Winkel A, Meszaros R, Reinsch S, Müller R, Travitzky N, Fey T, Greil P, Wondraczek L. Sintering of 3D-printed glass/HAp composites. J Am Ceram Soc. 2012;95:3387.
6. Bergmann C, Lindner M, Zhang W, Koczur K, Kirsten A, Telle R, Fischer H. 3D printing of bone substitute implants using calcium phosphate and bioactive glasses. J Eur Ceram Soc. 2010;30:2563.
7. Trombetta R, Inzana JA, Schwarz EM, Kates SL, Awad HA. 3D printing of calcium phosphate ceramics for bone tissue engineering and drug delivery. Ann Biomed Eng. 2017;45:23.
8. Taminger, KMB, Hafley RA. Electron beam freeform fabrication: a rapid metal deposition process. Proceedings of the 3rd annual automotive composites conference. 2003;9:10.
9. Brandl E, Baufeld B, Leyens C, Gault R. Additive manufactured Ti-6A1-4V using welding wire: comparison of laser and arc beam deposition and evaluation with respect to aerospace material specifications. Phys Procedia. 2010;5:595–606.
10. Acharya R, Bansal R, Gambone JJ, Kaplan MA, Fuchs GE, Rudawski NG, Das S. Additive manufacturing and characterization of René 80 Superalloy processed through scanning laser epitaxy for turbine engine hot-section component repair. Adv Eng Mater. 2015;17:942.
11. Carroll BE, Otis RA, Borgonia JP, Suh JO, Dillon RP, Shapiro AA, Hofmann DC, Liu ZK, Beese AM. Functionally graded material of 304L stainless steel and inconel 625 fabricated by directed energy deposition: characterization and thermodynamic modeling. Acta Mater. 2016;108:46.
12. Bobbio LD, Otis RA, Borgonia JP, Dillon RP, Shapiro AA, Liu ZK, Beese AM. Additive manufacturing of a functionally graded material from Ti-6Al-4V to invar: experimental characterization and thermodynamic calculations. Acta Mater. 2017;127:133.
13. Karunakaran KP, Suryakumar S, Pushpa V, Akula S. Low cost integration of additive and subtractive processes for hybrid layered manufacturing. Robot Comput Integr Manuf. 2010;26:490.
14. Flynn JM, Shokrani A, Newman ST, Dhokia V. Hybrid additive and subtractive machine tools - research and industrial developments. Int J Mach Tools Manuf. 2016;101:79.

15. Parthasarathy J, Starly B, Raman S, Christensen A. Mechanical evaluation of porous titanium (Ti6Al4V) structures with electron beam melting (EBM). J Mech Behav Biomed Mater. 2010;3(3):249–59.
16. Sochalski-Kolbus LM, Payzant EA, Cornwell PA, Watkins TR, Babu SS, Dehoff RR, Lorenz M, Ovchinnikova O, Duty C. Comparison of residual stresses in Inconel 718 simple parts made by Electron beam melting and direct laser metal sintering. Metall Mater Trans A Phys Metall Mater Sci. 2015;46:1419–32.
17. Townsend A, Senin N, Blunt L, Leach RK, Taylor JS. Surface texture metrology for metal additive manufacturing: a review. Precis. Eng. 2016;46:34–47.
18. Song B, Dong S, Liu Q, Liao H, Coddet C. Vacuum heat treatment of iron parts produced by selective laser melting: microstructure, residual stress and tensile behavior. Mater Des. 2014;54:727.
19. Wu AS, Brown DW, Kumar M, Gallegos GF, King WE. An experimental investigation into additive manufacturing-induced residual stresses in 316L stainless steel. Metall Mater Trans A: Phys Metall Mater Sci. 2014;45:6260.
20. Vaezi M, Yang S. Extrusion-based additive manufacturing of PEEK for biomedical applications. Virtual Phys Prototyping. 2015;10:123.
21. Baich L, Manogharan G, Marie H. Study of infill print design on production cost-time of 3D printed ABS parts. Int J Rapid Manufact. 2015;5:308–19.
22. Kang H-W, Lee SJ, Ko IK, Kengla C, Yoo JJ, Atala A. A 3D bioprinting system to produce human-scale tissue constructs with structural integrity. Nat Biotechnol. 2016;34:312.
23. Scheithauer U, Schwarzer E, Richter HJ, Moritz T. Thermoplastic 3D printing—an additive manufacturing method for producing dense ceramics. Int J Appl Ceram Technol. 2015;12:26.
24. Wu GH, Hsu SH. Review: polymeric-based 3D printing for tissue engineering. J Med Biol Eng. 2015;35:285.
25. Williams DF. On the nature of biomaterials. Biomaterials. 2009;30:5897.
26. Murr LE, Gaytan SM, Martinez E, Medina F, Wicker RB. Next generation orthopaedic implants by additive manufacturing using electron beam melting. Int J Biomater. 2012;2012:1.
27. Campoli G, Borleffs MS, Amin YS, Wauthle R, Weinans H, Zadpoor AA. Mechanical properties of open-cell metallic biomaterials manufactured using additive manufacturing. Mater Des. 2013;49:975–65.
28. Wang X, Xu S, Zhou S, Xu W, Leary M, Choong P, Qian M, Brandt M, Xie YM. Topological design and additive manufacturing of porous metals for bone scaffolds and orthopaedic implants: a review. Biomaterials. 2016;83:127–41.
29. Ratner BD, Hoffman AS, Schoen FJ, Lemons JE. Biomaterials science: an introduction to materials in medicine. New York: Academic Press; 2004.
30. Park J, Lakes RD. Biomaterials: an introduction. New York: Springer Science & Business Media; 2007.
31. Özel T, Bártolo PJ, Ceretti E, Gay JC, Rodriguez CA, Da Silva JVL, editors. Biomedical devices: design, prototyping, and manufacturing. New York: Wiley; 2016.
32. Simon JP, Fabry G. An overview of implant materials. Acta Orthop Belg. 1991;57(1):1–5.
33. Geetha M, Singh AK, Asokamani R, Gogia AK. Ti based biomaterials, the ultimate choice for orthopaedic implants—a review. Prog Mater Sci. 2009;54:397–425.
34. Zhong Y, Liu L, Wikman S, Cui D, Shen Z. Intragranular cellular segregation network structure strengthening 316L stainless steel prepared by selective laser melting. J Nucl Mater. 2016;470:170.
35. Zhong Y, Rännar L-E, Liu L, Koptyug A, Wikman S, Olsen J, Cui D, Shen Z. Additive manufacturing of 316L stainless steel by electron beam melting for nuclear fusion applications. J Nucl Mater. 2017;486:234.
36. Mower TM, Long MJ. Mechanical behavior of additive manufactured, powder-bed laser-fused materials. Mater Sci Eng A. 2016;651(0921–5093):198–213.
37. Gaytan SM, Murr LE, Martinez E, Martinez JL, MacHado BI, Ramirez DA, Medina F, Collins S, Wicker RB. Comparison of microstructures and mechanical properties for solid and mesh

cobalt-base alloy prototypes fabricated by electron beam melting. Metall Mater Trans A Phys Metall Mater Sci. 2010;41:3216–27.

38. Kim HR, Jang SH, Kim YK, Son JS, Min BK, Kim KH, Kwon TY. Microstructures and mechanical properties of Co-Cr dental alloys fabricated by three CAD/CAM-based processing techniques. Materials. 2016;9:596.

39. Murr LE, Quinones SA, Gaytan SM, Lopez MI, Rodela A, Martinez EY, Hernandez DH, Martinez E, Medina F, Wicker RB. Microstructure and mechanical behavior of Ti–6Al–4V produced by rapid-layer manufacturing, for biomedical applications. J Mech Behav Biomed Mater. 2009;2:20–32.

40. Niinomi M, Narushima T, Nakai M, editors. Advances in metallic biomaterials processing and applications. Berlin: Springer; 2015.

41. Okazaki Y, Rao S, Ito Y, Tateishi T. Corrosion resistance, mechanical properties, corrosion fatigue strength and cytocompatibility of new Ti alloys without Al and V. Biomaterials. 1998;19:1197–215.

42. Sing SL, Yeong WY, Wiria FE. Selective laser melting of titanium alloy with 50 wt% tantalum: microstructure and mechanical properties. J Alloys Compd. 2016;660:461.

43. Wauthle R, Van Der Stok J, Yavari SA, Van Humbeeck J, Kruth JP, Zadpoor AA, Weinans H, Mulier M, Schrooten J. Additively manufactured porous tantalum implants. Acta Biomater. 2015;14:217.

44. Vasilescu C, Drob S, Neacsu E, Rosca JM. Surface analysis and corrosion resistance of a new titanium base alloy in simulated body fluids. Corros Sci. 2012;65:431–40.

45. Matsuno H, Yokoyama A, Watari F, Uo M, Kawasaki T. Biocompatibility and osteogenesis of refractory metal implants, titanium, hafnium, niobium, tantalum and rhenium. Biomaterials. 2001;22:1253.

46. Wauthle R, Kruth J-P, Montero ML, Thijs L, Van Humbeeck J. New opportunities for using tantalum for implants with additive manufacturing. Eur Cells Mater. 2013;26:15.

47. Kurtz SM. PEEK biomaterials handbook. Oxford: William Andrew; 2011.

48. Campbell AA. Bioceramics for implant coatings. Mater Today. 2003;6:26–30.

49. Bartolo P, Kruth JP, Silva J, Levy G, Malshe A, Rajurkar K, Mitsuishi M, Ciurana J, Leu M. Biomedical production of implants by additive electro-chemical and physical processes. CIRP Ann Manuf Technol. 2012;61:635.

50. Mahmoud D, Elbestawi M. Lattice structures and functionally graded materials applications in additive manufacturing of orthopedic implants: a review. J Manufact Mater Process. 2017;1:13.

51. Taniguchi N, Fujibayashi S, Takemoto M, Sasaki K, Otsuki B, Nakamura T, Matsushita T, Kokubo T, Matsuda S. Effect of pore size on bone ingrowth into porous titanium implants fabricated by additive manufacturing: an in vivo experiment. Mater Sci Eng C. 2016;59:690–701.

52. Čapek J, Machová M, Fousová M, Kubásek J, Vojtěch D, Fojt J, Jablonská E, Lipov J, Ruml T. Highly porous, low elastic modulus 316L stainless steel scaffold prepared by selective laser melting. Mater Sci Eng C. 2016;69:631.

53. Lima DD, Mantri SA, Mikler CV, Contieri R, Yannetta CJ, Campo KN, Lopes ES, Styles MJ, Borkar T, Caram R, Banerjee R. Laser additive processing of a functionally graded internal fracture fixation plate. Mater Des. 2017;130:8.

54. Van Bael S, Chai YC, Truscello S, Moesen M, Kerckhofs G, Van Oosterwyck H, Kruth JP, Schrooten J. The effect of pore geometry on the in vitro biological behavior of human periosteum-derived cells seeded on selective laser-melted Ti6Al4V bone scaffolds. Acta Biomater. 2012;8(7):2824–34.

55. Wegst UGK, Bai H, Saiz E, Tomsia AP, Ritchie RO. Bioinspired structural materials. Nat Mater. 2015;14(1):23–36.

56. Challis VJ, Roberts AP, Grotowski JF, Zhang LC, Sercombe TB. Prototypes for bone implant scaffolds designed via topology optimization and manufactured by solid freeform fabrication. Adv Eng Mater. 2010;12:1106.

57. Sutradhar A, Park J, Carrau D, Nguyen TH, Miller MJ, Paulino GH. Designing patient-specific 3D printed craniofacial implants using a novel topology optimization method. Med Biol Eng Comput. 2016;54:1123.
58. Al-Tamimi AA, PRA F, Peach C, Cooper G, Diver C, Bartolo PJ. Metallic bone fixation implants: a novel design approach for reducing the stress shielding phenomenon. Virtual Phys Prototyping. 2017;12(2):141–51.
59. Marro A, Bandukwala T, Mak W. Three-dimensional printing and medical imaging: a review of the methods and applications. Curr Probl Diagn Radiol. 2016;45:2–9.
60. Iglesias A, Galvez A, Avila A. Immunological approach for full NURBS reconstruction of outline curves from noisy data points in medical imaging. IEEE/ACM Trans Comput Biol Bioinform. 2017.
61. Lewallen EA, Riester SM, Bonin CA, Kremers HM, Dudakovic A, Kakar S, Cohen RC, Westendorf JJ, Lewallen DG, van Wijnen AJ. Biological strategies for improved osseointegration and osteoinduction of porous metal orthopedic implants. Tissue Eng Part B Rev. 2015;21:218.
62. Chahine G, Koike M, Okabe T, Smith P, Kovacevic R. The design and production of Ti-6Al-4V ELI customized dental implants. JOM. 2008;60:50.
63. Ellingsen JE, Lyngstadaas SP. Bio-implant interface: improving biomaterials and tissue reactions. Boca Raton, FL: CRC; 2003.
64. Noble PC, Alexander JW, Lindahl LJ, Yew DT, Granberry WM, Tullos HS. The anatomic basis of femoral component design. Clin Orthop Relat Res. 1988;235(11):148–65.
65. Duda GN, Brand D, Freitag S, Lierse W, Schneider E. Variability of femoral muscle attachments. J Biomech. 1996;29(9):1185–90.
66. Churchill RS, Brems JJ, Kotschi H. Glenoid size, inclination, and version: an anatomic study. J Shoulder Elbow Surg. 2001;10(4):327–32.
67. Matsumura N, Ogawa K, Ikegami H, Collin P, Walch G, Toyama Y. Computed tomography measurement of glenoid vault version as an alternative measuring method for glenoid version. J Orthop Surg Res. 2014;9:17.
68. Boudarham J, Hameau S, Zory R, Hardy A, Bensmail D, Roche N. Coactivation of lower limb muscles during gait in patients with multiple sclerosis. PLoS One. 2016;11:e0158267.
69. Kim SH, Wise BL, Zhang Y, Szabo RM. Increasing incidence of shoulder arthroplasty in the United States. J Bone Joint Surg Am. 2011;93(24):249–2254.
70. Habermeyer P, Magosch P, Luz V, Lichtenberg S. Three-dimensional glenoid deformity in patients with osteoarthritis: a radiographic analysis. J Bone Joint Surg Am. 2006;88(6):1301–7.
71. Nowak DD, Bahu MJ, Gardner TR, Dyrszka MD, Levine WN, Bigliani LU, Ahmad CS. Simulation of surgical glenoid resurfacing using three-dimensional computed tomography of the arthritic glenohumeral joint: the amount of glenoid retroversion that can be corrected. J Shoulder Elbow Surg. 2009;18(5):680–8.
72. Dines DM, Gulotta L, Craig EV, Dines JS. Novel solution for massive glenoid defects in shoulder arthroplasty: a patient-specific glenoid vault reconstruction system. Am J Orthop (Belle Mead NJ). 2017;46(2):104.
73. Flurin PH, Janout M, Roche CP, Wright TW, Zuckerman J. Revision of the loose glenoid component in anatomic total shoulder arthroplasty. Bull NYU Hosp Jt Dis. 2013;71:68–76.
74. Hsu JE, Gee AO, Lucas RM, Somerson JS, Warme WJ, Matsen FA. Management of intraoperative posterior decentering in shoulder arthroplasty using anteriorly eccentric humeral head components. J Shoulder Elb Surg. 2016;25:1980.
75. Lewis GS, Conaway WK, Wee H, Kim HM. Effects of anterior offsetting of humeral head component in posteriorly unstable total shoulder arthroplasty: finite element modeling of cadaver specimens. J Biomech. 2017;53:78.
76. Anderson DD, Mosqueda T, Thomas T, Hermanson EL, Brown TD, Marsh JL. Quantifying tibial plafond fracture severity: absorbed energy and fragment displacement agree with clinical rank ordering. J Orthop Res. 2008;26:1046.

77. Egol KA, Chang EY, Cvitkovic J, Kummer FJ, Koval KJ. Mismatch of current intramedullary nails with the anterior bow of the femur. J Orthop Trauma. 2004;18(7):410–5.
78. Haynes RC, Pöll RG, Miles AW, Weston RB. Failure of femoral head fixation: a cadaveric analysis of lag screw cut-out with the gamma locking nail and AO dynamic hip screw. Injury. 1997;28:337.
79. Gardner MJ, Weil Y, Barker JU, Kelly BT, Helfet DL, Lorich DG. The importance of medial support in locked plating of proximal Humerus fractures. J Orthop Trauma. 2007;21:185–91.
80. Lewis GS, Caroom CT, Wee H, Jurgensmeier D, Rothermel SD, Bramer MA, Reid JS. Tangential bicortical locked fixation improves stability in Vancouver B1 periprosthetic femur fractures: a biomechanical study. J Orthop Trauma. 2015;29(10):364–70.
81. Tidwell JE, Roush EP, Ondeck CL, Kunselman AR, Reid JS, Lewis GS. The biomechanical cost of variable angle locking screws. Injury. 2016;47(8):1624–30.
82. Tank JC, Schneider PS, Davis E, Galpin M, Prasarn ML, Choo AM, Munz JW, Achor TS, Kellam JF, Gary JL. Early mechanical failures of the synthes variable angle locking distal femur plate. J Orthop Trauma. 2016;30(1):e7–e11.
83. Cronskär M, Rännar L-E, Bäckström M, Nilsson KG, Samuelsson B. Patient-specific clavicle reconstruction using digital design and additive manufacturing. J Mech Des. 2015;137:111418.
84. George M, Kevin Aroom BR, Harvey Hawes BG, Brijesh Gill BS, Love J. 3D printed surgical instruments: the design and fabrication process. World J Surg. 2017;41:314–9.
85. Takemoto M, Fujibayashi S, Ota E, Otsuki B, Kimura H, Sakamoto T, Kawai T, Futami T, Sasaki K, Matsushita T, Nakamura T, Neo M, Matsuda S. Additive-manufactured patient-specific titanium templates for thoracic pedicle screw placement: novel design with reduced contact area. Eur Spine J. 2016;25:1698.
86. Andersen RC, Nanos GP, Pinzur MS, Potter BK. Amputations in trauma. Skeletal trauma. 5th edition. Elsevier Saunders. 2015;2513–34.
87. Wong KC, Kumta SM, Geel NV, Demol J. One-step reconstruction with a 3D-printed, bio-mechanically evaluated custom implant after complex pelvic tumor resection. Comput Aided Surg. 2015;20:14.
88. Sanders G, Marks S, Dicaprio M. The treatment of femoral bone tumor using 3-D printing techniques: a mechanical analysis. Transactions of the 2014 Orthopedics Research Society Meeting. 2014;1109.
89. Fan H, Fu J, Li X, Pei Y, Li X, Pei G, Guo Z. Implantation of customized 3-D printed titanium prosthesis in limb salvage surgery: a case series and review of the literature. World J Surg Oncol. 2015;13:308.
90. Bibb R, Eggbeer D, Evans P, Bocca A, Sugar A. Rapid manufacture of custom-fitting surgical guides. Rapid Prototyping J. 2009;15(4):346–54.
91. Jardini AL, Larosa MA, Filho RM, Zavaglia CADC, Bernardes LF, Lambert CS, Calderoni DR, Kharmandayan P. Cranial reconstruction: 3D biomodel and custom-built implant created using additive manufacturing. J Cranio-Maxillofac Surg. 2014;42:1877.
92. Xu N, Wei F, Liu X, Jiang L, Cai H, Li Z, Yu M, Wu F, Liu Z. Reconstruction of the upper cervical spine using a personalized 3D-printed vertebral body in an adolescent with Ewing sarcoma. Spine. 2016;41:E50–4.

3D Printed Porous Bone Constructs

Wenjun Zheng, Qilin Wei, Xiaojie Xun, and Ming Su

Keywords Bone replacement · Three dimensional printing · Porous · Ceramics · Metals · Polymer

1 Introduction

Bone supports and protects organs in the body. Bone has three dimensional (3D) structures, where the spongy and porous inner part is surrounded by an outer part of low porosity. Bone diseases and fractures affect a myriad of people and are serious health concerns in population where aging is coupled with increased obesity and poor physical activity. As a metabolically active tissue, bone self-heals, but its ability is limited by ages, diseases, pathological conditions, and cannot repair large defects, which lead to bone fractures.

Gold standard for bone repair is bone graft, but bone grafting is limited by supply and disease transmission [1]. Engineered bone tissue is an alternate to conventional grafts. Ideal bone scaffolds should have right chemistry so that cells remain alive and function properly, three dimensional (3D) structures with sufficient strength to support organs, and porous microstructures to mimic extracellular matrix and allow diffusion of chemicals and growth factors into pore for cell growth and proliferation

W. Zheng
Department of Chemical Engineering, Northeastern University, Boston, MA, USA

Q. Wei
School of Chemistry and Chemical Engineering, Beijing Institute of Technology, Beijing, China

X. Xun
Wenzhou Institute of Biomaterials and Engineering, Chinese Academy of Science, Wenzhou Medical University, Zhejiang, China

M. Su (✉)
Department of Chemical Engineering, Northeastern University, Boston, MA, USA

Wenzhou Institute of Biomaterials and Engineering, Chinese Academy of Science, Wenzhou Medical University, Zhejiang, China
e-mail: m.su@northeastern.edu

© Springer International Publishing AG, part of Springer Nature 2018
B. Li, T. Webster (eds.), *Orthopedic Biomaterials*,
https://doi.org/10.1007/978-3-319-89542-0_3

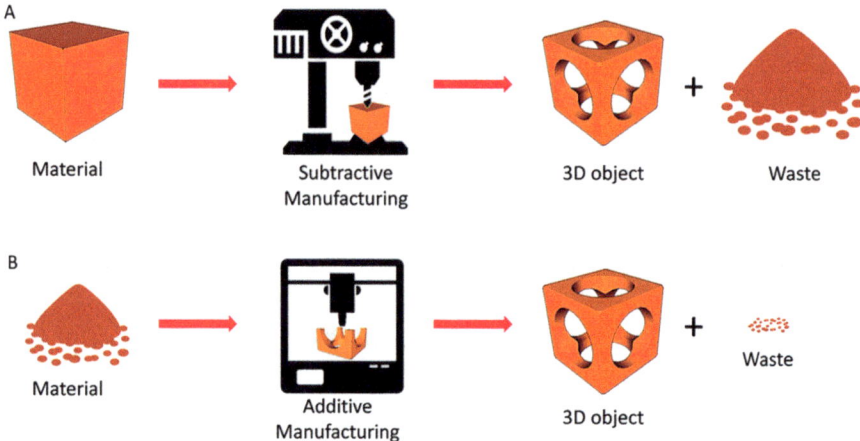

Fig. 1 Subtractive manufacturing (**a**) and additive 3D manufacturing (**b**)

[2, 3]. Bone scaffolds can be produced with subtractive techniques including chemical etching, drilling and electrical discharge machining, which are based on removal of excessive material from a bulk material until the desired shape and geometry are achieved, but these methods waste too much of raw materials, and cannot fully control pore size and interconnectivity of porous materials [4, 5]. Bone scaffolds can also be made using additive techniques (3D printing), in which materials are deposited in a control way to achieve desired shape and geometry (Fig. 1) [6], followed by a variety of pore forming approaches to create microstructures [7, 8]. Compared to subtractive techniques, the additive 3D printing techniques are powerful and efficient in producing bone constructs with complicated geometry that fits each patient.

2 3D Printing Techniques

A variety of 3D printing techniques has been developed such as selective laser sintering, stereo-lithography, fused filament fabrication, and inkjet and solvent casting [9–12]. A common nature of 3D printing techniques is that raw material is formed from a tiny nozzle in a pre-designed pattern at two dimension, and once a layer is complete, the nozzle (or sample) will be retracted for a controlled distance to build the second layer of material (Fig. 2) [7]. The raw materials can be in solid or liquid form, while the printed objects are in solid forms. For solid raw materials, a solid-liquid phase change induced by resistive heating or laser heating is needed before the material is deposited on receiving plate. For liquid raw materials, a solidification process involving either heat, ultraviolet radiation or chemicals is required to form 3D objects. Table 1 highlights major 3D printing techniques that can be used in producing bone constructs. All these methods can be used to generate customizable bone constructs with complicated geometry. Some of the techniques such as

Fig. 2 3D printing route

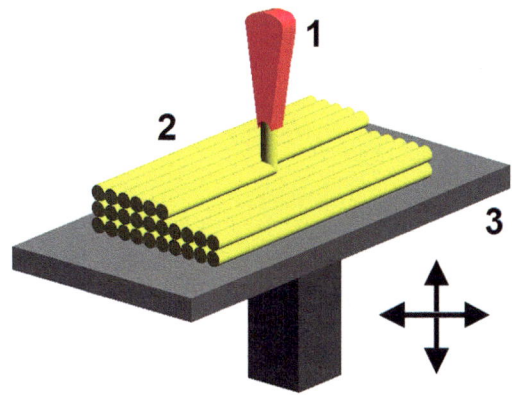

Table 1 3D printing technologies

Method	Strengths	Limitations
Selective laser sintering	High mechanical strengths	Grainy surface finish
	Living hinges and snap features possible	
Stereo-lithography	Fine details	Susceptible to sunlight and heat
	Smooth surface finish	
Binder jetting	Fast and multi-species print	Weak parts and rough surface finish
Poly-jet	High accuracy, multi-material capabilities	Low strength, sensitive to sunlight and heat
Fused deposition modeling	High mechanical strength and low cost	Poor surface finish
Injection molding	Broad material selection and high volume	High start-up cost and long lead time
Computer numeric control machining	All material, high tolerances	High equipment cost, thermoplastics only
Plastic forming	Large parts, affordable price	Limited shape and one side control
Plastic joining	All materials	Time consuming and labor intensive

selective laser sintering need to use scarifying powder materials, some need to work with liquid pre-polymer (stereo-lithography), and some need to use liquid binder to glue particulate raw materials, which limits the nature of materials that can be used as bone constructs [13]. In contrast, direct writing (ink-jet) method is limited by raw materials, but once established, this method would be the most convenient one [14]. Given the wide range of possible bone materials (metals, ceramics, and polymers), each technique has its own strengths and limitations, and the selection of technique will have to consider the material perspective such as material nature, melting temperature, reactivity, stability, and pore-forming ability.

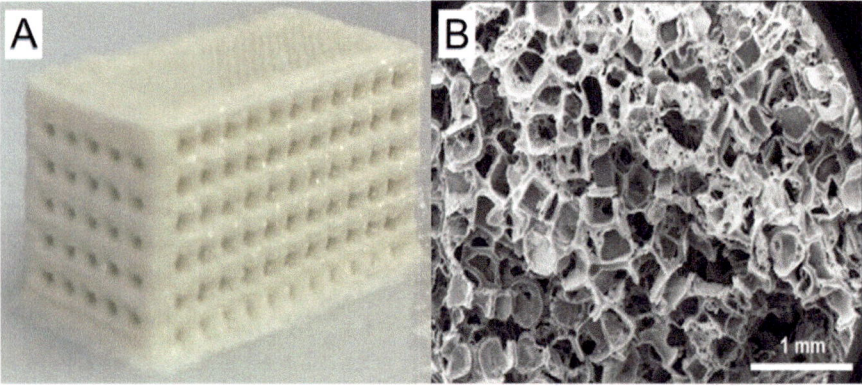

Fig. 3 3D printed ordered pores (**a**) and random pores (**b**)

3 Porous Materials for Cell Growth

Porous materials with pore sizes in hundreds of microns (100–400 μm) are suitable for cell migration and proliferation. The porous structures can be created by direct 3D printing, or by selective materials removal after forming 3D constructs. 3D printing techniques can produce uniformly ordered pores, but the pore size that is limited by precision of mechanical motion is over 200 μm, thus the total surface area of 3D constructs is low, which leads to less cells proliferated [8, 15]. On the other hand, selective material removal (particle leaching) can produce random porous structures with increased surface area for cell attachment and diffusion of culture media, but the finished material lacks desired 3D geometry to fit patient (Fig. 3) [16]. An ideal bone scaffold should have 3D printed large pores and material removal generated small pores, so that the scaffold can fit patient nicely, and enhance cell proliferation and penetration (Fig. 4) [17]. Since large pores are within the fabrication limit and can be readily made with 3D printers via software control, the following sections describe recent materials progresses in making 3D porous bone scaffolds using ceramics, metals and polymer materials.

4 3D Printing of Porous Ceramic Materials

Ceramic materials are used as bone substitutes due to biocompatibility and similar composition with bone [18, 19]. Among those, calcium phosphates such as hydroxy-apatite and tricalcium phosphate can be readily adsorbed when placed *in vivo*, and are used as coating materials to enhance cell attachment, and scaffolds for small bone growth to repair trauma or diseased induced osseous defects [20]. Although the materials are porous and have low density [21–25], their compression strengths (18.6 MPa) are lower compared with large bones due to porous structures. In order

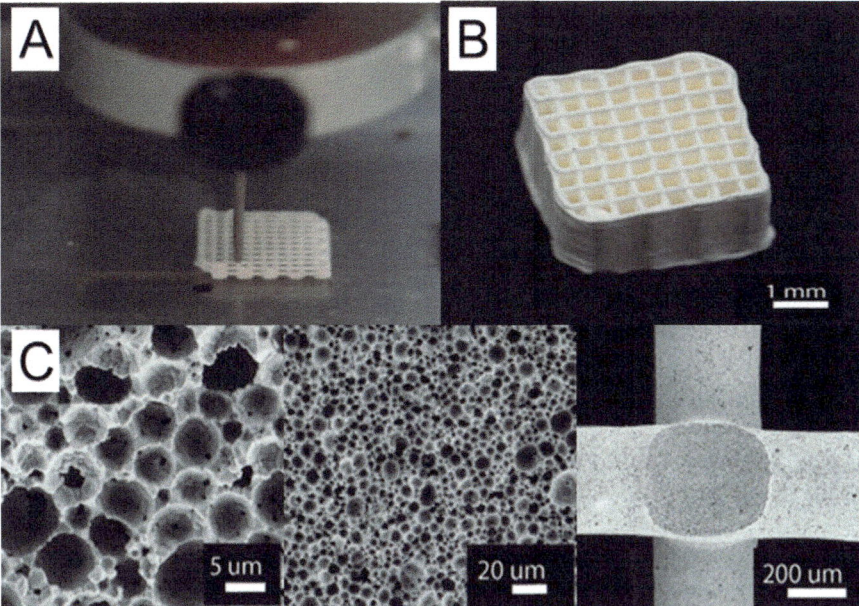

Fig. 4 3D printed dual-pore structure: (**a**) 3D printing process, (**b**) porous scaffold image, (**c**) image of internal pore structure

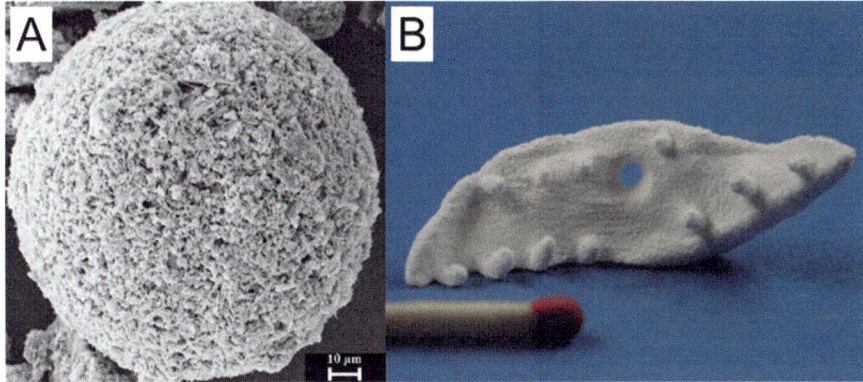

Fig. 5 3D printed ceramic structure: (**a**) ceramic granule, and (**b**) implant

to enhance mechanical strength, silica nanoparticles have been added into tricalcium phosphate when printing, and the printed materials have higher strengths compared to tricalcium phosphate (Fig. 5) [26]. The high temperature annealing of ceramic composite leads to solid-solid phase change, which causes slight shrinkage of the printed parts relative to 3D computer models [27]. Magnesium doped silicon carbide nanoparticles have been added to tricalcium phosphate before printing [28–30], the resulting ceramic material can maintain its original shape and strength even when immersed in a buffer solution. Meanwhile, the porous composites of polymer-ceramic materials can be processed to 3D scaffolds with minimum shrinkage and high compression strengths at low temperature [31, 32].

5 3D Printing of Porous Metal Materials

Metal bone materials are not brittle and used as large bone scaffold owing to high mechanical strength. 3D printing has been used to make bone scaffold using metal materials such as stainless steels, cobalt alloys, titanium and titanium alloys. Among these materials, CoCr alloy and stainless steel show high mechanical strength and corrosion resistance [33], but have higher stiffness and lower bone-implant contact than natural bones, which lead to bone tissue loss and implant failure. Although titanium and titanium alloy show better biocompatibility than others, they are still bioinert and their affinity to tissue is not always ideal, which lead to inflammation and low contact at implant-tissue interface. Making porous metallic materials or forming porous coating over metal implants can significantly enhance the tissue affinity of 3D printed metallic scaffold (Fig. 6) [34]. Surface modifications such as apatite coating, alkaline immersion and hydrothermal heating can be used to improve bioactivity of titanium and its alloy. Porous titanium shows higher affinity to tissue than those cultured on solid surface [35, 36], because the microstructures can enhance cell proliferation [37, 38]. Additionally, iron-manganese alloy has also been used as biodegradable bone scaffold material [39, 40].

Fig. 6 Porous titanium structure at (**a**) 1× and (**b**) 30×

6   3D Printing of Porous Polymer Materials

Polymer materials though have lower strength than metallic ones are easier to process and have higher biocompatibility. Polymer bone materials can degrade in body, and thus may not require surgical removal after implantation, which are ideal for children. Polymer bone materials can positively interact with cells, and improve cell adhesion, growth, migration and differentiation [41, 42]. An issue of porous polymer is low strength. For example, porous polylactic acid is brittle with less than 10% elongation at break. Introduction of additives such as hydroxyapatite and silica particles can enhance their strength. A particle size dependent enhancement of mechanical strength has been identified in porous polylactic acid, where hydroxyapatite nanoparticles can enhance strength and inhibit crack propagation in porous polylactic more than hydroxyapatite microscope [43, 44]. A directional strength enhancement has also been found with nanoparticle additives, likely due to alignment of nanoparticles when polymer-nanoparticle composites are injected from tiny nozzle [45]. Bioactive silica particles have added in polylactic acid to produce bone scaffolds with hierarchical pore architecture, which show a compressive strength of 200 times higher than polyurethane foam [46]. Cell culture results confirm good biocompatibility of scaffolds (Fig. 7) [47]. *In vivo* animal results confirm bone regeneration ability and degradation of materials [48].

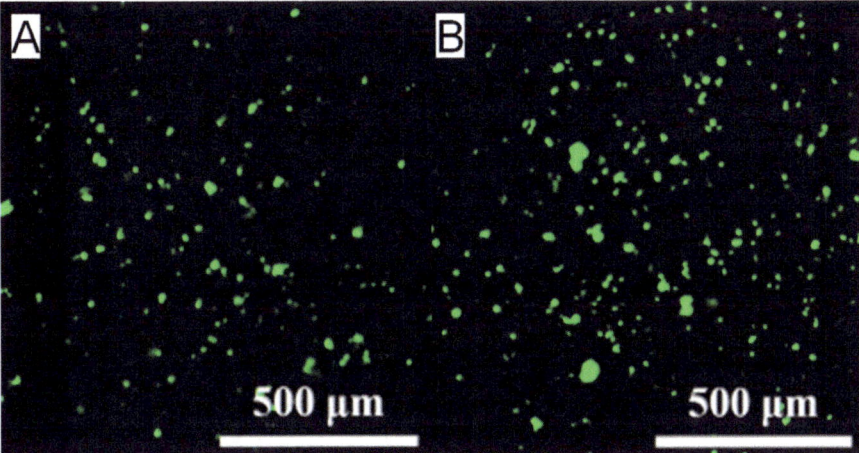

Fig. 7 Bone cells adhesion on (**a**) polylactic acid, and (**b**) silica-modified polylactic acid

7 Conclusions

A variety of 3D printing techniques have been used to make bone scaffold materials based on ceramics, metals and alloys, and polymeric materials. Generating controlled porosity in engineering bone materials can enhance cell attachment and cell proliferation, leading to better recovery of patient. The microstructures of bone scaffold (porosity, pore size and size distribution) can be produced by selectively removing pore forming reagents from 3D printed materials. Given the simultaneous control of macroscopic shape and dimension, 3D printed porous materials could become a promising selection for bone tissue regeneration.

References

1. Nandi SK, Roy S, Mukherjee P, Kundu B, De DK, Basu D. Orthopaedic applications of bone graft & graft substitutes: a review. Indian J Med Res. 2010;132(1):15–30.
2. Hutmacher DW, Schantz JT, Lam CXF, Tan KC, Lim TC. State of the art and future directions of scaffold-based bone engineering from a biomaterials perspective. J Tissue Eng Regen Med. 2007;1(4):245–60.
3. Hollister SJ. Porous scaffold design for tissue engineering. Nat Mater. 2005;4(7):518.
4. Giannitelli S, Accoto D, Trombetta M, Rainer A. Current trends in the design of scaffolds for computer-aided tissue engineering. Acta Biomater. 2014;10(2):580–94.
5. Hollister SJ, Bergman TL. Biomedical applications of integrated additive/subtractive manufacturing. Additive/Subtractive Manufacturing Research and Development in Europe, vol. 1001. 2004. p. 55.
6. Ambrosi A, Pumera M. 3D-printing technologies for electrochemical applications. Chem Soc Rev. 2016;45(10):2740–55.
7. Bose S, Vahabzadeh S, Bandyopadhyay A. Bone tissue engineering using 3D printing. Mater Today. 2013;16(12):496–504.
8. Chia HN, Wu BM. Recent advances in 3D printing of biomaterials. J Biol Eng. 2015;9(1):4.
9. Williams JM, Adewunmi A, Schek RM, Flanagan CL, Krebsbach PH, Feinberg SE, Hollister SJ, Das S. Bone tissue engineering using polycaprolactone scaffolds fabricated via selective laser sintering. Biomaterials. 2005;26(23):4817–27.
10. Kim K, Yeatts A, Dean D, Fisher JP. Stereolithographic bone scaffold design parameters: osteogenic differentiation and signal expression. Tissue Eng Part B. 2010;16(5):523–39.
11. Hutmacher DW, Schantz T, Zein I, Ng KW, Teoh SH, Tan KC. Mechanical properties and cell cultural response of polycaprolactone scaffolds designed and fabricated via fused deposition modeling. J Biomed Mater Res Part A. 2001;55(2):203–16.
12. Mozafari M, Moztarzadeh F, Rabiee M, Azami M, Maleknia S, Tahriri M, Moztarzadeh Z, Nezafati N. Development of macroporous nanocomposite scaffolds of gelatin/bioactive glass prepared through layer solvent casting combined with lamination technique for bone tissue engineering. Ceram Int. 2010;36(8):2431–9.
13. Tan K, Chua C, Leong K, Cheah C, Cheang P, Bakar MA, Cha S. Scaffold development using selective laser sintering of polyetheretherketone–hydroxyapatite biocomposite blends. Biomaterials. 2003;24(18):3115–23.
14. Saijo H, Igawa K, Kanno Y, Mori Y, Kondo K, Shimizu K, Suzuki S, Chikazu D, Iino M, Anzai M. Maxillofacial reconstruction using custom-made artificial bones fabricated by inkjet printing technology. J Artif Organs. 2009;12(3):200–5.

15. Bose S, Roy M, Bandyopadhyay A. Recent advances in bone tissue engineering scaffolds. Trends Biotechnol. 2012;30(10):546–54.
16. Senatov F, Niaza K, Zadorozhnyy MY, Maksimkin A, Kaloshkin S, Estrin Y. Mechanical properties and shape memory effect of 3D-printed PLA-based porous scaffolds. J Mech Behav Biomed Mater. 2016;57:139–48.
17. Minas C, Carnelli D, Tervoort E, Studart AR. 3D printing of emulsions and foams into hierarchical porous ceramics. Adv Mater. 2016;28(45):9993–9.
18. Burg KJ, Porter S, Kellam JF. Biomaterial developments for bone tissue engineering. Biomaterials. 2000;21(23):2347–59.
19. Vallet-Regi M, González-Calbet JM. Calcium phosphates as substitution of bone tissues. Prog Solid State Chem. 2004;32(1):1–31.
20. Bose S, Tarafder S. Calcium phosphate ceramic systems in growth factor and drug delivery for bone tissue engineering: a review. Acta Biomater. 2012;8(4):1401–21.
21. Seitz H, Rieder W, Irsen S, Leukers B, Tille C. Three-dimensional printing of porous ceramic scaffolds for bone tissue engineering. J Biomed Mater Res Part B. 2005;74(2):782–8.
22. Warnke PH, Seitz H, Warnke F, Becker ST, Sivananthan S, Sherry E, Liu Q, Wiltfang J, Douglas T. Ceramic scaffolds produced by computer-assisted 3D printing and sintering: characterization and biocompatibility investigations. J Biomed Mater Res Part B. 2010;93(1):212–7.
23. Vail N, Swain L, Fox W, Aufdlemorte T, Lee G, Barlow J. Materials for biomedical applications. Mater Des. 1999;20(2):123–32.
24. Lee G, Barlow J. In: Selective laser sintering of calcium phosphate powders. Proceedings of the solid freeform fabrication symposium, Austin, TX, 1994; pp. 191–7.
25. Lee G, Barlow J. In: Selective laser sintering of bioceramic materials for implants. Proceedings of the solid freeform fabrication symposium, Austin, TX, 1993; pp. 376–80.
26. Bergmann C, Lindner M, Zhang W, Koczur K, Kirsten A, Telle R, Fischer H. 3D printing of bone substitute implants using calcium phosphate and bioactive glasses. J Eur Ceram Soc. 2010;30(12):2563–7.
27. Schickle K, Zurlinden K, Bergmann C, Lindner M, Kirsten A, Laub M, Telle R, Jennissen H, Fischer H. Synthesis of novel tricalcium phosphate-bioactive glass composite and functionalization with rhBMP-2. J Mater Sci Mater Med. 2011;22(4):763–71.
28. Shao H, He Y, Fu J, He D, Yang X, Xie J, Yao C, Ye J, Xu S, Gou Z. 3D printing magnesium-doped wollastonite/β-TCP bioceramics scaffolds with high strength and adjustable degradation. J Eur Ceram Soc. 2016;36(6):1495–503.
29. Xie J, Shao H, He D, Yang X, Yao C, Ye J, He Y, Fu J, Gou Z. Ultrahigh strength of three-dimensional printed diluted magnesium doping wollastonite porous scaffolds. MRS Commun. 2015;5(4):631–9.
30. Sun M, Liu A, Shao H, Yang X, Ma C, Yan S, Liu Y, He Y, Gou Z. Systematical evaluation of mechanically strong 3D printed diluted magnesium doping wollastonite scaffolds on osteogenic capacity in rabbit calvarial defects. Sci Rep. 2016;6:34029.
31. Taboas J, Maddox R, Krebsbach P, Hollister S. Indirect solid free form fabrication of local and global porous, biomimetic and composite 3D polymer-ceramic scaffolds. Biomaterials. 2003;24(1):181–94.
32. Schek RM, Taboas JM, Segvich SJ, Hollister SJ, Krebsbach PH. Engineered osteochondral grafts using biphasic composite solid free-form fabricated scaffolds. Tissue Eng. 2004;10(9–10):1376–85.
33. Fousová M, Kubásek J, Vojtěch D, Fojt J, Čapek J. 3D printed porous stainless steel for potential use in medicine, IOP Conference Series: Materials Science and Engineering. Bristol: IOP Publishing; 2017. p. 012025.
34. Lewallen EA, Jones DL, Dudakovic A, Thaler R, Paradise CR, Kremers HM, Abdel MP, Kakar S, Dietz AB, Cohen RC. Osteogenic potential of human adipose-tissue-derived mesenchymal stromal cells cultured on 3D-printed porous structured titanium. Gene. 2016;581(2):95–106.
35. De Peppo G, Palmquist A, Borchardt P, Lennerås M, Hyllner J, Snis A, Lausmaa J, Thomsen P, Karlsson C. Free-form-fabricated commercially pure Ti and Ti_6Al_4V porous scaffolds

support the growth of human embryonic stem cell-derived mesodermal progenitors. Sci World J. 2012;2012:1.

36. Lewallen EA, Riester SM, Bonin CA, Kremers HM, Dudakovic A, Kakar S, Cohen RC, Westendorf JJ, Lewallen DG, Van Wijnen AJ. Biological strategies for improved osseo-integration and osteoinduction of porous metal orthopedic implants. Tissue Eng Part B. 2014;21(2):218–30.

37. Liu X, Chu PK, Ding C. Surface modification of titanium, titanium alloys, and related materials for biomedical applications. Mater Sci Eng R Rep. 2004;47(3):49–121.

38. Elias C, Lima JH, Valiev R, Meyers M. Biomedical applications of titanium and its alloys. JOM. 2008;60(3):46–9.

39. Chou D-T, Wells D, Hong D, Lee B, Kuhn H, Kumta PN. Novel processing of iron–manganese alloy-based biomaterials by inkjet 3-D printing. Acta Biomater. 2013;9(10):8593–603.

40. Hong D, Chou D-T, Velikokhatnyi OI, Roy A, Lee B, Swink I, Issaev I, Kuhn HA, Kumta PN. Binder-jetting 3D printing and alloy development of new biodegradable Fe-Mn-ca/mg alloys. Acta Biomater. 2016;45:375–86.

41. Rezwan K, Chen Q, Blaker J, Boccaccini AR. Biodegradable and bioactive porous polymer/inorganic composite scaffolds for bone tissue engineering. Biomaterials. 2006;27(18):3413–31.

42. Wei G, Ma PX. Macroporous and nanofibrous polymer scaffolds and polymer/bone-like apatite composite scaffolds generated by sugar spheres. J Biomed Mater Res A. 2006;78(2):306–15.

43. Serra T, Planell JA, Navarro M. High-resolution PLA-based composite scaffolds via 3-D printing technology. Acta Biomater. 2013;9(3):5521–30.

44. Serra T, Ortiz-Hernandez M, Engel E, Planell JA, Navarro M. Relevance of PEG in PLA-based blends for tissue engineering 3D-printed scaffolds. Mater Sci Eng C. 2014;38:55–62.

45. Cox SC, Thornby JA, Gibbons GJ, Williams MA, Mallick KK. 3D printing of porous hydroxy-apatite scaffolds intended for use in bone tissue engineering applications. Mater Sci Eng C. 2015;47:237–47.

46. Wu C, Luo Y, Cuniberti G, Xiao Y, Gelinsky M. Three-dimensional printing of hierarchical and tough mesoporous bioactive glass scaffolds with a controllable pore architecture, excellent mechanical strength and mineralization ability. Acta Biomater. 2011;7(6):2644–50.

47. Yin H-M, Qian J, Zhang J, Lin Z-F, Li J-S, Xu J-Z, Li Z-M. Engineering porous poly (lactic acid) scaffolds with high mechanical performance via a solid state extrusion/porogen leaching approach. Polymers. 2016;8(6):213.

48. Cowan CM, Aghaloo T, Chou Y-F, Walder B, Zhang X, Soo C, Ting K, Wu B. MicroCT evaluation of three-dimensional mineralization in response to BMP-2 doses in vitro and in critical sized rat calvarial defects. Tissue Eng. 2007;13(3):501–12.

Biopolymer Based Interfacial Tissue Engineering for Arthritis

Krishanu Ghosal, Rohit Khanna, and Kishor Sarkar

Keywords Arthritis · Interfacial tissue · Osteochondral tissue · Tissue interface · Biopolymer · 3D scaffold · Tissue regeneration · Gradient scaffold · Growth factor delivery · Orthopedic application

1 Introduction

Osteoarthritis (OA) and rheumatoid arthritis (RA) are common joint diseases that damage both the articular cartilage and the underlying subchondral bone. It was estimated that over 39 million Europeans and more than 20 million Americans are suffering from osteoarthritis according to the report published in 2008 and this number is expected to double by the year 2020 [1]. Conventional treatments like periosteal grafts, subchondral drilling or microfracture, autograft transplantation/mosaicplasty, allograft transplant and lavage are used to treat osteoarthritis [2–5]. Among these, only autograft and allograft transplantation show promising results for cartilage-bone repair. But, donor site morbidity, degenerative changes and graft rejection limit its long term therapeutic effects on patients [6, 7]. Therefore, long-term clinical outcomes with minimal complications may be obtained through implantation of engineered osteochondral grafts using a patient's own cells and osteochondral scaffolds to regenerate or repair cartilage-bone defects.

Tissue engineering (TE) is considered as an emerging field that has shown great promise for the generation of functional tissues like cartilage and bone by merging three principle components such as cells, biodegradable scaffolds and growth factors *in vitro* followed by subsequent implantation *in vivo* [8–12]. But, conventional tissue engineering approaches have challenges regenerating the osteochondral

K. Ghosal · K. Sarkar (✉)
Gene Therapy and Tissue Engineering Lab, Department of Polymer Science and Technology, University of Calcutta, Kolkata, India
e-mail: kspoly@caluniv.ac.in; http://www.kishorgttl.com

R. Khanna
Department of Mechanical Engineering, University of Texas at San Antonio, San Antonio, TX, USA

© Springer International Publishing AG, part of Springer Nature 2018   67
B. Li, T. Webster (eds.), *Orthopedic Biomaterials*,
https://doi.org/10.1007/978-3-319-89542-0_4

interface due to presence of progressive gradient materials composition and complex physiological properties [13–16]. In addition, due to the limitation in the capacity of articular cartilage to self-repair, it is essential to develop approaches based on engineered biomaterials. In this regard, interfacial tissue engineering (ITE) has become more attractive in tissue engineering due to its capability to regenerate the complex bi- or multiphasic nature of defects in different tissue types like bone-cartilage, and muscle-tendon complex tissue interfaces where complex material composition and mechanical properties are present. It mainly consists of two approaches including the generation of the tissue interface and the specific tissues which transform from one tissue type to another tissue. Generally, 3D cylindrical osteochondral scaffolds or plugs consisting of both cartilage and subchondral bone constructs are used in ITE.

The development of scaffolds is one of the primary challenges in tissue engineering to mimic the natural extracellular matrix (ECM). The primary function of the scaffold is to provide physical support to growing cells and to mimic the tissue specific environment in order to differentiate and subsequently create functional tissue regeneration, as scaffolds degrade at a controlled rate without leaving toxic effects [17–19]. Despite the development of metals, ceramics, and inorganic material based scaffolds; polymeric biomaterials have gained tremendous attention for scaffold design due to their ease of synthesis and modification as required, biodegradability, biocompatibility and less immunogenic nature [20–22]. Therefore, the objective of this chapter is to discuss the anatomy of the osteochondral interface, the current status of conventional and interfacial tissue engineering and biopolymers used for ITE, and to present the clinical status and future perspectives of ITE.

2   Anatomy of Osteochondral Tissue Interface

Osteochondral tissue consists mainly of articular cartilage and subchondral bone as shown in Fig. 1. Articular cartilage is further subdivided by the superficial zone, middle zone, deep zone and calcified cartilage. The upper 10% of the articular cartilage is the superficial zone having a small amount of proteoglycans with low permeability. The next 45% down is the middle zone consisting of higher contents of proteoglycans with a low number of cells. The orientation of the collagen fibers at this zone makes it highly compressive in nature which allows the cartilage to recovery from articular surface impacts. The perpendicular orientation of collagen fibrils and cells to the articular cartilage surface makes the deep zone (last 45%) have greater compressive strength than the middle zone as shown in Fig. 2.

Osteochondral tissue and its interface possesses different material composition and mechanical properties. Articular cartilage mainly consists of small amounts of chondrocyte cells surrounded by an ECM which includes collagen, proteoglycans, water and some minor proteins. Water with 70–80% is the major part in articular cartilage followed by 60–70% and 30% collagen and proteoglycan, respectively.

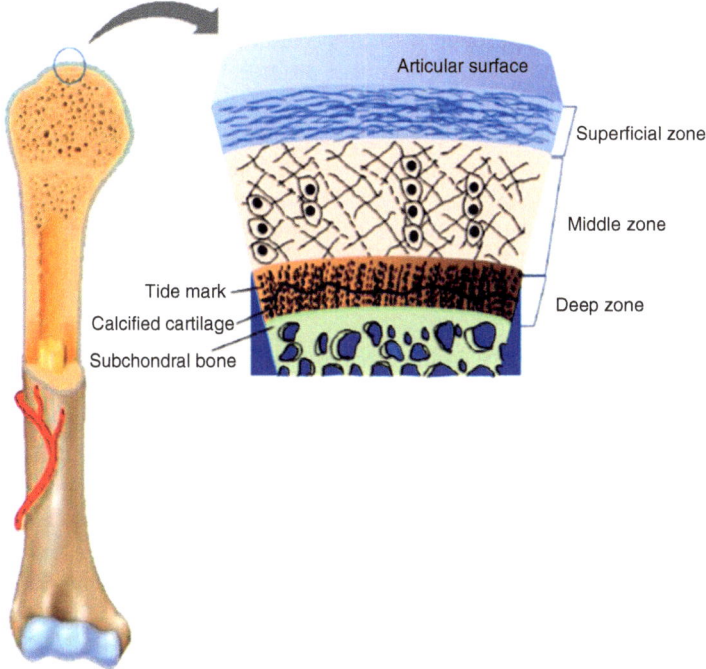

Fig. 1 Cross section of osteochondral tissue with different cartilage zones and subchondral bone. Modified from [23] with permission from Elsevier

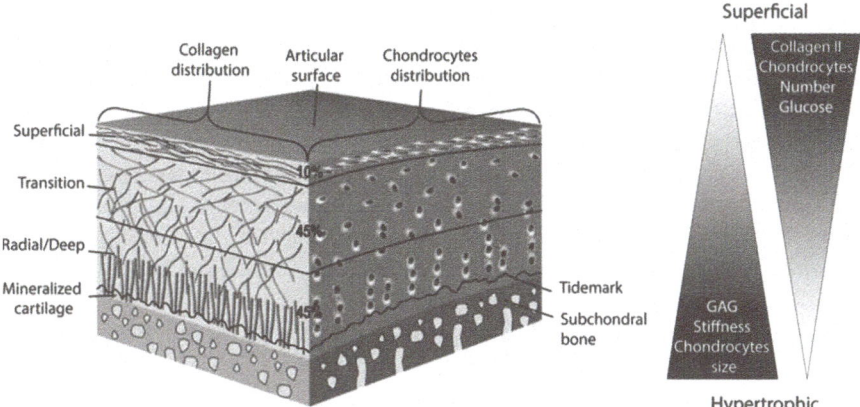

Fig. 2 Schematic representation of osteochondral tissue with gradient representation of collagen, chondrocyte cell size and number, and cartilage stiffness. Reprint from [28] with permission from Wiley

Proteoglycans are high molecular weight macromolecules consisting of a glycos-aminoglycan (GAG) core protein with chondroitin sulfate and keratan sulfate. The concentration of water gradually decreases from the surface to the deep zone of articular cartilage and the reverse trend is found with proteoglycans. Collagen is mainly a fibrous protein with collagen type II as the main component along with other types of collagen in a smaller amount. The collagen fibers are mainly respon-sible for an improvement of structural and elastic strength of articular cartilage [24]. It is believed that collagen type X mainly assists in the mineralization at the inter-face of articular cartilage and underlying bone [25]. More than 90% of the organic mass of bone consists of collagen type I which is a triple helix chain having two identical $\alpha 1(I)$-chains and one $\alpha 2(I)$-chain.

The boundary between the cartilage and the underlying subchondral bone is sep-arated by calcified cartilage and the interface between calcified cartilage and the deep zone is called the *tidemark*. The osteochondral interface contains hypertrophic chondrocytes penetrating calcified cartilage. Chondrocytes are generally small in size having a diameter of 13 μm with a surface area of approximately 821 μm^2 and a volume of 1748 μm^3. The size of the cells remains almost the same throughout the cartilage zone. But the volume increases up to 20 times when the cells are in a hypertrophic state with the production of collagen X [26, 27]. The interface is struc-turally weaker compared to that of the transition of different cartilage zones due to the absence of collagen fibers. Osteochondral tissue consists of different axial gra-dients of composition and ECM organization along its structure as shown in the right side of Fig. 2.

The subchondral bone mainly consists of collagen type I and hydroxyapatite (HA). Approximately, $85 \pm 3\%$ of HA is present in subchondral bone i.e. this amount of HA is present at the bottom portion as shown in above figure (left side figure) and the HA content gradually decreases with going toward upside ($65 \pm 2\%$ of HA in the calcified cartilage) and completely disappear in the superficial zone of articular car-tilage [29]. In the same way, the amount of GAGs which is a measure of stiffness and chondrocyte size form a linear gradient from subchondral bone towards articu-lar cartilage. Conversely, a linear gradient of collagen type II and chondrocyte num-ber is observed from the superfacial zone of articular cartilage towards subchondral bone. In accordance to composition, mechanical properties also varies from the superficial zone to subchondral bone. The compressive modulus of the superficial, middle and deep zones are 0.079, 2.1 and 320 MPa, respectively [13, 30]. Whereas, subchondral bone is composed of HA crystals, collagen type I and water with a compressive modulus of 5.7 GPa [31].

3   Conventional Vs. Interfacial Tissue Engineering

Despite the early success of tissue engineering, researchers have faced challenges in regenerating tissues specifically at the interfacial injured tissue site, due to the pres-ence of complex material composition and different mechanical properties [32, 33].

Current approaches exploit the use of bioactive or bio-resorbable scaffolds, with the objective of developing physical and biological functionality after implantation [34]. The next step is the utilization of passive *in vitro* recellularization prior to implantation. The outcome of this approach is limited since cell differentiation, proliferation, and tissue remodeling do not progress physiologically [35]. Functional tissue engineering is an alternative well established approach that would provide proper *in vitro* propagated cells to cellularize matrices, together with suitable physical conditioning, to develop tissue functionality prior to implantation.

Synthetic biomaterials or chemically crosslinked xenograft tissues are generally employed in conventional therapies for the repair or regeneration of tissues [10]. But several drawbacks, such as calcification, hardening and degeneration limit its therapeutic clinical application in patients [36]. Over the last few decades, researchers used three major tissue engineering approaches including *in situ* tissue regeneration where freshly isolated or cultured cells implanted directly, *in vitro* tissues assembled using cells and scaffolds followed by implantation for tissue regeneration. Isolation of individual cells or small cellular aggregates from the recipient or a donor followed by *in vitro* culture and finally injection into the injured tissue directly are the main strategies of direct implantation tissue engineering. Bioresorbable or bioactive natural or synthetic scaffolds directly implanted in a patient to exploit the body's natural ability to regenerate *in situ* tissue, has been successfully employed for repairing heart valves, small diameter vascular grafts, ligaments and tendons, bladder and surgical patches [37]. However, a variable patient response in terms of resorption, recellularization and regeneration are the main drawbacks of this approach and may result in improper tissue growth with poor mechanical and biological properties *in vivo* and subsequently failure.

So, the conventional tissue engineering approach is not suitable for the regeneration of cartilage to bone, tendon to bone, or ligament to bone simultaneously with their appropriate biological function at the interfacial injured site. To overcome these issues, an interfacial tissue engineering approach has been recently used to regenerate different types of soft-hard tissues simultaneously with a gradual change in mechanical, biological and chemical properties at the interfacial injured sites [30, 38–41]. Different approaches so far adopted for the fabrication of 3D scaffolds for interfacial tissue engineering are shown in Fig. 3. In monophasic construction as shown in Fig. 3 (II), the subcondral bone forming scaffold is cultured in chondrogenic cells to form neocartilage tissue on the top of the scaffold [42]. Various biomaterials such as Bioglass®, calcium sulphate and calcium phosphate, chitosan, collagen, poly(L-lactic acid) or PLLA etc. have been used to form this type of osteochondral construct. Wang et al. [43] used PLLA, poly(DL-lactide) and collagen-HA composites as a bone forming scaffold. They obtained superior neocartilage formation on collagen-HA composites after culturing with chondrocytes in a static bioreactor for 15 weeks. In another study, β-tricalcium phosphate (β-TCP)-based subchondral bone scaffolds were developed by Guo et al. [44].

After neocartilage formation on the 3D construct, they implanted the osteochondral construct in a sheep model and the osteochondral defect was completely filled by neocartilage tissue after 24 weeks of implantation. In a biphasic construct, there

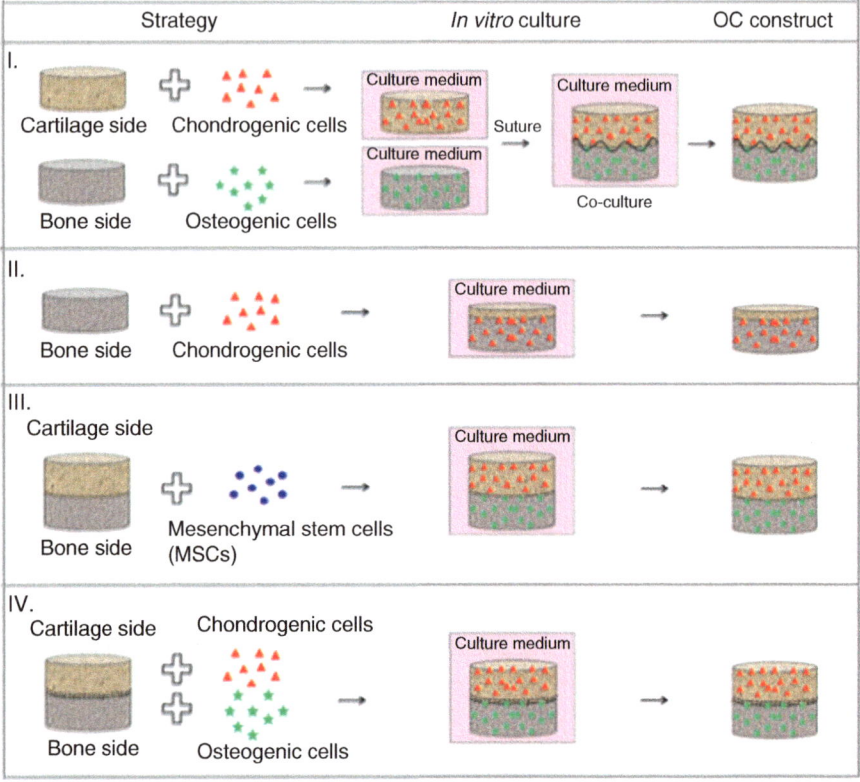

Fig. 3 Different approaches for the fabrication of 3D scaffolds used in interfacial tissue engineering. Reprint from [45] with permission from Springer

are three different approaches as shown in Fig. 3 (I), (III) and (IV). In the first approach, two separate osteogenic and chondrogenic scaffolds are cultured separately with respective cells followed by combining them with adhesives or sutures. On the other approaches, the biphasic scaffolds are cultured with only stem cells or with osteogenic and chondrogenic cells to form an osteochondral construct. A biphasic scaffold consisting of a collagen sponge at the upper layer for cartilage formation and bottom layer with PLGA/collagen composites for subcondral bone formation was fabricated by Chen et al. [46] for osteochondral tissue regeneration. After culturing the scaffold with canine MSCs, the scaffold was implanted in osteo-chondral defects in the knee of canines and osteochondral tissue was formed after 16 weeks of implantation. In another study, Gotterbarm et al. [47] regenerated deep osteochondral defects of mini pigs using porous β-TCP (for bone formation) and fibrous collagen type I/III layer (for cartilage formation) biphasic scaffolds with the addition of a growth factor mixture (GFM). It was found that the bone formation occurred on the TCP layer after 6 weeks of implantation whereas cartilage was formed after 12 weeks of implantation on the collagen layer.

4 Polymeric Biomaterials for Interfacial Tissue Engineering

Collagen, starch, chitosan, alginate, hyaluronic acid, silk, poly(3-hydroxybutyrate-co-3-hydroxyhexanoate) (PHBHHx) are very well known natural biopolymers that can support cell adhesion and guide cell differentiation to specific tissue lineages due to the presence of specific molecular domains in its structure [48]. As discussed earlier, osteochondral defects consist of two different types of tissues and their chemical and physiological properties are different from each other, so it is difficult to mimic the interfacial defect site with only one single polymer. To mimic this type of interfacial tissue regeneration, either two different polymers/polymer-nanoparticle composites are used or single polymer consists of different types of growth factors, which can promote the differentiation of two different types of cells which are present in the interfacial defect site. A list of different polymers in combination with cell stimulus is shown in Table 1.

Natural polymers can be easily functionalized into various functional groups, containing specific molecular patterns in the polymer chain, like RGD which can enhance various cellular activities, including cell adhesion, communication and differentiation [48]. As an example, alginate based scaffolds coupled with gelatin can facilitate both cell adhesion and differentiation. Alginate scaffolds crosslinked with gelatin have been demonstrated to enhance cell adhesion, proliferation while supporting the differentiation of MSCs into chondrogenic and osteogenic cell lineages [71]. Due to structural similarity of chitosan with GAG, chitosan has been widely studied for cartilage tissue regeneration [72–75]. In this concern, Chen et al. made a bilayered chitosan-gelatin scaffold, loaded with plasmid DNA that can induce both osteogenesis and chondrogenesis in a systematic manner. This bilayered polymeric scaffold proved to be effective for osteochondral repair. *In vitro* studies showed that the bilayered scaffold can simultaneously support articular cartilage and subchondral bone regeneration in a rabbit knee osteochondral defect model [64]. In another study, Abarrategi et al. fabricated different chitosan scaffolds with different molecular weights and deacetylation degree. The optimum chitosan scaffold with 17.9 wt% mineral content (calcium carbonate), lowest molecular weight (11.49 KDa) and lowest deacetylation degree (83%) showed well-structured subchondral bone and cartilage tissue regeneration [76]. A novel HA/chitosan scaffold was developed by Reis and co-workers for osteochondral tissue regeneration [77]. They fabricated the scaffold by combining a sintering and a freeze drying technique. After seeding goat bone marrow stem cells (GBMCs) separately on the scaffold, the cells were differentiated into osteogenic and chondrogenic cell linages after 14 and 21 days, respectively. *In vitro* physiological and biological data suggested that the HA/chitosan bilayered scaffold is very effective for the regeneration of osteochondral defects.

Collagen is another important biopolymer, an abundant protein in animal tissues and maintains structural integrity of the ECM. Due to major organic constituents of the bone and cartilage ECM, collagen is considered a perfect biomaterial for bone, cartilage or osteochondral tissue engineering [78–81]. In a recent study, O'Brien

Table 1 Notable targeted interfacial tissue engineering approaches for orthopedic interfacial tissue regeneration

Material used	Scaffold geometry	Fabrication technique	Stimuli delivery	Targeted interfacial tissue	References
PU/TCP	2-layer	Vacuum impregnation	None	Cortical-cancellous bone	[49]
PLGA/TCP	3-layer	Multinozzle low temperature deposition and particulate leaching	None	Cartilage-bone	[50, 51]
Polyglactin/ PL-GA/ bioglass	3-layer	Heat sintering	None	Ligament-bone	[15, 52, 53]
Collagen I/ HAP	2-layer 3-layer 3-layer	Solid freeform fabrication Separate layer synthesis by deposition technique Precipitation followed by crosslinking	None	Cortical-cancellous bone Cartilage-bone Cartilage-bone	[54–56]
Chitosan/HA	3-layer	Lyophilization	Drug delivery (TCH)	Any defect at interface	[57]
PU/PE/CEL2	2-layer	Glazing and powder pressing technique	None	Cortical-cancellous bone	[58]
PCL/ collagen-I/ TCP	2-layer	Fused diffusion and electrospinning	None	Cartilage-bone	[59]
Collagen-I	2-layer (core-shell)	Solution suspension technique	None	Cartilage-bone	[60]
OPF/gelatin	2-layer	Gelatin microsphere embedded hydrogel	Growth factor delivery (BMP-2 and IGF-1)	Cartilage-bone	[61]
OPF/ bPEI-HA	2-layer	bPEI-HA/DNA complex embedded hydrogel	Gene delivery (RUNX and SOX trio gene)	Cartilage-bone	[62]
PLL/PCL/ chitosan	2-layer	Porous nanofiber by electrospinning containing chitosan gel	Growth factor delivery (BMP-2)	Cartilage-bone	[63]
Chitosan/ gelatin/HA	2-layer	Attaching separate layer with fibrin glue	Gene delivery (TGF-β1 and BMP-2 gene)	Cartilage-bone	[64]

(continued)

Table 1 (continued)

Material used	Scaffold geometry	Fabrication technique	Stimuli delivery	Targeted interfacial tissue	References
PGA/PLA/PCL/HA	2-layer	Computer-aided design and manufacturing (CAD/CAM) technology	None	Articular cartilage-bone	[65]
PEOT-co-PBT/PCL/PLA	Gradient	Computer-aided design and manufacturing (CAD/CAM) technology	None	Cartilage-bone	[66]
PLGA/CS/NaHCO₃/β-TCP	Gradient	Ethanol fused microsphere based 3D-scaffold	None	Cartilage-bone	[67]
MPDMS/PEG-DA	Gradient	Photo polymerization	None	Ligament-bone	[68]
PLGA	Gradient	Ethanol fused microsphere based 3D-scaffold	Growth factor delivery (BMP-2 and TGF-β_1)	Cartilage-bone	[69]
PLGA/silk	Gradient	Microsphere gradient by sonication in alginate hydrogel	Growth factor delivery (HRP, rhBMP-2 and rhIGF-I)	Cartilage-bone	[70]

PU polyurethane; *TCP* tricalcium phosphate; *PLGA* poly(lactic-co-glycolic acid); *HA* hydroxyapatite; *PE* polyethylene; *CEL2* bioactive glass; *PCL* polycaprolactone; *OPF* oligo(polyethylene glycol) fumarate; *bPEI* branched polyethyleneimine; *HA* hyaluronic acid; *PLL* poly-L-lysine; *PGA* poly(glycolic acid); *PLA* Poly(lactic acid); *PEOT-co-PBT* poly(ethylene oxide terephthalate)/poly(butylene terephthalate) copolymer; *CS* chondroitin sulphate; *MPDMS* methyl phenyldimethoxysilane; *PEG-DA* poly(ethylene glycol) diacrylate

and co-workers fabricated a cell-free multi-layered scaffold for layer specific regeneration of functional osteochondral tissue for carpine joints. Histological analysis revealed that the scaffold supports formation of both well-structured subchondral bone and cartilage tissue after 12 months of scaffold implantation with an anatomical tidemark [81]. Getgood et al. also made a biphasic collagen/GAG scaffold for osteochondral tissue regeneration. They loaded the scaffold with rhFGF18 or BMP-7 growth factors. *In vivo* results in a sheep model with osteochondral defects showed that the scaffold significantly improved osteochondral repair [82]. Apart from the growth factor or gene signaling for differentiation of progenitor cell to osteogenesis or chondrogenesis, nanoparticles also helped to differentiate toward osteogenesis or chondrogenesis although the actual mechanism behind this nanoparticle induce differentiation still remains unrevealed. Recently, Yi and co-workers have been trying to find out the mechanism of osteogenic differentiation of MSC by gold nanoparticles [83]. They proposed that gold nanoparticles induce mechanical stress on the MSCs and resulted the activation of the protein kinase (MAPK) signaling pathway by activating of p38 mitogen, which regulates the expression of relevant genes for osteogenic differentiation. Vanadium loaded collagen scaffolds differentiated bone marrow progenitor cells to osteo-chondrogenesis without any toxic effect. Better biocompatibility (adhesion, growth and osteoblastic and chondrocyte differentiation) was obtained with vanadium loaded membranes than unloaded collagen scaffolds [84].

Recently, silk has gained tremendous attention as an alternative natural biopolymer for tissue engineering applications owing to its superior mechanical properties [85]. Saha et al. reported osteochondral tissue regeneration using silk fibroin scaffolds from Mulberry and Nonmulberry silkworms. After 1–2 months of *in vitro* culture in osteo- and chondro-inductive media, non-mulberry silkworms pre-seeded with human bone marrow stromal cells showed well distinguished areas with chondrocyte-like cells, whereas mulberry silkworms pre-seeded with human bone marrow stromal cells formed osteoblast-like cells. An *in vivo* study established that neo-osteochondral tissue was generated on cell-free multi-layer silk scaffolds absorbed with TGF-β3or recombinant human bone morphogenetic protein-2 [86]. Levingstone et al. [87], fabricated a multi-layered scaffold for the repair of osteochondral defects in a rabbit model. The scaffold material consisted of 3 different layers and the bone layer contained collagen I and HA powder, the intermediate layer consisted of both collagen I and collagen type II and the superficial layer consisted of collagen I, collagen II and GAG hyaluronic acid. *In vivo* implantation of the multiphasic scaffold in a rabbit model with osteochondral defects showed tissue regeneration with zonal organization and repaired the subchondral bone defect with the formation of a tidemark. Bertoni et al., synthesized a biphasic scaffold consisting of fibroin and alginate for bone regeneration and collagen and alginate for cartilage formation [88]. They demonstrated that both collagen and fibroin can promote differentiation of stem cells into an osteogenic cell lineage whereas alginate promotes chondrogenic differentiation. Kalpan and co-workers developed a osteochondral construct having a gradient in growth factors for interfacial tissue engineering at osteochondral defect sites [70]. At first, they made a cylindrical like shape with alginate gels and then incorporated PLGA and silk fibroin microspheres into it, in a reversed gradient. The microspheres were loaded with two different growth factors, bone morphogenic protein 2 (rhBMP-2) and insulin like growth factor-I (rhIGF-I). They found that silk microspheres are much more efficient to deliver rhBMP-2, while they were less efficient to deliver rhIGF-I compared with PLGA microspheres only. After 5 weeks of seeding with hMSCs in proper osteogenic and chondrogenic medium, hMSC differentiated into an osteogenic and chondrogenic cell linage along with a concentration gradient of growth factors.

Alginate is another important polysaccharide next to chitosan for tissue engineering applications. Coluccino et al. [89] reported recently highly porous alginate scaffolds for osteochondral tissue regeneration. For the chondral layer, they mixed TGF in citric acid with an alginate/alginate sulphate solution whereas HA granules were added into the alginate solution for osteoconductive surfaces. *In vitro* results demonstrated that the designed scaffolds were capable for osteochondral tissue formation with suitable functional and mechanical properties. Khanarian et al. [90] used a hybrid scaffold of alginate hydrogel and HA for the regeneration of the osteochondral interface. They observed that the HA containing part of the scaffold promoted the formation of the proteoglycan and type II collagen-rich matrix when it was seeded with chondrocytes. More significantly, the elevated biosynthesis resulted in a substantial increase in both compressive and shear moduli relative to the mineral free control.

Several advantages of natural biopolymers such as, immunogenicity, lack of large scale production and difficulty in purification, can limit their clinical application [91]. To overcome these problems with natural polymers, researchers have developed biodegradable synthetic biopolymers including PLGA, PCL, poly(propylene fumarate), polyorthoesters, polydioxanone, polyphosphazenes, and polyanhydrides [92, 93]. The main advantages of synthetic biodegradable polymers rely in their chemistries, ease of processing, and controllable molecular weight distribution that can be tailored depending upon their end applications [94]. Cao et al. [95] fabricated PCL three dimensional (3D) porous osteochondral scaffolds by using a fused deposition technique followed by seeding osteogenic cells in one half of the scaffold and chondrogenic cells in rest of the scaffold. After culturing in a co-culture medium, two different types of ECMs at two different portions of the scaffold were formed. In addition to the 3D porous scaffold, polymeric nanofibers also showed promising results in osteochondral tissue regeneration [96, 97]. A biomimetic PCL nanofibrous scaffold having insulin and β-glycerophosphate concentration gradients was prepared by Erisken et al. [98]. After culturing human adipose-derived stromal cells on the graded scaffold for over 8 weeks, chondrogenic differentiation was observed at insulin rich locations of the scaffold whereas mineralization occurred at the β-glycerophosphate site.

Dormer et al. [16] developed PLGA microspheres based 3D gradient scaffolds containing bone morphogenetic protein-2 (BMP-2) and transforming growth factor-β_1 (TGF-β) in reverse gradients. After *in vitro* culture of human umbilical cord mesenchymal stromal cells (hUCMSCs) for 6 weeks, a gradient dependent ECM was formed. They found that in comparison with blank controls, the growth factor loaded scaffold demonstrated greater GAG production, alkaline phosphatase activity, collagen content and respective chondrogenic and osteogenic gene expression. They proposed that engineered signal gradients may be useful for interfacial osteochondral tissue regeneration. Similarly, Gupta et al. [67] reported a high molecular weight PLGA microsphere based gradient scaffold having chondroitin sulfate and tricalcium phosphate in reverse gradients for osteochondral defect repair. They noticed that the mechanical properties of the scaffold initially depend on the composition of the encapsulating material whereas in the later stage, the mechanical properties rely on the degradation of the polymeric scaffold and the newly formed ECM by seeded cells. Initial results suggested that raw material encapsulation on a per cell basis might be dominated by a medium-regulated environment, although it was observed that *in vivo* differentiation of the cells would be controlled by the surrounding native environment of the implant site. In another study, Solchaga and co-workers [99] reported on the restoration of osteochondral defects with hyaluronan and polyester based scaffolds. They used two different polyesters, PLGA and PLLA based polymers and two hyaluronan-based polymers (ACP™ and

HYAFF®-11) for their study. They found that the degradation rate of the scaffolds is a crucial factor for the repairing process. Hyaluronan-based scaffolds showed better cell infiltration which leads to faster tissue regeneration. Degradation of the ACP™ leads to rapid bone formation, however, the slow degradation of HYAFF®-11 prolongs the existence of the cartilage.

5   Design Considerations for Interfacial Tissue Engineering

In this section, we will discuss the design considerations and elegant design methodologies adopted for interfacial tissue engineering to regenerate the osteochondral tissue interface. Success of interfacial tissue engineering not only depends on scaffold design but also depends on biodegradability and mechanical properties, surface topography, microstructure, scaffold porosity, pore geometry and orientation of the scaffolds. Mainly, two techniques such as stratified and gradient techniques have been used to fabricate scaffolds for interfacial tissue engineering applications and the techniques are described briefly below.

5.1   Stratified Scaffold Design

The first generation stratified scaffold design for osteochondral tissue engineering consists of two well distinguished cartilage and bone regions attached together using either sealants or sutures [100, 101]. In an attempt, Novakovic and co-workers [100] separately prepared cartilage constructs by culturing bovine calf articular chondrocytes on PLGA meshes and bone-like constructs by bovine calf periosteal cells on PLGA/polyethylene glycol blend foams. After that, biphasic scaffolds were prepared by suturing the distinct constructs together after 1 or 4 weeks of cell culture. The integration between two different scaffolds at the cartilage-bone interface was observed after another 4 weeks of culture. Alhadlaq et al. [102] designed a bilayered osteochondral construct with a condyle shape composed of polyethylene glycol-diacrylate hydrogel. The upper hydrogel layer of the construct consisted of MSC derived chondrocytes whereas the bottom layer consisted of MSC derived osteoblasts. After 12 weeks of implantation, two distinct osseous and cartilaginous regions were observed in the defect site. These pioneering studies on osteochondral tissue engineering showed how to facilitate multi-tissue formation on a multi-phase stratified scaffold, the next challenging task is how to fit osteochondral tissue interface into this constructs/scaffolds. In this concern, many researchers are working on stratified scaffold designs that can mimic the structural environment of the natural osteochondral interface. For example, Lu et al. [103] developed a 3D-osteoblast chondrocyte co-culture on a biomimetic hybrid of a hydrogel and a polymer-bioactive glass composite. The novel osteochondral construct was composed of three regions: gel only, gel/composite interface and composite only region. It was observed that chondrocyte and osteoblast co-cultures on this microsphere based scaffold system supported the development of distinct continuous chondrocyte and osteoblast regions, with good integration between these two regions. Not only that, the multiphase scaffold with varied calcium phosphate content promoted the growth of multiple matrix zones: a mineralized collagen matrix with osteoblasts, an interfacial matrix rich in GAG+ collagen, and a GAG rich chondrocyte region. Gibson and co-workers [104] reported a new method "liquid phase co-synthesis" that produces porous, layered scaffolds that can mimic the structure and composition of articular cartilage on one side of the scaffold, subchondral bone on the other side of the scaffold and there was a continuous gradual interface between these two. Their designed layered scaffold can be inserted into the osteochondral defect site without the need of any types of glue, sutures or screws with a highly interconnected porous structure throughout the entire osteochondral defect.

The insertion of the anterior cruciate ligament (ACL) into bone is a typical example of complex soft tissue to hard tissue interfacial tissue engineering. Which consists of spatial variations in cell type, and matrix composition resulting in three distinct regions (the ligament, fibrocartilage, and bone [105]) whereas the fibrocartilage section is further divided into mineralized and non-mineralized regions. It is well established that matrix organization at the interfacial tissue site is optimized to sustain both compressive and tensile stress simultaneously [106]. As an example, fibrocartilage is often confined in anatomical regions when exposed to compressive stress [106]. This type of site specific tunable mechanical properties facilitate a smooth transition in strain throughout the insertion and deliver valuable cues for ligament-to-bone interfacial scaffold design.

The multi tissue alteration at the interfacial site of ligament to bone at the ACL to bone joint represents a challenging task for interfacial tissue engineering. Until now, most of the recent studies focus on the restoration of the anatomic ACL-bone interface [107–110]. Cooper et al. [107] designed a multiphasic scaffold, based on polymeric nanofibers composed of PLGA using a 3D braiding technology. The resultant scaffold demonstrated optimal pore diameters of 175–233 μm for ligament growth, as well as it showed suitable mechanical properties, same as the native ligament. Paxton et al. [111] engineered a stratified scaffold to regenerate the bone-ligament interface using polyethylene glycol diacrylate with HA and the cell adhesive peptide RGD. Their designed multiphase scaffold showed significant improvement over a single phase ACL graft. Lu and co-workers designed for the first time a continuous triphasic scaffold, which consists of three distinct phases [113]. Phase A designed for soft tissue, phase B for interfacial region and phase C for bone regeneration. They engineered each phase with optimal composition and geometry which is suitable for ligament to bone interfacial tissue engineering. They used a double emulsion method to develop the microspheres and continuous triphasic scaffolds were formed by a sintering process above the polymer glass transition temperature. In another work, they designed a triphasic scaffold that can mimic the ligament to bone interface [15]. They used a polyglactin, PLGA (85,15) copolymer and bioactive glass. They also fabricated and characterized a mechanoactive scaffold consisting of a composite of poly-α-hydroxyester nanofibers and sintered microspheres. They observed that their scaffold induced compression of the patellar tendon grafts resulting in matrix remodeling and up regulated expression of fibrocartilage related markers such as aggrecan, TGFβ3, Type II-collagen, etc. These results suggests that their polymeric nanofiber based stratified scaffold can be used for the formation of anatomic fibrocartilage at the interfacial enthesis on ACL reconstruction grafts [112].

Similar to the ligament to bone interface, tendon to bone interfacial tissue engineering is also an example of a critical type of tissue engineering. Although tendon to bone and ligament to bone interface reconstruction are biochemically and physiologically almost similar, the tissue engineering strategy should differ in terms of stress loading environment, mineral distribution, and surgical repair methods that are applied during the operation, which would also influence the healing response.

The devastating pain of rotator cuff tears coupled with the high percentage of failure of conventional repairing strategies [113, 114] draw attention for the clinical need of integrative solutions for tendon to bone regeneration. To overcome this problem, several research groups have designed biopolymer based scaffolds in such a way so it can form an anatomic insertion site similar to the tendon to bone interface. Moffat et al. [115] prepared PLGA nanofiber scaffolds that can repair rotator cuff. They showed that nanofiber orientation play an important role for fibroblast morphology, and alignment as well as integrin expression. Furthermore, scaffold mechanical properties also depend on the fiber orientation. They also engineered a composite nanofiber system composed of PLGA and hydroxyapatite nanoparticles that can regenerate both non-mineralized and mineralized fibrocartilage regions of the supraspinatus insertion site [116].

It was found that the stratified 3D scaffold design technique is suitable for interfacial tissue regeneration whereas most of applications require improvement in mechanical properties or tissue integration, although the fabrication technique is very simple and is able to mimic the specific tissue interface like the cortical-cancellous interface. In addition to this, the stratified design technique is also dependent on material type, physical, chemical and topological properties of the materials. To overcome these problems, tissue engineers have chosen a continuous gradient technique for interfacial tissue engineering applications.

5.2 Gradient Scaffold Design

Gradient scaffolds design is a new, emerging and promising approach for interfacial tissue engineering. These novel types of scaffolds posses either a gradient of composition [117, 118], growth factors [119] or genes [120]. The gradient strategy can replicate the complex transition of chemical as well as mechanical properties of the interfacial regions and offer regional control over a complex native interface. Compared with the previously discussed stratified scaffold, gradient scaffolds consist with reasonably gradual and continuous change in both mechanical and compositional properties.

Detamore and his co-workers reported uniform PLGA (50,50 lactic acid, glycolic acid, acid-end group) microsphere scaffolds with continuous and gradual change in macroscopic gradients in stiffness [118]. They produced structural gradients by incorporating a nanomaterial. Structural gradients were produced by incorporating a high stiffness nanomaterial into portions of the microspheres during the scaffold manufacturing process. Similarly, Erisken et al. [117] designed functionally graded electrospun scaffolds from biodegradable polymers to regenerate the cartilage-to-bone interfacial tissue engineering. The scaffold consists of PCL and tricalcium phosphate nanocomposite with a gradient in calcium phosphate concentration. From the *in vitro* results they showed that MC3T3 osteoblast cells formed a gradient in the calcified matrix within 4 weeks. Li et al. [119] also engineered nanofiber scaffolds with a mineral content gradient to regenerate the tendon to bone interface. They fabricated a calcium phosphate coating on a mat of gelatin-coated PCL nanofibers by incubating the scaffolds in a ten times concentrated simulated body fluid (SBF). This gradient in mineral content resulted in a variation of scaffold stiffness and affected the total number of adhered MC3T3 cells to the nanofibrous scaffold.

Philips et al. [121] reported a novel alternative approach to design gradient scaffolds which can mimic the bone soft tissue interface. They engineered the scaffold by a simple, one-step seeding process of primary dermal fibroblasts onto polymeric scaffolds containing a graded distribution of immobilized Runx2 retrovirus. *In vivo* results showed that by controlling the spatial distribution of Runx2, fibroblasts can be stimulated to deposit a mineralized matrix on the scaffold. Berkland and co-workers designed microsphere based seamless scaffolds with gradients of growth factors. They prepared uniform size microspheres using a Precision Particle Fabrication technique. The scaffolds were assembled into a cylindrical glass mold and subsequently attached using ethanol treatments to form a continuous scaffold. This process offers a unique advantage over other conventional sintering processes, which will require quite high temperatures, which can affect the bioactivity of the growth factors as they are incorporated into the scaffolds [122]. Ramalingam et al. [123] fabricated nanofiber scaffolds with gradients in composition for interfacial tissue engineering. They simultaneously electrospun two types of nanofibers in an overlapping pattern to create a nanowoven mat with a composition gradient. PCL was used for the fabrication of nanofibers and they varied the concentration of osteoconductive amorphous calcium phosphate nanoparticles to form a gradient. *In vitro* results demonstrated that adhesion and proliferation of osteogenic cells on nanofiber gradient scaffolds increased considerably on those portions of the scaffold that contained higher percentages of calcium phosphate nanoparticles.

In summary, it can be postulated that compared with stratified scaffolds, gradient based scaffolds have the ability to mimic the orthopedic interfaces more precisely. The smooth transition in composition resulted in gradients of functional properties which offers efficient load transfer between distinct types of tissues. In other words, design strategies and design parameters must be prioritized for interfacial tissue engineering. From this point of view, multiphase stratified structures represent a simpler approach whereas gradient based scaffolds are more complex for fabrication but can mimic functional properties as well as mechanical properties inherent at the interfacial tissue site.

6 Present Clinical Status of Interfacial Tissue Engineering

Injury at the interfacial tissue site can occur at any age and is frequently seen in young athletes. Curl et al. [124] found that greater than 60% of patients who possess knee arthroscopies suffered from Grade III or Grade IV lesions where the harshness of this type of damage varies significantly. Many times, injury at the articular cartilage site leads to the osteoartharaitis. Not only for young athletes, elderly often deal with osteoartharitis. The pain associated with this type of interfacial tissue site destroys the quality of a patient's life. There are approximately 500,000 osteochondral and cartilage repair events done in the USA alone. These numbers are increasingly rapidly due to lifestyle and aging of the population. The costs associated with this repairing process is predicted to be greater than $1.5 billion in the near future [125]. A list of conventional clinical methods and advanced tissue engineering strategies for the treatment of defects at the interfacial tissue site can be subcategorized into the following methods:

1. Conventional clinical methods

   (a) Palliative treatment methods
   (b) Reparative treatment methods
   (c) Restorative treatment methods

2. Tissue engineering methods

   (a) Single phase scaffolds
   (b) Multiphase scaffolds
   (c) Gradient scaffolds

However, despite so many *in vitro* and *in vivo* animal model studies, there are no reported data/studies on humans to treat arthritis or any type of injury at an interfacial tissue site by interfacial tissue engineering.

7 Future Perspectives of Interfacial Tissue Engineering in Orthopedic Applications

The field of tissue engineering, especially interfacial tissue engineering, is the latest and rapidly growing field. In the last decade, osteochondral constructs have been implanted in various animals including rats, rabbits, pigs, goats, sheep, horses and dogs for osteochondral tissue regeneration at the defect site, although the size of defect varies from animal size. But the success of the interfacial tissue engineering concept for osteochondral tissue regeneration in small animals is one step closer to clinical trials on humans. Although current efforts are focused on the development of the optimum scaffold designs and fabrication for interfacial tissue engineering in orthopedic applications, we predict that the future will turn towards the identification of the most significant strategies of interfacial tissue engineering to treat arthritis patients. However, competition with other conventional clinical approaches make interfacial tissue engineering in orthopedic applications well warranted and holds a significant promise in the near future although a significant number of challenges and limitations still exist.

8 Conclusion

With the advancement of technologies, interfacial tissue engineering is continuing to progress at a fast rate to design more smart scaffolds for the regeneration of the osteochondral interface tissue. Different scaffold fabrication techniques, like multiphasic or gradient scaffolds, containing signaling cues for chondro- or osteogenic differentiation to mimic osteochondral tissue are continuously developing to find new ways to improve osteochondral scaffolds and their potential for long-term *in vivo* studies for the generation of osteochondral tissue. *In vivo* experiments should be conducted on larger animals with large defect sizes to optimize the formation of required collagen types, material composition and mechanical properties since their functionality will be useful towards clinical trials for humans.

Acknowledgement This work was funded by the Science and Engineering Research Board (SERB), Department of Science and Technology (DST), Govt. of India, (EEQ/2016/000712 and ECR/2016/002018/ES). KG and KS also acknowledge Sayantan Tripathy to support during writing this chapter.

References

1. Csaki C, Schneider PShakibaei M. Mesenchymal stem cells as a potential pool for cartilage tissue engineering. Ann Anat Anatomischer Anzeiger. 2008;190:395–412.
2. Hangody L, Kish G, Karpati Z, et al. Arthroscopic autogenous osteochondral mosaicplasty for the treatment of femoral condylar articular defects a preliminary report. Knee Surg Sports Traumatol Arthrosc. 1997;5:262–7.
3. Carranza-Bencano A, García-Paino L, Padron JA, et al. Neochondrogenesis in repair of full-thickness articular cartilage defects using free autogenous periosteal grafts in the rabbit. A follow-up in six months. Osteoarthritis Cartilage. 2000;8:351–8.
4. Sledge SL. Microfracture techniques in the treatment of osteochondral injuries. Clin Sports Med. 2001;20:365–78.
5. Livesley P, Doherty M, Needoff M, et al. Arthroscopic lavage of osteoarthritic knees. Bone Joint J. 1991;73:922–6.
6. Paul J, Sagstetter A, Kriner M, et al. Donor-site morbidity after osteochondral autologous transplantation for lesions of the talus. J Bone Joint Surg Am. 2009;91:1683–8.
7. Hangody L, Dobos J, Baló E, et al. Clinical experiences with autologous osteochondral mosaicplasty in an athletic population a 17-year prospective multicenter study. Am J Sports Med. 2010;38:1125–33.
8. Amini AR, Laurencin CTNukavarapu SP. Bone tissue engineering: recent advances and challenges. Crit Rev Biomed Eng. 2012;40:363–408.
9. Parveen S, Krishnakumar KSahoo S. New era in health care: tissue engineering. J Stem Cells Regener Med. 2006;1:8–24.

10. Ikada Y. Challenges in tissue engineering. J R Soc Interface. 2006;3:589–601.
11. Langer RVacanti J. Tissue engineering. Science. 1993;260:920–6.
12. Lu HH, Subramony SD, Boushell MK, et al. Tissue engineering strategies for the regeneration of orthopedic interfaces. Ann Biomed Eng. 2010;38:2142–54.
13. Yang PJTemenoff JS. Engineering orthopedic tissue interfaces. Tissue Eng Part B Rev. 2009;15:127–41.
14. Stoop R. Smart biomaterials for tissue engineering of cartilage. Injury. 2008;39:77–87.
15. Spalazzi JP, Doty SB, Moffat KL, et al. Development of controlled matrix heterogeneity on a triphasic scaffold for orthopedic interface tissue engineering. Tissue Eng. 2006;12:3497–508.
16. Dormer NH, Singh M, Wang L, et al. Osteochondral interface tissue engineering using macroscopic gradients of bioactive signals. Ann Biomed Eng. 2010;38:2167–82.
17. Streuli C. Extracellular matrix remodelling and cellular differentiation. Curr Opin Cell Biol. 1999;11:634–40.
18. Gailit JClark RA. Wound repair in the context of extracellular matrix. Curr Opin Cell Biol. 1994;6:717–25.
19. Kleinman HK, Philp DHoffman MP. Role of the extracellular matrix in morphogenesis. Curr Opin Biotechnol. 2003;14:526–32.
20. Griffith L. Polymeric biomaterials. Acta Mater. 2000;48:263–77.
21. Chandra RRustgi R. Biodegradable polymers. Prog Polym Sci. 1998;23:1273–335.
22. Angelova NHunkeler D. Rationalizing the design of polymeric biomaterials. Trends Biotechnol. 1999;17:409–21.
23. Nukavarapu SPDorcemus DL. Osteochondral tissue engineering: current strategies and challenges. Biotechnol Adv. 2013;31:706–21.
24. Eyre DRWJ-J. Collagen structure and cartilage matrix integrity. J Rheumatol Suppl. 1995;43:82–5.
25. Aigner T, Reichenberger E, Bertling W, et al. Type X collagen expression in osteoarthritic and rheumatoid articular cartilage. Virchows Arch B. 1993;63:205.
26. Hunziker E, Quinn THäuselmann H-J. Quantitative structural organization of normal adult human articular cartilage. Osteoarthritis Cartilage. 2002;10:564–72.
27. Farnum C, Lee R, O'Hara K, et al. Volume increase in growth plate chondrocytes during hypertrophy: the contribution of organic osmolytes. Bone. 2002;30:574–81.
28. Di Luca A, Van Blitterswijk CMoroni L. The osteochondral interface as a gradient tissue: from development to the fabrication of gradient scaffolds for regenerative medicine. Birth Defects Res C Embryo Today. 2015;105:34–52.
29. Zhang Y, Wang F, Tan H, et al. Analysis of the mineral composition of the human calcified cartilage zone. Int J Med Sci. 2012;9:353–60.
30. Keeney MPandit A. The osteochondral junction and its repair via bi-phasic tissue engineering scaffolds. Tissue Eng Part B Rev. 2009;15:55–73.
31. Arvidson K, Abdallah B, Applegate L, et al. Bone regeneration and stem cells. J Cell Mol Med. 2011;15:718–46.
32. Qu D, Mosher CZ, Boushell MK, et al. Engineering complex orthopaedic tissues via strategic biomimicry. Ann Biomed Eng. 2015;43:697–717.
33. Erisken C, Kalyon DMWang H. Viscoelastic and biomechanical properties of osteochondral tissue constructs generated from graded polycaprolactone and beta-tricalcium phosphate composites. J Biomech Eng. 2010;132:091013.
34. Scarritt ME, Pashos NCBunnell BA. A review of cellularization strategies for tissue engineering of whole organs. Front Bioeng Biotechnol. 2015;3:43.
35. Garreta E, Oria R, Tarantino C, et al. Tissue engineering by decellularization and 3D bioprinting. Mater Today. 2017;20:166–78.
36. Maitz MF. Applications of synthetic polymers in clinical medicine. Biosurf Biotribol. 2015;1:161–76.
37. Ko IK, Lee SJ, Atala A, et al. In situ tissue regeneration through host stem cell recruitment. Exp Mol Med. 2013;45:e57.

38. Grayson WL, Chao P-HG, Marolt D, et al. Engineering custom-designed osteochondral tissue grafts. Trends Biotechnol. 2008;26:181–9.
39. Hutmacher DW. Scaffolds in tissue engineering bone and cartilage. Biomaterials. 2000;21:2529–43.
40. Klein TJ, Malda J, Sah RL, et al. Tissue engineering of articular cartilage with biomimetic zones. Tissue Eng Part B Rev. 2009;15:143–57.
41. Lu HJiang J. Interface tissue Engineeringand the formulation of multiple-tissue systems. Tissue Eng I. 2006;102:91–111.
42. Allan K, Pilliar R, Wang J, et al. Formation of biphasic constructs containing cartilage with a calcified zone interface. Tissue Eng. 2007;13:167–77.
43. Wang X, Grogan SP, Rieser F, et al. Tissue engineering of biphasic cartilage constructs using various biodegradable scaffolds: an in vitro study. Biomaterials. 2004;25:3681–8.
44. Guo X, Wang C, Duan C, et al. Repair of osteochondral defects with autologous chondrocytes seeded onto bioceramic scaffold in sheep. Tissue Eng. 2004;10:1830–40.
45. Nooeaid P, Salih V, Beier JP, et al. Osteochondral tissue engineering: scaffolds, stem cells and applications. J Cell Mol Med. 2012;16:2247–70.
46. Chen G, Sato T, Tanaka J, et al. Preparation of a biphasic scaffold for osteochondral tissue engineering. Mater Sci Eng C. 2006;26:118–23.
47. Gotterbarm T, Richter W, Jung M, et al. An in vivo study of a growth-factor enhanced, cell free, two-layered collagen–tricalcium phosphate in deep osteochondral defects. Biomaterials. 2006;27:3387–95.
48. Mano JReis R. Osteochondral defects: present situation and tissue engineering approaches. J Tissue Eng Regen Med. 2007;1:261–73.
49. Hsu Y, Turner IMiles A. Fabrication of porous bioceramics with porosity gradients similar to the bimodal structure of cortical and cancellous bone. J Mater Sci Mater Med. 2007;18:2251–6.
50. Liu C, Han ZCzernuszka J. Gradient collagen/nanohydroxyapatite composite scaffold: development and characterization. Acta Biomater. 2009;5:661–9.
51. Liu L, Xiong Z, Yan Y, et al. Multinozzle low-temperature deposition system for construction of gradient tissue engineering scaffolds. J Biomed Mater Res B Appl Biomater. 2009;88:254–63.
52. Spalazzi JP, Dagher E, Doty SB, et al. In vivo evaluation of a multiphased scaffold designed for orthopaedic interface tissue engineering and soft tissue-to-bone integration. J Biomed Mater Res A. 2008;86:1–12.
53. Wang INE, Shan J, Choi R, et al. Role of osteoblast–fibroblast interactions in the formation of the ligament-to-bone interface. J Orthop Res. 2007;25:1609–20.
54. Wahl DA, Sachlos E, Liu C, et al. Controlling the processing of collagen-hydroxyapatite scaffolds for bone tissue engineering. J Mater Sci Mater Med. 2007;18:201–9.
55. Kon E, Delcogliano M, Filardo G, et al. A novel nano-composite multi-layered biomaterial for treatment of osteochondral lesions: technique note and an early stability pilot clinical trial. Injury. 2010;41:693–701.
56. Tampieri A, Sandri M, Landi E, et al. Design of graded biomimetic osteochondral composite scaffolds. Biomaterials. 2008;29:3539–46.
57. Teng SH, Lee EJ, Wang P, et al. Functionally gradient chitosan/hydroxyapatite composite scaffolds for controlled drug release. J Biomed Mater Res B Appl Biomater. 2009;90:275–82.
58. Vitale-Brovarone C, Baino FVerné E. Feasibility and tailoring of bioactive glass-ceramic scaffolds with gradient of porosity for bone grafting. J Biomater Appl. 2010;24:693–712.
59. Ho STB, Hutmacher DW, Ekaputra AK, et al. The evaluation of a biphasic osteochondral implant coupled with an electrospun membrane in a large animal model. Tissue Eng A. 2009;16:1123–41.
60. Mimura T, Imai S, Kubo M, et al. A novel exogenous concentration-gradient collagen scaffold augments full-thickness articular cartilage repair. Osteoarthritis Cartilage. 2008;16:1083–91.

61. Holland TA, Bodde EW, Baggett LS, et al. Osteochondral repair in the rabbit model utilizing bilayered, degradable oligo (poly (ethylene glycol) fumarate) hydrogel scaffolds. J Biomed Mater Res A. 2005;75:156–67.

62. Needham CJ, Shah SR, Dahlin RL, et al. Osteochondral tissue regeneration through polymeric delivery of DNA encoding for the SOX trio and RUNX2. Acta Biomater. 2014;10:4103–12.

63. Ferrand A, Eap S, Richert L, et al. Osteogenetic properties of electrospun Nanofibrous PCL scaffolds equipped with chitosan-based Nanoreservoirs of growth factors. Macromol Biosci. 2014;14:45–55.

64. Chen J, Chen H, Li P, et al. Simultaneous regeneration of articular cartilage and subchondral bone in vivo using MSCs induced by a spatially controlled gene delivery system in bilayered integrated scaffolds. Biomaterials. 2011;32:4793–805.

65. Ding C, Qiao Z, Jiang W, et al. Regeneration of a goat femoral head using a tissue-specific, biphasic scaffold fabricated with CAD/CAM technology. Biomaterials. 2013;34:6706–16.

66. Di Luca A, Longoni A, Criscenti G, et al. Surface energy and stiffness discrete gradients in additive manufactured scaffolds for osteochondral regeneration. Biofabrication. 2016;8:015014.

67. Gupta V, Mohan N, Berkland CJ, et al. Microsphere-based scaffolds carrying opposing gradients of chondroitin sulfate and Tricalcium phosphate. Front Bioeng Biotechnol. 2015;3:96.

68. Hou Y, Schoener CA, Regan KR, et al. Photo-cross-linked PDMSstar-PEG hydrogels: synthesis, characterization, and potential application for tissue engineering scaffolds. Biomacromolecules. 2010;11:648–56.

69. Dormer NH, Singh M, Zhao L, et al. Osteochondral interface regeneration of the rabbit knee with macroscopic gradients of bioactive signals. J Biomed Mater Res A. 2012;100:162–70.

70. Wang X, Wenk E, Zhang X, et al. Growth factor gradients via microsphere delivery in biopolymer scaffolds for osteochondral tissue engineering. J Control Release. 2009;134:81–90.

71. Petrenko YA, Ivanov R, Petrenko AY, et al. Coupling of gelatin to inner surfaces of pore walls in spongy alginate-based scaffolds facilitates the adhesion, growth and differentiation of human bone marrow mesenchymal stromal cells. J Mater Sci Mater Med. 2011;22:1529–40.

72. Croisier FJérôme C. Chitosan-based biomaterials for tissue engineering. Eur Polym J. 2013;49:780–92.

73. da Silva MA, Crawford A, Mundy J, et al. Chitosan/polyester-based scaffolds for cartilage tissue engineering: assessment of extracellular matrix formation. Acta Biomater. 2010;6:1149–57.

74. Malafaya P, Pedro A, Peterbauer A, et al. Chitosan particles agglomerated scaffolds for cartilage and osteochondral tissue engineering approaches with adipose tissue derived stem cells. J Mater Sci Mater Med. 2005;16:1077–85.

75. Malafaya PBReis RL. Bilayered chitosan-based scaffolds for osteochondral tissue engineering: influence of hydroxyapatite on in vitro cytotoxicity and dynamic bioactivity studies in a specific double-chamber bioreactor. Acta Biomater. 2009;5:644–60.

76. Abarrategi A, Lópiz-Morales Y, Ramos V, et al. Chitosan scaffolds for osteochondral tissue regeneration. J Biomed Mater Res A. 2010;95:1132–41.

77. Oliveira JM, Rodrigues MT, Silva SS, et al. Novel hydroxyapatite/chitosan bilayered scaffold for osteochondral tissue-engineering applications: scaffold design and its performance when seeded with goat bone marrow stromal cells. Biomaterials. 2006;27:6123–37.

78. Chajra H, Rousseau C, Cortial D, et al. Collagen-based biomaterials and cartilage engineering. Application to osteochondral defects. Biomed Mater Eng. 2008;18:33–45.

79. Galois L, Freyria A, Grossin L, et al. Cartilage repair: surgical techniques and tissue engineering using polysaccharide-and collagen-based biomaterials. Biorheology. 2004;41:433–43.

80. Shim J-H, Kim AJ, Park JY, et al. Effect of solid freeform fabrication-based polycaprolactone/poly (lactic-co-glycolic acid)/collagen scaffolds on cellular activities of human adipose-derived stem cells and rat primary hepatocytes. J Mater Sci Mater Med. 2013;24:1053–65.

81. Levingstone TJ, Ramesh A, Brady RT, et al. Cell-free multi-layered collagen-based scaffolds demonstrate layer specific regeneration of functional osteochondral tissue in caprine joints. Biomaterials. 2016;87:69–81.

82. Getgood A, Henson F, Skelton C, et al. Osteochondral tissue engineering using a biphasic collagen/GAG scaffold containing rhFGF18 or BMP-7 in an ovine model. J Exp Orthop. 2014;1:13.

83. Yi C, Liu D, Fong C-C, et al. Gold nanoparticles promote osteogenic differentiation of mesenchymal stem cells through p38 MAPK pathway. ACS Nano. 2010;4:6439–48.

84. Cortizo AM, Ruderman G, Mazzini FN, et al. Novel vanadium-loaded ordered collagen scaffold promotes osteochondral differentiation of bone marrow progenitor cells. Int J Biomater. 2016;2016:1.

85. Melke J, Midha S, Ghosh S, et al. Silk fibroin as biomaterial for bone tissue engineering. Acta Biomater. 2016;31:1–16.

86. Saha S, Kundu B, Kirkham J, et al. Osteochondral tissue engineering in vivo: a comparative study using layered silk fibroin scaffolds from mulberry and nonmulberry silkworms. PLoS One. 2013;8:e80004.

87. Levingstone TJ, Thompson E, Matsiko A, et al. Multi-layered collagen-based scaffolds for osteochondral defect repair in rabbits. Acta Biomater. 2016;32:149–60.

88. Bertoni L, Zavatti M, Resca E, et al. Bilayered scaffolds colonized with dental pulp stem cells for osteochondral tissue engineering application. Ital J Anat Embryol. 2015;120:101.

89. Coluccino L, Stagnaro P, Vassalli M, et al. Bioactive TGF-β1/HA alginate-based scaffolds for osteochondral tissue repair: design, realization and multilevel characterization. J Appl Biomater Funct Mater. 2016;14:e42–52.

90. Khanarian NT, Jiang J, Wan LQ, et al. A hydrogel-mineral composite scaffold for osteochondral interface tissue engineering. Tissue Eng A. 2011;18:533–45.

91. O'Shea TM, Miao X. Bilayered scaffolds for osteochondral tissue engineering. Tissue Eng Part B Rev. 2008;14:447–64.

92. Dhandayuthapani B, Yoshida Y, Maekawa T, et al. Polymeric scaffolds in tissue engineering application: a review. Int J Polym Sci. 2011;2011:1.

93. Liu X, Holzwarth JMMa PX. Functionalized synthetic biodegradable polymer scaffolds for tissue engineering. Macromol Biosci. 2012;12:911–9.

94. Shafiee A, Soleimani M, Chamheidari GA, et al. Electrospun nanofiber-based regeneration of cartilage enhanced by mesenchymal stem cells. J Biomed Mater Res A. 2011;99:467–78.

95. Cao T, Ho K-HTeoh S-H. Scaffold design and in vitro study of osteochondral coculture in a three-dimensional porous polycaprolactone scaffold fabricated by fused deposition modeling. Tissue Eng. 2003;9:103–12.

96. Yunos DM, Ahmad Z, Salih V, et al. Stratified scaffolds for osteochondral tissue engineering applications: electrospun PDLLA nanofibre coated bioglass®-derived foams. J Biomater Appl. 2013;27:537–51.

97. Stodolak-Zych E, Menaszek E, Szatkowski P, et al. Carbon nanofibers (CNF) as scaffolds for osteochondral tissue regenerative medicine. Front Bioeng Biotechnol. 2016.

98. Erisken C, Kalyon DM, Wang H, et al. Osteochondral tissue formation through adipose-derived stromal cell differentiation on biomimetic polycaprolactone nanofibrous scaffolds with graded insulin and Beta-glycerophosphate concentrations. Tissue Eng A. 2011;17:1239–52.

99. Solchaga LA, Temenoff JS, Gao J, et al. Repair of osteochondral defects with hyaluronan-and polyester-based scaffolds. Osteoarthritis Cartilage. 2005;13:297–309.

100. Schaefer D, Martin I, Shastri P, et al. In vitro generation of osteochondral composites. Biomaterials. 2000;21:2599–606.

101. Gao J, Dennis JE, Solchaga LA, et al. Tissue-engineered fabrication of an osteochondral composite graft using rat bone marrow-derived mesenchymal stem cells. Tissue Eng. 2001;7:363–71.

102. Alhadlaq AMao JJ. Tissue-engineered osteochondral constructs in the shape of an articular condyle. JBJS. 2005;87:936–44.

103. Lu HH, El-Amin SF, Scott KD, et al. Three-dimensional, bioactive, biodegradable, polymer–bioactive glass composite scaffolds with improved mechanical properties support collagen synthesis and mineralization of human osteoblast-like cells in vitro. J Biomed Mater Res A. 2003;64:465–74.

104. Harley BA, Lynn AK, Wissner-Gross Z, et al. Design of a multiphase osteochondral scaffold III: fabrication of layered scaffolds with continuous interfaces. J Biomed Mater Res A. 2010;92:1078–93.
105. Benjamin M, Evans ECopp L. The histology of tendon attachments to bone in man. J Anat. 1986;149:89.
106. Ma Z, Kotaki M, Inai R, et al. Potential of nanofiber matrix as tissue-engineering scaffolds. Tissue Eng. 2005;11:101–9.
107. Cooper JA, Lu HH, Ko FK, et al. Fiber-based tissue-engineered scaffold for ligament replacement: design considerations and in vitro evaluation. Biomaterials. 2005;26:1523–32.
108. Cooper JA, Sahota JS, Gorum WJ, et al. Biomimetic tissue-engineered anterior cruciate ligament replacement. Proc Natl Acad Sci. 2007;104:3049–54.
109. Inoue N, Ikeda K, Aro HT, et al. Biologic tendon fixation to metallic implant augmented with autogenous cancellous bone graft and bone marrow in a canine model. J Orthop Res. 2002;20:957–66.
110. Itoi E, Berglund LJ, Grabowski JJ, et al. Tensile properties of the supraspinatus tendon. J Orthop Res. 1995;13:578–84.
111. Paxton JZ, Donnelly K, Keatch RP, et al. Engineering the bone–ligament interface using polyethylene glycol diacrylate incorporated with hydroxyapatite. Tissue Eng A. 2008;15:1201–9.
112. Spalazzi JP, Vyner MC, Jacobs MT, et al. Mechanoactive scaffold induces tendon remodeling and expression of fibrocartilage markers. Clin Orthop Relat Res. 2008;466:1938–48.
113. Coons DABarber FA. Tendon graft substitutes—rotator cuff patches. Sports Med Arthrosc Rev. 2006;14:185–90.
114. Derwin KA, Baker AR, Spragg RK, et al. Commercial extracellular matrix scaffolds for rotator cuff tendon repair: biomechanical, biochemical, and cellular properties. JBJS. 2006;88:2665–72.
115. Moffat KL, Kwei AS-P, Spalazzi JP, et al. Novel nanofiber-based scaffold for rotator cuff repair and augmentation. Tissue Eng A. 2008;15:115–26.
116. Moffat KL, Levine WN, Lu HH. In vitro evaluation of rotator cuff tendon fibroblasts on aligned composite scaffold of polymer nanofibers and hydroxyapatite nanoparticles. Transactions of the 54th Orthopaedic Research Society. 2008.
117. Erisken C, Kalyon DMWang H. Functionally graded electrospun polycaprolactone and β-tricalcium phosphate nanocomposites for tissue engineering applications. Biomaterials. 2008;29:4065–73.
118. Singh M, Sandhu B, Scurto A, et al. Microsphere-based scaffolds for cartilage tissue engineering: using subcritical CO_2 as a sintering agent. Acta Biomater. 2010;6:137–43.
119. Li X, Xie J, Lipner J, et al. Nanofiber scaffolds with gradations in mineral content for mimicking the tendon-to-bone insertion site. Nano Lett. 2009;9:2763–8.
120. Liu H, Yang L, Zhang E, et al. Biomimetic tendon extracellular matrix composite gradient scaffold enhances ligament-to-bone junction reconstruction. Acta Biomater. 2017;56:129–40.
121. Phillips JE, Burns KL, Le Doux JM, et al. Engineering graded tissue interfaces. Proc Natl Acad Sci. 2008;105:12170–5.
122. Singh M, Morris CP, Ellis RJ, et al. Microsphere-based seamless scaffolds containing macroscopic gradients of encapsulated factors for tissue engineering. Tissue Eng Part C Methods. 2008;14:299–309.
123. Ramalingam M, Young MF, Thomas V, et al. Nanofiber scaffold gradients for interfacial tissue engineering. J Biomater Appl. 2013;27:695–705.
124. Curl WW, Krome J, Gordon ES, et al. Cartilage injuries: a review of 31,516 knee arthroscopies. Arthroscopy. 1997;13:456–60.
125. McNickle AG, Provencher MTCole BJ. Overview of existing cartilage repair technology. Sports Med Arthrosc Rev. 2008;16:196–201.

Performance of Bore-Cone Taper Junctions on Explanted Total Knee Replacements with Modular Stem Extensions: Mechanical Disassembly and Corrosion Analysis of Two Designs

Pooja Panigrahi, Kyle Snethen, Kevin G. Schwartzman, Jorg Lützner, and Melinda K. Harman

Keywords Orthopedics · Biomaterials · *In vivo* · Total knee replacement · Modularity · Morse taper · Bore-cone taper junction · Modular stem · Titanium alloy · Cobalt chromium alloy · Corrosion · Fretting · Mechanical testing · Implant retrieval

1 Introduction

Modern joint prostheses commonly incorporate modular junctions in the design to provide intraoperative flexibility and a manageable inventory of components needed to address various bone defects [1–3]. Modularity increases the surgeon's options related to component size and materials [4–6] during primary arthroplasty [7], as well as during revision arthroplasty [8]. Contemporary hip prostheses usually have at least one modular taper junction, while knee prostheses typically incorporate modular taper junctions only when stem extensions are necessary, often in complicated revision procedures [1, 9]. It is likely that concerns associated with modularity are better documented for hip prostheses [10] due to the higher prevalence of modular junctions in total hip prostheses.

While the benefits of modularity are well established for joint prostheses, there are lingering concerns about localized corrosion at taper junctions. Galvanic corrosion is thought to play a role in modular taper junctions when other forms of corrosion are also involved [11–13], though in other cases this may not be a major factor

P. Panigrahi · K. Snethen · K. G. Schwartzman · M. K. Harman (✉)
Bioengineering Department, Clemson University, Clemson, SC, USA
e-mail: harman2@clemson.edu

J. Lützner
Orthopedic and Trauma Surgery Department, University Hospital Carl Gustav Carus, Dresden, Germany

© Springer International Publishing AG, part of Springer Nature 2018
B. Li, T. Webster (eds.), *Orthopedic Biomaterials*,
https://doi.org/10.1007/978-3-319-89542-0_5

that leads to surface degradation [14, 15]. The geometries of taper junctions themselves lead to crevice environments where chloride ions can reach sufficiently high concentrations to result in accelerated localized corrosion [16]. This type of damage has been observed in several retrieval studies of modular hip prosthesis tapers [17, 18]. In addition to corrosion, the release of metal debris due to wear and fretting specifically affects modular taper surfaces [19]. Bore and cone tapers are press-fit during manual intraoperative assembly and ideally form a taper-lock that resists relative movement between the components until disassembly during revision surgery [20, 21]. The force required to disassemble a taper-lock junction is an established measure of the strength and stability of taper-lock junctions [22–24]. However, evidence of displacements less than 100 μm parallel to the taper axis have been noted on modular taper surfaces after *in vitro* [25–27] and *in vivo* [28, 29] cyclic loading. This micromotion results in mechanical disruption of the taper surfaces [19, 30], and the thermodynamic corrosion reaction is able to occur freely in areas where the passive film is disrupted [29, 31]. The high prevalence of fretting corrosion in modular joint prostheses at the time of revision surgery persists in recent reports [32, 33]. Finally, taper junctions are subject to a variety of compressive and tensile stresses during *in vivo* function, and it is possible that the stresses themselves or the resulting strained material leads to localized corrosion [34, 35].

Clinical consequences from corrosion at modular junctions may include elevated serum levels of metal ions and systemic metal allergy, aseptic local tissue reactions, instability at the taper interface, and metallosis [36]. While a few case reports [37, 38] have been able to link the corrosion observed on retrievals with *in vivo* consequences, most large-scale retrieval studies with corrosion results [29, 33, 39, 40] have found that a broad range of clinical outcomes exist, with the most common reasons for revision including excessive pain, infection, loosening, and polyethylene wear. Recent review articles on the clinical consequences of taper corrosion suggests that the risks of modularity need to be better understood and there is a need to define realistic monitoring protocols [41, 42]. Retrieval studies coupled with mechanical disassembly could identify the designs' and patient factors' effects on corrosion.

The objective of this study was to evaluate bore-cone taper junctions of explanted total knee replacements (TKR) designed with modular stem extensions, documenting forces necessary to disassemble the modular stems, the amount and distribution of surface corrosion observed on the bore-cone taper junctions, and to determine any associations between corroded modular tapers and patient demographics. Previously established photogrammetric [30] and profilometric [43] methods were adapted to characterize the taper surfaces and complete the surface corrosion measurements. The amount of surface corrosion was compared across two different knee prosthesis designs, disassembly forces, anatomic location, alloy couplings, patient demographics, and clinical records. Since the effects of these variables were found to be significant in prior studies on corrosion or fretting micromotion in modular hip prostheses [10, 13, 18, 19], it was hypothesized that some or all of these variables are associated with the amount of corrosion observed in modular knee prostheses.

2 Materials and Methods

2.1 Implant Retrieval and Archiving

Knee prosthesis components were collected consecutively from consenting patients after revision knee arthroplasty by the surgeon co-author (JL) between 2011–2014 and archived in an IRB-approved explant registry (Technical University Dresden EK348112009; Clemson University IRB2008–308). Out of 316 retrieved knee prosthesis components collected, 60 femoral or tibial components included a modular stem extension. Thirty-eight prosthesis components met the following inclusion criteria: (1) The prosthesis was explanted by the same surgical team at the same clinical center; (2) Components belonged to a TKR design with options for using modular stem extensions on the femoral and tibial components; (3) The TKR design was either the *RT Plus* (Plus Orthopedics) or *NexGen RH* (Zimmer) design (Table 1), as necessitated by the need for at least 3 prostheses of a single design for statistical reasons; (4) All metal components were available for analysis, including stems, set-screws, through-bolts, sleeves, tibial baseplates, femoral articulating components, and augmentation blocks. Key design differences for the included TKR were the orientation of the bore-cone taper junctions (bore located on the component versus bore located on the modular stem) and the use of similar or mixed metal combinations at the taper junction (Fig. 1).

The patient clinical history and surgical reports for each retrieved prosthesis was abstracted from available medical records and summarized (Table 2). The included TKR were explanted from 21 patients (14 female and 7 male) after an average duration of 3.5 ± 2.9 (range, 0.1–11.2) years *in vivo*. Average patient age and body mass index (BMI) at the time of revision was 71.7 ± 8.4 (range, 55.4–88.6) years and 30.9 ± 4.6 (range, 19.8–37.2), respectively. All patients had endured one or more prior knee arthroplasty procedures (range, 2–9), except one who received a TKR with stem extensions as a primary surgery due to posttraumatic osteoarthritis. Reasons for revision included infection (11 patients), loosening (6 patients), peri-prosthetic fracture (3 patients), and unreported (1 patient). Upon removal during revision surgery, the explanted components were assigned a unique identifying number and handled according to standard protocols for implant retrieval. Each TKR was fixed in formalin for >48 h, rinsed in tap water, air dried, packaged in absorbent cloth, and then sealed in labeled plastic bags and shipped to the lab. Upon receipt, each component was cleaned and disinfected by sonication and rinsing in a mild detergent (Liquinox, Alconox Inc., White Plains, NY), deionized water, and ethyl alcohol.

Each knee prosthesis, including either a tibial baseplate, tibial stem extension, and screws/augments, or a femoral condylar replacement, femoral stem, and screws/augments, was disassembled as recommended by ISO 7206–10 and ASTM F2009 [44, 45] using a servohydraulic test system (Instron, Norwood, MA) with a 25 kN load cell. A test setup was designed [46] to apply a pure axial tensile load to the stem extension while avoiding off-axis moments (Fig. 2). A preload condition of 150 N

Table 1 Taper design features of TKR with modular stem extensions

Parameters		Design A	Design B
Manufacturer information	Manufacturer	Plus Orthopedics[a]	Zimmer
	Brand	RT Plus modular knee	NexGen RH modular knee
Taper orientation	Stem junction	Bore	Cone
	Component junction	Cone	Bore
Taper material	Bore	TiAlV or CoCrMo	CoCrMo
	Cone	CoCrMo	TiAlV
Mechanical locking of taper	Conical or threaded	Conical	Conical
	Machine line depth (bore[b])	2.0 ± 0.2 μm	1.3 ± 0.1 μm
	Machine line depth (cone[b])	3.1 ± 0.4 μm	1.9 ± 0.3 μm
	Set screws	1	2, on femoral components only
	Through-bolt	No	On tibial components only
	Stem extension assembly instructions	Component placed on assembly block. Automatic hammer placed on stem and used to impact the stem three times. Set screw is tightened with screwdriver [52].	Stem extension manually inserted to achieve "snug" fit. While stem extension is protected, a two-pound mallet is used to strike it solidly one time. Set screws or through-bolt are tightened with screwdriver [53]
Experimental design	Femoral components	14	6
	Tibial components	14	4
	n (total)	28	10

[a]Acquired by Smith & Nephew in 2007
[b]Measured from non-corroded regions on retrievals

was specified followed by a ramp tensile load applied to the stem axis at a constant displacement rate of 0.04 mm/s. The load and displacement was recorded continuously until the loading force registered less than 100 N indicating disassembly and test completion. The maximum disassembly force for the taper-lock junction was recorded for each component. After disassembly, exposed taper surfaces were cleaned and disinfected by the same sonication protocol. Due to the large difference in sample sizes for the two included taper designs, all Design B tapers and a matched cohort of Design A tapers from patients with comparable demographics (statistically similar age and BMI) were selected for further analysis.

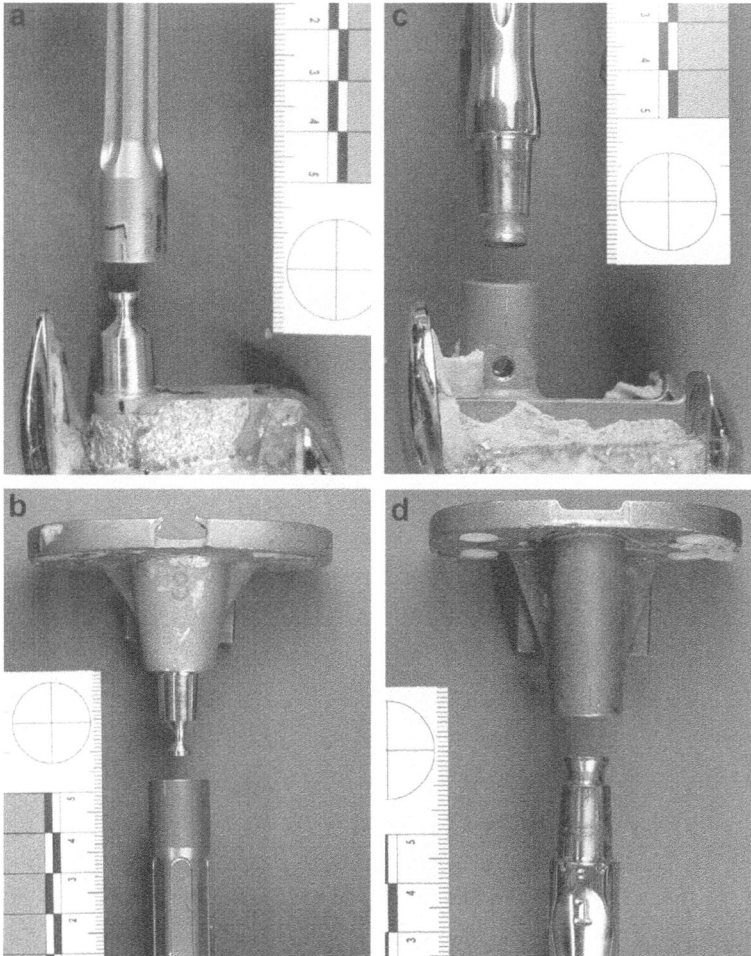

Fig. 1 Explanted TKR components representative of Design A tapers had the male cone tapers located on the (**a**) femoral and (**b**) tibial components and the female bore tapers on the modular stems, with material combinations consisting of CoCrMo cones and TiAlV or CoCrMo bores. Explanted TKR components representative of Design B tapers had the female bore tapers located on the (**c**) femoral and (**d**) tibial components and the male cone tapers on the modular stems, with material combinations consisting of TiAlV cones and CoCrMo bores

2.2 Assessment of Surface Corrosion Area

Surface damage on the cone tapers was evaluated using an optical microscope (K400P, Motic Inc., Xiamen, China) and a photogrammetric measurement method adapted from Harman, et al. [30] to determine the total taper junction surface area with visual evidence of gross corrosion, hereafter referred to as "surface corrosion area". In this study, "surface corrosion areas" had evidence of features consistent

Table 2 Summary of knee prostheses and patient demographics

Prosthesis number	Patient number	Sex/age at retrieval	BMI	Time *in vivo* (years)	Design	Component	Bore taper material	Cone taper material
A1	1	F/68.5	24.4	1.5	A	Femoral	TiAlV	CoCrMo
A2					A	Tibial	TiAlV	CoCrMo
A3	2	M/74.2	26.4	1.0	A	Femoral	CoCrMo	CoCrMo
A4					A	Tibial	CoCrMo	CoCrMo
A5	3	F/75.1	31.1	4.8	A	Femoral	TiAlV	CoCrMo
A6					A	Tibial	TiAlV	CoCrMo
A7	4	M/71.6	33.7	0.1	A	Femoral	TiAlV	CoCrMo
A8					A	Tibial	TiAlV	CoCrMo
A9	5	F/85.1	19.8	2.9	A	Femoral	CoCrMo	CoCrMo
A10					A	Tibial	TiAlV	CoCrMo
A11	6	F/76.9	31.2	1.9	A	Femoral	CoCrMo	CoCrMo
A12					A	Tibial	CoCrMo	CoCrMo
A13	7	F/69.0	33.3	3.8	A	Femoral	TiAlV	CoCrMo
A14	8	M/69.8	29.9	7.7	A	Femoral	TiAlV	CoCrMo
A15					A	Tibial	TiAlV	CoCrMo
A16	9	F/74.5	33.5	3.6	A	Femoral	TiAlV	CoCrMo
A17					A	Tibial	TiAlV	CoCrMo
A18	10	M/n.a.	n.a.	n.a.	A	Tibial	TiAlV	CoCrMo
A19	11	M/56.8	36.3	2.2	A	Femoral	CoCrMo	CoCrMo
A20					A	Tibial	TiAlV	CoCrMo
A21	12	F/55.4	31.2	1.6	A	Femoral	TiAlV	CoCrMo
A22					A	Tibial	TiAlV	CoCrMo
A23	13	F/56.9	35.2	0.7	A	Femoral	TiAlV	CoCrMo
A24					A	Tibial	TiAlV	CoCrMo
A25	14	F/70.5	26.6	0.9	A	Femoral	TiAlV	CoCrMo
A26					A	Tibial	TiAlV	CoCrMo
A27	15	F/76.9	33.7	0.2	A	Femoral	TiAlV	CoCrMo
A28					A	Tibial	TiAlV	CoCrMo
B1	16	F/74.8	37.2	6.6	B	Femoral	CoCrMo	TiAlV
B2					B	Tibial	CoCrMo	TiAlV
B3	17	M/69.9	34.6	5.0	B	Femoral	CoCrMo	TiAlV
B4					B	Tibial	CoCrMo	TiAlV
B5	18	M/66.8	28.4	6.4	B	Femoral	CoCrMo	TiAlV
B6					B	Tibial	CoCrMo	TiAlV
B7	19	F/75.5	35.7	2.5	B	Femoral	CoCrMo	TiAlV
B8					B	Tibial	CoCrMo	TiAlV
B9	20	F/88.6	30.4	6.3	B	Femoral	CoCrMo	TiAlV
B10	21	F/77.1	25.3	11.2	B	Femoral	CoCrMo	TiAlV

Fig. 2 The experimental test setup for measuring disassembly force was designed using a shaft collar (**a**) attached to the load cell (**b**) to apply a pure axial tensile load to the stem extension with the component base potted in bismuth alloy (**c**) and fixed to the test surface. Unconstrained dovetail sliders (**d**) aided alignment of the stem axis (**e**) with the applied force vector and countered any off-axis moments during testing

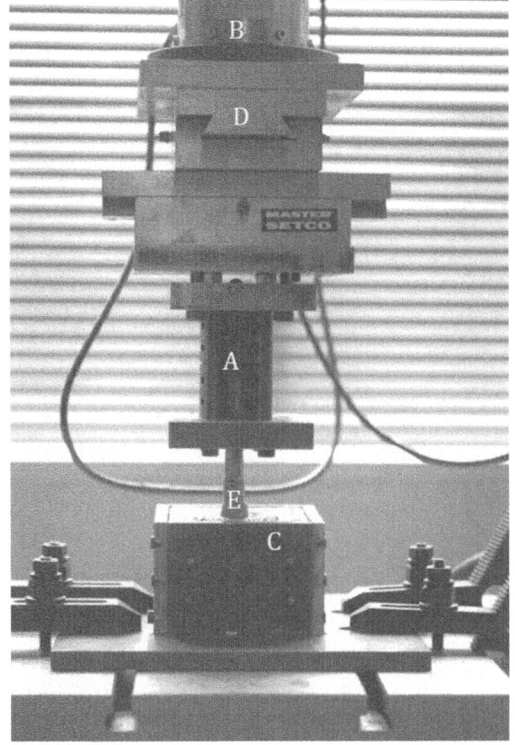

with corrosion and were characterized by a white frosted appearance, discoloration, burnishing, or corrosion products that obscured three or more consecutive machine lines. High-resolution calibrated photographs from each of the four anatomic side-views of the cone taper (anterior, posterior, medial, and lateral) were acquired using a digital camera (D60 with MicroNikkor lens, Nikon, Melville, NY) (Fig. 3a). Open-source software (ImageJ, NIH) was used to calibrate the image, measure diameters and taper junction length, and to manually segment the boundaries of the 90° quadrant at six- 15° increments, assuming a conical taper geometry (Fig. 3b). Three users were trained to visually identify surface corrosion regions using low magnification optical microscopy and manually digitize the regions for measurement, and the intra- and inter- user accuracy was recorded based on the user training described in Harman, et al. [30]. Measured regions in each 15° segment (Fig. 3c) were corrected to account for the projection of the conical three-dimensional geometry to a two-dimensional photograph. Surface corrosion areas were normalized as a percentage of the total surface area of each taper junction or quadrant.

A profilometric imaging protocol was developed as an additional method of measuring the surface corrosion area, and validated against the photogrammetric method. Three-dimensional profilometric scans were acquired at 40 points uniformly distributed on each taper junction surface with a white light interferometer (NPFlex Optical Interferometer, Bruker, Billerica, MA) using a 10X super-long

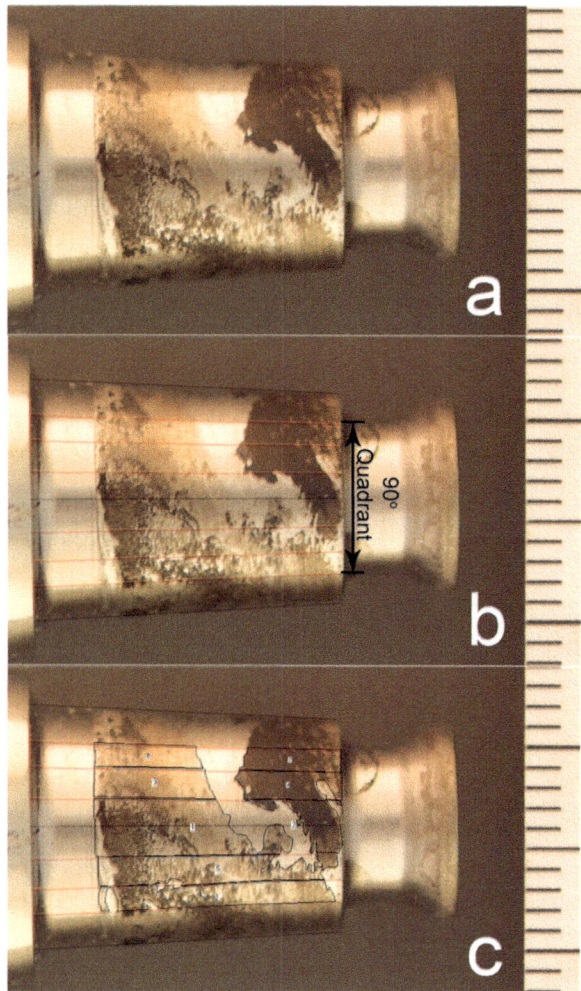

Fig. 3 High-resolution photograph of cone taper (**a**) with 15° radial increment grid overlay of a 90° anterior quadrant (**b**) and manual user selection of surface corrosion areas within each 15° region (**c**)

working distance objective lens: 5 equidistant points were scanned on each of the 4 main (anterior, medial, posterior, lateral) and 4 intermediate (anterior-medial, posterior-medial, anterior-lateral, posterior-lateral) anatomic views (Fig. 4). Each point was described as either "grossly corroded" or "not grossly corroded", based on the three-dimensional scans (Fig. 5). Points with gross corrosion had had one or more of these distinguishing features: more than three consecutive machine lines obscured by surface damage, pitting damage on the scale of 50 μm or larger, or distinct fretting in addition to other signs of corrosion. Points with no gross corrosion had a machined surface with no microscopically discernable defects, had

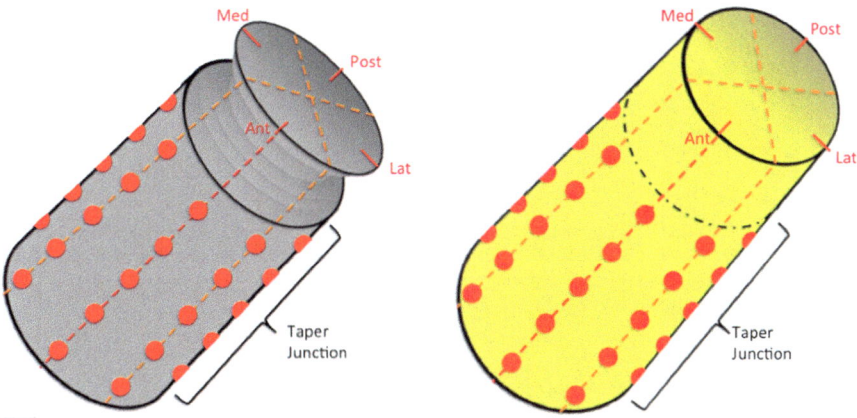

Fig. 4 Schematic of 40-point profilometric scans on cone tapers (left) and PVS molds (right) of bore taper surfaces

mechanically flattened peaks, or had microscopic defects below 50 µm in size. These 40-point measurements of the taper junction were compared against the photogrammetric measurement results to ensure the similarity of surface corrosion area measurements obtained by either method.

The surface corrosion area on the bore tapers was evaluated using noncontact profilometric imaging on high resolution polyvinyl siloxane (PVS) molds of the otherwise inaccessible bore taper surface. Light-viscosity PVS (416A/416B, Doje's Forensic Supplies, Ocoee, FL) was applied using a dual-cartridge applicator with a mixing tip at the base of each bore taper, and anchored at the base with a plastic grate as described by Panigrahi et al. [47]. PVS molds were marked for anatomic location and removed manually after 60 seconds of curing. While the PVS molds did not replicate any discoloration on the original bore taper surfaces, the method was validated to reproduce fine (0.5–3 µm) surface features, confirming that the corroded topography was preserved [47] (Fig. 6). Similar to the cone taper surface, the 40-point profilometric imaging protocol (Fig. 4) was used to determine the surface corrosion area on the PVS molds as a percentage of the taper junction surface, and by extension, on the bore taper surfaces.

2.3 Damage Mode Characterization

In addition to surveying each taper surface to determine the absence or presence of corrosion at each region, the mode(s) of corrosion evident on each taper surface were determined based on the gross visual localization of corrosion products, and microscopic evidence of fretting. Crevice corrosion was characterized by a ring of corrosion products, likely titanium or chromium oxides, at the edge of the taper junction (Fig. 7a). Fretting corrosion was characterized by regions of discoloration

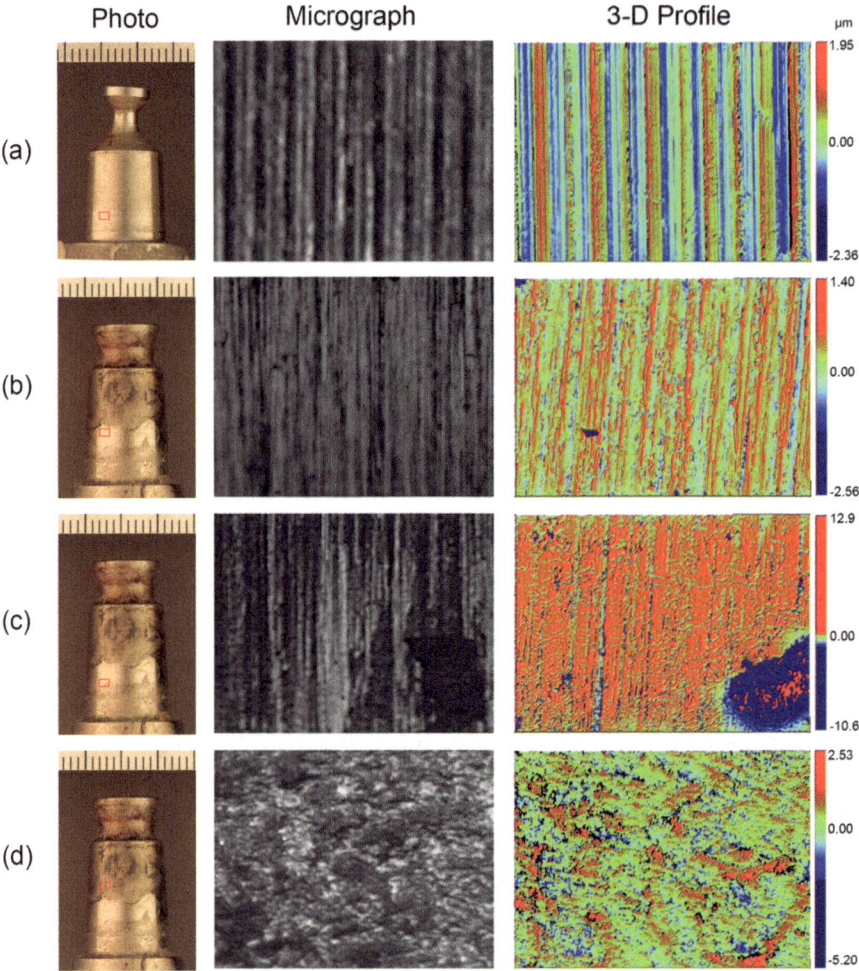

Fig. 5 Representative three-dimensional profiles of taper regions that were not grossly corroded (**a**, **b**) and grossly corroded (**c**, **d**). Grossly corroded regions had evidence of macroscopic pits (**c**), build-up of corrosion products (**d**), or at least three consecutive machine lines obscured by corrosive damage (**d**). Regions that were considered to not have gross corrosion had undamaged machine lines with no build-up of corrosion products (**a**) or had microscopic pits under 50 μm (**b**)

and irregular regions of oxide build-up across the taper junction length, often in the same areas with fretting asperity wear marks 25–100 μm perpendicular to circumferential machine lines (Fig. 7b). If evidence of both types of corrosion was present, this was noted.

After these nondestructive characterization techniques, representative sections of bore and cone tapers on Design A and B prostheses were taken normal to the taper axis. Tapers were mounted in a conductive embedding medium (Technovit 7200, Kulzer, Hanau, Germany), sectioned with a diamond blade precision saw (Isomet

Fig. 6 Light viscosity PVS molds were used to reproduce inaccessible bore taper surfaces of explanted modular TKRs. A bore taper surface with no gross corrosion is pictured in the top row; bore taper surface with gross corrosion is pictured in the bottom row. Red rectangles indicate corresponding regions on bore tapers and PVS molds. Colored profile legends are in μm

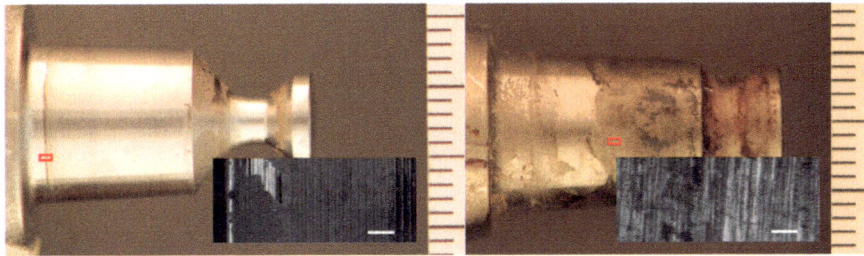

Fig. 7 Evidence representative of crevice (**a**) and fretting (**b**) corrosion on modular tapers of knee prostheses. Crevice corrosion was characterized by a white or discolored frosted ring at the base of the taper junction. Fretting corrosion was characterized by large burnished or discolored regions with fretting scars (horizontal in example micrograph) perpendicular to the circumferential machine lines (vertical in example micrograph). Micrograph scale bars are 100 μm

5000, Buehler, Lake Bluff, IL), and polished sequentially with silicon carbide grit paper and colloidal diamond paste on polishing cloths. Interfaces (between the metal and any oxide products) and surfaces (metal without corrosion oxide buildup) were examined for any evidence of corrosion cracks or pits with a scanning electron microscope (S3400, Hitachi, Tokyo, Japan).

2.4   Data Analysis

The effects of categorical and numerical variables on the surface corrosion area were determined using several non-parametric statistical tests utilizing commercially available statistics software (MiniTab 15, Minitab Inc., State College, PA). A preliminary validation of the statistical similarity between the photogrammetric and profilometric imaging methods was established with a paired-sample Wilcoxon test on the surface corrosion area measured on cone tapers. The effect of taper design (A or B), patient's sex (M or F), metal combination (similar or mixed) of Design A tapers, component (femoral or tibial), crevice corrosion evidence (no or yes), and fretting corrosion evidence (no or yes) on the disassembly force and surface corrosion area were determined individually with Mann-Whitney ranked-sum tests. Paired-sample Wilcoxon tests were performed on the surface corrosion area on the bore and cone of the same component, surface corrosion area on medial and lateral quadrants of each cone taper, surface corrosion area on anterior and posterior quadrants of each cone taper, and surface corrosion area on the femoral and tibial modular tapers on the same TKR. Spearman's Rho correlations were determined between time *in vivo* and surface corrosion area, between patient's BMI and surface corrosion area, between patient's age at retrieval and surface corrosion area, between number of previous surgeries and surface corrosion area, and between disassembly force and surface corrosion area. To determine any dependencies among categorical results (design, evidence of crevice corrosion, evidence of fretting corrosion), chi square independence tests were completed on contingency and expected values tables.

3   Results

The 40-point profilometric imaging method of approximating the surface corrosion area was found to be statistically similar to the photogrammetric method ($p = 0.570$). This implies that the criteria distinguishing regions with and without gross corrosion results in similar measurements for both imaging techniques. Since PVS molds are able to replicate macroscopic and microscopic surface features applicable to corrosion analysis of tapers [35, 47], and since the method of measuring the gross corrosion on the taper junction area does not affect the results, comparison of photogrammetric measurements on cone tapers and approximation based on indirect profilometric analysis of bore tapers was deemed valid. Inter- and intra- user measurement accuracy were acceptable (±4.45% and ± 1.34%, respectively).

Taper design (A or B) had a significant effect on disassembly force, with significantly higher disassembly forces recorded for Design A tapers compared to Design B tapers ($p < 0.001$). The median disassembly force was 3987 N for Design A tapers and 393 N for Design B tapers. Hand disassembly occurred upon removal of the through bolts and/or set screws for 3.6% (1/28) of Design A tapers and 38% (3/8) of

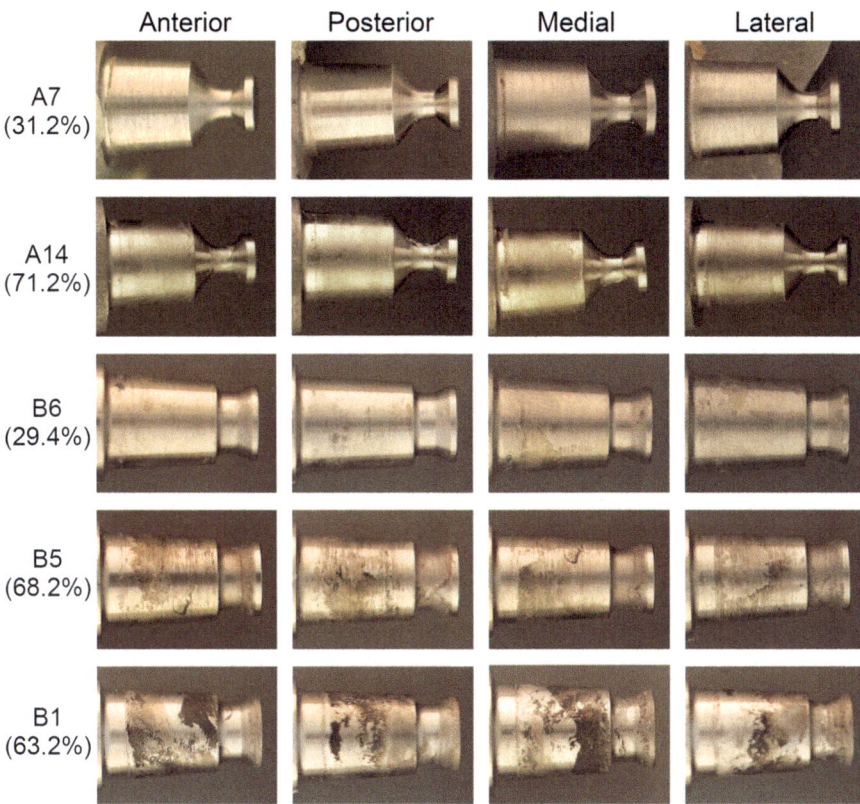

Fig. 8 Gross photos of Design A and Design B cone tapers with varying percentages (in parentheses) and distributions of surface corrosion area

Design B tapers. Disassembly force for one tibial component with Design B tapers was not reported because of failure to remove the through-bolt prior to testing.

In general, the disassembled cone tapers displayed a notable variety in surface corrosion area and visual localization of corrosion (Fig. 8). There were significant differences ($p < 0.05$) in surface corrosion area when comparing the variables of taper design (A or B), fretting corrosion (no/yes), and anatomic location (anterior/posterior) (Table 3). Differences in metal combination (similar/mixed), component (femoral/tibial, overall and paired), crevice corrosion (no/yes), and patient's sex (male/female) were not significantly associated with surface corrosion areas (Table 4). There was a significant moderate negative correlation between disassembly force and surface corrosion area (Table 3, Fig. 9). A significant correlation was not detected between the patient demographics (time *in vivo*, number of previous surgeries, age at retrieval, and BMI) and surface corrosion area (Table 4).

Regions with obscured machine lines due to buildup of corrosion products were examined further to determine the extent of surface corrosion. While the mean peak-to-valley height of these regions was 4.3 ± 1.4 μm, imaging of the taper cross

Table 3 Significant factors affecting surface corrosion area and clinical outcomes

Comparison	Statistical test	n	Result	p value
Design factors				
Design A vs. design B *	Mann-Whitney	17, 9	$\text{Median}_A = 7.50\%$ $\text{Median}_B = 67.47\%$	0.0047
No fretting corrosion vs. fretting corrosion *	Mann-Whitney	15, 11	$\text{Median}_{NoFC} = 5.46\%$ $\text{Median}_{FC} = 68.16\%$	0.0001
Design and fretting corrosion **	Chi Square independence	26	Dependent	0.000015
Anterior vs. posterior (paired) *	Paired-sample Wilcoxon	26	$\text{Median}_{Ant\text{-}Post} = -1.864\%$	0.016

*Difference in medians is statistically significant at $p < 0.05$
**Variables are statistically dependent at $p < 0.05$

Table 4 Other statistically insignificant factors considered to affect surface corrosion area

Comparison	Statistical test	n	Result	p value
Design factors				
Similar vs. mixed metals (design A only)	Mann-Whitney	5, 12	$\text{Median}_{Similar} = 5.46\%$ $\text{Median}_{Mixed} = 11.25\%$	0.6353
Femoral components vs. tibial components	Mann-Whitney	15, 11	$\text{Median}_{Femoral} = 15.00\%$ $\text{Median}_{Tibial} = 16.92\%$	0.8762
No crevice corrosion vs. crevice corrosion	Mann-Whitney	7, 19	$\text{Median}_{NoCC} = 15.00\%$ $\text{Median}_{CC} = 17.35\%$	0.5249
Design and crevice corrosion	Chi Square independence	26	Independent	0.592
Bore vs. cone (paired)	Wilcoxon	26	$\text{Median}_{Bore\text{-}Cone} = 4.377\%$	0.134
Medial vs. lateral (paired)	Wilcoxon	26	$\text{Median}_{Med\text{-}Lat} = -0.9445\%$	0.163
Femoral vs. tibial (on same TKR, paired)	Wilcoxon	11	$\text{Median}_{Fem\text{-}Tib} = 13.10\%$	0.230
Patient factors				
Male patients vs. female patients	Mann-Whitney	10, 17	$\text{Median}_{Male} = 30.26\%$ $\text{Median}_{Female} = 12.27\%$	0.1472
Age at retrieval vs. surface corrosion area	Spearman's rho correlation	26	$\rho = -0.111$	0.589
Patient BMI vs. surface corrosion area	Spearman's rho correlation	26	$\rho = 0.216$	0.289
Time *in vivo* vs. surface corrosion area—Design A	Spearman's rho correlation	17	$\rho = 0.307$	0.231
Time *in vivo* vs. surface corrosion area—Design B	Spearman's rho correlation	9	$\rho = 0.131$	0.736
Number of previous surgeries vs. surface corrosion area	Spearman's rho correlation	26	$\rho = 0.093$	0.651

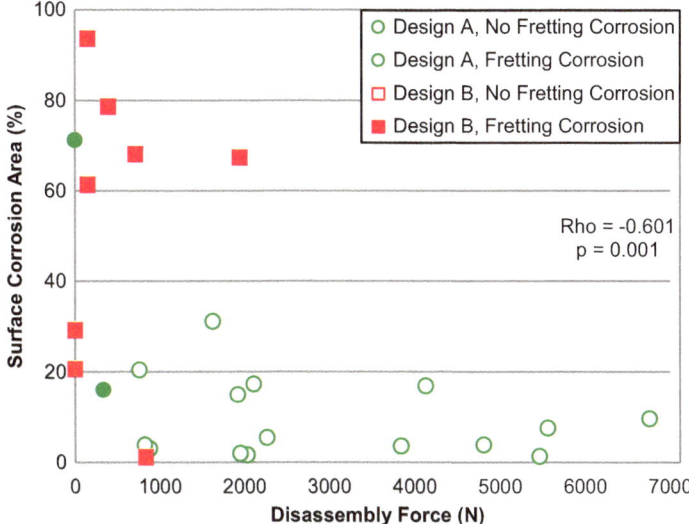

Fig. 9 Scatterplot of modular taper disassembly force and surface corrosion area: negative correlation of ranked values indicates that taper junctions with a higher surface corrosion area disassembled under lower loads

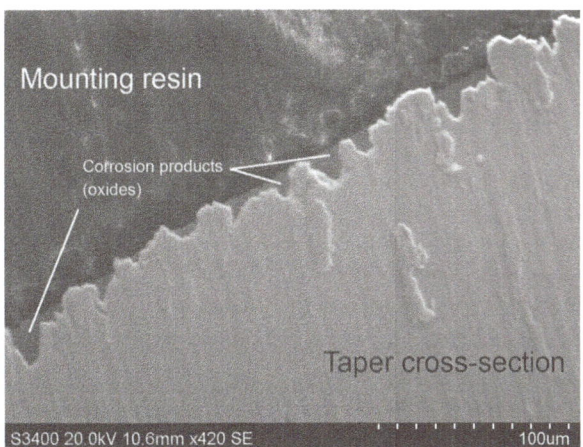

Fig. 10 Scanning electron microscope image of Design B taper with cross-section taken perpendicular to taper axis. Corrosion pits 0–20 μm deep were present beneath accumulated corrosion products in fretting corrosion regions

sections demonstrated that pits below the corrosion oxide surface were over 10 μm deep (Fig. 10). Fretting micromotion in these areas may have initiated pitting corrosion behavior that altered the taper junction interface.

4 Discussion

4.1 Effects of Design and Modes of Corrosion

The design of the modular knee prostheses was found to be one of the most significant factors (p = 0.0047) that influenced the surface corrosion area. Since the two models of TKR in this study differed in manufacturer, overall prosthesis design, taper geometry, machining specifications, assembly method, and use of set screws and through-bolts, it is likely that some or all of these factors affect the amount of surface corrosion that occurs in a relatively short amount of time. A previous multi-center retrieval study on modular TKR corrosion found that the taper geometry (threaded or conical) had a very significant (p < 0.001) effect on the degree of corrosion measured by a semi-quantitative grading scale [33]. The present study adds that within the category of conical taper designs, other design factors also contribute significantly to the surface corrosion area. Specifically, Design B featured mixed metal combinations (CoCrMo components and Ti-6Al-4 V stems) with the orientation of the bore-cone taper junctions configured with bore taper located on the component and the conical taper located on the stem. It is of particular interest that several components and surgical tools associated with the Design B in the present study were recalled by the United States FDA in 2013 and 2014; the reasons cited include a loose connection between the tibial baseplate and tibial stem extension, as well as the potential for the through-bolt threads being "out of specification" [48, 49]. Results from this study suggest that femoral component taper junctions, in addition to the tibial components mentioned in the FDA recalls, are also problematic.

There was a strong dependence between fretting corrosion and design and significantly larger surface corrosion areas on tapers that experienced fretting corrosion. Therefore, it is likely that the larger surface corrosion areas on Design B can be attributed to its higher susceptibility to fretting corrosion. Several design-based factors may contribute to micromotion during *in vivo* service that lead to fretting corrosion. Prior work has demonstrated that the assembly force and technique has a direct effect on the Morse taper strength of hip tapers [50, 51], and the two designs in the present study incorporate different assembly techniques per the manufacturers' labeling [52, 53]. Mixed reports on modular hip tapers suggest that conical taper angle may [59] or may not [54] have a significant effect on the amount of fretting or corrosion observed. Prior *in vitro* work on taper contact area and taper surface roughness suggest these factors may also contribute to the prevalence of passive film disruption and corrosion [55].

The prevalence of tapers having a surface corrosion area over 10% was 15/26 (57.7%) in the present study. This varied from data reported in the only large-scale modular TKR retrieval study [33], in which 184/198 (92.9%) of modular tapers had a Goldberg corrosion score of 2 or greater, indicative of over 10% surface corrosion [56]. The prevalence of tapers with fretting corrosion (11/26, 42.3%) or surface corrosion area over 50% (8/26, 30.8%) was more similar to the previous study [33] reporting 36/108 (33.3%) of conical tapers had a Goldberg corrosion score of 4, indicative of over 50% surface corrosion [56]. It should be noted that the previous study incorporated a semi-quantitative grading scale to promote rapid comparisons among a larger total sample size, assuming that the mechanism of any damage observed was both fretting and crevice corrosion [33]. Additionally, that previous study incorporated modular TKR from at least 6 manufacturers and designs, while the present study only included retrieved prostheses from 2 designs. Since design is such a significant factor that affects fretting corrosion, it is not surprising that reports vary among different retrieval archives.

4.2   Effects of Patient Factors and Anatomical Location

Patient age at revision, BMI, sex, and the *in vivo* duration did not have a significant effect on the surface corrosion areas in this study. Prior work on modular knee retrievals also found that patient weight, patient age, and implantation time did not correlate with semi-quantitative corrosion scores [33]. One benefit of the photogrammetric method over semi-quantitative grading scales was the ability to compare trends in the localization of corrosion. In this study, the paired differences in surface corrosion area between the anterior quadrant and the posterior quadrant were deemed significant, while no significant differences were observed between the medial and lateral quadrants or between the femoral and tibial components. Since the femur experiences anterior-posterior bending during normal gait patterns and tensile stresses may lead to increased corrosion [35], it is thus unsurprising that a difference was observed. It is important, however, to note that the median difference in surface corrosion area was only 1.9% and this difference may not be clinically significant.

4.3   Mechanical Disassembly and Surface Corrosion Area

Unlike the data available for modular total hip prostheses, mechanical disassembly of explanted TKR with modular stem extensions is not widely reported [46]. The disassembly forces measured for Design A tapers were comparable to the reported average magnitudes (3000 N to 6900 N) measured during mechanical testing of explanted total hip prostheses with modular tapers at the head/neck [57] and neck/stem junctions [58]. In contrast, the lower disassembly forces measured for Design B tapers were considerably less than the average magnitudes from total hip prostheses measured immediately after manual assembly at the time of surgery (1000 N to

1700 N) [24] or after retrieval [57, 58]. Comparable data from testing performed immediately after manual assembly of TKR is unavailable.

Using the mechanical disassembly data, it was possible to consider the mechanical response to corroded taper surfaces. Mixed conclusions from prior studies suggest that corroded taper surfaces can either be associated with difficult disassembly (very high disassembly forces) [57], or with taper instability (spontaneous disassembly) [32]. This study found that lower disassembly forces were associated with high surface corrosion areas ($\rho = -0.601$, $p = 0.001$). Due to the limitations of a retrieval study, it cannot be determined with certainty which event (taper corrosion or mechanical instability) caused the other. However, due to the fretting evidence associated with high surface corrosion areas and the low disassembly forces (Fig. 9), it seems likely that mechanical instability contributed to fretting, which lead to high surface corrosion areas.

4.4   Limitations

While this study addressed the limitations of not distinguishing fretting and crevice corrosion when using semi-quantitative visual methods to evaluate modular knee prostheses [33], other limitations inherent to retrieval studies apply to the current findings. The retrieved knee components in this study were explanted during revision surgery for infection, loosening, and periprosthetic fracture. Thus, the incidence of fretting/crevice corrosion and the surface corrosion areas reported in this study are not indicative of all knee prostheses of these two designs; they are only representative of clinically failed devices. Since modular stem TKRs are usually indicated for patients with poor bone stock following a failed primary arthroplasty or for patients with excessive trauma and bone fracture, it is possible that the high surface corrosion areas are due in part to the physiological changes that come with a severely compromised joint. While these results may not be representative of the performance of these modular TKR designs in healthier bone, they bring attention to the consequences of metal release in the patient group already suffering from poorest joint health and function.

5   Conclusion

The performance of bore-cone taper junctions for two TKR designs with modular stem extensions was differentiated using mechanical disassembly combined with photogrammetric and profilometric techniques for measuring surface corrosion areas. Taper design has a significant effect on disassembly force, the surface corrosion area or the likelihood of fretting corrosion, while anatomical locations and patient factors have little effect. Higher surface corrosion areas were associated with lower disassembly forces at the taper junction. This study demonstrates that high surface corrosion areas can be generated within short durations of *in vivo* function in patients with modular stem TKR and poor clinical outcomes.

Given the prevalence of fretting corrosion and negative correlation between disassembly forces and corrosion area, TKR designs that incorporate modularity using bore-cone taper junctions should carefully consider taper fit to achieve effective mechanical stability and avoid fretting. Taper junctions can provide a self-locking connection given appropriate manufacturing tolerances for creating an interference fit and application of a suitable assembly force. Based on analysis of the modular TKR designs in this study, use of supplemental fixation (e.g. through-bolts and set screws) for improving the mechanical locking of bore-cone taper junctions was not supported. Several parameters specific to TKR design (femoral anatomic location, bore on component taper orientation) showed a trend toward greater surface corrosion area and warrant additional study under more controlled experimental conditions.

References

1. Barrack RL. Modularity of prosthetic implants. J Am Acad Orthop Surg. 1994;2(1):16–25.
2. Sporer SM, Paprosky WG. Femoral fixation in the face of considerable bone loss: the use of modular stems. Clin Orthop Relat Res. 2004;429:227–31.
3. Whittaker JP, Dharmarajan R, Toms AD. The management of bone loss in revision total knee replacement. J Bone Joint Surg Br. 2008;90-B(8):981–7.
4. Lecerf G, Fessy MH, Phillippot R, Massin P, Giraud F, Flecher X, Girard J, Merti P, Marchetti E, Stindel E. Femoral offset: anatomical concept, definition, assessment, implications for pre-operative templating and hip arthroplasty. Orthop Traumatol Surg Res. 2009;95(3):210–9.
5. Sakai T, Sugano N, Nishii T, Haragushi K, Ochi T, Ohzono K. Optimizing femoral anteversion and offset after total hip arthroplasty, using a modular femoral neck system: an experimental study. J Orthop Sci. 2000;5(5):489–94.
6. Knahr K. Total hip arthroplasty: tribological considerations and clinical consequences. In: Puhl W, Bentley G, Klaus-Peter G, editors. EFORT reference in orthopaedics and traumatology. Berlin: Springer Science & Business Media; 2013.
7. Harris WH. A new total hip implant. Clin Orthop Relat Res. 1971;81:105–13.
8. Kopec MA, Pemberton A, Milbrandt JC, Allan G. Component version in modular total hip revision. Iowa Orthop J. 2009;29:5–10.
9. Scott CEH, Biant LC. The role of the design of tibial components and stems in knee replacement. J Bone Joint Surg Br. 2012;94-B(8):1009–15.
10. Marlowe DE, Parr JE, Mayor MB, editors. Selected Technical Papers 1301: modularity of orthopedic implants. Philadelphia, PA: American Society for Testing and Materials (ASTM); 1997.
11. Collier JP, Surprenant VA, Jensen RE, Mayor MB. Corrosion at the interface of cobalt-alloy heads on titanium-alloy stems. Clin Orthop Relat Res. 1991;(271):305-312.
12. Cook SD, Barrack RL, Clemow AJT. Corrosion and wear at the modular interface of uncemented femoral stems. J Bone Joint Surg Br. 1994;76(1):68–72.
13. Cook SD, Barrack RL, Baffes GC, Clemow AJ, Serekian P, Dong N, Kester MA. Wear and corrosion of modular interfaces in total hip replacements. Clin Orthop Relat Res. 1994;(298):80–8.
14. Kummer FJ, Rose RM. Corrosion of titanium/cobalt-chromium alloy couples. J Bone Joint Surg Am. 1983;65(8):1125–6.
15. Lucas LC, Buchanan RA, Lemons JE. Investigations on the galvanic corrosion of multialloy total hip prostheses. J Biomed Mater Res. 1981;15(5):731–47.

16. Marcus P, Mansfield FB. Analytical methods in corrosion science and engineering. Boca Raton, FL: CRC; 2005.
17. Willert HG, Broback LG, Buchhorn GH, Jensen PH, Koster G, Lang I Ochsner P, Schenk R. Crevice corrosion of cemented titanium alloy stems in total hip replacements. Clin Orthop Relat Res. 1996;333:51.
18. Gilbert JL, Buckley CA, Jacobs JJ. In vivo corrosion of modular hip prosthesis components in mixed and similar metal combinations. The effect of crevice, stress, motion, and alloy coupling. J Biomed Mater Res. 1993;27:1533–44.
19. Viceconti M, Baleani M, Squarzoni S, Toni A. Fretting wear in a modular neck hip prosthesis. J Biomed Mater Res. 1997;35:207–16.
20. Fernandez J, Miller GJ, Mauldin CM. Modular hip prosthesis. US Patent 6319286 B1. 2001.
21. McCarthy JC, Bono JV, O'Donnell PJ. Custom and modular components in primary total hip replacement. Clin Orthop Relat Res. 1997;344:162.
22. Nganbe M, Khan U, Louati H, Speirs A, Beaule PE. In vitro assessment of strength, fatigue durability, and disassembly of Ti6Al4V and CoCrMo necks in modular total hip replacements. J Biomed Mater Res B Appl Biomater. 2011;97B(1):132–8.
23. Bishop RA, Sellenschloh K, Morlock MM. Strength of the taper lock at the stem-neck junction in hip replacement. Proceedings of the 58th Annual Meeting of the Orthopaedic Research Society, 2012, p. 1048.
24. Pallini F, Cristofolini L, Traina F, Toni A. Modular hip stems: determination of disassembly force of a neck-stem coupling. Artif Organs. 2007;31(2):166–70.
25. Goldberg JR, Gilbert JL. In vitro corrosion testing of modular hip tapers. J Biomed Mater Res B. 2003;64:78–93.
26. Flemming C, Brown SA. Mechanical testing for fretting corrosion of modular total hip tapers. In: Kambic HE, Yokobori AT, editors. Biomaterials' mechanical properties, Issue, vol. 1173; 1994. p. 156–66.
27. Swaminathan V, Gilbert JL. Fretting corrosion of CoCrMo and Ti6Al4V interfaces. Biomaterials. 2012;33(22):5487–503.
28. Brown SA, Flemming C, Kawalec JS. Fretting corrosion accelerates crevice corrosion of modular hip tapers. J Appl Biomater. 1995;6:19–26.
29. Kop AM, Swarts E. Corrosion of a hip stem with a modular neck taper junction. J Arthroplasty. 2009;24:1019–23.
30. Harman MK, Baleani M, Juda K, Viceconti M. Repeatable procedure for evaluating taper damage on femoral stems with modular necks. J Biomet Mater Res B. 2011;99:431–9.
31. Geringer J, Forest B, Combrade P. Fretting-corrosion of materials used as orthopaedic implants. Wear. 2005;259:943–51.
32. Meyer H, Mueller T, Goldau G, Chamaon K, Ruetschi M, Lohmann CH. Corrosion at the cone/taper interface leads to failure of large-diameter metal-on-metal total hip arthroplasties. Clin Orthop Relat Res. 2012;470(11):3101–8.
33. Arnholt CM, DW MD, Tohfafarosh M, Gilbert JL, Rimnac CM, Kurtz SM. Implant Research Center Writing Committee, Klein G, Mont MA, Parvizi J, Cates HE, Lee GC, Malkani A, Kraay M. Mechanically assisted taper corrosion in modular TKA. J Arthroplasty. 2014;29(9 Suppl):205–8.
34. Lanting BA, Teeter MG, Vasarhelyi EM, Ivanov TG, Howard JL, Naudie DD. Correlation of corrosion and biomechanics in the retrieval of a single modular neck total hip arthroplasty design. J Arthroplasty. 2015;30(1):135–40.
35. Panigrahi P, Poursaee A, Harman MK. Corrosion behavior of medical-grade Ti-6Al-4V exposed to tensile loads. Proceedings of the 61st Annual Meeting of the Orthopaedic Research Society, 2015.
36. Wassef AJ, Schmalzried TP. Femoral taperosis: an accident waiting to happen? Bone Joint J. 2013;95-B(11 Suppl):3–6.
37. McMaster WC, Patel J. Adverse local tissue response lesion of the knee associated with Morse taper corrosion. J Arthroplasty. 2013;28(2):375.e5–8.
38. Shulman RM, Zywiel MG, Gandhi R, Davey JR, Salonen DC. Trunnionosis: the latest culprit in adverse reactions to metal debris following hip arthroplasty. Skeletal Radiol. 2015;44:433–40.

39. Higgs GB, Hanzlik JA, MacDonald DW, Gilbert JL, Rimnac CM, Kurtz SM. Is increased modularity associated with fretting and corrosion damage in metal-on-metal total hip arthroplasty devices? A retrieval study. J Arthroplasty. 2013;28(Suppl 1):2–6.
40. Kurtz SM, Kocagoz SB, Hanslik JA, Underwood RJ, Gilbert JL, MacDonald DW, Lee G-C, Mont MA, Kraay MJ, Klein GR, Parvazi J, Rimnac CM. Do ceramic femoral heads reduce taper fretting corrosion in hip arthroplasty? A retrieval study. Clin Orthop Relat Res. 2013;471(10):3270–82.
41. Berry DJ, Abdel MP, Callaghan JJ. What are the current clinical issues in wear and tribocorrosion? Clin Orthop Relat Res. 2014;472(12):3659–64.
42. Esposito CI, Wright TM, Goodman SB, Berry DJ. What is the trouble with trunnions? Clin Orthop Relat Res. 2014;472(12):3652–8.
43. Munir S, Walter WL, Walsh WR. Variations in the trunnion surface topography between different commercially available hip replacement stems. J Orthop Res. 2015;33:98–105.
44. ISO 7206–10:2003(E). Implants for surgery—partial and total hip-joint prostheses—Part 10: Determination of resistance to static load of modular femoral heads.
45. ASTM F2009-00. Standard test method for determining the axial disassembly force of taper connections of modular prostheses.
46. Snethen K, Henson K, Lutzner J, Kirschner S, Harman M. Mechanical disassembly of retrieved long-stem total knee replacement with taper modularity. Annual Meeting & Exposition of the Society for Biomaterials; 2014.
47. Panigrahi P, Schwartzman KG, Harman MK. Polyvinyl siloxane molds for nondestructive surface feature metrology of failed joint prostheses. J Fail Anal Prev. 2015;15:266–71.
48. Class 2 Recall: NexGen complete knee solutions stemmed Tibial component Precoat. FDA medical device recalls database no. Z-0480-2014. 2013. http://www.accessdata.fda.gov/scripts/cdrh/cfdocs/cfRES/res.cfm?id=123053.
49. Class 2 Recall: NexGen complete knee solution MIS Total knee procedure stemmed Tibial component, Precoat. FDA medical device recalls database no. Z-1938-2014. 2014. http://www.accessdata.fda.gov/scripts/cdrh/cfdocs/cfRES/res.cfm?id=127988.
50. Pennock AT, Schmidt AH, Bourgeault CA. Morse-type tapers: factors that may influence taper strength during total hip arthroplasty. J Arthroplasty. 2002;17(6):773–8.
51. Rehmer A, Bishop NE, Morlock MM. Influence of assembly procedure and material combination on the strength of the taper connection at the head-neck junction of modular hip endoprostheses. Clin Biomech. 2012;27(1):77–83.
52. RT-PLUS Modular: constrained rotating total knee prosthesis: surgical technique "Intramedullary Application". Plus Orthopaedics literature no. 1313-e-Ed; 2007.
53. Zimmer NexGen Rotating Hinge Knee Primary/Revision Surgical Technique. Zimmer literature no. 97-5880-002-00 Rev. 3; 2009.
54. Padgett DE, Stoner K, Nassif N, Nawabi D, Wright T, Elpers M. The effect of taper geometry on large head MOM THA taper-trunnion damage. Bone Joint J. 2013;95-B(Suppl 34):472.
55. Panagiotidou A, Meswania J, Hua J, Muirhead-Allwood S, Hart A, Blunn G. Enhanced wear and corrosion in modular tapers in total hip replacement is associated with the contact area and surface topography. J Orthop Res. 2013;31(12):2032–9.
56. Goldberg JR, Gilbert JL, Jacobs JJ, Bauer TW, Paprosky W, Leurgans S. A multicenter retrieval study of the taper interfaces of modular hip prostheses. Clin Orthop Relat Res. 2002;401:149–61.
57. Lieberman JR, Rimnac CM, Garvin KL, Klein RW, Salvati EA. An analysis of the head-neck taper interface in retrieved hip prostheses. Clin Orthop Relat Res. 1994;300:162–7.
58. Csernica RM, Harman MK, Baleani M, Tozzi G, Erani P, Stea S, Toni A. Mechanical disassembly and taper damage assessment of retrieved femoral stems with modular necks. Proceedings of the 59th annual meeting of the Orthopaedic Research Society; 2013.
59. Padgett DE, Stoner K, Nassif N, Nawabi D, Wright T, Elpers M. The effect of taper geometry on large head MOM THA taper-trunnion damage. Bone Joint J. 2013;95-B(Suppl 34):472.

Wear Simulation Testing for Joint Implants

Peter Liao

Keywords Joint simulator · Wear simulation · Bio-tribology · Joint arthroplasty · Displacement-controlled simulators · Force-controlled simulators · Standards in wear simulation · Hip implant wear · Knee implant wear · Joint lubrication · Patient activities and implant wear · Adverse testing conditions

1 Introduction: Why Joint Simulator?

Total joint replacement has become a common procedure today. There were about 1.4 million hip and knee implant procedures performed on inpatients in the United States in 2016, according to the estimate from the Millennium Research Group (MRG) of Toronto, Ontario, with about 550,000 hip replacement procedures and 850,000 knee replacements [1]. With various implant materials and designs, it is important to have suitable methods for assessing the simulated performance of these replacement parts prior to patient clinical trials. Complex joint simulators have been developed and are one tool to assess implants via non-clinical testing, to such an extent that international standards exist. Joint simulators have been used in evaluating the wear resistance of joint implants, reproducing a clinical outcome, and/or exploring various simulated uses [2–4]. The formal definition of wear is the process of interaction between surfaces, which causes the deformation and removal of materials on the surface because of mechanical action between the sliding faces. Wear also refers to dimensional loss. The processes of wear are studied in the field of tribology.

Wear simulation provides one of several means to evaluate the laboratory performance of total joint replacements. Technology continues to advance and as the number of independently controlled axes increases there are more opportunities to employ test machines that provide capabilities to simulate more parameters of the clinical use of the implants. This chapter reviews the type of tribological joint simulators and how they have been used in different applications. Examples will be given

P. Liao (✉)
DePuy Synthes Joint Reconstruction, Warsaw, IN, USA
e-mail: pliao@its.jnj.com

© Springer International Publishing AG, part of Springer Nature 2018
B. Li, T. Webster (eds.), *Orthopedic Biomaterials*,
https://doi.org/10.1007/978-3-319-89542-0_6

on what has been achieved. Also, the benefits, limitation and the future of simulation will be discussed.

2 What Is a Joint Simulator?

A joint simulator consists of three major portions: a test station that holds the test samples in an environmental chamber that simulates the human hip/knee joints, the mechanics that apply the loading and kinematics through the test stations, and a software control system that generates desired inputs to the mechanics and receives feedback through data acquisition (Fig. 1). An environmental chamber is where the wear testing is conducted and test samples are held in fixtures and enclosed in a capsule to mimic a clinical condition for the implants including anatomical orientation. Diluted bovine serum is usually used as the joint lubricant because it has a similar composition to human joint fluid, and, it generates similar wear mechanisms as *in vivo* wear [5]. Most of the testing has been conducted at room temperature, while some advanced simulators have the capability to maintain at a desired temperature level replicating the human body at approximately 37°C. The load and motion that is applied to the test samples are driven by a variety of mechanisms based on simulator design, these are typically hydraulically, electromechanically or pneumatically with control by computer software. A typical load curve consists of a stance and a swing phase with approximately a 60–40% ratio (Fig. 2a). During the stance phase, the first peak force occurs when the heel strikes on the floor and the second peak force occurs at the moment of toe off. The computer software generates displacement and/or force to the links of the simulator, resulting in a displacement-controlled or force-controlled effect on the test samples (Fig. 2b). All these make the simulator a useful tool for evaluating implant performance under various testing conditions.

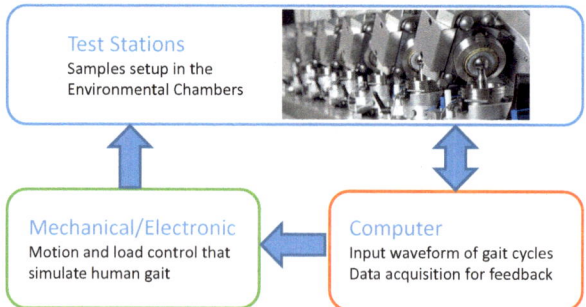

Fig. 1 A joint simulator

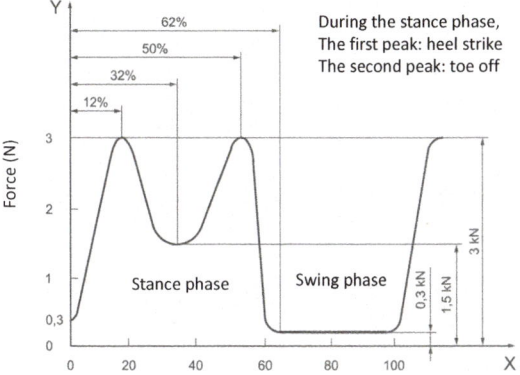

X axis represents Time, as a percentage of a gait cycle [from ISO 14242-1: 2012(E)]

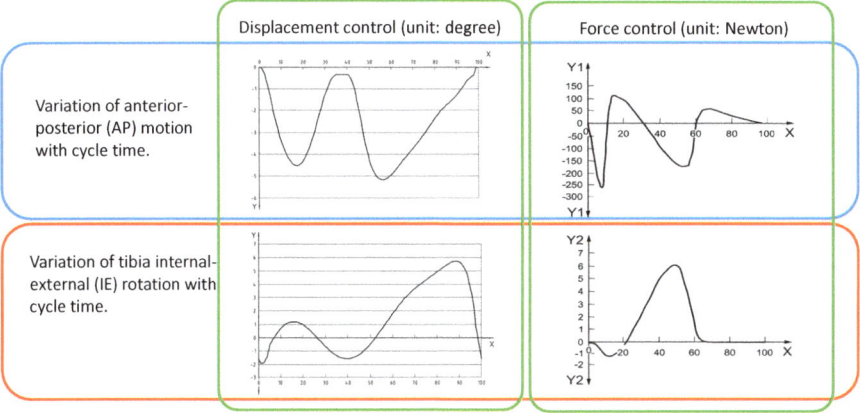

X axis represents Time, as a percentage of a gait cycle [from ISO 14243-3:2004(E) and ISO14243-1:2009(E)]

Fig. 2 (**a**) Typical load curve in a Gait cycle. (**b**) Displacement vs. force control

3 Types of Joint Simulators

There are various types of simulators made by different manufacturers. The intention here is to introduce the basic types that accepted by research institutes and test labs.

(a) Pin-on-Disk

The pin-on-disk (POD) or pin-on-plate (POP) style simulator is the simplest form of testing method to study the wear behavior of two contacted materials/components moving against each other under certain loads (Fig. 3). The POD simulator is primarily used for screening various material combinations or surface related treatments [6]. It takes less time and resource to run than a full joint simulation. It can be used as a simplified knee model for simulating the knee condyle (pin) on the tibia plateau (disk). Some machines have dynamic loading curves, with pin/plate

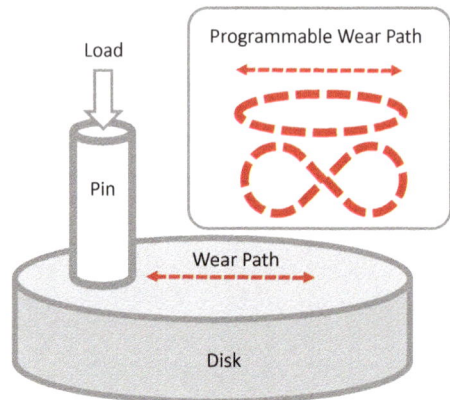

AMTI 6-station POD (Water Town, MA, USA)   Schematics of a POD

Fig. 3 Pin-on-Disk (POD) simulator

motion control for complex relative motions. These motion patterns vary from a back and forth reciprocating motion to square patterns and figure eight patterns. Overall, the lubrication mechanisms of POD/POP are different from that of a full joint simulation, due to the contact geometry.

(b) Orbital Bearing Motion Simulator

The orbital bearing motion (OBM) simulator has a ball-in-the-socket configuration of a hip implant and is used in hip joint simulation (Fig. 4) [7–9]. The original setup was a stationary femoral head on the top of an insert/cup that was secured on an angle block of 22.5-degrees. The angle block was driven mechanically by a motor, resulting in a total of 45-degrees of rocking motion in abduction/adduction (Ab/Ad) and flexion/extension (F/E) direction. There is no internal/external (I/E) rotation in this type of simulator. The machines have a hydraulic dynamic loading control for gait cycles.

(c) Anatomic Position Simulator

The anatomic position simulators intend to mimic the realistic usage and orientation of implants. An anatomic hip simulator has the acetabular cup positioned above the femoral head and stem surrogate, and includes dynamic loading, and motion control in F/E, Ab/Ad and I/E rotations. Each direction has a reasonable range of angles (Fig. 5) [10]. For an anatomic knee simulator, the motion is described in terms of anterior/posterior motion, internal/external rotation and flexion/extension rotation and includes dynamic loading (Fig. 6).

Shore Western 12-station hip simulator
(Monrovia, CA, USA)

MTS 8-station hip simulator
(Eden Prairie, MN, USA)

Fig. 4 Orbital bearing machine (OBM)

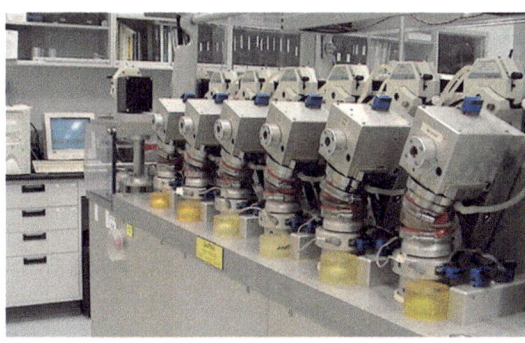

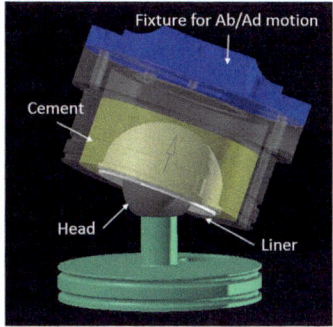

AMTI 12-station hip simulator (Water Town, MA, USA)

Drawing of a 1-station setup

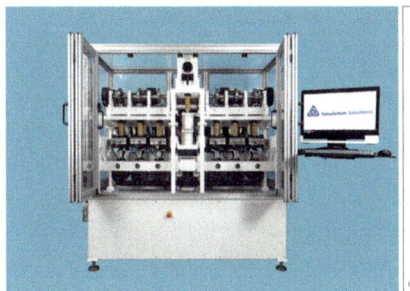

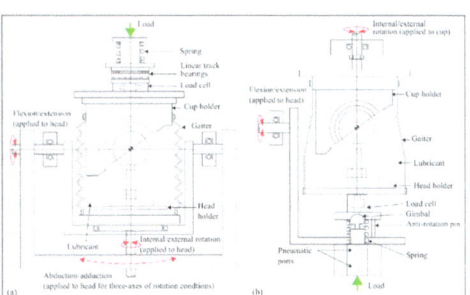

PROSIM 6-station hip simulator
(Simulation Solutions, Stock Port, UK)

Test cell schematics of (a) the electromechanical hip joint
simulator and (b) the pneumatic hip joint simulator.[10]

Fig. 5 Anatomic position simulator

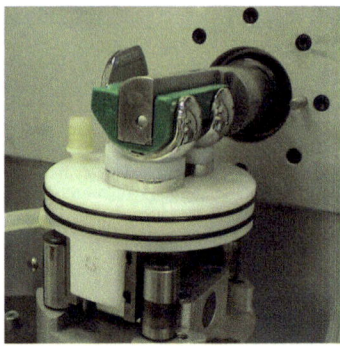 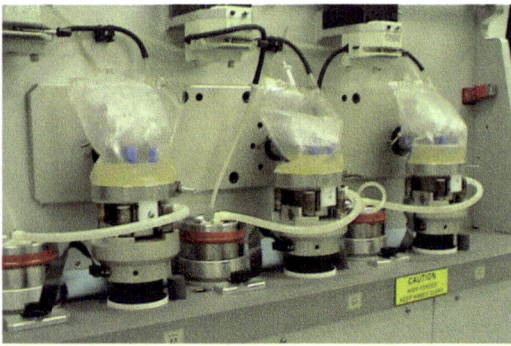

A knee test station (open for observation)   A knee wear simulation (AMTI) – picture shows three stations, samples are inside the IV bags, serum as lubricant that is circulated through a reservoir with temperature control

Fig. 6 Knee simulator

4 Current Wear Simulation Standards

Standardization of testing method may provide the ability to compare between laboratories. The current standard for POD is ASTM F732. The ASTM standards for hip and knee wear simulation have been aligned with the ISO standards. The most adopted standards for hip wear simulation are the ISO 14242 series, and those for knee simulations are the ISO 14243 series (Table 1).

The part 1 of both hip (ISO 14242) and knee (ISO 14243) wear standards define a physiological loading and kinematics in a gait cycle, with a specific relative sample orientation and lubricant recommendations. The test length is five million gait cycles (running at 1 gait cycle per second), with a hypothesis of one million cycles equivalent to 1 year of clinical steps [17]. The requirement of the report is also described in detail.

The part 2 of both ISO 14242 and ISO 14243 focus on the wear quantification method. Both standards describe the gravimetric method, including sample cleaning and preparation. The initial weight of a sample is measured and compared to the weight of the sample at different testing cycles. The weight change is further corrected by the weight gain due to fluid uptake during testing, which is achieved by using control specimens submerged in the same lubricant during testing. The corrected weight loss and total cycles determine the wear rate. For ISO 14242 part 2, it also covers the volumetric method, the initial sample surface was scanned by a coordinate measurement machine (CMM) and compared to the scanned data with a computer algorithm at different testing cycles to determine volume loss [18].

The ISO 14242 part 3 for hip wear standard is developed for an OBM style simulator, as a result of alignment with ASTM F1714 (Standard Guide for Gravimetric Wear Assessment of Prosthetic Hip-Designs in Simulator Devices).

Table 1 Standards for hip and knee wear simulation

	Current standards for hip and knee wear simulation
POD	ASTM F732–00(2006)
Hip	ISO 14242 implants for surgery - wear of total hip-joint prostheses [11–13] ISO 14242–1 (2012E): Loading and displacement parameters for wear-testing machines and corresponding environmental conditions for test ISO 14242–2 (2016E): Methods of measurement ISO 14242–3 (2009E): Loading and displacement parameters for orbital bearing type wear testing machines and corresponding environmental conditions for test
Knee	ISO 14243 implants for surgery - wear of total knee-joint prostheses [14–16] ISO 14243–1 (2009E): Loading and displacement parameters for wear-testing machines with load control and corresponding environmental conditions for test ISO 14243–2 (2009E): Methods of measurement ISO 14243–3 (2004E): Loading and displacement parameters for wear-testing machines with displacement control and corresponding environmental conditions for test

The ISO 14243 part 3 of the knee wear standard is for a displacement-controlled simulator, as a result of alignment with ASTM F1715 (Standard Guide for Wear Assessment of Prosthetic Knee Designs in Simulator Devices).

There are functional machine limitations for different simulators and there has been tremendous effort expended in the development of standard test methods that may be used around the world. Researchers have been working to expand the capabilities of simulator use from the standardized tests to advanced or unique kinematics and loading to develop a means to generate higher amounts of wear to help differentiate between materials or attempting to derive outcomes that may simulate various clinical situations or to mimic a measurement from explanted components. These advanced research techniques continue to evolve and provide possible standardization through ASTM and ISO organizations. Examples of published research include: edge loading, neck-liner impingement, separation/subluxation of implants, mal-orientation, third-body wear, and aging of bearing materials [19–22].

5 The Achievement of Wear Simulation

A simulation provides a more detailed understanding of the implant behavior under a certain testing scenario. It allows comparison of the relative performance of different implant materials or designs, under standard or non-standard gait and load cycles. A comparison side-by-side testing on the simulator is usually preferred for showing the performance of a target group to a control group. The following are examples of simulation use:

(a) Wear of total joint replacement is directly depending on the articulating bearing materials. Hip simulators have been used to evaluate hard-on-soft bearings, such as metal-on-polyethylene (MoP), ceramic-on-polyethylene (CoP), and

hard-on-hard bearings, such as metal-on-metal (MoM), ceramic-on-ceramic (CoC), and ceramic-on-metal (CoM) [23–26].
(b) Simulation testing can be used to compare designs such as effect of head size, liner thickness, diametrical clearances in hips, or fixed vs. mobile tibia inserts in knees [24, 27–29]. Sometimes a modified fixture is required to hold test samples with extreme sizes or for a specific test purpose.

To mimic the human daily activities, a simulator can be programmed to perform a certain sequence of movements. This is achieved by creating a repeated sequence of input waveforms, the start-and-stop walking, stair climbing or jogging [30, 31].

With the trend of younger patients and increased level of activities, extreme or hypothetical aggressive testing scenarios have been developed (micro-separation, rim loading, impingement, or 3rd body wear) to challenge the performance of the implants [32, 33]. Some testing methods are becoming standards, while a lot of the protocols were developed based on the research questions and could be case-dependent.

6 The Limitation of Wear Simulation

The human joint is functioning in a multi-variable environment. The patient's weight, activity levels, joint fluid quality and/or surgical techniques all influence the performance of the implants. A simulation test is a simplified model with controlled variables. It therefore does not fully represent *in vivo* performance in some cases; for examples:

(a) Patients' weight and activity levels vary from person to person in reality, while the loading and kinematics in the ISO/ASTM standards are based on the averaged population.
(b) Bovine serum is used as a test lubricant. But, the protein concentration and types of proteins in the serum may vary from batch to batch. Both factors affect wear of the implants [5]. Development of a universal lubricant may reduce the impact of batch-variation and standardization of testing methods.
(c) The frictional heating at the bearing surfaces in a simulator due to continuous testing cycles may affect the implant wear [34]. This non-stop motion is rare in human daily activity and could cause overheating in bearing surfaces that's not clinical relevant. It is reasonable to control the temperature of the wear station, or, use a start-stop gait cycles in wear simulation.
(d) Impingement, high cup angle, or misalignment due to a suboptimum outcome of a surgical technique may be simulated for exploratory purposes. The test setup may be extreme and beyond the clinical condition, to observe the performance limit. Interpret the simulated outcome under extreme conditions with caution, because the data may be valuable for comparing the effect of certain variables, but not for predicting the clinical performance.

7 Conclusions

The joint simulator opens a window to the *in vivo* performance of total joint replacement. Although currently there are limitations in mimicking the real-life activities, a joint simulator provides a controlled environment for evaluating bearing materials, new design concepts, or reproducing a clinical outcome. With the improvement of sensor technology (instrumented implants or gait analysis), the measured joint load and motion will not only benefit the computer-assisted simulation, but also achieve a more realistic simulation [35] (Table 2).

Table 2 Type of wear simulators and comparison

Types	Advantages	Disadvantages	Future improvement
Pin-on-disk	The test samples are prepared as pins and disks. It's a good screening device for evaluating the wear performance of bearing combination	May not represent the wear performance of final design due to geometry	Add sensing technology in the test chamber that provides feedback (temperature, lubricant pH value, or others) for wear modeling
Orbital bearing motion	The test samples are setup as a ball in a socket configuration with an "inverted position" (seen as an upside-down person). It's a good screening device for wear evaluation. Some labs use the simulator for 3rd body wear testing because wear debris tend to accumulated in the bearing surfaces due to gravity	It doesn't have the flexibility for the users to change kinematics, except the load profile. This is because the orbital bearing motion was driven by the mechanical design in the simulator	The simulator may be developed as a tool for observing wear mechanism, creating wear models, and establishing a wear library for joint materials
Anatomic position simulator	The state-of-the-art simulator design allows the test samples to be setup as clinical use. The load profile and joint kinematics are programmable to simulate various scenarios	The high cost of the simulator. Elaborate work in maintenance and calibration are required	With the feedback from sensing technologies, the researcher can take advantage of the programmable features and conduct more realistic wear simulation

References

1. Mendenhall S. Hip and knee implant review. Orthop Netw News. 2017;28(3):1–24.
2. Affatato S, et al. Tribology and total hip joint replacement: current concepts in mechanical simulation. Med Eng Phys. 2008;30(10):1305–17.
3. Zietz C, et al. Wear testing of total hip replacements under severe conditions. Expert Rev Med Devices. 2015:1–18.
4. Fisher J, Dowson D. Tribology of total artificial joints. Proc Instn Mech Engrs. 1991;205:73–9.
5. Liao Y-S, Benya P, McKellop H. Effect of protein lubrication on the wear properties of materials for prosthetic joints. J Biomed Mater Res. 1999;48(4):465–73.
6. McKellop HA, et al. Polyethylene wear in prosthetic joints. In: Dowson D, Wright V, editors. Evaluation of artificial joints. London: FS Moore Ltd; 1977. p. 109–34.
7. Herrera L, et al. Hip simulator evaluation of the effect of femoral head size on sequentially cross-linked acetabular liners. Wear. 2007;263(7–12):1034–7.
8. Medley JB, et al. Kinematics of the MATCO hip simulator and issues related to wear testing of metal-metal implants. Proc Inst Mech Eng H. 1997;211(1):89–99.
9. Liao Y-S, McNulty D, Hanes M. Wear rate and surface morphology of UHMWPE cups are affected by the serum lubricant concentration in a hip simulation test. Wear. 2003;255(7–12):1051–6.
10. Ali M, et al. Influence of hip joint simulator design and mechanics on the wear and creep of metal-on-polyethylene bearings. Proc Inst Mech Eng H. 2016;230(5):389–97.
11. ISO 14242-1. Implants for surgery—wear of total hip joint prostheses—Part 1: loading and displacement parameters for wear-testing machines and corresponding environmental conditions for test; 2012.
12. ISO 14242-2. Implants for surgery—wear of total hip joint prostheses—Part 2: methods of measurement; 2016.
13. ISO 14242-3. Implants for surgery—wear of total hip joint prostheses—Part 3: loading and displacement parameters for orbital bearing type wear testing machines and corresponding environmental conditions for test; 2009.
14. ISO 14243-1. Implants for surgery—wear of total knee-joint prostheses—loading and displacement parameters for wear-testing machines with load control and corresponding environmental conditions for test; 2009.
15. ISO 14243-2. Implants for surgery—wear of total knee-joint prostheses—methods of measurement; 2009.
16. ISO 14243-3. Implants for surgery—wear of total knee-joint prostheses—loading and displacement parameters for wear-testing machines with displacement control and corresponding environmental conditions for test; 2004.
17. Schmalzried TP, et al. Quantitative assessment of walking activity after Total hip of knee replacement. J Bone Joint Surg. 1998;80-A(1):54–9.
18. Raimondi MT, Sassi R, Pietrabissa R. A method for the evaluation of the change in volume of retrieved acetabular cups. Proc Inst Mech Eng H. 2000;214(6):577–87.
19. ISO/CD 14242-4. Implants for surgery—wear of total hip-joint prostheses—Part 4: testing hip prostheses under variations in component positioning which results in direct edge loading: variation in cup inclination and medial-lateral centres offset; 2017.
20. ASTM, F3047M-15. Standard guide for high demand hip simulator wear testing of hard-on-hard articulations.
21. ASTM F2582-14: Standard Test Method for impingement of acetabular prostheses.
22. ASTM F2003-02. Standard practice for accelerated aging of ultra-high molecular weight polyethylene after gamma irradiation in air; 2015.
23. McKellop H, et al. Friction and wear properties of polymer, metal, and ceramic prosthetic joint materials evaluated on a multichannel screening device. J Biomed Materials Res. 1981;15:619–53.

24. Whitaker D, et al. Effect of gamma irradiation and head size on the wear of moderately cross-linked UHMWPE inserts with EtO sterilisation in a hip simulation study. Bone Joint J Orthop Proc. 2013;95B(Supp 34):586.

25. Wimmer MA, et al. Wear mechanisms in metal-on-metal bearings: the importance of tribo-chemical reaction layers. J Orthop Res. 2010;28(4):436–43.

26. Williams S, et al. Ceramic-on-metal hip arthroplasties: a comparative in vitro and in vivo study. Clin Orthop Relat Res. 2007;465:23–32.

27. Shen F-W, Lu Z, McKellop HA. Wear versus thickness and other features of 5-Mrad cross-linked UHMWPE acetabular liners. Clin Orthop Relat Res. 2011;469(2):395–404.

28. Wang A, Essner A, Klein R. Effect of contact stress on friction and wear of ultra-high molecular weight polyethylene in total hip replacement. Proc Inst Mech Eng H. 2001;215(2):133–9.

29. McEwen HMJ, et al. The influence of design, materials and kinematics on the in vitro wear of total knee replacements. J Biomech. 2005;38(2):357–65.

30. Wimmer MA, et al. Knee flexion and daily activities in patients following total knee replacement: a comparison with ISO standard 14243. Biomed Res Int. 2015;2015:7.

31. Hadley M, et al. Development of a Stop-Dwell-Start (SDS) protocol for in vitro wear testing of metal-on-metal total hip replacements. Phoenix, AZ: ASTM; 2012.

32. Liu F, Williams S, Fisher J. Effect of microseparation on contact mechanics in metal-on-metal hip replacements—a finite element analysis. J Biomed Mater Res B Appl Biomater. 2015;103(6):1312–9.

33. Partridge S, et al. Evaluation of a new methodology to simulate damage and wear of polyethylene hip replacements subjected to edge loading in hip simulator testing. J Biomed Mater Res B Appl Biomater. 2017.

34. Liao Y-S, et al. The effect of frictional heating and forced cooling on the serum lubricant and wear of UHMW polyethylene cups against cobalt-chromium and zirconia balls. Biomaterials. 2003;24(18):3047–59.

35. Fitzpatrick CK, et al. Validation of a new computational 6-DOF knee simulator during dynamic activities. J Biomech. 2016;49(14):3177–84.

Mechanical Stimulation Methods for Cartilage Tissue Engineering

Stefan Balko, Joanna F. Weber, and Stephen D. Waldman

Keywords Cartilage · Chondrocytes · Mechanical stimulation · Mechanotransduction · Mechanobiology · Compression · Indentation · Shear · Tension · Multiaxial loading · Vibrations · Stochastic resonance

1 Cartilage Anatomy

Articular cartilage is located at the distal ends of the bones in all articulating joints, providing a gliding surface for the bones to articulate on. It is composed of four major components: collagen, proteoglycans, water and chondrocytes [1–3]. Collagens are fibrillar proteins that provide tensile properties to the cartilage and make the tissue resistant to mechanical forces. Proteoglycans are negatively charged glycosaminoglycans (GAG) covalently bonded to a protein core that attract cations and water into the collagen network which in turn causes swelling and increases tension within the collagen mesh further adding to the resistance to compressive forces. Water, which is the largest component of articular cartilage, contributes between 65–80% of the wet weight. Lastly, chondrocytes are the specialized cartilage cells distributed throughout the extracellular matrix (ECM) of the tissue, comprising roughly only 5% of the tissue volume and although they play no direct role in resisting mechanical forces, they are responsible for tissue maintenance and control the synthesis and degradation of both collagen and proteoglycans.

Articular cartilage has four different zones, each with a unique composition and organization allowing for a complex structure which is able to not only act as a smooth gliding surface but also is able to withstand large loads while not compromising joint function. The 4 different zones are: (i) superficial zone; (ii) middle zone; (iii) deep zone; and (iv) calcified zone. The articulating surface is the

S. Balko · J. F. Weber · S. D. Waldman (✉)
Chemical Engineering, Faculty of Engineering and Architectural Science, Ryerson University, Toronto, ON, Canada

Li Ka Shing Knowledge Institute, St. Michael's Hospital, Toronto, ON, Canada
e-mail: swaldman@ryerson.ca; http://www.waldmanlab.com

© Springer International Publishing AG, part of Springer Nature 2018
B. Li, T. Webster (eds.), *Orthopedic Biomaterials*,
https://doi.org/10.1007/978-3-319-89542-0_7

superficial zone, which protects against tangential shear forces and comprises approximately 10–20% of the overall volume. There is a denser population of chondrocytes which take on a flatter, elongated shape and the collagen fibers in this zone are oriented parallel to the articulating surface. This layer is responsible for the majority of the tensile properties of cartilage, allowing it to resist shear, tensile, and compressive forces. In this zone, resident chondrocytes also synthesize superficial zone protein (also known as lubricin or proteoglycan 4), a specialized proteoglycan which accumulates at the surface and is secreted into the synovial fluid of the joint, acting as a boundary lubricant. The zone just below the superficial zone is the middle zone, which comprises about 40–60% of the overall volume. It is a transitional zone, providing a bridge between the superficial and deep zone. The collagen fibers in this zone are organized obliquely and are thicker than in the superficial zone. Proteoglycans are also present in this zone making it the first resistor to compressive forces. The chondrocytes in this layer are more sparsely distributed and have a spherical shape. The deep zone is the zone just below the middle zone and comprises of approximately 30% of the volume. The collagen fibers in this zone are oriented perpendicular to the surface, having the largest diameter fibrils of all the zones. The deep zone also has the largest content of proteoglycans, and combined with the size and orientation of the collagen fibers, makes the deep zone the largest resistor to compressive forces. The chondrocytes are organized in columns oriented perpendicular to the surface. A notable feature of the deep zone is the tide mark, which separates the three previous zones from the calcified zone. The calcified zone is responsible for ensuring the cartilage is adhered to the subchondral bone. In this zone, the population of chondrocytes is very limited. The unique orientation of chondrocytes in each zone leads to the specific organization and maintenance of the overall ECM of the tissue [1, 3, 4].

2   Cartilage as a Material

From a materials standpoint, cartilage is a porous, viscoelastic material and has three key phases: (i) a solid phase which is predominately a strong collagen mesh with proteoglycans interwoven in; (ii) a fluid phase, which is comprised of water; and (iii) an ion phase, which has many dissolved electrolytes with positive and negative charges [5]. It is the combination of these three phases that allows articular cartilage to withstand large loads imposed on it by the human body. The most notable theory on how to describe the relationship between the stress and strain of articular cartilage is the biphasic theory, originally introduced by Mow et al. [5–7]. This theory states that three major internal forces act within loaded cartilage tissue: (i) the stress developed in the solid phase (collagen and proteoglycan woven network); (ii) the pressure developed in the liquid phase; and (iii) the drag force acting on each phase as they pass through one and other [5–7]. Later, the triphasic theory was introduced by Lai et al. which builds on the biphasic model of articular cartilage, but also takes into account the Donnan osmotic pressure, which is created by the imbalance of mobile counter-ions between the inner proteoglycan molecules and the outer solution [5, 8]. The interaction of these phases lead to a complex coupled mechano-electrochemical environment in which chondrocytes are exposed to multiple stimuli. These stimuli include mechanical forces, fluid flow, hydrostatic pressure, an osmotic pressure gradient, electric current, and electric potential differences within the ECM itself.

Since cartilage is characterized as a viscoelastic material, there are three important time dependant phenomena that relate to it: creep, hysteresis, and stress relaxation. Creep is the tendency for a material to permanently deform under constant stress. Hysteresis is the phenomenon in which previous loading influences the behavior of the tissue. Stress relaxation is the tendency for a material, under constant strain, to slowly decrease in stress until an equilibrium is reached [2]. In regards to healthy cartilage found in humans, the aggregate modulus and Poisson's ratio is between 0.5–0.7 MPa and 0.07–0.1, respectively [9].

Articular cartilage is constantly exposed to mechanical forces inside the human body which are essential to the growth of healthy human cartilage. Areas of the joint which are load-bearing have cartilage which is thicker and mechanically stronger than those areas which are non-load-bearing [1, 10]. Articular cartilage must be able to withstand the applied forces or it will start to deteriorate which can lead to severe joint pain and eventually osteoarthritis. There are a number of mechanisms, which have been proposed to be involved in the transduction of mechanical forces to biochemical signals in chondrocytes. As cartilage deforms certain effects are generated, such as interstitial fluid flow, electrical potentials, increased osmotic pressure, and decreased pH. Changes in pH have been linked to the changes in proteoglycan and collagen metabolism during compression. Osmotic stress has been shown to change the mechanical properties of chondrocytes, further altering the deformation rate of the tissue under a given load. The flow of ions in and out of the cell is also affected by deformation via stretch activated ion channels. Integrins also play a crucial role during mechanotransduction as they act as the primary bridge between the cell and its ECM. Salter el al. have shown that mechanical stimulation results in an influx of Ca^{2+} via stretch activated channels and a multitude of signal transduction events through integrin receptors [11–14]. In addition to this, the cytoskeleton and the nucleus of the chondrocyte play a role in transducing mechanical forces into different signals to alter gene expression and secreted constituents of the ECM [15, 16]. All of this leads to the fact that, although the exact mechanisms of mechanotransduction are unclear, mechanical loading has a direct influence on tissue formation. This has lead engineers and scientists to study mechanical forces and mechanical stimulation to induce cartilaginous tissue growth *in vitro*.

3   Cartilage Tissue Engineering

Since cartilage has an inability to repair and heal itself, a need has arisen to find
solutions to the problem of deteriorating articular cartilage. Through engineering
principles and techniques, tissue engineered constructs have been created *in vitro* to
be implanted into the defect site to resurface the affected joint. In order to engineer
viable tissue constructs, three general considerations are required: (i) cell sourcing;
(ii) scaffold design; and (iii) growth stimulus. In regards to cell sourcing, the cell
found in articular cartilage is the chondrocyte. Although chondrocytes are the primary cells involved in cartilage tissue formation, there are different means by which
one can obtain chondrocytes. Chondrocytes can be directly harvested from existing
cartilaginous tissues in the body. Alternatively, stem cells from different sources can
be differentiated into chondrocytes. Thus far clinically, chondrocytes have typically
served as the primary cell source for articular cartilage repair. Chondrocytes do have
their limitations though, such as dedifferentiation during *in vitro* expansion and the
limited availability of healthy autologous chondrocytes [17, 18]. To avoid the issues
related to chondrocyte sourcing, extensive research has been conducted into the use
of stem cells, including: embryonic (ESCs), mesenchymal (MSCs), and adipose-
derived (ASCs) stem cells, as a cell source. MSCs and ASCs are multipotent stem
cells that can be easily isolated from many mesenchymal tissues and have the ability
to undergo chondrogenesis given the correct physiochemical cues, making it a popular choice for cell sourcing and cellular therapy [19, 20]. ESCs have the trait of
unlimited proliferation and can essentially differentiate into any type of cell, making it extremely promising in tissue regeneration and cell sourcing. As research into
the use of ESCs develops, there are ethical concerns as well as many unknown
safety concerns that may limit their ability for use.

When designing tissue engineered constructs it is also important to take into
consideration scaffold design. Scaffolds can provide a 3D structure to support cell
growth and proliferation, ECM deposition, and tissue regeneration. Scaffolds must
be biocompatible to minimize the host response, be biodegradable to allow for
replacement with newly grown tissue, have suitable porosity to allow for cellular
proliferation and interconnectivity, and possess the proper mechanical properties to
support tissue growth under mechanical loads, all while encouraging the growth of
newly formed tissue [21]. In order to meet all these requirements, the biomaterial
chosen for the scaffold must be chosen carefully. Generally speaking, scaffolds are
designed using a natural or synthetic based biomaterial. Natural biomaterials are a
popular choice for scaffolds due to their biocompatibility. Specifically, carbohydrate-
based hyaluronic acid, agarose, alginate, chitosan, protein based collagen or fibrin
are currently used [21]. For synthetic biomaterials, polymers are used because of
their ease of fabrication and the ability to tailor the surface and bulk properties. The
most popular synthetic polymers for scaffold design are poly-lactic acid, polyglycolic acid, and their copolymer poly-lactic-co-glycolic acid [17]. Alternative to
using scaffolds, there are several scaffold-free techniques for tissue engineered constructs, such as pellet culture, aggregate cultures, and self-assembling techniques

[17]. Scaffold-free techniques benefit from being free of any excess material that would otherwise be used to create the scaffold which would need to degrade to allow for new tissue formation to replace it. In addition to this, a scaffold material may potentially induce other problems, such as stress shielding and toxicity that scaffold-free designs do not suffer from.

Lastly, when designing tissue engineered constructs the incorporation of growth stimuli are crucial in ensuring the tissue is formed properly and in a time-efficient manner. Growth factors have been studied extensively since the 1950's as a means to encourage tissue growth *in vitro*. Many different growth factors have been used and incorporated into tissue engineered constructs to elicit differentiation, proliferation and synthesis [17, 22, 23]. Alongside growth factors, there are other biophysical stimulation methods which are extensively studied. The three most common are mechanical stimulation, electrical stimulation, and magnetic stimulation. The use of magnetic stimulation in the treatment of diseases has been of great interest for a long time and because of it, there is a large breadth of literature covering the use of biomagnetism. Although the field of study is robust, much of it is met with skepticism [24]. Similar to magnetic simulation, the use of electrical stimulation is a well-established clinical therapy readily available with promising uses in cartilage tissue engineering [25]. This article focuses on the application of mechanical stimulation to tissue engineered constructs, but it is highly encouraged that the reader refers to the vast literature available in both magnetic stimulation and electrical stimulation in order to get a broader understanding of the different aspects of biophysical stimuli.

Articular cartilage is exposed to mechanical loads under normal physiological conditions. With the desire to make constructs that better represent what is found in native healthy cartilage, the environment in which cartilage grows *in vivo* was assessed and the application of mechanical forces to the cartilage constructs *in vitro* was determined to be a viable method to create a healthier stronger construct [1, 10, 13, 17, 26–28]. Applying mechanical stimulation to cartilage constructs *in vitro* allows for better tissue growth, uniformly organized tissue constituents and better mechanical properties, similar to that of native cartilage. Studies have shown that in general, low to moderate magnitude loads applied at frequencies on the order of 1 Hz substantially enhances the expression and synthesis of matrix proteins [13]. Although mechanical forces applied to cartilage have been extensively studied, the means by which the cells sense these mechanical signals and affect change remains poorly understood [13]. In the human body, articular cartilage is exposed to stresses between 3 and 10 MPa and because of this, the focus has been on applying forces in this physiological range [29].

There are currently many different methods for applying mechanical stimulation to cartilage constructs, but each of the methods can generally be broken down into one of two following categories: static and dynamic loading. Static loading refers to a constant force being applied to a tissue engineered construct over a given period of time. Static conditions for mechanical stimulation have been studied through many variations of force application, the majority of them falling under the categories of hydrostatic pressure and direct compression. The two general methods for applying hydrostatic pressure are by compressing a gas phase that transmits the load through the medium to the cells and, a less complicated approach, is by compressing the fluid phase. In general, studies assessing the effects of static hydrostatic pressure on chondrocytes cultured in monolayer have been shown to have no effect or a negative effect on tissue growth. Jortikka et al. showed that applying 5 MPa statically for 20 h had no effect on sulfated GAG incorporation [30]. Smith et al. found that applying 10 MPa for 4 h actually decreased the collagen mRNA levels present in the tissue [31].

In regards to static compression, the most common method for applying a static compressive load is to directly apply the load to the construct through a platen loading surface. It has been shown, similar to the findings in static hydrostatic pressure, to have no beneficial effect to tissue growth. Static compression inhibited matrix synthesis at compressive loads of 0.05–1.0 MPa and strains of 50–60%. Prolonged static compression has also been shown to induce matrix consolidation and hinders diffusional transport of macromolecules. Compression at 50% diminishes total protein and sulfated GAG synthesis by 35% and 57% respectively when compared to the uncompressed controls. It also decreased the percentage of protein retention by 30% [32, 33]. Buschmann et al. also found that applying a constant compression force (up to 50% total strain) produced little or no change in the biosynthesis of the cartilage construct [34]. This observation of no effect, or even a supressing effect, on tissue growth has become the consensus when talking about static loading conditions, and is the reason as to why the field now primarily focuses on dynamic loading conditions.

Static tension has also been investigated, again showing small to negative effects, depending on the duration of load as well as the intermittency of load application. Fan et al. showed that biaxial static tensile loading applied for 30 min, 3 times a week for 4 weeks was able to upregulate proteoglycan content and tissue thickness without altering mechanical properties. In the same study though they observed that stimulating constructs 30 min a day every day resulted in no change as compared to control groups [35].

The mechanisms by which static loading affects cartilage growth are similar to that of dynamic loading and may be categorized as: (i) cell deformation; (ii) transport-related; (iii) physicochemical; and (iv) cell-matrix interactions [34]. It has been suggested that the availability of nutrients (oxygen, glucose, growth factors), as well as transport related mechanisms are not solely responsible for the inhibition of biosynthesis during static compression. The physiochemical environment is also altered due to an increase in fixed charge density via a decrease in hydration with compression. The increase in fixed charge density upregulates the intracellular concentrations of cations, thereby increasing osmotic pressure and reducing the pH. Reduced pH has been shown to also reduce biosynthesis. In regards to cell-matrix interaction, Buschmann et al. have shown that specific connections between the chondrocyte and the pericellular matrix are required for cellular response to mechanical stimuli. Constructs with longer culture times were more responsive to the static compression as opposed to those with shorter culture periods, indicating the development of the necessary transduction pathways at the cell-matrix interface [17, 29, 32, 34, 36].

As mentioned earlier, the study of mechanical stimulation has become focused primarily on dynamic loading scenarios as they have been proven to effectively increase tissue growth and proliferation as well as increasing the mechanical properties of the construct itself.

4 Dynamic Loading Scenarios for Mechanical Stimulation

Dynamic conditions refer to the application of a force which changes over time. Usually in a cyclical pattern, the force will be applied at a certain frequency over a given time period. Dynamic mechanical stimulation has been shown to have the greatest effect on tissue engineering constructs thus far, showing the highest growth rates as well as stronger tissue constructs with better mechanical properties.

Dynamic conditions of mechanical stimulation have been studied through a variety of methods of force application. The most notable of these applications, and subsequently the methods that have been studied the most, are: (i) compression; (ii) tension; (iii) shear; (iv) friction; and (v) vibration. Although there are different methods of application, it has been shown many times over that a dynamic form of mechanical stimulation provides the greatest growth potential for cartilage constructs *in vitro*. Each method has its benefits and will be discussed individually to give a better understanding of how each is applied and the overall effect they have.

4.1 Compression

Dynamic compression applied to tissue engineered constructs has been the focus for studies in compressive mechanical stimulation for quite some time. The most common method for applying a dynamic compressive load is to directly apply the load to the construct through a loading surface and alter the load via a sinusoidal wave form. Compressive loading falls under three categories: (i) confined compression; (ii) unconfined compression; and (iii) indentation.

The different parameters of interest in terms of applying the stimulation, which have also been the focus of optimization studies, have been frequency (or duty cycle), duration, and strain or force amplitude used. Typically, frequencies ranging from 0.0001 to 3 Hz, strains from 0.1 to 35%, loads from 0.1 to 24 MPa and durations lasting hours to weeks have been examined at various cycles and waveforms, attempting to find an optimal configuration while staying in the realm of physiological conditions [17, 32, 34, 37].

4.1.1 Confined Compression

Confined compression refers to the construct being radially confined during the application of force. Special chambers are created to inhibit the transverse strain of constructs during force application, while allowing for uniaxial compression to still occur. Because of confinement, hydrostatic pressure plays a larger role in the force being applied, depending on the frequency of the compressive force. Soltz and Ateshian were able to show that as the frequency of dynamic load approaches 0.00044 Hz, the magnitude and phase of fluid pressurization matched that of the applied stress [38]. Studies in confined compression seem to further elucidate the crucial role of interstitial fluid pressurization in the load bearing capabilities of cartilage [32, 38–40]. In terms of tissue growth, Davisson et al. showed a dramatic increase in sulfated GAG and protein synthesis for confined dynamic compression at 0.1 Hz with a 50% static compression offset, showing results that agree with the literature in terms of dynamic compression increasing matrix synthesis [32]. In general, confined compression requires a more rigorous compression apparatus as it requires a chamber designed to inhibit transverse strain and higher precision in construct shape consistency. As such, it seems that the preference is to study dynamic compression in unconfined parameters.

4.1.2 Unconfined Compression

Unconfined compression, in contrast to confined compression, refers to the construct being free to expand radially during force application. The majority of research in dynamic compression falls this category. As a construct is compressed, it expands radially, which introduces a transverse strain into the construct in addition to the axial strain. This transverse strain has the ability to induce further mechanotransduction and help tissue growth. Also, with unconfinement, hydrostatic pressure still increases, but not exponentially as it does with confined compression.

Numerous short-term and long-term studies have applied unconfined compression protocols to cartilage constructs using hydrogels or microporous scaffolds, differentiated, undifferentiated, or de-differentiated cells to stimulate cell differentiation, proliferation, biosynthetic activity and functional ECM development [41]. In regards to short-term studies, Wong et al. showed that 45 h of unconfined cyclical compression increased protein synthesis by 50% above control values [42]. Sah et al. has shown that depending on the frequency, the biosynthetic response of cartilage would either decrease or increase. At a frequency of 0.0001–0.001 Hz, compressions up to 4% total strain had little effect but at a frequency of 0.01 Hz, compressions of 1.1–4.5% total strain caused a 40% increase and a 30% increase in collagen II and GAG synthesis respectively [43]. Through the use of 3D scaffolds and dynamic compression stimulus, Démarteau et al. showed that cartilage construct response does not depend directly on the stage of cell differentiation [44].

The development of functional ECM can be further appreciated when looking at long-term studies that apply unconfined compressive loading. Through a 2-month long study, Mauck et al. showed that constructs loaded with intermittent loading (at 10% deformation, at a 1 Hz frequency, 1 h on 1 h off, 3 h per day 3 days per week) had a 2-fold increase in material properties relative to the control, even though the GAG content was similar between both the control and the stimulated group. This suggests that the organization and assembly of the ECM was regulated by the dynamic loading, allowing for better mechanical properties [45]. Mauck et al. have also reported a three-fold increase in the equilibrium aggregate modulus in argarose-seeded constructs which were dynamically stimulated as compared to their free-swelling controls. It was also noted that a significant difference in stiffness occurred in the last week of growth as compared to the first 3 weeks, further eluding to the role of the ECM on material properties and response to stimulation [46].

As stated earlier, the mechanisms by which the dynamic compression affects the growth of cartilage can be categorized as: (i) cell deformation; (ii) transport-related fluid flow and cell-protein interaction; (iii) physicochemical; and (iv) cell-matrix interactions. Most notably, it has been shown that constructs cultured for longer periods of time respond to the stimulation to a greater degree than constructs with shorter culture periods. This indicates that the cell-matrix interactions play a crucial role in the biosynthetic response to compression stimulation. It has also been shown that GAG accumulation in cartilage constructs only occurs if the GAG content prior to compression is sufficiently high, once again indicating that the ECM plays a significant role in supporting the biosynthesis of the cartilage construct [34, 44, 47]. With regards to the application of the force, Suh et al. has reported that when cartilage constructs undergo dynamic compressive loading, the ECM goes through a repeated compression-expansion cycle, causing an oscillating positive-negative hydrostatic pressure together with interstitial fluid flow, which in turn leads to tissue biosynthesis [48].

Although dynamic loading has shown many benefits for both short-term and long-term culture periods, there are still drawbacks that arise. One of the drawbacks is the desensitization of the constructs to the stimulation itself, over longer periods of time. Weber & Waldman have shown through the examination of the durations of dynamic compressive loading, a minimum amount of stimulation was required to elicit an anabolic response, but desensitization could quickly be reached with an increase in loading cycles [2]. In addition to this, although compression has been shown to increase tissue growth and mechanical properties, the overall stiffness of constructs has still yet to reach that of native cartilage. Although the properties can be altered, not all mechanical properties are affected in the same manner. Kelly et al. demonstrated that, although the dynamic loading increased the bulk properties, the overall profile of construct properties in the axial direction were qualitatively the same as free swelling controls over the course of 42 days. That being the case, the Poisson's ratio did increase, hinting at an improved collagen network [49].

4.1.3 Indentation

Indentation stimulation is a special type of compressive stimulation where the size of the stimulating platen is smaller than the surface of the construct [50]. Indentation stimulation, therefore, results in a complex loading profile throughout the construct where directly under the indenter experiences direct compression while the surrounding area remains uncompressed. Through special analysis techniques which allow for the correlation to spatial mapping (e.g. autoradiography), this type of stimulation allows for the comparison of directly stimulated, indirectly stimulated, and unstimulated conditions. Parkkinen et al. noted differences in the amount of proteoglycan synthesis due to direct vs. indirect stimulation with different trends observed in each area under several loading magnitudes. Depth-dependent variations were also observed due to the different loading magnitudes [50].

The complex loading state created by indentation stimulation is difficult to define for constructs with irregular geometry and non-uniform material properties, thus it is not a common mode of stimulation. More often, indentation (or double indentation) is used for mechanical property testing rather than stimulation [51, 52].

4.2 Tension

The physiological loading condition for articular cartilage *in vitro* is seen as a combination of primarily hydrostatic pressure, compression, shear, and a small degree of tension. Because of this, tension is generally overlooked in terms of a method for mechanical stimulation. That being said, there is still a library of research, albeit smaller than for compression and shear, on tensile forces used for mechanical stimulation. Tension can be split into two main categories for discussion: (i) uniaxial and (ii) biaxial or multiaxial.

4.2.1 Uniaxial

Uniaxial tension refers to the tensile force being applied in a single axis. As with compression studies, the duration, frequency, and force applied all play a crucial role when investigating the effects of uniaxial tensile loading. Vanderploeg et al. have demonstrated that uniaxial dynamic tension with an oscillation period of 48 and 68 h, matrix synthesis was inhibited as compared to the control. It was also noted that after the 68 h stimulation, chondrocytes adopted a morphology similar to that of a fibrochondrocytic phenotype, potentially shedding light on the potential reason for failure of articular cartilage repair due to different strain environments occurring at healing regions [53]. Wong et al. showed that cyclical tensile strains upregulated matrix degrading enzymes, such as MMP-13. MMP-13 is a member of the matrix metalloproteinase family with preferential activity to degrade collagen type II [54]. A similar finding was observed by Mawatari et al., showing a down regulation of aggrecan (the major proteoglycan of cartilage), collagen type II and SOX9 (master chondrogenic transcription factor), all of which are gene markers for chondrogenic growth [55]. Although, in general it seems that tensile loading has mainly detrimental effects, Vanderploeg et al. also investigated the effect of tensile strain on zonal cartilage, finding that although the middle and deep zones showed no increase in GAG synthesis, the superficial zone did [56]. This potentially relates back to the superficial zone having cartilage and collagen oriented parallel to the surface for the purpose of resisting tensile and shear forces.

4.2.2 Biaxial or Multiaxial

Biaxial tension refers to applying the tensile force in two or more axes. With the introduction of the Flexcell™ system, biaxial tension became a popular mode of mechanical stimulation as it enabled the ability to apply the tensile force equixially in a simple and controllable manner. The Flexcell system uses an expanding and contracting membrane upon which a cartilage construct is fixated and the membrane is expanded or collapsed applying a uniaxial strain throughout the construct.

In terms of loading regimes, frequencies ranging from 0.03–2.5 Hz, durations up to 96 h, and strains up to 23% have been applied [57]. Again, as parameters varied, so did the results obtained from these experiments. In terms of collagen synthesis, Thomas et al. showed an increase in collagen II and aggrecan synthesis with a cyclical tensile loading regime of 7.5% strain at 1 Hz for 30 min [58]. In terms of gene expression, Chen et al. was able to show that cyclical tensile strain, applied at 6% strain at 0.25 Hz altered chondrocyte gene expression [59]. Cyclical tensile strain has also been shown to be an anti-inflammatory signal as well as significantly surpassing IL-1β-dependant mRNA induction for multiple proteins responsible for the initiation of cartilage degradation [60]. In general, tensile loading has been shown to have mixed results.

4.3 Shear

The loading that occurs in articulating joints is complex and has multiple forces acting on the joints at all times. One specific force that is generally low in magnitude is that of a shear force acting on the articulating surface. Although synovial fluid reduces the friction factor, emulating this shear force *in vitro* has been examined to determine its effects on cartilage constructs [17]. When looking at the architecture of cartilage tissue, as previously showed, the superficial layer has chondrocytes and collagen oriented parallel to the surface as to protect against tensile and shear forces [3]. The two main avenues that are explored in regards to shear stress are: hydrodynamic and mechanical shear. The first, hydrodynamic shear, is usually the result of a fluid flowing either over top of the cartilage construct or directly through the construct. There are many different systems that have been designed with the purpose of applying continuous fluid flow to constructs housed in a bioreactor. The second method, mechanical shear, is achieved through the direct application of shear forces on the construct.

4.3.1 Hydrodynamic Shear

In relation to hydrodynamic shear, the shear force is generated through the flow of fluid over the cartilage construct. It can also be generated by flowing through the constructs as well, which is known as perfusion. Many bioreactors have been developed to allow for fluid flow over cultured constructs as it has been generally accepted that fluid flow will increase nutrients available via an increase in mass transport. That being said, shear forces are still being generated at the point of contact between the flowing fluid and the cartilage construct which in turn have a direct result and mechanical properties and growth constituents of the constructs.

Raimondi et al. demonstrated an increase in collagen type II and type I after cartilage constructs were cultured in a perfusion bioreactor for 2 weeks. More specifically, a perfusion pressure of 1 mPa showed the highest effect on collagen synthesis [61]. Bueno et al. found an increase in collagen formation with increased shear stresses, as well as an overall increase in mechanical properties. Using their novel wavy-walled bioreactor they were able to alternate the fluid flow uniformity as well as the shear stresses applied to the constructs, showing an increase in thickness with increased shear due to an increase in GAG accumulation at the core of the constructs [62]. Gemmiti et al. has shown, through the use of a dual-chambered flow bioreactor, that flow mediated stress increased type II collagen without any significant increase in collagen type I, a marker of dedifferentiation. Furthermore, they were able to show an increase in tissue modulus of the tissue from control groups at 1.5–2.5 MPa after flow regimes were carried out [63].

Fluid flow-induced shear does have its potential drawbacks though. If shear forces are too high, or if fluid velocity is too high, ECM constituents can be cleaved off and washed away and there is the potential for cleavage of cells from their scaffold [61–63].

4.3.2 Mechanical Shear

Mechanical shear is a result of tangentially moving a platen surface against a cartilage construct and generating a deformation. In order to achieve the shear force, a static compressive force is pre-loaded to ensure enough friction between the platen surface and the construct surface exists such that there is no relative movement between the two contacting surfaces. Fitzgerald et al. showed that dynamic shear increased the biosynthesis of cartilage, preferentially collagen type II, at the surface of the constructs [64]. Jin et al. have shown that by applying a dynamic shear load at a frequency range of 0.01–1.0 Hz and a strain range of 1–3%, the synthesis of protein and proteoglycans increased by 50% and 25% respectively when compared to controls [65]. Frank et al. was also showed that using a 0.1 Hz frequency and 1% applied shear strain increases the ECM synthesis. The small amplitude simple shear deformation induces low levels of fluid flow which are associated with metabolic stimulation [66]. Nugent et al. showed an increase in proteoglycan 4 released at the surface, which is one of the constituents found in the synovial capsule of articulating joints and is crucial for proper joint lubrication [67].

A long-term study done by Waldman et al. showed that intermittent applications of dynamic shear forces improved the growth of the cartilage tissue and the mechanical properties. Stimulated at 2% shear strain at a frequency of 1 Hz for 400 cycles every other day, both collagen and proteoglycan synthesis increased by 40 and 35% respectively over a 4 week period. A threefold increase in compressive load-bearing capacity and a sixfold increase in stiffness was also shown, with the maximum equilibrium stress reaching 16 kPa and the maximum equilibrium modulus of 112 kPa [68].

Although shear deformation shows a positive effect in tissue growth, some researchers view it as a poor method of stimulus for cartilage tissue since it has been shown to increase interleukin-6 and nitric oxide levels which are catabolic mediators and similar to indicators observed in the development of osteoarthritis [17]. In relation to this, Fitzgerald et al. showed an up regulation of COX-2, a known inflammatory mediator that stimulates proteoglycan degradation and causes an increase in protease (enzyme responsible for the degradation of matrix constituents) transcription [64].

4.4 Friction

Surface sliding friction is an important mediator for the development and maintenance of the zonal differences in the cartilage matrix, specifically, the development of the superficial zone morphology and production of lubricating surface molecules (i.e. proteoglycan 4). Friction stimulation differs from shear stimulation in that there is relative motion between the two contacting surfaces. Frictional stimulation apparatuses have been developed with the intention to simulate this sliding motion seen in physiologic joint movement [69–71]. These can be divided into two main types: full surface stimulation and moving point-of-contact stimulation.

Full-surface stimulation can be achieved by applying a rotating apparatus to the surface of the construct. Fukuda et al. showed that by applying gliding friction through a rotating glass apparatus on the surface of agarose-cartilage constructs, the surface region increased both collagen II and GAG secretion when compared to the middle region of the construct, leading to an anisotropic structure similar to what is found in native cartilage [69]. Grad et al. were also able to show an increase in proteoglycan 4 released at the surface when applying a rotational ball to the surface of cartilage constructs [70].

Point-of-contact stimulation can be accomplished by reciprocating a small platen or indenter across the surface of the construct. Kaupp et al. showed that this type of stimulation was capable of increasing superficial zone specific constituents such as proteoglycan 4 and biglycan and also increasing collagen II expression at the surface of the constructs [71].

4.5 Vibration

Vibration stimulation of cells *in vitro* is a novel approach that many studies are examining. The two main categories are: high-frequency ultrasonic and lower frequency mechanical vibrations.

4.5.1 High-Frequency Ultrasonic Vibration

The application of ultrasound requires the vibrations to be transduced through a medium before being applied to the cells themselves. Different methods of achieving this have been studied, with the most common approach having transducers transmitting the ultrasonic waves through a coupling medium to the culture plates. Different forms of ultrasound have been investigated in terms of low and high intensity. Thakurta et al. have shown that ultrasound increases the proliferation of chondrocytes, maintains the chondrocyte phenotype in scaffolds over 21 days and selectively enhances the gene expression of TGF-β3 over TGF-β1 (chondrogenic growth factors) [72]. Parvizi et al. were able to show an increase in proteoglycan synthesis and aggrecan mRNA expression through the use of a low intensity-pulsed ultrasound stimulation [73]. Similarly, Noriega et al. showed an increase in collagen type II and mRNA expression in chondrocytes seeded in 3D scaffolds stimulated with a 5 MHz ultrasonic wave applied for 51 s, twice a day [74]. Further studies from the same group also showed an up regulation of both the ROCK-I and Rho-A when the same application of ultrasound was applied. Both ROCK-I and Rho-A are genes known to regulate cytoskeleton formation in chondrocytes, more specifically the formation of actin stress fibres [75].

4.5.2 Lower-Frequency Mechanical Vibrations

A common mechanism used for mechanical vibrations is to have a stage on which the cell culture plate rests and apply the mechanical vibration with a modular piezoelectric device through the stage itself. Jankovitch et al. showed that a 0.3 g amplitude, 30 Hz vibration increased chondrogenic differentiation *in vitro* as well as upregulated SOX9 expression and downregulated MMP-13 activity [76]. Kaupp et al. obtained similar results, showing an increase in cell proliferation at 1 g and 350 Hz. They also showed that over a longer culture period (1 week), there was in increase in ECM accumulation which led to a decrease in the effectiveness of the stimulation [77].

Although vibrational therapies show a promising effect, it is still unsure as to their use in long-term applications to chondrocyte biosynthesis and tissue growth.

5 General Drawbacks of Mechanical Stimulation

An overview of the overall outcomes achieved through each type of stimulus can be seen in the table below. The table below does not include static loading parameters because, as mentioned earlier, it is the general consensus that static mechanical stimuli inhibit tissue growth and are overall detrimental to tissue engineered constructs (Table 1).

In regards to the different techniques, each method has the ability to increase tissue growth and cartilage-specific ECM production in constructs depending on the loading parameters that are defined. However, this parameter specificity, which is often unique to each type of stimulus if not to each experimental model, is one of the major obstacles of mechanical stimulation. There are a vast number of variables that are in play when applying mechanical stimulation to a tissue engineered construct. Some of these include the magnitude of the load, the duration of the load, whether it is one application of a load or several applications in a span of a week or a month, and the list goes on. When surveying the literature, there is a plethora of different loading regimes being used within each subcategory of mechanical stimulation, each reporting varying results. There is no definitive loading scenario for each specific type of stimulus that is the best when compared to all the others. If mechanical stimulation is to be viewed as a method to improve cartilage tissue growth to the point where it can be industrialized, parameter optimization must be investigated. Currently, relatively few studies optimize loading regimes and often researchers tend to rely on previously reported loading protocols.

Another drawback to mechanical stimulation is the fact that the cells in tissue engineered constructs can become desensitized. This has been most commonly seen in compression stimulation, but is still relevant in all other forms of stimulation. As chondrocytes undergo a specific loading regime, the mechanical forces activate signaling pathways which result in chondrogenic ECM synthesis. Over the course of the stimulation, certain parts of the pathway (e.g. receptors) may become fatigued, leading to a desensitization to the mechanical force which subsequently leads to a detrimental effect on ECM production [2, 78]. If the stimulation could have a negligible or detrimental effect in the development of the constructs, then its value, in terms of a tissue growth stimulus, is not worth the effort.

Although mechanical stimulation can be a robust method for increasing tissue growth and mechanical properties of cartilage constructs, it still falls short from that of native healthy cartilage. Promising results have been made, but native articular cartilage has better mechanical properties than that found in any tissue engineered construct [79, 80]. In order for tissue engineering to be seen as a viable option for replacing damaged cartilage, methods have to allow for cartilage to be created with properties that are equivalent to native cartilage otherwise this may potentially

Table 1 Comparison of single mode loading scenarios

Mode	Pros	Cons
Confined compression	– Increased GAG synthesis	– Requires geometrically consistent constructs to fit in specific compression chambers – Desensitization to mechanical stimulus – Large forces cause permanent deformation to constructs [32, 38–40]
Unconfined compression	– Increased GAG and collagen synthesis – Increased mechanical properties	– Desensitization to mechanical stimulus over longer periods of time – Long-term effects are not as promising as short-term studies show [2, 34, 41–49]
Indentation	– Increased GAG synthesis	– Complex loading profile is difficult to predict on irregular shapes and non-uniform materials [50–52]
Uniaxial tension	– Increased GAG synthesis in superficial zone cartilage	– Down regulated of cartilage growth transcription factors – Down regulated cartilage matrix constituents – Upregulated degradative enzyme secretion [53–56]
Biaxial/ multiaxial tension	– Increased GAG and collagen type II production – Upregulated aggrecan	– Upregulated degradative enzyme secretion [57–60]
Hydrostatic shear	– Increased collagen type II – Increased mechanical properties (Young's Modulus) – Increased nutrient flow through fluid flow and mass transport	– Large shear forces are detrimental and cause cleaving of constituents [61–63]
Mechanical shear	– Increased GAG and collagen type II production – Increased compressive load bearing capacity	– Upregulated proinflammatory mediators and degradative enzymes [17, 64–68]
Friction	– Increased GAG and collagen type II production – Increased proteoglycan 4 secretion – Oriented surface chondrocytes	– Effects are limited to surface and subsurface area of constructs [69–71]
Ultrasonic vibration	– Increased proliferation of chondrocytes – Upregulated cartilage growth transcription factors – Upregulated chondrocyte cytoskeleton formation	– Long-term effects need to be investigated [72–75]

(continued)

Table 1 (continued)

Mode	Pros	Cons
Mechanical vibration	– Upregulated cartilage growth transcription factors – Downregulated degradative enzymes – Increased cartilage ECM growth	– Desensitization to stimulation over longer periods of time [76, 77]

negatively influence the quality of the repair. A possible reason for the general deficiency in mechanical properties may actually relate to the fact that, although mechanical stimulation upregulates tissue growth, it also upregulates the degradative enzymes associated with cartilage turn-over. Alongside an anabolic response post mechanical stimulation, a catabolic response is also observed. Although degradation is required for tissue remodeling, over-expression may potentially lead to weaker ECM structures over extended periods of mechanical stimulation [81]. This is an area that is typically overlooked as the short-term studies as they tend to focus solely on immediate collagen and GAG synthesis.

There are still many issues that must be overcome before a mechanical stimulation becomes an effective and efficient technique for increasing properties of cartilage constructs. However, approaches have been put forward which aim to overcome some of these drawbacks. Mixed mode loading is a more recent field of study, which applies a mixed variation of the mechanical stimulation types, in an attempt to not only make more mechanically robust constructs but to mimic the complex loading that is seen *in vivo* and attempt to overcome drawbacks with singular forms of loading.

6   Mixed Mode Loading

As studies progress in the field of mechanical stimulation applied to cartilage *in vitro*, it is becoming apparent that a more complex mode of mechanical stimulation is required to achieve not only better tissue growth and mechanical properties, but also to make up for drawbacks found in single mode stimulation protocols. For this reason, combining different loading parameters together is an approach that has recently been a focus of study. The major areas of mixed mode loading fall under two categories: compression-shear and compression-vibration. In general, it has been shown that dynamic compression has many great aspects to it that contribute to the *in vitro* growth of strong healthy cartilage constructs. By adding a secondary form of stimulation in addition to the compression, even more promising results have been achieved which open the door for even further exploration into mixed loading platforms for mechanical stimulation.

6.1 Compression and Shear

Two of the main forces exerted on articular cartilage are compression shear, and recently, this form of mixed loading has been explored with promising results. Waldman et al. investigated a mixed loading protocol of 5% dynamic compression with 5% shear strain, finding an increase in collagen by 76% and proteoglycan by 73% when compared to controls. They also showed a threefold increase in compressive modulus and 1.75-fold increase in shear modulus in constructs that were stimulated using the dynamic-shear loading over a 4 week period [82]. Grad et al. explored using a multi-motion compression-shear apparatus to provide two forms of shear as well as dynamic compressive loading to cartilage constructs to better resemble the motion in the joints of the human body. They found an increase in proteoglycan 4 and cartilage oligomeric matrix protein mRNA expression as well as an increase in collagen type II, aggrecan and tissue inhibitor of metalloproteinase-3 mRNA expression. They also noticed an increase in proteoglycan 4 and cartilage oligomeric matrix protein released to the culture media [70, 83]. Stoddart et al. also explored a compression-shear loading regime showing an increase in collagen type II and aggrecan expression within an hour of applying a 0.5 N load, and observing a peak at 2 h of loading. This increase translated into an increase of up to 60% in GAG content of the constructs after 4 days of intermittent cyclical loading [84]. Although the outcomes are quite promising, the issue of whether or not the constructs maintain their enhanced properties after the stimulation has yet to be investigated.

6.2 Compression and Vibration

A new and novel approach that is currently being explored is the application of a combined dynamic compression-vibration loading, termed stochastic resonance. Stochastic resonance refers to a white noise, or vibrational load, superimposed on top of a dynamic compressive load. Stochastic resonance has been observed in many biological systems, from molecular level DNA transcription systems in gene expression to whole body level devices developed for regular use to maintain blood pressure, blood oxygenation, and balance [2]. Originally, stochastic resonance was investigated as a method to inhibit the desensitization of chondrocytes to dynamic mechanical compression *in vitro*. Weber & Waldman showed improved cellular sensitivity to mechanical loading and increased matrix synthesis between 20 and 60% over short-term culture. Stochastic resonance also limited the load-induced desensitization by maintaining sensitivity under desensitized loading conditions [85].

Further work to examine the long-term effects of stochastic resonance on younger and older cell populations has also been done. Cell-agarose constructs were prepared with primary bovine chondrocytes from two age groups. Constructs were then subjected to mechanical stimuli coinciding with the media exchange cycle, 3 times

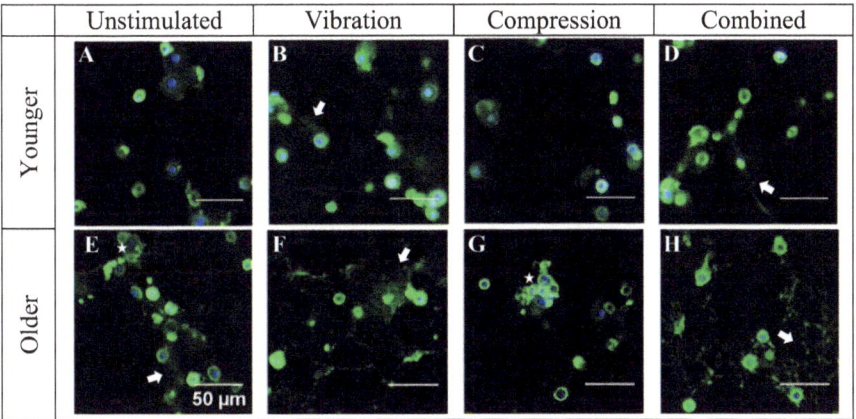

Fig. 1 Collagen type VI (FITC) immuno-staining of agarose-encapsulated bovine cells from older and younger animals. Arrows indicate the development of interconnected collagen VI matrix, stars indicate cell clusters. Scale bar = 50 μm

per week for up to 4 weeks. Immunohistochemistry staining for collagen type VI (Fig. 1) showed an increased development of an interconnected matrix linking adjacent cells in constructs treated with the vibratory stimulus (both alone and in conjunction with the dynamic compressive stimulus). Collagen VI is a unique collagen with a beaded micro-filament structure which forms a distinctive network in the ECM [86]. In adult cartilage, collagen VI is concentrated in the pericellular matrix where it connects the cells to the ECM through integrin binding and plays a crucial role in transducing mechanical loads from the tissue to the cells [87, 88]. The presence of an interconnected collagen VI matrix in the tissue engineered constructs, while recapitulating the distribution pattern of collagen VI in the growth plate during development and endochondral ossification, also suggests that the cells within the construct may be more sensitive to an applied global strain such as dynamic mechanical compression. Furthermore, the development of cell clusters in the constructs created from older cells was observed. Cell clustering is indicative of osteoarthritis in native cartilage and may indicate pathological tissue development in engineered cartilage constructs [89]. In the constructs treated with the vibratory stimulus (again, both alone and in conjunction with compression) a reduction in the number of cell clusters was noted.

Currently, the effect of stochastic resonance on passaged cells is being studied. As only few cells can be reasonably harvested from an individual, passaging is an effective method to expand the cell population to be able to create an engineered construct. However, as passaged cells tend to not be as synthetically active as primary (freshly isolated) cells, several different methods have been explored to redifferentiate these cells, including mechanical stimulation. Preliminary data on agarose-encapsulated passaged bovine chondrocytes (passage 4) subjected to combined compressive-vibration stimuli showed increases in both collagen and

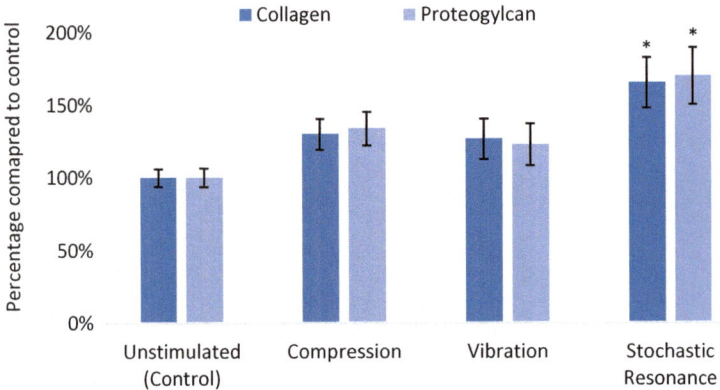

Fig. 2 Bovine constructs (P4) subjected to mechanical stimuli. Synthesis measured by radioisotope incorporation and DNA measured by PicoGreen assay. Data normalized to control, presented as mean ± standard error of the mean, n > 9/group; *p < 0.05 significant difference from control

proteoglycan synthesis (relative to DNA content) compared to unstimulated controls (p < 0.05) indicating the utility of this method (Fig. 2).

7 Future Directions

As previously discussed, there needs to be a better solution to repairing damaged or diseased cartilage and the approach of tissue engineered constructs shows potential. In regards to tissue engineered cartilage constructs, the correct properties are required in order to ensure success of the construct once implanted *in vivo*. Mechanical stimulation has been shown to be a very effective approach at increasing the mechanical properties and the growth of cartilage constituents in constructs *in vitro*, and has potential to allow for cartilage constructs created *in vitro* to resemble healthy native cartilage. However, there is still a long way to go to achieve this goal.

As noted earlier, engineered constructs typically fall short in mechanical properties when compared to native healthy cartilage. This is one of the biggest issues that will need to be addressed. If the constructs do not have the same properties as the surrounding native cartilage, this may hinder their ability to survive post-implantation. A potential approach to solving this problem is by having a stimulation protocol that combines a multitude of stimulatory factors, such as growth factors, mechanical stimulation, electrical stimulation, magnetic stimulation, etc. By combining a multitude of stimuli, which have their individual ability to increase growth potential of cartilage, an additive effect might be seen in terms of mechanical strength achievable *in vitro*.

An issue that has not been discussed yet, but is still crucial is that of anatomy and shape. Although mechanical stimulation methods are evolving, they are usually applied to uniformly shaped constructs that are small in size, which do not represent the anatomy at the defect site. While recent efforts to engineer cartilage constructs have been incorporating anatomical shape are underway, the ability to mechanically stimulate these constructs effectively needs to be investigated. Currently used methods may not readily work as the irregularities in construct shape will results in regional differences in the applied stresses/strains and ultimately differences in ECM biosynthesis.

Finally, once all these other issues are addressed and cartilage constructs are able to reach that of healthy native cartilage, in order to create products readily available for surgeons, manufacturing requirements will need to be addressed, some of which include: (i) upscaling of the cartilage constructs to create an ample supply; (ii) upscaling of the stimuli apparatus in order to ensure correct stimulation of each construct; and (iii) quality control to ensure each construct is receiving correct stimuli. These manufacturing issues, although quite a long time from being a true concern, still need to be discussed and investigated.

The future of mechanical stimulation for cartilage tissue engineered constructs looks bright with many promising results thus far showing the potential of using the mechanical stimuli to create and develop strong, healthy constructs *in vitro*. Although further experimentation is still required, the concept of repairing damaged articular cartilage may soon be resolved with the introduction of cartilage tissue engineered constructs into the marketplace.

References

1. Hall AC, Horwitz ER, Wilkins RJ. The cellular physiology of articular cartilage. Exp Physiol. 1996;81(3):535–45.
2. Weber JF. The sensitivity of articular chondrocytes to dynamic mechanical stimulation. Ph.D. dissertation. Dept. Mech. and Mat. Eng., Queen's University, 2015.
3. Sophia Fox AJ, Bedi A, Rodeo SA. The basic science of articular cartilage: structure, composition, and function. Sports Health. 2009;1(6):461–8.
4. Brady MA, Waldman SD, Ethier CR. The application of multiple biophysical cues to engineer functional neo-cartilage for treatment of osteoarthritis (part I: cellular response). Tissue Eng Part B Rev. 2015;21(1):1–19.
5. Lu XL, Mow VC. Biomechanics of articular cartilage and determination of material properties. Med Sci Sports Exerc. 2008;40(2):193–9.
6. Mow VC, Holmes MH, Michael Lai W. Fluid transport and mechanical properties of articular cartilage: a review. J Biomech. 1984;17(5):377–94.
7. Mow VC, Kuei SC, Lai WM, Armstrong CG. Biphasic creep and stress relaxation of articular cartilage in compression? Theory and experiments. J Biomech Eng. 1980;102(1):73–84.
8. Lai WM, Hou JS, Mow VC. A triphasic theory for the swelling and deformation behaviors of articular cartilage. J Biomech Eng. 1991;113(3):245–58.
9. Athanasiou KA, Rosenwasser MP, Buckwalter JA, Malinin TI, Mow VC. Interspecies comparisons of Insitu intrinsic mechanical-properties of distal femoral cartilage. J Orthop Res. 1991;9(3):330–40.
10. Stockwell RA. Cartilage failure in osteoarthritis: relevance of normal structure and function. A review. Clin Anat. 1991;4(3):161–91.

11. Guilak F, Jones WR, Ting-Beall HP, Lee GM. The deformation behavior and mechanical properties of chondrocytes in articular cartilage. Osteoarthr Cartil. 1999;7(1):59–70.
12. Gray ML, Pizzanelli AM, Grodzinsky AJ, Lee RC. Mechanical and physicochemical determinants of the chondrocyte biosynthetic response. J Orthop Res. 1988;6(6):777–92.
13. Shieh AC, Athanasiou KA. Principles of cell mechanics for cartilage tissue engineering. Ann Biomed Eng. 2003;31(1):1–11.
14. Salter DM, Nuki G, Wright MO. Integrin—Interleukin-4 mechanotransduction pathways in human chondrocytes. Clin Orthop Relat Res. 2001;391:49–60.
15. Buschmann MD, Hunziker EB, Kim YJ, Grodzinsky AJ. Altered aggrecan synthesis correlates with cell and nucleus structure in statically compressed cartilage. J Cell Sci. 1996;109(Pt 2):499–508.
16. Guilak F. Pression-induced changes in Cyte Nucleu. J Biomech. 1995;28(12):1529–41.
17. Zhang L, Hu J, Athanasiou KA. The role of tissue engineering in articular cartilage repair and regeneration. Crit Rev Biomed Eng. 2009;37(1–2):1–57.
18. Darling EM, Athanasiou KA. Rapid phenotypic changes in passaged articular chondrocyte subpopulations. J Orthop Res. 2005;23(2):425–32.
19. Wakitani S, et al. Mesenchymal cell-based repair of large, full-thickness defects of articular cartilage. J Bone Joint Surg Am. 1994;76(4):579–92.
20. Yoo J, Barthel T, Nishimura K. The chondrogenic potential of human bone- marrow-derived mesenchymal progenitor cells. J Bone Jt Surg. 1998;80A(12):1745–59.
21. Getgood A, Brooks R, Fortier L, Rushton N. Articular cartilage tissue engineering. J Bone Jt Surg. 2009;91-B(5):565–76.
22. Takeichi M. The factor affecting the spreading of chondrocytes upon inorganic substrate. J Cell Sci. 1973;13(1):193–204.
23. Corvol MT, Malemud CJ, Sokoloff L. A pituitary growth-promoting factor for articular chondrocytes in monolayer culture. Endocrinology. 1972;90(1):262–71.
24. Haddad JB, Obolensky AG, Shinnick P. The biologic effects and the therapeutic mechanism of action of electric and electromagnetic field stimulation on bone and cartilage: new findings and a review of earlier work. J Altern Complement Med. 2007;13(5):485–90.
25. Snyder MJ, Wilensky JA, Fortin JD. Current applications of electrotherapeutics in collagen healing. Pain Physician. 2002;5(2):172–81.
26. Ryan JA, Eisner EA, DuRaine G, You Z, Reddi AH. Mechanical compression of articular cartilage induces chondrocyte proliferation and inhibits proteoglycan synthesis by activation of the ERK pathway: implications for tissue engineering and regenerative medicine. J Tissue Eng Regen Med. 2009;3(2):107–16.
27. Natenstedt J, Kok AC, Dankelman J, Tuijthof GJ. What quantitative mechanical loading stimulates in vitro cultivation best? J Exp Orthop. 2015;2(1):15.
28. Brown TD. Techniques for mechanical stimulation of cells in vitro: a review. J Biomech. 2000;33(1):3–14.
29. Elder BD, Athanasiou KA. Hydrostatic pressure in articular cartilage tissue engineering: from chondrocytes to tissue regeneration. Tissue Eng Part B Rev. 2009;15(1):43–53.
30. Jortikka MO, et al. The role of microtubules in the regulation of proteoglycan synthesis in chondrocytes under hydrostatic pressure. Arch Biochem Biophys. 2000;374(2):172–80.
31. Smith RL, et al. In vitro stimulation of articular chondrocyte mRNA and extracellular matrix synthesis by hydrostatic pressure. J Orthop Res. 1996;14(1):53–60.
32. Davisson T, Kunig S, Chen A, Sah R, Ratcliffe A. Static and dynamic compression modulate matric metabolism in tissue engineered cartilage. J Orthop Res. 2002;20:842–8.
33. Sah RL, Grodzinsky AJ, Plaas AH, Sandy JD. Effects of tissue compression on the hyaluronate-binding properties of newly synthesized proteoglycans in cartilage explants. Biochem J. 1990;267(3):803–8.
34. Buschmann MD, Gluzband YA, Grodzinsky AJ, Hunziker EB. Mechanical compression modulates matrix biosynthesis in chondrocyte/agarose culture. J Cell Sci. 1995;108(Pt 4):1497–508.

35. Fan JCY, Waldman SD. The effect of intermittent static biaxial tensile strains on tissue engineered cartilage. Ann Biomed Eng. 2010;38(4):1672–82.
36. Grodzinsky AJ, Levenston ME, Jin M, Frank EH. Cartilage tissue remodeling in response to mechanical forces. Annu Rev Biomed Eng. 2000;2:691–713.
37. Lin WY, et al. The study of the frequency effect of dynamic compressive loading on primary articular chondrocyte functions using a microcell culture system. Biomed Res Int. 2014;2014(1):1–11.
38. Soltz MA, Ateshian GA. Interstitial fluid pressurization during confined compression cyclical loading of articular cartilage. Ann Biomed Eng. 2000;28:150–9.
39. Heiner AD, Martin JA. Cartilage responses to a novel triaxial mechanostimulatory culture system. J Biomech. 2004;37(5):689–95.
40. Gu WY, Yao H, Huang CY, Cheung HS. New insight into deformation-dependent hydraulic permeability of gels and cartilage, and dynamic behavior of agarose gels in confined compression. J Biomech. 2003;36(4):593–8.
41. Lee C, Grad S, Wimmer M, Alini M. The influence of mechanical stimuli on articular cartilage tissue engineering. Top Tiss Eng. 2006;2(2):1–32.
42. Wong M, Siegrist M, Cao X. Cyclic compression of articular cartilage explants is associated with progressive consolidation and altered expression pattern of extracellular matrix proteins. Matrix Biol. 1999;18(4):391–9.
43. Wall A, Board T. Biosynthetic response of cartilage explants to dynamic compression. Class Pap Orthop. 2014:427–9.
44. Démarteau O, et al. Dynamic compression of cartilage constructs engineered from expanded human articular chondrocytes. Biochem Biophys Res Commun. 2003;310(2):580–8.
45. Mauck RL, Wang CCB, Oswald ES, Ateshian GA, Hung CT. The role of cell seeding density and nutrient supply for articular cartilage tissue engineering with deformational loading. Osteoarthr Cartil. 2003;11(12):879–90.
46. Mauck RL, et al. Functional tissue engineering of articular cartilage through dynamic loading of chondrocyte-seeded agarose gels. J Biomech Eng. 2000;122(3):252–60.
47. Babalola OM, Bonassar LJ. Parametric finite element analysis of physical stimuli resulting from mechanical stimulation of tissue engineered cartilage. J Biomech Eng. 2009;131(6):61014.
48. Suh JK. Dynamic unconfined compression of articular cartilage under a cyclic compression load. Biorheology. 1996;33(4–5):289–304.
49. Kelly TAN, Ng KW, Wang CCB, Ateshian GA, Hung CT. Spatial and temporal development of chondrocyte-seeded agarose constructs in free-swelling and dynamically loaded cultures. J Biomech. 2006;39(8):1489–97.
50. Parkkinen JJ, Lammi MJ, Helminen HJ, Tammi M. Local stimulation of proteoglycan synthesis in articular cartilage explants by dynamic compression in vitro. J Orthop Res. 1992;10(5):610–20.
51. Hayes WCC, Keer LMM, Herrmann G, Mockros LFF. A mathematical analysis for indentation tests of articular cartilage. J Biomech. 1972;5(5):541–51.
52. Hori RY, Mockros LF. Indentation tests of human articular cartilage. J Biomech. 1976;9(4):259–68.
53. Vanderploeg EJ, Imler SM, Brodkin KR, García AJ, Levenston ME. Oscillatory tension differentially modulates matrix metabolism and cytoskeletal organization in chondrocytes and fibrochondrocytes. J Biomech. 2004;37(12):1941–52.
54. Wong M, Siegrist M, Goodwin K. Cyclic tensile strain and cyclic hydrostatic pressure differentially regulate expression of hypertrophic markers in primary chondrocytes. Bone. 2003;33(4):685–93.
55. Mawatari T, Lindsey DP, Harris AHS, Goodman SB, Maloney WJ, Smith RL. Effects of tensile strain and fluid flow on osteoarthritic human chondrocyte metabolism in vitro. J Orthop Res. 2010;28(7):907–13.

56. Vanderploeg EJ, Wilson CG, Levenston ME. Articular chondrocytes derived from distinct tissue zones differentially respond to in vitro oscillatory tensile loading. Osteoarthr Cartil. 2008;16(10):1228–36.
57. Bleuel J, Zaucke F, Brüggemann GP, Niehoff A. Effects of cyclic tensile strain on chondrocyte metabolism: a systematic review. PLoS One. 2015;10(3):1–25.
58. Thomas RS, Clarke AR, Duance VC, Blain EJ. Effects of Wnt3A and mechanical load on cartilage chondrocyte homeostasis. Arthritis Res Ther. 2011;13(R203):10.
59. Chen C, et al. Cyclic equibiaxial tensile strain alters gene expression of chondrocytes via histone deacetylase 4 shuttling. PLoS One. 2016;11(5):e0154951.
60. Xu Z, Buckley MJ, Evans CH, Agarwal S. Cyclic tensile strain acts as an antagonist of IL-1 beta actions in chondrocytes. J Immunol. 2000;165(1):453–60.
61. Raimondi MT, et al. Engineered cartilage constructs subject to very low regimens of interstitial perfusion. Biorheology. 2008;45(3–4):471–8.
62. Bueno EM, Bilgen B, Barabino GA. Hydrodynamic parameters modulate biochemical, histological, and mechanical properties of engineered cartilage. Tissue Eng Part A. 2009;15(4):773–85.
63. Gemmiti CV, Guldberg RE. Shear stress magnitude and duration modulates matrix composition and tensile mechanical properties in engineered cartilaginous tissue. Biotechnol Bioeng. 2009;104(4):809–20.
64. Fitzgerald JB, Jin M, Grodzinsky AJ. Shear and compression differentially regulate clusters of functionally related temporal transcription patterns in cartilage tissue. J Biol Chem. 2006;281(34):24095–103.
65. Jin M, Frank EH, Quinn TM, Hunziker EB, Grodzinsky AJ. Tissue shear deformation stimulates proteoglycan and protein biosynthesis in bovine cartilage explants. Arch Biochem Biophys. 2001;395(1):41–8.
66. Frank EH, Jin M, Loening AM, Levenston ME, Grodzinsky AJ. A versatile shear and compression apparatus for mechanical stimulation of tissue culture explants. J Biomech. 2000;33(11):1523–7.
67. Nugent GE, Aneloski NM, Schmidt TA, Schumacher BL, Voegtline MS, Sah RL. Dynamic shear stimulation of bovine cartilage biosynthesis of proteoglycan 4. Arthritis Rheum. 2006;54(6):1888–96.
68. Waldman SD, Spiteri CG, Grynpas MD, Pilliar RM, Kandel RA. Long-term intermittent shear deformation improves the quality of cartilaginous tissue formed in vitro. J Orthop Res. 2003;21(4):590–6.
69. Fukuda K, et al. Relationship between dynamic stress field and ECM production in regenerated cartilage tissue. In: 2016 International Symposium on Micro-NanoMechatronics and Human Science (MHS), 2016, pp. 3–5.
70. Grad S, Gogolewski S, Alini M, a Wimmer M. Effects of simple and complex motion patterns on gene expression of chondrocytes seeded in 3D scaffolds. Tissue Eng. 2006;12(11):3171–9.
71. Kaupp JA, Tse MY, Pang SC, Kenworthy G, Hetzler M, Waldman SD. The effect of moving point of contact stimulation on chondrocyte gene expression and localization in tissue engineered constructs. Ann Biomed Eng. 2013;41(6):1106–19.
72. Guha Thakurta S, Kraft M, Viljoen HJ, Subramanian A. Enhanced depth-independent chondrocyte proliferation and phenotype maintenance in an ultrasound bioreactor and an assessment of ultrasound dampening in the scaffold. Acta Biomater. 2014;10(11):4798–810.
73. Parvizi J, Wu CC, Lewallen DG, Greenleaf JF, Bolander ME. Low-intensity ultrasound stimulates proteoglycan synthesis in rat chondrocytes by increasing aggrecan gene expression. J Orthop Res. 1999;17(4):488–94.
74. Noriega S, Mamedov T, Turner JA, Subramanian A. Intermittent applications of continuous ultrasound on the viability, proliferation, morphology, and matrix production of chondrocytes in 3D matrices. Tissue Eng. 2007;13(3):611–8.

75. Noriega S, Hasanova G, Subramanian A. The effect of ultrasound stimulation on the cytoskeletal organization of chondrocytes seeded in three-dimensional matrices. Cells Tissues Organs. 2012;197(1):14–26.
76. Gauthier BJ. The effects of mechanical vibration on human chondrocytes in vitro. MS thesis, Marquette University, 2016.
77. a Kaupp J, Waldman SD. Mechanical vibrations increase the proliferation of articular chondrocytes in high-density culture. Proc Inst Mech Eng H. 2008;222(5):695–703.
78. Fanning PJ, Emkey G, Smith RJ, Grodzinsky AJ, Szasz N, Trippel SB. Mechanical regulation of mitogen-activated protein kinase signaling in articular cartilage. J Biol Chem. 2003;278(51):50940–8.
79. Duda GN, et al. Mechanical quality of tissue engineered cartilage: results after 6 and 12 weeksin vivo. J Biomed Mater Res. 2000;53(6):673–7.
80. Waldman SD, Grynpas MD, Pilliar RM, Kandel RA. The use of specific chondrocyte populations to modulate the properties of tissue-engineered cartilage. J Orthop Res. 2003;21(1):132–8.
81. De Croos JNA, Dhaliwal SS, Grynpas MD, Pilliar RM, Kandel RA. Cyclic compressive mechanical stimulation induces sequential catabolic and anabolic gene changes in chondrocytes resulting in increased extracellular matrix accumulation. Matrix Biol. 2006;25(6):323–31.
82. Waldman SD, Couto DC, Grynpas MD, Pilliar RM, Kandel RA. Multi-axial mechanical stimulation of tissue engineered cartilage : review. Eur Cell Mater. 2007;13(613):66–74.
83. Grad S, Lee CR, Wimmer MA, Alini M. Chondrocyte gene expression under applied surface motion. Biorheology. 2006;43:259–69.
84. Stoddart MJ. Enhanced matrix synthesis in de novo, scaffold free cartilage-like tissue subjected to compression and shear. J Anat. 2006;189:503–5.
85. Weber JF, Waldman SD. Stochastic resonance is a method to improve the biosynthetic response of chondrocytes to mechanical stimulation. J Orthop Res. 2016;34(2):231–9.
86. Cescon M, Gattazzo F, Chen P, Bonaldo P. Collagen VI at a glance. J Cell Sci. 2015;128(19):3525.
87. Keene DR, Engvall E, Glanville RW. Ultrastructure of type VI collagen in human skin and cartilage suggests an anchoring function for this filamentous network. J Cell Biol. 1988;107(5):1995–2006.
88. Söder S, Hambach L, Lissner R, Kirchner T, Aigner T. Ultrastructural localization of type VI collagen in normal adult and osteoarthritic human articular cartilage. Osteoarthr Cartil. 2002;10(6):464–70.
89. Lotz MK, Otsuki S, Grogan SP, Sah R, Terkeltaub R, D'Lima D. Cartilage cell clusters. Arthritis Rheum. 2010;62(8):2206–18.

Mechanically Assisted Electrochemical Degradation of Alumina-TiC Composites

Hetal U. Maharaja and Guigen Zhang

Keywords Alumina-TiC · Ceramic composite · Abrasion · Low load · Oxidation · Electrochemical · Brushing · Mechanically assisted electrochemical degradation · Oxidative wear · Microploughing

1 Introduction

Metal alloys such as Ti-6Al-4V and CoCrMo, known for their biocompatibility and mechanical strength, are common biomaterials for total hip replacements (THR) [1]. The surface of these metallic biomaterials often forms a passive oxide film providing resistance against corrosion or electrochemical degradation. Electrochemical degradation is a charge-transfer process in which metallic materials react and/or interact with the aqueous environment. Such a process always includes oxidation of metallic components and the concomitant reduction of active species in the aqueous environment somewhere else on the implant surface. Whether an electrochemical reaction will occur or not under certain given conditions is governed by its thermodynamic favorability (Gibbs free energy of reaction, ΔG) which is a function of the type and state of reactants, surface property, reaction kinetics, chemical species and their concentration, and temperature, among others [2].

The passive oxide film on metal alloys is typically a few nanometers thick and provides protection against the dissolution of metals into metal ions and electrons

H. U. Maharaja
Department of Bioengineering, Clemson University, Clemson, SC, USA

Institute for Biological Interfaces of Engineering, Clemson University, Clemson, SC, USA

G. Zhang (✉)
Department of Bioengineering, Clemson University, Clemson, SC, USA

Department of Electrical and Computer Engineering, Clemson University, Clemson, SC, USA

Institute for Biological Interfaces of Engineering, Clemson University, Clemson, SC, USA

Department of Biomedical Engineering, University of Kentucky, Lexington, KY, USA
e-mail: guigen.bme@uky.edu

149

when exposed to aqueous environments. High mechanical loads and relative motion between two contacting surfaces of articulating load-bearing joints could cause disruption of this oxide film that exposes the underlying metal alloy resulting in oxidation and the release of metal ions into the surrounding synovial fluid [3–5]. The released metal ions can possibly illicit inflammatory responses, leading to osteolysis and eventual aseptic implant loosening from failure of the osteointegration process [6, 7].

Mechanically assisted electrochemical degradation is an electrochemical process triggered by a mechanical condition such as abrasion, micromotion or fretting between two articulating surfaces, leading to damage of the protective oxide film. The undesirable electrochemical activity of the biomaterial may be compounded by the release of wear debris that can further damage the protective oxide film and compromise the integrity of the implant surface. In tribology terms, it is often regarded as a tribo-electrochemical wear process.

Ceramic biomaterials are known to provide superior mechanical strength, wear resistance, hardness, chemical inertness than their metallic counterparts [8, 9]. For example, alumina-based ceramic composites demonstrate reduced friction as articulating joints and improved wear resistance compared with metal-metal or metal-polymer combinations [10]. With a Vicker's hardness of more than 2000, alumina (Al_2O_3) can be polished to a smooth surface. It possesses high wettability providing better adhesion to lubricating fluid, enabling it to have a much lower wear rate as articulating components, some 0.025 µm/yr to 4 µm/yr in comparison with 100 µm/yr for metal-polyethylene articulating pairs [1]. Alumina is also one of the most thermodynamically stable oxides of aluminum and is less susceptible to degradation by usual oxidation making it highly biocompatible [11].

There have been many advances in alumina manufacturing processes since the 1970s. Today alumina used for total hip arthroplasties has fewer impurities, smaller grain sizes, higher density and improved fracture toughness (by the addition of zirconia and other oxides) [10, 12]. However, due to their inherent brittleness, ceramics are not able to sustain high impact or non-uniform loads [1] and show little to no plastic deformation under extreme mechanical situations, and they are prone to micro-fracture under abrasive conditions. The resulting fragments or debris, even if not harmful to the host tissues, could act as third body particles to accelerate wear damage. Some studies have shown prominent inflammatory responses to ceramic wear debris [13, 14] requiring revision surgeries due to aseptic loosening for ceramic-on-ceramic hip prosthesis [15].

The intrinsic brittleness of alumina can be reduced by the addition of hard reinforcements like metal carbides, nitrides and oxides. Monolithic alumina when enhanced by metal refractory ceramic reinforcements like titanium carbide (TiC), titanium carbonitride (Ti(C,N)), tungsten-titanium composite carbide ((W,Ti)C) yield a composite with increased flexural strength, fracture toughness, hardness and an improved friction coefficient [16–18]. The alumina-TiO_2 nanocomposite exhibited lower wear volume and better mechanical properties with a 10 mol% TiO_2 addition [19]. Likewise, mechanical properties and wear behavior of alumina-TiN as a potential biomaterial has been explored too [8]. Titanium carbide (TiC) added to alumina increases hardness, toughness and more importantly wear resistance of just plain alumina [18, 20]. Due to superior hardness, TiC coatings on titanium substrates enhance resistance to tribochemical wear and offer corrosion resistance to an underlying metal substrate. It also improves osseointegration by stimulating the growth of osteoblasts and their proliferation, offering a biocompatible interface to metallic or polymer substrates [21].

Besides improvement in mechanical and wear properties of alumina by the transition metal carbides and nitrides, it is important to consider the electrochemical behavior of a reinforcing material when evaluating chemical and tribological stability of an alumina composite. Even alumina, although an insulator [22], does not participate in the electrochemical processes, studies have shown that its wear resistance decreases in an aqueous environment due to its hydrophilic nature by reacting with water to form aluminum hydroxide in basic and acidic environments at elevated temperatures [23, 24]. Moreover, these transition metal refractory ceramics mentioned earlier demonstrate metal-like conductivity, enabling charge transfer during electrochemical reactions [25–28], hence electrical and chemical implications of their additions to alumina need to be carefully assessed.

Besides its extreme hardness, TiC is a conductive ceramic with resistivity of 0.003–0.008 Ω-m (vs Cu -1.72×10^{-8} Ω-m) and can partake in electrochemical processes in a chemical environment [22]. TiC, when used as reinforcement for alumina to improve wear resistance, retains its conductive nature [20] and this property is in fact favorable in magnetic recording disk drive application where the metal-like conductivity of TiC is beneficial to dissipate charge build up due to frictional contact [29]. Electrochemical nature of TiC was highlighted in studies [30, 31] that have shown anodic dissolution of TiC in aggressive chemical conditions and [32] where TiC nanowires exhibited enhanced electrocatalytic properties allowing facile electron transfer and redox activity. It is known from a study [29] that alumina-TiC composites have high oxidation resistance in air and nitrogen environments for temperatures up to 350 °C. However, alumina-TiC composites also undergo oxidative wear in dry conditions. Bare alumina-TiC has been observed to release CO_2 as a byproduct of tribochemical wear of TiC at 120 °C under dry sliding wear [33] and that oxygen chemisorption and carbon oxidation is catalyzed by alumina-TiC. Such a tribochemical wear mechanism is likely to be enhanced under an aqueous environment that allows continuous electrochemical interaction.

Considering improved mechanical strength, wear resistance of alumina-TiC and biocompatibility of TiC itself, alumina-TiC composites may be regarded as an appealing biomaterial for load bearing implants. However, the tendency of TiC to facilitate charge transfer in electrochemical processes and reactivity of alumina in a wet environment pose a need to evaluate the electrochemical behavior of alumina-TiC composites. Hence the electrochemical activity of alumina-TiC in an aqueous chemical environment, especially when compounded by destructive mechanical processes like high impact loads, abrasion, cyclic fatigue, frictional wear, needs to be evaluated. Such an evaluation would enable a better material design of the ceramic composite for biomedical applications.

Table 1 Experimental methods used in the study

	Tests
Quantitative/semi-quantitative	(i) Electrochemical methods: (a) Open circuit current (ZRA) (b) Electrochemical impedance spectroscopy
	(ii) Chemical analysis: (a) Inductively coupled plasma mass spectrometry (ICP-MS) (b) Xray photoelectron spectroscopy (XPS)
Qualitative	Scanning Electron microscopy (SEM)

In this study, we focus our efforts on investigating the effect of TiC reinforcement on the electrochemical degradation of alumina-TiC composites. We aim to elucidate the degradation mechanism involved by studying the electrochemical response of alumina-TiC composites when it is abraded in an aqueous environment and assess corresponding alterations in surface chemistry and appearance with quantitative and qualitative techniques, respectively. The experimental methods employed to understand the interplay of mechanical and electrochemical processes are as follows (Table 1):

2   Methods and Materials

To examine the effect of gentle abrasion on the electrochemical behavior of alumina-TiC composites, we built an experimental apparatus allowing us to abrade the composite with a brush in an aqueous environment while simultaneously measuring electrochemical response from the composite. Doing so will enable us to study the spontaneous oxidative processes induced by brushing abrasion and establish interdependence among different parameters of abrasion, including temperature, brushing acceleration and speed, electrochemical potential and current. Further characterization of the degradation of alumina composites by brushing abrasion is achieved by comparing microstructural damage due to abrasion in a dry environment.

2.1   Brushing Abrasion Setup

The setup built for the brushing tests is shown in Fig. 1. Its center piece is a motorized overhead stirrer (Eurostar power control-visc, IKA Works, Wilmington, NC) for rotational-motion of brush about the vertical axis. Brushing abrasion of an alumina-TiC composite sample is accomplished with a nylon brush, attached to the end of the motor shaft. The acceleration and speed of the stirrer motor is controlled using the labworldsoft 5 program (IKA Works, Wilmington, NC). This setup also simultaneously measures the electrochemical response of alumina-TiC composites

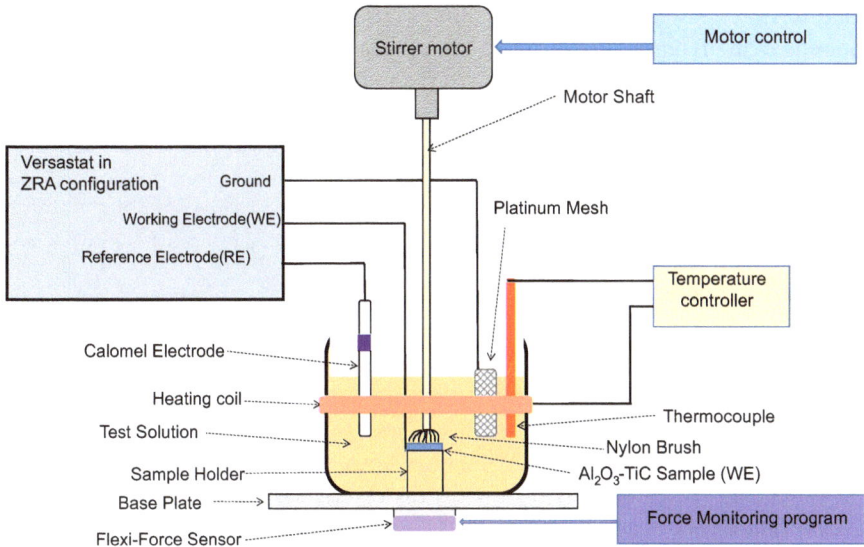

Fig. 1 Experiment apparatus set up for brushing abrasion

to brushing abrasion with a potentiostat (VersaStat MC, Princeton Applied Research, Oak Ridge, TN). A heating unit in the set up comprises of a J-type thermocouple, heating tape (BIH051020L, BriskHeat, Columbus, OH) and a digital temperature controller ITC-106 (Inkbird, Shenzhen, China) is used to control the temperature of the test solution (electrolyte) throughout the duration of a test. The contact load between the brush and the composite sample is monitored and controlled using a force sensor (flexiforce sensor: A201, Tekscan, South Boston, MA). In a typical brushing test, a nylon brush is brought into contact with the surface of the alumina-TiC composite sample and a contact force of 45 gm-f (0.44 N) is achieved by the force monitoring unit and by adjusting the base plate on linear translation stage. This force may fluctuate about this constant set value during brushing due to scattered contact of rotating bristles of the brush. A ramp scheme is designed in the labworldsoft to control the acceleration of the motor from rest to a preset maximum brushing speed, governing the rate and degree of abrasion.

2.2   Sample Preparation

For every test, commercially available alumina-TiC (70%/30%) samples, 10 × 10 mm in size, were first cleaned ultrasonically for 10 min in ethanol followed by rinsing in deionized (DI) water. An electrical connection with the composite sample was established using a copper tape and the test sample was used as the working electrode (WE). All sides of the sample except the top surface and copper tape to be exposed to the test solution were coated with lacquer to minimize unwanted interferences from copper. The prepared sample was kept at a fixed position on a sample holder with its top surface facing up. A nylon brush in the form of a bundle of bristles with a polyethylene base was used for brushing about the vertical axis. The nylon brush was cleaned by sonicating in ethanol for 10 min followed by rinsing in DI water.

2.3   Electrochemical Measurements

Electrochemical measurements were made using a VersaStat MC. A Saturated Calomel Electrode (SCE) was used as the reference electrode (RE), a platinum mesh as the counter electrode for potentiostatic (applied potential) and potentiodynamic tests and as a ground lead (GND) for open circuit condition (free potential) measurements. Micro90 (pH~9.5), a corrosive organic solution diluted to 0.3% (vol/vol) in DI water and heated to 75 °C, was used as an electrolyte unless stated otherwise. To monitor the brushing induced electrochemical current response without applying any electrical potential (free potential), a potentiostat set in a Zero Resistance Ammeter (ZRA) configuration was used in a three-electrode setting.

2.4   Brushing Abrasion Testing

Before starting the brushing abrasion, the open circuit potential (OCP) of the alumina-TiC sample was allowed to stabilize for 10 min while recording the electrochemical current. Brushing acceleration and speeds were controlled by adjusting the ramping times for the stirrer motor to reach maximum abrasion speed from rest. Following the brushing test, the composite sample was rinsed with DI water and ethanol and stored for surface analysis.

By using the same test parameters and setup as described above for every run of an experiment, a systematic study of brushing abrasion was performed to characterize the mechanically assisted degradation of alumina-TiC composites and identify factors that affect the degradation process.

2.4.1   Effect of Brushing Acceleration and Speed

Since acceleration i.e. the ramping time to reach maximum brushing speed controls the amount of rotational force imparted to the surface features under abrasion, the effect of ramping time was studied under three different ramping schemes: 10, 40 and 70 s to ramp the rotational speed of brush from rest to maximum speed of 800 revolutions per minute (rpm). Total test duration was about 17 min with 10 min for OCP stabilization, 5 min of brushing, and some remaining time for motion actuation and slowdown. Two maximum brushing speeds were used: 500 and 1200 rpm. After the brushing tests, the resulting surfaces were imaged and analyzed for morphological changes. This set of experiment was designed to correlate electrochemical responses with the abrasion ramping time and speed.

2.4.2   Effect of Temperature

Temperature is an important factor that governs the thermodynamic favorability of electrochemical processes, rate of reaction and conductivity of the test solution. To understand the effect of temperature, the prolonged exposure of alumina-TiC composites to an aqueous alkaline environment was carried out in a heated (75 °C) and room (25 °C) temperature. The total duration of these tests was 2 h with brushing abrasion for 6 min (in three consecutive brushing cycles of 2 min each). The brushing abrasion parameters were kept the same: a contact force of 45 gm-f, ramping time of 10 s, and maximum brushing speed of 800 rpm.

2.4.3   Effect of Environment

To ascertain if the material degradation mechanism is an abrasion assisted electrochemical process and not just a tribological process, brushing abrasion in dry conditions (no electrolyte) was also performed. The resulting microstructural damage of the dry-test samples was examined and compared with the wet-test samples. Aside from the dry and wet difference, other experimental settings were kept unchanged at a contact force of 45 gm-f, ramping time at 10 s, and maximum brushing speed of 800 rpm. In this way, the surface damage incurred would be mainly due to brushing abrasion because the electrochemical interactions of TiC with the aqueous environment were eliminated in dry abrasion.

2.5   Electrochemical Impedance Study

Electrochemical impedance spectroscopy (EIS) tests were undertaken using the same setup described earlier to characterize oxide formation and change of sample-electrolyte interface properties. Impedance scans were taken before and after potentiostatic conditions (anodic and cathodic) and abrasion tests in Micro90 at 75 °C. Potentiostatic tests were performed to verify the propensity and stability of the oxide formation particularly on the TiC domain as alumina does not participate in the charge transfer processes. The value for the applied potential in potentiostatic tests was chosen from active cathodic and anodic regions of potentiodynamic tests which coincided with average OCP values (-250 mV vs. SCE) observed in free potential mode during brushing abrasion. The experimental design for EIS analysis is shown in Table 2 below.

Table 2 Experimental conditions for EIS study

Experiment	Duration	Temperature	Brushing abrasion	Applied potential vs. SCE
Anodic biasing	43 min	75 °C	No	250 mV
Cathodic biasing	43 min	75 °C	No	−250 mV
No brushing-no biasing	43 min	75 °C	No	No
Brushing-no biasing	10 min OCP stabilization +30 min brushing +3 min to restore	75 °C	Yes	No

Three runs of impedance test were performed before and after the application of each experimental condition listed in Table 2 to study the altered sample-electrolyte interface. In every impedance scan, alternating current (AC) perturbation with magnitude of 50 mV was applied at open circuit condition in the frequency range of 1 Hz–50 kHz. For brushing abrasion experiments, 10s of ramping time with a maximum speed of 800 rpm was applied after 10 min of OCP stabilization. Charge transfer resistance (R_{ct}), the parameter of interest in this experimental design, was obtained with ZView by fitting an equivalent circuit model to the Nyquist plot acquired. Any alterations in charge transfer resistance before and after a test condition would give us clues about possible changes in oxide film on the TiC domain.

2.6 Surface Characterization

Abrasive alterations in surface morphology were examined through comparison of images obtained from Scanning Electron Microscopy (SEM) (SU6600 and S4800, Hitachi High Technologies, Tokyo, Japan) for abraded samples and pristine samples. Changes caused by different parameters of brushing abrasion tests were analyzed. Note that with less energetic secondary electrons reflected from sample surfaces, images taken at low kV (0.7 kV) will reveal more superficial information than 5 kV, and that at the lower kV, the contrast will be reversed and alumina matrix domain will appear as darker regions under 0.7 kV rather than lighter under 5 kV.

2.7 Chemical Analysis

As the mode of abrasion employed in this study is of gentle nature, to obtain measurable alterations in surface chemistry, a much longer duration of brushing abrasion was employed. To accelerate the surface chemistry changes by brushing abrasion, continuous brushing in heated Micro90 at 1000 rpm (ramping time of 10 s) for 2.5 h was performed following 10 min of OCP stabilization. Abraded regions were marked under optical microscope for chemical analysis by X-ray Photoelectron Spectroscopy (XPS). Elemental scans for titanium were performed to observe changes in oxidation states after brushing abrasion. Atomic percentages of each element (Ti, C, Al and O) from XPS scans were obtained to gain a preliminary understanding of changes in surface chemistry due to the electrochemical process activated by brushing abrasion of alumina-TiC.

The test solution after brushing abrasion was also analyzed for any traces of titanium oxide or alumina particles released as debris during the prolonged brushing. ICP-MS (Inductively Coupled Plasma Mass Spectrometry) was employed to determine elemental titanium and aluminum concentration of test solution. Untested Micro90 solution was also analyzed as a control.

3   Results and Discussion

3.1   Electrochemical Response to Brushing Abrasion

Under OCP conditions, the composite shows a baseline current of nearly 1 μA in the heated Micro90 environment, suggesting a dynamic electrochemical process on the sample surface. These measurable electrochemical interactions during OCP stabilization period signify that the electro-active TiC domain of the composite interacts with the test solution. After OCP stabilization, brushing abrasion results in typical 'passive layer breakdown' behavior as evident in both the OCP and current responses which are commonly observed with metal and metal alloys [3, 34, 35] as well as metal-ceramic composites [36]. In response to abrasion after OCP stabilization, electrochemical current shows a sharp increase with a concurrent negative drop in OCP as shown in Fig. 2, indicating activation of oxidative reactions on TiC. The decay of both current and OCP suggests a re-passivation process occurring to remedy the disruption of the oxide barrier on the sample surface. This re-passivation behavior may not completely form a compact non-porous film due to continuing abrasion when the brushing motion is going on.

As the brushing stops, the current decays to its original rate of stabilization. Such behavior shows that the rate of electrochemical reaction on the TiC domain is increased due to brushing abrasion. Parameters like ΔV and ΔI depicted in Fig. 2 are the differential values of OCP and electrochemical current respectively from baseline, that provide a quantitative measure of the electrochemical activity of alumina-TiC initiated by abrasion.

Magnitudes of the electrochemical current and potential vary with brushing abrasion parameters like ramping speed and maximum brushing speed. As seen in Fig. 3a, current response to brushing with a ramping time of 10s is the highest with the greatest average ΔV (0.022 V) and ΔI (17.3 μA) as plotted in Fig. 3b. Clearly, a shorter ramping time generates a larger rotational acceleration hence exerting greater abrasive forces on a sample surface than a longer ramping time, leading to

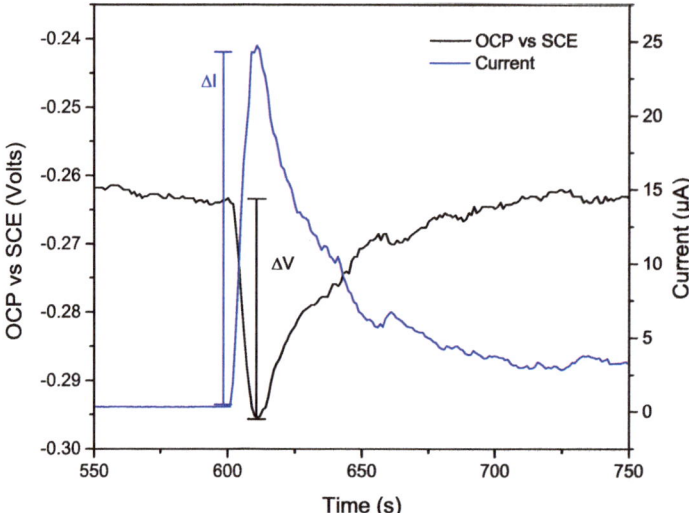

Fig. 2 Typical electrochemical response of alumina-TiC to brushing abrasion in Micro 90 at 75 °C

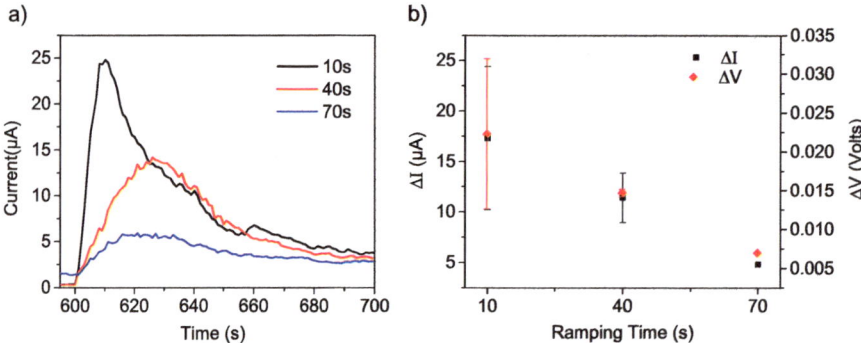

Fig. 3 Effect of ramping time on (**a**) Current response to brushing abrasion at 10 s, 40 s, 70 s and (**b**) Average ΔV and ΔI values for n = 3 in Micro90 at 75 °C

more morphological damage. Similarly, a larger maximum speed generates a higher current and causes more surface damage. For example, as seen in Fig. 4a, b, the case of 1200 rpm maximum speed results in higher delta values for OCP and electrochemical current ($\Delta V = 0.035$ V,$\Delta I = 26.19$ μA) than the case of 500 rpm ($\Delta V = 0.017$ V, $\Delta I = 17.92$ μA). These facts suggest that the abrasion of alumina-TiC activates an oxidative electrochemical process, resulting in increased chemical interactions with an aqueous environment and the corresponding electrochemical response depends on abrasion parameters.

At elevated temperature of 75 °C, the baseline current and the peak value in the electrochemical response curve are much higher than at room temperature, as shown

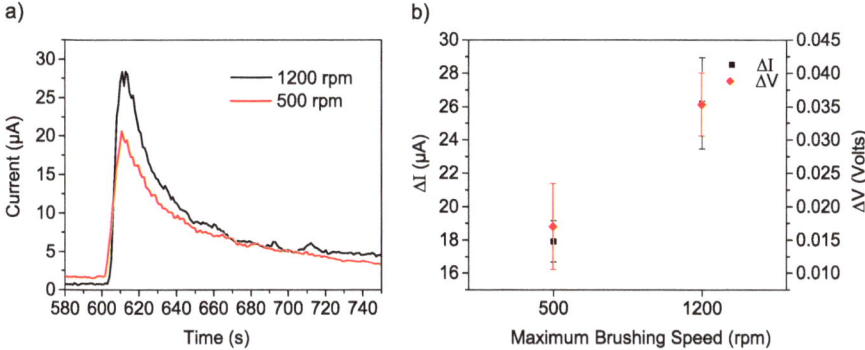

Fig. 4 Effect of maximum brushing speed on (**a**) Current response to abrasion at 1200 rpm and 500 rpm and (**b**) Average ΔV and ΔI values for n = 3 in Micro90 at 75 °C with a ramping time of 10 s

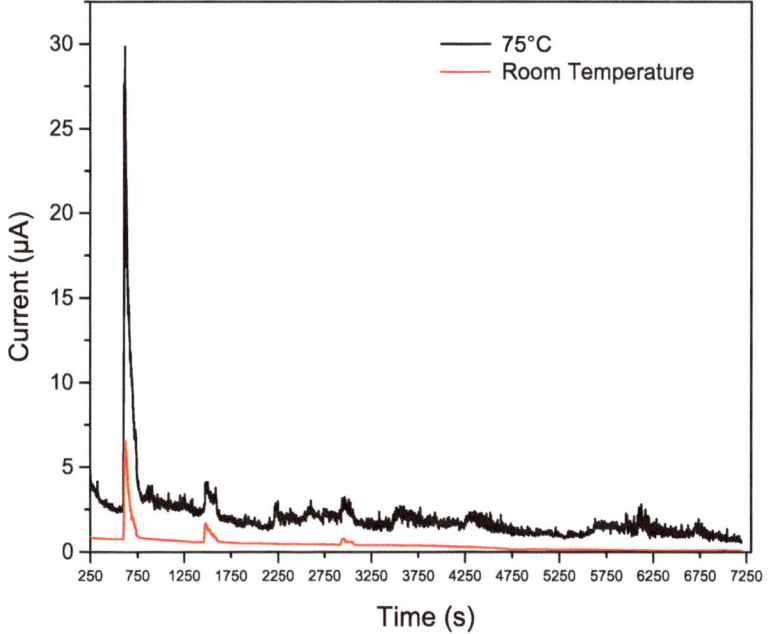

Fig. 5 Effect of temperature on current response to brushing abrasion (3 cycles) in Micro 90

in Fig. 5. This indicates that the degradation mechanism triggered by abrasion is an electrochemical process involving an oxidation reaction whose thermodynamic favorability is enhanced at higher temperatures. Moreover, at a higher temperature, the increased conductivity of a solution could also play a role by making more charged species available to enable faster reaction kinetics on the TiC domain. Among the three brushing cycles, the current response to the first cycle of brushing is the highest and successive current responses reduce in magnitude. This reduction in current could be attributed to several reasons. It could be due to the hysteresis loosening of the bristles after each brushing motion, or it may be due to a loss of material by abrasion in each brushing cycle, leading to less brushing contact in subsequent brushing cycles.

3.2   Surface Characterization

SEM images of the sample surfaces after the abrasion test given in Fig. 6a–f show that the abraded samples have been brushed off [37] exhibiting circular 'ploughing' [38–40] marks, likely caused by material removal along a curvilinear track. Under a low magnification (at 100X or lower) these marks appear as concentric rings, consistent with the rotational brushing trajectory (Fig. 6a–c). At a higher magnification, the ploughing marks appear as dark and bright bands. Under closer inspections, these dark bands are formed due to a greater amount of surface wear than brighter bands.

The appearance of these circular bands is affected by the way brush bristles spread on the sample based on the initial contact force between the sample and brush at the beginning of the test. Darker regions show a greater degree of morphological damage and material removal, mostly on the alumina domains than the brighter band region, in which the alumina exhibited much lesser damage. At a 18,000X magnification (Fig. 6g), the alumina domain is white and TiC black. In comparison with a pristine sample (Fig. 6h), we clearly see the grain boundary wear and material removal on the alumina domain on the brushed sample (Fig. 6g).

Surface damage as observed in Fig. 6 suggests the susceptibility of the alumina-TiC composite to abrasive wear incurred along with electrochemical activation (Fig. 2) under gentle abrasion condition. Keep in mind that alumina domains appear white in higher kV SEM images from SU6600 and the TiC domain is black, but the contrast reverses in lower kV SEM images from S4800.

As seen in Fig. 7, the case of a 10 s ramping time which corresponds to the highest current response induces the most severe surface damage than the two other slower cases. High brushing speed causes a similar outcome: more severe surface damage under 1200 rpm than under 500 rpm as shown in (Fig. 8). The TiC grain boundaries show more wear giving a smeared boundary appearance [37]. As we see in these images, the overall damage is of the same 'microploughing' type. The variation of brushing abrasion parameters is manifested in the severity of damage induced in grain boundary region, with the most severe damage seen for the 10 s case followed by 40 s and 70 s cases. Similarly, a lower maximum speed (500 rpm) causes less damage than a higher maximum speed (1200 rpm). These revealed relationships between electrochemical current response and brushing acceleration and maximum brushing speed and the induced surface damage confirm that the degradation mechanism is an abrasion assisted electrochemical process and the degree of

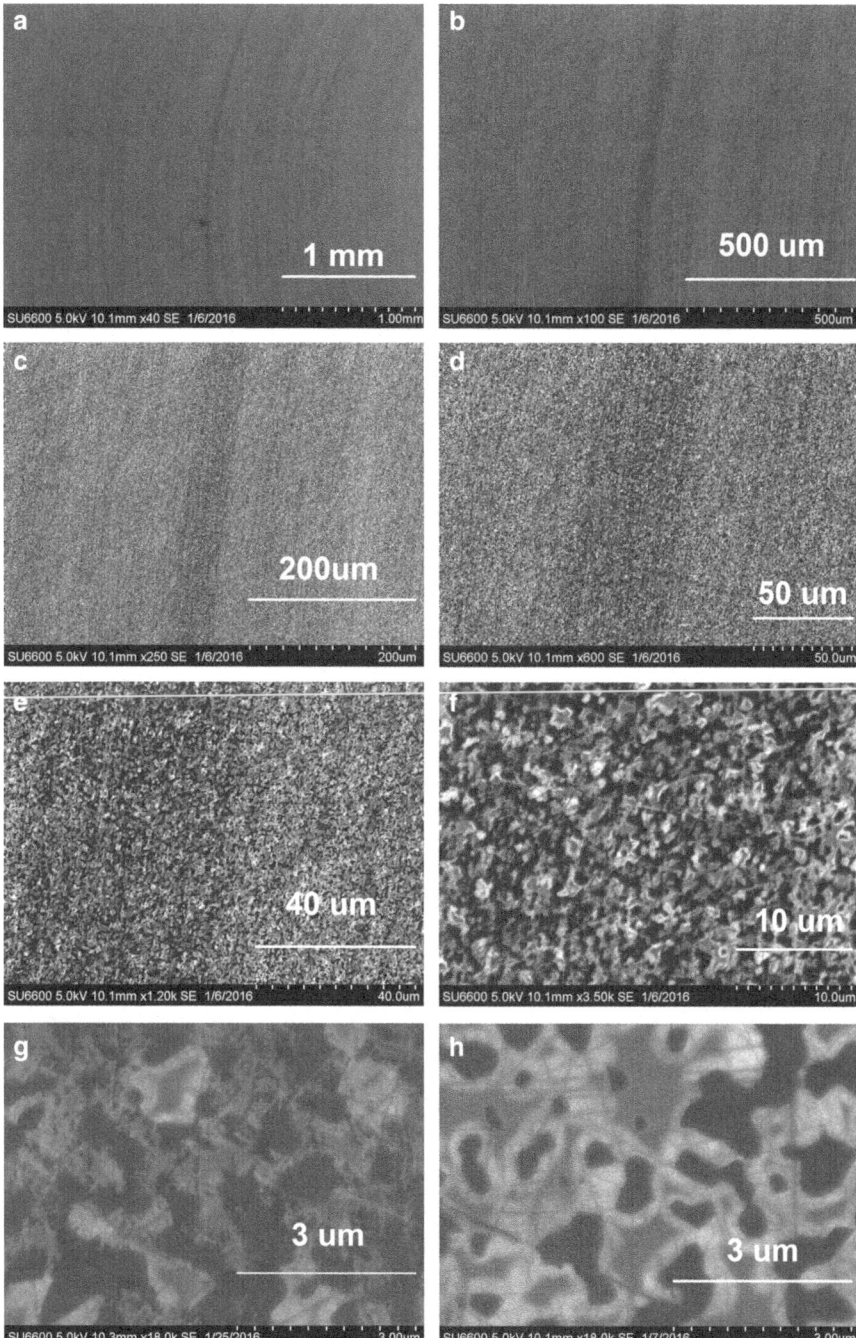

Fig. 6 SEM images (SU6600) of alumina-TiC sample after brushing abrasion in a heated environment at magnifications of (**a**) 40×, (**b**) 100×, (**c**) 250×, (**d**) 600×, (**e**) 1200×, and (**f**) 3500×; 18,000× images of the microstructure of (**g**) Brushed and (**h**) Pristine samples

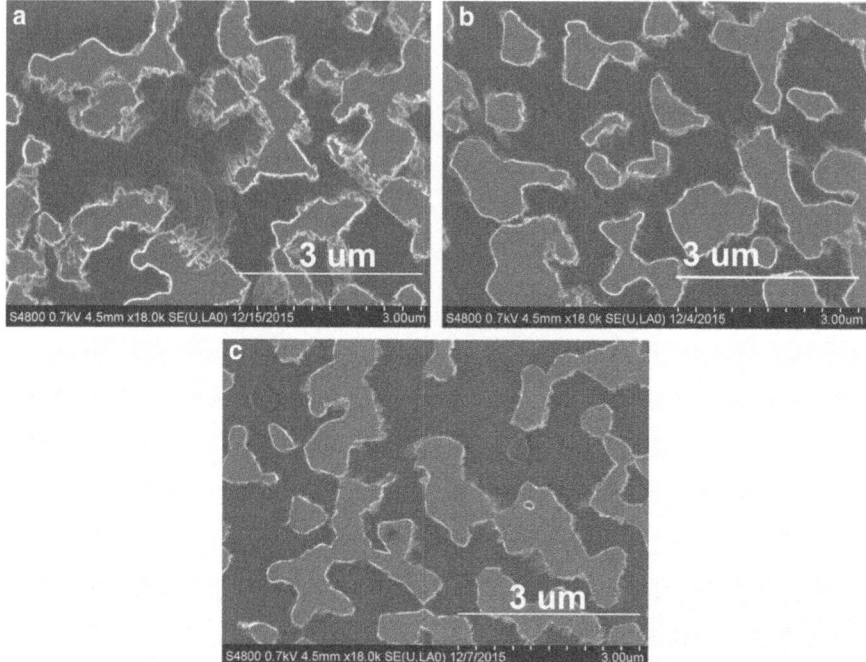

Fig. 7 Low kV SEM images (S4800) of alumina-TiC showing differences in morphological damage at ramping speeds of (**a**) 10 s, (**b**) 40 s and (**c**) 70 s to maximum brushing speed of 800 rpm

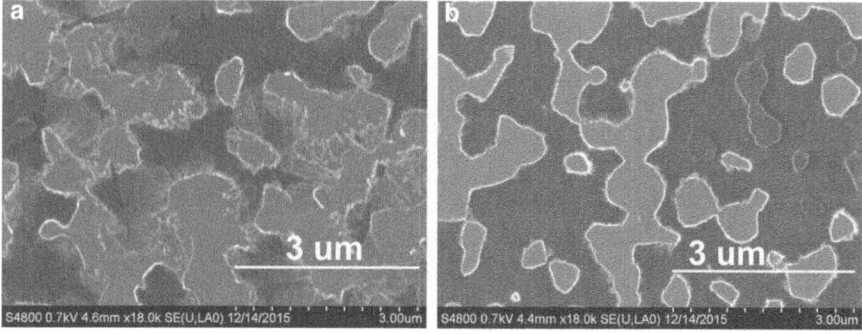

Fig. 8 Low kV SEM images (S4800) of alumina-TiC showing differences in morphological damage at different maximum brushing speeds of (**a**) 1200 rpm and (**b**) 500 rpm with ramping time of 10 s

electrochemical interaction of TiC with the environment depends on the magnitude of parameters controlling the mechanical process of abrasion.

As apparent in Fig. 9, grain boundaries are more intact at room temperature and alumina domains show much lesser wear in the room-temperature condition than in

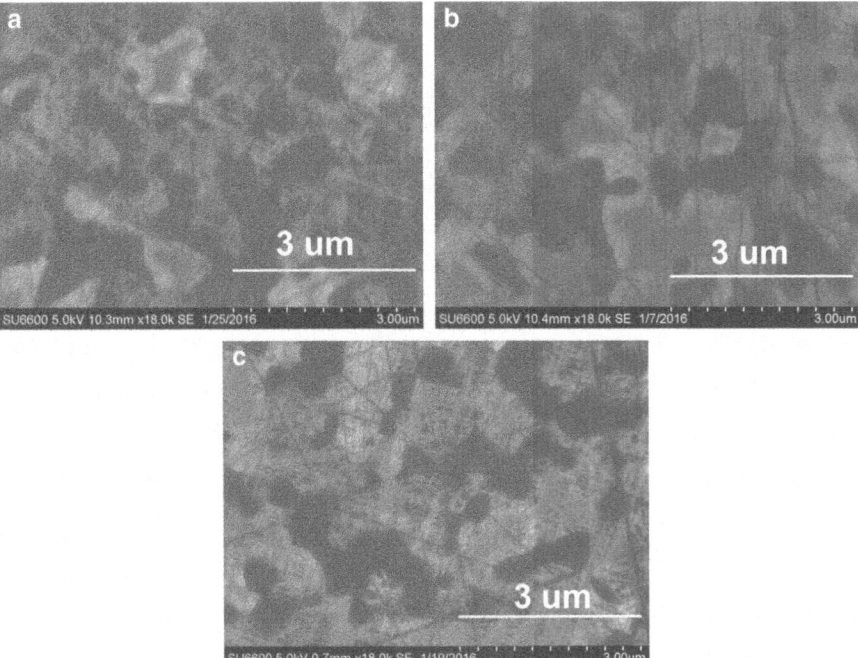

Fig. 9 Comparison of SEM images (SU6600) of alumina-TiC composite brushed in (**a**) Wet (Micro90) and heated, (**b**) Wet (Micro90) and room temperature (Rt) and (**c**) Dry Rt environment

a heated condition. This morphological damage corresponds well to the current responses obtained (Fig. 5) at these at two different temperatures.

While comparing the surface of dry brushed samples with wet brushed samples, the material removal of the alumina domain was much less in the dry brushed samples with no damage in the grain boundary region. Damage incurred on alumina domain in a wet environment, especially near grain boundary region could be due to a greater chemical reactivity of alumina to aqueous environment possibly driven by a reaction with water to form hydroxide. At a higher temperature, not only is the rate of electrochemical interactions of TiC with an aqueous environment higher, the susceptibility of alumina domain to abrasive wear is also increased.

3.3 Chemical Analysis

XPS analysis of the brushed samples revealed alterations in surface chemistry caused by the oxidative electrochemical process on the TiC domain. Brushed and pristine samples showed the similar elemental composition but their atomic percentages were different (Table 3).

Table 3 Atomic percentages of brushed and pristine samples from XPS measurements

Sample/element		Atomic percent (%)			
		Ti	C	O	Al
Pristine	Average	3.42	38.8433	38.96	18.77
	Std. dev	0.98	7.21	3.98	2.37
Brushed	Average	7.01	28.93	47.28	16.77
	Std. dev	0.88	2.02	1.23	0.66
Statistical significant difference		$p < 0.05$	$p < 0.1$	$p < 0.05$	No difference

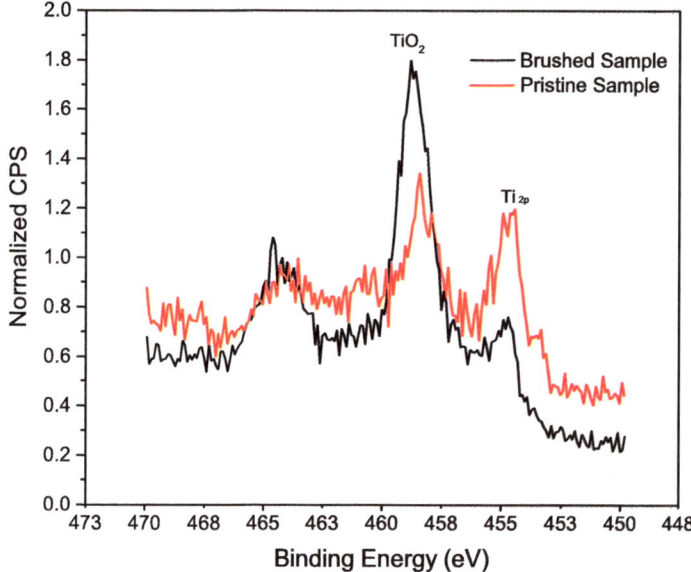

Fig. 10 Elemental scans from XPS analysis for brushed and pristine samples

As seen in Table 3, brushed samples show a reduced amount of carbon atoms in the scanned regions along with a higher percentage of titanium and oxygen atoms than the pristine sample. The altered atomic percentages indicate a loss of carbon atoms and acquisition of oxygen on the surface. Further, the regional elemental scans plotted in Fig. 10 for titanium show the relative percentage of the $Ti_{2p3/2}$ bonded to Ti and the $Ti_{2p\ 3/2}$ bonded to oxygen in TiO_2, where a normalized count per second (cps) is obtained with respect to a common peak at approximately 464.4 eV for both samples.

The pristine sample shows a default TiO_2 peak at 458.3 eV with a $Ti_{2p3/2}$ peak at 454 eV representing a Ti-Ti bond. For the brushed sample, the number of Ti bonded atoms to oxygen increased by almost two times as marked by a higher peak at 458.3 eV corresponding to $Ti_{2p3/2}O_2$ formation. Thus, chemical analysis through XPS suggests abrasion induced electrochemical oxidation of TiC to TiO_2 accompanied by a release of some carbon-based product in aqueous environments.

a)   b)

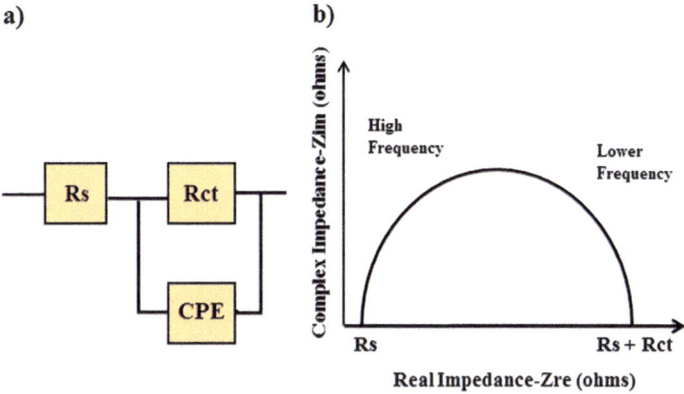

Fig. 11 (a) Equivalent circuit for Randles cell and (b) Corresponding Nyquist plot

Results of the ICP-MS analysis show the presence of aluminum at 57 ppb for the tested solution and an undetectable level for the untested solution. For titanium, the amount is below the detection limits for both solutions. Higher aluminum concentration in the tested solution indicate that even with gentle abrasion the composite will release wear particles into the test solution, though the precise chemical state, e.g., whether alumina or aluminum hydroxide, is unknown. These particles may have been immediately swept away from sample surface and brushed along the surface. If the loose particles are alumina, they could have resulted in ploughing of the sample surface.

3.4 Electrochemical Impedance Data Analysis

Electrochemical impedance spectroscopy data can provide crucial information about the state of the oxide film at the composite-solution interface. An oxide film formed on a surface often exhibits resistive (frequency-independent) and capacitive (frequency dependent) behavior. The Nyquist plot is an effective way to characterize the charge transfer processes. In a Nyquist plot (often a semicircle), fast kinetic controlled reactions are represented in a high frequency region at the left end, and slow diffusion and mass transfer controlled reactions are captured in a low frequency region at the right end, as depicted in Fig. 11b. With the equivalent Randles circuit shown in Fig. 11a, containing a constant phase element (CPE), charge transfer resistance of oxide films (R_{ct}) and solution resistance (R_s), we can determine the various parameters through statistical fitting of the circuit model to the Nyquist plots. Here R_s measures the resistance present in solution between the reference electrode and working electrode, which is affected by ionic concentration, temperature, type of ions and area of electrode. R_{ct} is the resistance of the oxide film and electrode-electrolyte interface to charge transfer, and it varies with the type of

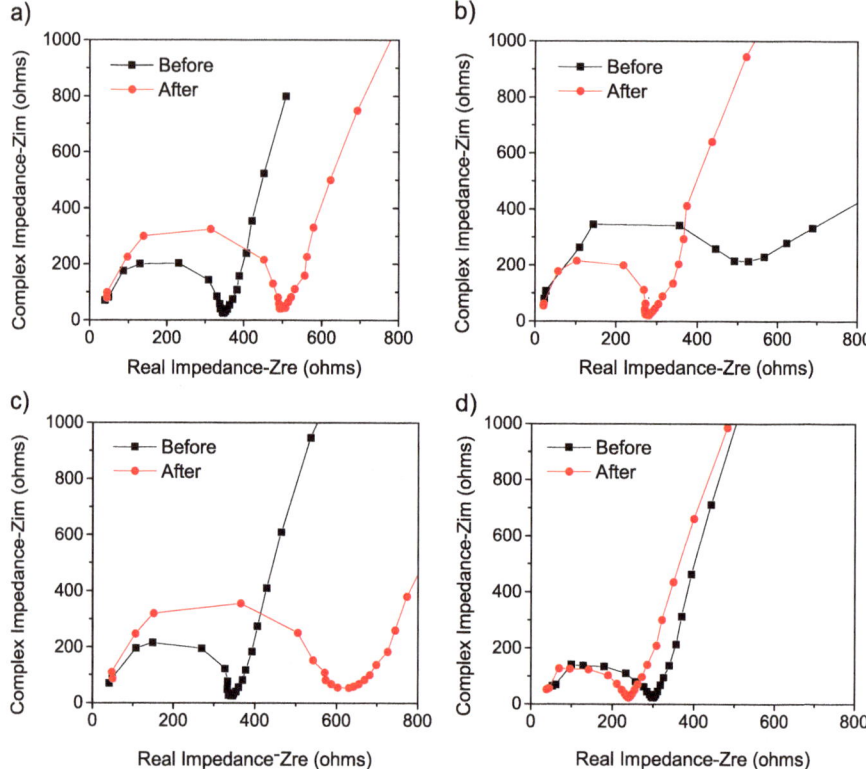

Fig. 12 Nyquist plot obtained before and after each experiment. (**a**) Anodic biasing (250 mV vs. SCE), (**b**) Cathodic biasing (−250 mV vs. SCE), (**c**) No brushing-no biasing, and (**d**) Brushing-no biasing

reactions, conditions of oxide film (compact or porous with defects), temperature, electrode potential and concentration of reactant species. CPE is a non-ideal representation of capacitive behavior of an electrical double layer on an electrode-electrolyte interface [41]. For analyzing EIS data, the alumina domain is considered not to be participating in the charge transfer processes during electrochemical interactions.

R_{ct} values before and after each test run in experimental conditions mentioned in Table 2 were obtained by fitting of the equivalent circuit model to the inner semi-circle (high frequency region) of the Nyquist data obtained (Fig. 12). Frequency dependent behavior of the phase between the applied input and measured output signal and impedance at electrode interface in Bode plot (not shown) was marked by a one time constant, which is representative of a single R-C component like equivalent circuit for a Randles cell.

After anodic biasing (250 mV vs. SCE), the R_{ct} value increased, indicating formation of a stable and compact barrier to charge transfer processes verifying the tendency of the TiC domain to form TiO_2 in an aqueous environment. Similar

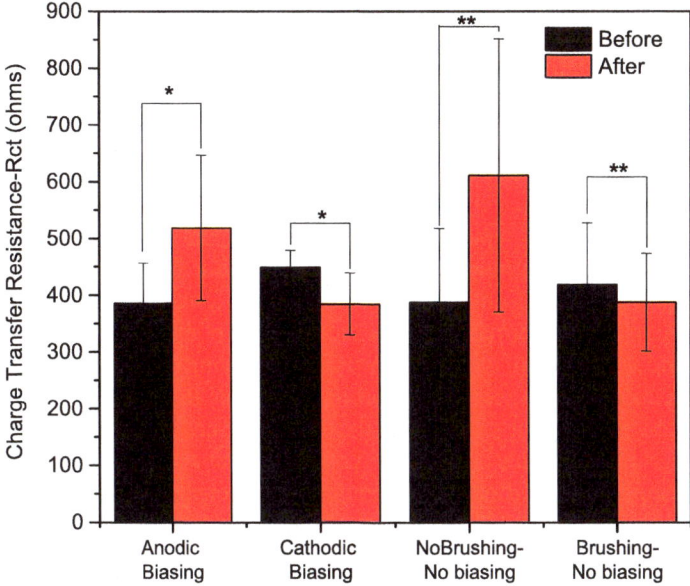

Fig. 13 Charge transfer resistance R_{ct} values for composite sample before and after each experimental test condition in Micro90 at 75 °C (*significant difference ($p < 0.05$) & **significant difference ($p \leq 0.1$)

behavior was observed when the sample was just exposed to the test solution under a no-biasing condition without brushing abrasion. However, after cathodic biasing (-250 mV vs. SCE), R_{ct} values reduced. Cathodic potentials are known to deteriorate the stability of the TiO_2 film [3] that results in an increased amount of electrochemical interaction of TiC with the environment and enhanced charge transfer rate at the interface. The brushing with a no-biasing condition, which favors oxide film formation under undisturbed conditions, produces reduced R_{ct} as seen in Fig. 13 after brushing abrasion.

3.5 Understanding the Degradation Mechanism of Alumina-TiC Composite

In the current study, the alumina-TiC composite underwent 'gentle abrasion' and the fundamental degradation mechanism can be categorized as tribo-electrochemical wear. The oxidative current response and ploughing damage on the composite surface are all indicators of tribo-electrochemical wear induced by brushing abrasion. There is no visible wear on TiC grains due to its higher hardness [22] and wear resistance. However, TiC likely undergoes oxidation [30, 31] to form TiO_2 and CO_2 as per the following reaction:

$$TiC + 4H_2O \rightarrow TiO_2 + CO_{2(aq)} + 8H^+_{(aq)} + 8e^- \qquad (1)$$

Upon contact with an aqueous environment and when left undisturbed, a passive oxide layer would form on TiC, as the R_{ct} value plotted in Fig. 13 indicates. The onset of brushing could lead to the removal of micro-asperities on the alumina domain along with the rupture of a passive layer on TiC thus activating a burst of electrochemical oxidative processes on conductive TiC as shown in Eq. (1). While the TiC domains would repassivate and form TiO_2, the ongoing brushing would keep the electrochemical interactions of TiC active due to continuous abrasion of any possible oxide layer formed. Under this situation, a stable oxide layer can hardly exist. Aside from TiO_2 formation, the reduction in carbon percentage in XPS results point toward the release of carbon atoms into the solution to form CO_2 as per Eq. (1) due to oxidation and replacement of carbon by oxygen in the TiC crystal lattices.

In general, gentle wear process like the one performed in this study would cause less damage [23]. Through qualitative comparison of the surface morphology of abraded samples under different environment conditions, it is clear that the alumina domain showed a greater amount of material 'chipped away' in a wet and heated environment than at wet room temperature and dry brushing conditions. Alumina experiences tribological and chemical degradation. The hydrophilic nature of alumina causes it to readily hydrate to aluminum hydroxide when in contact with water and this reaction is thermodynamically more prevalent at higher temperatures [42, 43]. Strong reactivity to water and resulting greater surface plasticity of alumina in a wet environment makes it more susceptible to wear in a wet environment than dry conditions [23]. During abrasion, the hydroxide formed that may not be adherent is abraded away along with chipping of brittle alumina domain as supported by ICP-MS analysis. The low stress during brushing abrasion leaves released debris unconstrained [44] and they could be immediately swept away from the sample surface and brushed along creating circular trajectories further abrading the alumina matrix. Thus, debilitated wear resistance of the alumina domain in a heated aqueous alkaline environment plays a crucial role in the degradation of the composite.

Relating this understanding to the evidence from altered charge transfer resistance, the wear debris released also interrupts the passive TiO_2 layer formation on the TiC domain during continuous brushing. It is important to note that the electrochemical current response may have negligible or no contributions from alumina even though it could be chemically reacting with an aqueous environment. However, XPS chemical analysis of the brushed sample for 2 h does show traces of TiO_2. This indicates that prolonged brushing and exposure to the solution does result in TiO_2 formation in worn out regions, however, its structure may not be uniform and compact to offer any resistance to oxidative charge transfer processes because of brushing abrasion. A similar reduction in R_{ct} after abrasion due to damage in the protective layer was observed by others [45, 46].

4   Conclusions

This study has shed new insights into the interplay of abrasion and electrochemical degradation of alumina-TiC ceramic composite in an aqueous environment. An oxidative electrochemical process on TiC is activated by brushing abrasion while alumina also undergoes abrasive wear. It is established in this chapter that:

Brushing abrasion causes electrochemical activation of alumina-TiC composite, forming TiO_2 on the composite surface due to oxidation of TiC in aqueous environment.

- Abrasive damage occurs near the grain boundaries with traces of "microploughing" on alumina domains.
- Electrochemical response to brushing abrasion and the corresponding surface damage are affected by abrasion parameters like acceleration and speed of the abrasion and the temperature of test environments.
- An elevated temperature enhances thermodynamic favorability and reaction rate of TiC oxidation and enables faster charge transfer.
- A wet and heated environment increases susceptibility of abrasion damage in alumina domain in comparison with a dry condition.
- Wear debris released from alumina abrasion may hinder the formation of a protective TiO_2 film.

It is crucial to consider that abrasion mechanisms employed in the study are gentle and under extremely low load conditions and such mechanically assisted electrochemical degradation mechanism is certainly to be aggravated under a greater load in aggressive ionic environments present in the biological milieu.

Acknowledgments The authors would like to acknowledge the support from the Department of Bioengineering and the Institute for Biological Interfaces of Engineering at Clemson University. Funding support for this work is provided by a storage media company through a research agreement with Clemson University under contract number 146422 and by the Institute for Biological Interfaces of Engineering at Clemson University.

References

1. Jazrawi LM, Kummer FJ, Di Cesare PE. Hard bearing surfaces in total hip arthroplasty. Am J Orthop (Belle Mead NJ). 1998;27(4):283–92.
2. Bard AJ, Faulkner LR, Leddy J, Zoski CG. Electrochemical methods: fundamentals and applications, vol. 2. New York: Wiley; 1980. p. 44–82.
3. Gilbert JL, Mali SA. Medical implant corrosion: electrochemistry at metallic biomaterial surfaces. In: Degradation of implant materials. New York: Springer; 2012. p. 1–28.
4. Royhman D, Patel M, Runa MJ, et al. Fretting-corrosion behavior in hip implant modular junctions: the influence of friction energy and pH variation. J Mech Behav Biomed Mater. 2016;62:570–87.
5. Swaminathan V, Gilbert JL. Fretting corrosion of CoCrMo and Ti6Al4V interfaces. Biomaterials. 2012;33(22):5487–503.

6. Cooper HJ, Della Valle CJ, Berger RA, et al. Corrosion at the head-neck taper as a cause for adverse local tissue reactions after total hip arthroplasty. J Bone Joint Surg. 2012;94(18):1655–61.
7. Cooper HJ, Urban RM, Wixson RL, et al. Adverse local tissue reaction arising from corrosion at the femoral neck-body junction in a dual-taper stem with a cobalt-chromium modular neck. J Bone Joint Surg Am. 2013;95(10):865–72.
8. Begin-Colin S, Mocellin A, Stebut JV, et al. Alumina and alumina–TiN wear resistance in a simulated biological environment. J Mater Sci. 1998;33(11):2837–43.
9. Lashneva VV, Kryuchkov YN, Sokhan SV. Bioceramics based on aluminum oxide. Glas Ceram. 1998;55(11):357–9.
10. Garino J, Rahaman MN, Bal BS. The reliability of modern alumina bearings in total hip arthroplasty. Semin Arthroplasty. 2006;17(3):113–9.
11. Skinner HB. Ceramic bearing surfaces. Clin Orthop Relat Res. 1999;369:83–91.
12. Willmann G. Ceramic femoral head retrieval data. Clin Orthop Relat Res. 2000;379:22–8.
13. Boutin P, Christel P, Dorlot JM, et al. The use of dense alumina–alumina ceramic combination in total hip replacement. J Biomed Mater Res. 1988;22(12):1203–32.
14. Nine MJ, Choudhury D, Hee AC, et al. Wear debris characterization and corresponding biological response: artificial hip and knee joints. Materials. 2014;7(2):980–1016.
15. Mittelmeier H, Heisel J. Sixteen-years' experience with ceramic hip prostheses. Clin Orthop Relat Res. 1992;282:64–72.
16. Jianxin D, Zeliang D, Jun Z, et al. Unlubricated friction and wear behaviors of various alumina-based ceramic composites against cemented carbide. Ceram Int. 2006;32(5):499–507.
17. Fei YH, Huang CZ, Liu HL, et al. Mechanical properties of Al 2 O 3–TiC–TiN ceramic tool materials. Ceram Int. 2014;40(7):10205–9.
18. Guu YY, Lin JF, Ai CF. The tribological characteristics of titanium nitride, titanium carbonitride and titanium carbide coatings. Thin Solid Films. 1997;302(1–2):193–200.
19. Lee SW, Morillo C, Lira-Olivares J, et al. Tribological and microstructural analysis of Al2O3/TiO 2 nanocomposites to use in the femoral head of hip replacement. Wear. 2003;255(7):1040–4.
20. Cai KF, McLachlan DS, Axen N, et al. Preparation, microstructures and properties of Al2O3–TiC composites. Ceram Int. 2002;28(2):217–22.
21. Brama M, Rhodes N, Hunt J, et al. Effect of titanium carbide coating on the osseointegration response in vitro and in vivo. Biomaterials. 2007;28(4):595–608.
22. Shackelford JF, Han YH, Kim S, Kwon SH. CRC materials science and engineering handbook. Boca Raton, FL: CRC; 2016.
23. Sahoo P, Davim JP. Tribology of ceramics and ceramic matrix composites. In: Tribology for scientists and engineers. New York: Springer; 2013. p. 211–31.
24. Oda K, Yoshio T. Hydrothermal corrosion of alumina ceramics. J Am Ceram Soc. 1997;80(12):3233–6.
25. Lauwers B, Kruth JP, Liu W, Eeraerts W, Schacht B, Bleys P. Investigation of material removal mechanisms in EDM of composite ceramic materials. J Mater Process Technol. 2004;149(1):347–52.
26. Landfried R, Kern F, Burger W, Leonhardt W, Gadow R. Development of electrical discharge Machinable ZTA ceramics with 24 vol% of TiC, TiN, TiCN, TiB2 and WC as electrically conductive phase. Int J Appl Ceram Technol. 2013;10(3):509–18.
27. Avasarala B, Haldar P. Electrochemical oxidation behavior of titanium nitride based electrocatalysts under PEM fuel cell conditions. Electrochim Acta. 2010;55(28):9024–34.
28. Meijs S, Fjorback M, Jensen C, Sørensen S, Rechendorff K, Rijkhoff NJ. Electrochemical properties of titanium nitride nerve stimulation electrodes: an in vitro and in vivo study. Front Neurosci. 2015;9:268.
29. Zhang L, Koka RV. A study on the oxidation and carbon diffusion of TiC in alumina–titanium carbide ceramics using XPS and Raman spectroscopy. Mater Chem Phys. 1998;57(1):23–32.
30. Cowling RD, Hintermann HE. The corrosion of titanium carbide. J Electrochem Soc. 1970;117(11):1447–9.

31. Cowling RD, Hintermann HE. The anodic oxidation of titanium carbide. J Electrochem Soc. 1971;118(12):1912–6.
32. Kiran V, Srinivasu K, Sampath S. Morphology dependent oxygen reduction activity of titanium carbide: bulk vs. nanowires. Phys Chem Chem Phys. 2013;15(22):8744–51.
33. Ramirez AG, Kelly MA, Strom BD, et al. Carbon-coated sliders and their effect on carbon oxidation wear. Tribol Trans. 1996;39(3):710–4.
34. Contu F, Elsener B, Böhni H. Corrosion behaviour of CoCrMo implant alloy during fretting in bovine serum. Corros Sci. 2005;47(8):1863–75.
35. Barril S, Debaud N, Mischler S, et al. A tribo-electrochemical apparatus for in vitro investigation of fretting–corrosion of metallic implant materials. Wear. 2002;252(9–10):744–54.
36. Bratu F, Benea L, Celis JP. Tribocorrosion behaviour of Ni–SiC composite coatings under lubricated conditions. Surf Coat Technol. 2007;201(16):6940–6.
37. Jianxin D, Tongkun C, Zeliang D, et al. Tribological behaviors of hot-pressed Al 2 O 3/TiC ceramic composites with the additions of CaF 2 solid lubricants. J Eur Ceram Soc. 2006;26(8):1317–23.
38. Jahanmir S. Wear transitions and tribochemical reactions in ceramics. Proc Inst Mech Eng J J Eng Tribol. 2002;216(6):371–85.
39. Yingjie L, Xingui B, Keqiang C. A study on the formation of wear debris during abrasion. Tribol Int. 1985;18(2):107–11.
40. Lee GY, Dharan CKH, Ritchie RO. A physically-based abrasive wear model for composite materials. Wear. 2002;252(3):322–31.
41. Gamry Instruments. Basics of electrochemical impedance spectroscopy. Gamry Instruments: 20Primer; 2006. p. 202006.
42. Gee MG. The formation of aluminium hydroxide in the sliding wear of alumina. Wear. 1992;153(1):201–27.
43. Gates RS, Hsu M, Klaus EE. Tribochemical mechanism of alumina with water. Tribol Trans. 1989;32(3):357–63.
44. Gates JD. Two-body and three-body abrasion: a critical discussion. Wear. 1998;214(1):139–46.
45. Azzi M, Szpunar JA. Tribo-electrochemical technique for studying tribocorrosion -behavior of biomaterials. Biomol Eng. 2007;24(5):443–6.
46. Vitry V, Sens A, Kanta AF, et al. Wear and corrosion resistance of heat treated and as-plated duplex NiP/NiB coatings on 2024 aluminum alloys. Surf Coat Technol. 2012;206(16):3421–7.

Part II
Biology and Clinical and Industrial Perspectives

Biomaterials in Total Joint Arthroplasty

Lindsey N. Bravin and Matthew J. Dietz

Keywords Total joint arthroplasty · Total joint replacement · Biomaterials ·
Polyethylene · Ultra-high molecular weight polyethylene · Polyethylene wear ·
Ceramic · Zirconium · Cobalt chrome · Metallosis · Bearing surfaces · Implant
manufacturing

1   Introduction

Total joint arthroplasty, or total joint replacement, is a successful procedure used
when the native joint tissue no longer functions appropriately or when pain due to
degradation of the joint surface limits a person's daily activities. From osteoarthritis
to osteonecrosis, there are multiple indications to perform a total joint arthroplasty.
No matter the diagnosis, the procedure's goal is to provide the patient with pain
control, stability and a clinically functional joint for return to their daily life.
Regardless of age, affected patients have decreases in activities ranging from the
inability to participate in sports to the inability to mobilize for activities of daily
living because of joint destruction. Patients must fail nonoperative management
prior to becoming a candidate for total joint arthroplasty. Therefore, most patients
are on long-term anti-inflammatory medications. They modify their daily activities
sometimes to the extreme of not using entire floors of their house just to avoid stairs.
Most of them have reached a point where they have to use a cane, walker or wheel-
chair for assistance with mobilization. Some patients have such limitations of their
motion that they are unable to sit in a chair or car for any length of time. These limitations typically cause a substantial decrease in the patient's quality of life. Patients
who undergo a total hip or total knee arthroplasty (THA, TKA) have successful
recoveries that allow return to work and normal daily activities affecting society in

Electronic supplementary material: The online version of this chapter (doi:10.1007/978-3-
319-89542-0_9) contains supplementary material, which is available to authorized users.

L. N. Bravin · M. J. Dietz (✉)
Department of Orthopedics, West Virginia University School of Medicine,
Morgantown, WV, USA
e-mail: mdietz@hsc.wvu.edu

© Springer International Publishing AG, part of Springer Nature 2018   175
B. Li, T. Webster (eds.), *Orthopedic Biomaterials*,
https://doi.org/10.1007/978-3-319-89542-0_9

a positive way. The societal benefit of both THA and TKA are significant, with approximately $33,000 net societal savings per THA performed and a net societal savings of $19,000 per TKA performed [1, 2]. When compared to preoperative baseline, an updated study projected the estimated incremental cost-effectiveness ratio for THA to be $40,000 per quality-adjusted life years (QALY) and, for TKA, to be $43,107 per QALY [3]. Projections of future primary and revision total joint arthroplasties are estimated to significantly increase from 2005 to 2030. According to Kurtz et al.'s prediction, the number of primary THAs performed will increase from 208,600 in 2005 to 572,000 in 2030. Primary TKAs are predicted to increase from 450,000 in 2005 to 3.48 million procedures in 2030 [4]. Although these predicted values are based on the use of the Nationwide Inpatient Sample, efforts to produce implants with improved longevity will highly impact the rates of revision. The revision procedures documented in 2005 are predicted to double by the year 2026 for THA and by 2015 for TKA [4]. Currently, projected case numbers are outgrowing the available medical and economical resources [4–6].

Total joint arthroplasty affords patients the means to achieve a better lifestyle with a return to most of their daily activities. The success of total joint arthroplasty has continued to improve over time with better prosthetic designs as well as better patient selection for the surgical procedure. When considering the definition of success, there are both patient and surgical factors. Most patients report success as a fulfillment of their preoperative expectations, with pain relief and mobility being at the top of their list. When assessing patients pre- and postoperatively, there are several approaches to quantify success. Based on a patient's subjective and objective perception of their health status and function, surgeons are able to assess the patient's improvement postoperatively. Patient Reported Outcome Measurements (PROMs) collected by some of the largest national databases in the United States have shown that there is improvement in pain relief, functional status and activity level after a total joint arthroplasty. According to a study using a Hospital for Special Surgery (HSS) Hip or Knee Replacement Expectations Survey, patients' preoperative expectations are very high, identifying 11 of 18 items (for THA) and 10 of 19 items (for TKA) as "important." These items included pain, walking status, psychological state and daily activities (both essential and nonessential). Surgeon expectations were typically lower than the patients' expectations preoperatively [7]. This study also corroborated results from previous studies that showed patient satisfaction postoperatively was better for THA than for TKA [7–9]. In addition, patient characteristics play a large role in opportunity for a successful outcome. There is evidence that obesity and depression are risk factors for poor outcomes [10, 11]. If a surgeon can set patient expectations to appropriately prepare them for a successful outcome, the satisfaction rate will, in turn, improve. In the small proportion of patients who do not have successful results, these databases and PROMs allow surgeons to re-evaluate the cause of suboptimal outcomes. Albeit a small cohort of patients, the information obtained from unsuccessful outcomes paves the way for future improvements in implant design and function, surgical techniques and patient management pre- and postoperatively.

When patient satisfaction is taken out of the picture, there are several factors that must be considered to successfully reconstruct a particular joint. A joint arthroplasty should replicate the function of the particular joint of interest. It should also be accepted without consequence by its host and last long enough to limit the number of subsequent revision procedures. Therefore, success of a joint arthroplasty is measured according to the stability, sterility and survivability of the implant. Research is continually yielding new and improved material combinations that successfully satisfy these three expectations. Current material combinations have demonstrated increasing success over the years but shortcomings still exist.

2 Stability

The ability of a material to withstand the joint reactive forces and the cyclic loads of a particular joint is determined by its fatigability and wear characteristics. The implant must provide enough strength to withstand these loads while still providing elasticity to prevent stress shielding at the bone-implant interface. In addition, the structural design has to conform to the anisotropic properties of the bone to allow for load sharing during the early phase of implant integration. The fixation technique also plays a significant role in the stability of an implant and can include bone ingrowth, bone ongrowth or cementation. These added characteristics of the material play a role in increasing its structural stability when utilized appropriately by the surgeon. The ultimate goal is to increase the strength of the implant while decreasing its incidence of failure (periprosthetic fracture or catastrophic implant failure). Initial implant integration occurs through mechanical loading of the bone to stimulate bone-implant integration. Inflammatory responses to infection and wear debris, micro-motion at the bone-implant interface and inappropriate mechanical loads (too great or too little) on an implant can negatively impact bone quality causing resorption. Late failures typically occur due to fatigue failure at material interfaces (bone-implant, bone-cement or cement-implant), physical integrity of the implant itself and bearing surface wear [12]. It has taken decades of failures to determine why each one occurs and how to prevent these failures in each generation of material production. Discussion later in this chapter will focus on the individual biomaterial advancements that have been sought to overcome each failure mode.

Malalignment leading to excessive wear or stress on the bearing surface can also lead to catastrophic failure of the bearing component. For instance, a rotational mismatch between the femoral and tibial component when using a bearing insert with a cam and post may result in fracture of the post (Fig. 1). The acetabular liner bearing surface also sees undue stress with excessive abduction or anteversion of the acetabular shell leading to catastrophic rim breakage of the liner.

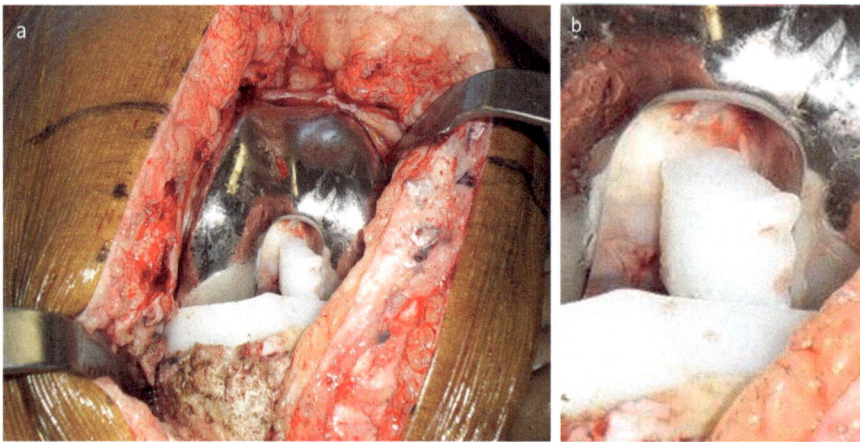

Fig. 1 This is a revision left total knee arthroplasty with a standard medial parapatellar approach performed for a fractured post of a posterior stabilized polyethylene insert with improved wear characteristics with decreased mechanical strength. This was a recognized complication of early polyethylenes utilized for posterior stabilized implants.

3   Sterility

Sterilization of each implant is necessary from an operative standpoint but has side effects that can produce less than optimal results from a materials standpoint. The goal is sterilization techniques that have the highest wear resistance profile while maintaining the mechanical toughness of the material. Limitations of free radical production and oxidation are direct contributors to increasing the wear resistance of a material. In addition, there are certain materials that contain properties that can elicit an allergic response in some patients. Unlike the inert nature of polyethylene, nickel can stimulate an allergic response, as well as a systemic response to metal ions and metal debris from traumatic and normal wear resulting in revision surgeries and results that are less than ideal. Limiting the host response to implant material is another characteristic to consider when trying to determine the optimal implant combination. Furthermore, the residue left on implants as a result of machining and sterilization can affect the biologic activity in the host by preventing bony ingrowth or ongrowth which can lead to aseptic loosening. An example is the Sulzer Inter-Op TM acetabular component (Sulzermedica, acquired by Zimmer Biomet, Warsaw, IN), recalled in December 2001 for failure of ingrowth due to residue found on certain packaged components [13]. The documented failure rate of this cup as estimated by Sulzer Orthopedics is 13% (2353 implants failed out of 17,500 implanted). A prospective study of patients with this particular shell reported a 33% overall failure rate (30% failure of recalled acetabular components) due to aseptic loosening [13]. Bonsignore further studied the consequence of machine oil residue on osseointegration and osteoblast function finding a decreased ability to attach, spread, grow, differentiate and mineralize [14]. This local effect leads to decreased pullout measures and decreased bone-implant contact percentage [14]. Events such as these have led to the addition of steps during sterilization to avoid adverse effects.

4 Survivability

The number of joint arthroplasties in our population of patients aged 50 to 70 years old is increasing with few answers as to what material will outlast the lifetime of these patients. The desire is to limit the number of revisions a patient will experience. However, the age at the first arthroplasty and activity level desired by that patient necessitates an implant survival range greater than 20 years. Biomaterials act differently according to their chemical makeup, geometric design and articulation with the adjacent implant material. Although each material at some point displays abrasive wear, adhesive wear and fatigue failure, the primary wear characteristics are different for each material [12]. In addition to the survivability of normal wear characteristics, the role of the material to take on the burden of infection resistance also affects its survivability. There are several moves in research to find materials that can decrease bacterial adhesion to the implant, which will, in turn, decrease the number of revisions for periprosthetic joint infections. Changes in machining techniques, manufacturing processes and sterilization techniques have all improved the wear characteristics and the survivability of the implants *in vivo*. The most recent development is the production of a drug eluting highly cross-linked polyethylene that maintains its biomechanical properties despite eluting a higher concentration of antibiotics for a longer duration than the gold standard treatment using antibiotic cement spacers [15].

5 Bearing Surfaces: Polyethylene

5.1 Polyethylene Then

The use of polyethylene dates to the 1950s with Drs. John Charnley and GK McKee using what was classified as high-density polyethylene for total hip arthroplasty [16, 17]. Over the past several decades, research has changed the way polyethylene is manufactured, machined, stabilized and sterilized. The available resin powders vary in their molecular structure giving each resin slightly differing properties including molecular weight and viscosity, which have a direct relationship to the mechanical strength of the material and its wear resistance. The powder resin is formed into a solid polyethylene bar stock by ram extrusion or directly molded into its implant shape using compression molding techniques. Historically, compression molding has improved linear and volumetric wear rates compared to ram extruded polyethylene with linear wear rates of 0.05 mm/year vs. 0.11 mm/year, respectively [18].

Their volumetric wear rates were studied in combination with sterilization technique using either gamma irradiation in air or radiation in an inert gas. Volumetric wear was better for compression molding irradiated in inert gas (52.12 mm³/year) compared to the ram extruded polyethylene irradiated in inert gas or with gamma irradiation in air (62.32 mm³/year and 66.09 mm³/year, respectively) [19]. It is thought that the machining process involved with ram extrusion, as well as the decreased temperatures used to manufacture the implants, produces changes in the surface of the polyethylene both macroscopically and microscopically, supporting the effects of machining on fusion defects both prior to implantation and after retrieval [20]. It is known that the early process of ultra-high molecular weight polyethylene (UHMWPE) implants displayed more wear characteristics in knees than in hips, and were worse if machined from ram extruded bar stock [21]. Knees distribute forces over the polyethylene insert differently than hips, making wear characteristics more pronounced over shorter times. The acceptable amount of wear in a hip prosthesis polyethylene liner demonstrates a significantly higher and unacceptable wear rate profile in the knee. In addition to consolidation, radiation cross-linking or chemical cross-linking was used to induce higher amounts of cross-linking within the polyethylene. Gamma irradiation provided increased cross-linking with the disadvantage of increased free radical production. Sterilization in gas plasma or ethylene oxide avoids immediate and long-term oxidative degradation by limiting the free radical production but with no improvement in wear resistance. In addition, the increase in cross-linking by radiation has a linear correlation with decreasing the mechanical properties of the material. Oonishi compared high dose radiation to ethylene oxide sterilization and showed a steady-state linear wear of 0.006 mm/year compared to 0.098 mm/year, respectively [22]. In the late 1990s, thermal stabilization was combined with the cross-linking techniques to improve the wear and oxidation resistance of UHMWPE. An increase in the oxidation resistance of UHMWPE by quenching the majority of the post-irradiation residual free radicals was shown to maintain the mechanical property performance standards specified by the American Society for Testing Materials (ASTM), International Standards Organization (ISO) and the US Food and Drug Administration (FDA) [23]. Remelting further decreased free radical production and measurable oxidation; however, recrystallization and crystallinity decreased, further reducing the fatigue strength of the material [24]. With both remelting and annealing, there are reports of rim fracture that cause clinical concerns [25, 26]. In the wake of improvement in the above processes, the packaging of the implants in air penetrable encasements allowed for shelf life oxidation of any residual free radicals and material degradation prior to implantation.

The goal of continued clinical research is to provide longer lasting implants with fewer revision procedures. It is well documented that the incidence of osteolysis and the need for revision are both directly correlated to the amount of volumetric wear [27, 28]. The clinical failures and inadequacies of the conventional polyethylene as well as the first-generation UHMWPE brought our attention to the need for improved wear rates and furthermore, improved processing techniques. In the lab, this translates into a goal of increased wear resistance, decreased oxidation potential and elimination of the material's free radicals while maintaining its mechanical properties. Unfortunately, the processes that are available and currently in use have a risk/benefit ratio with improving one of these characteristics while hindering the others.

5.2 Polyethylene Now

Since the 1950s, the molecular weight has increased from high-density polyethylene to ultra-high molecular weight polyethylene with an average molecular weight of 3–6 million g/mol [29]. Maximum bulk impact strength and abrasive wear resistance are obtained between molecular weights of 2.4–3.3 million g/mol with a nonlinear correlation to the intrinsic viscosity of the polyethylene. The wear characteristics appear to plateau around 3.3 million g/mol [29–31]. With polyethylene resin now at an optimum molecular weight, consolidation needed improvement to decrease the porosity and microscopic impurities and to prevent fusion defects during the process of either ram extrusion or compression molding. As compared to the first-generation UHMWPE, recent data show little difference in the density, tensile yield, ultimate tensile strength and elongation to failure measurements between compression molding and ram extrusion [29] which is likely due to improvements in medical grade resin production and the evolution of converters used for consolidation [29]. Increased research has led to more modern techniques such as Hot Isostatic Pressing (HIPing). Zimmer Biomet (Warsaw, IN) has become the sole proprietor of the HIPing process in making their product ArCom [29]. This hybrid process uses high temperature pressure with argon gas and compression molding, followed by either a turning or milling operation.

Packaging and sterilization processes have evolved to increase cross-linking within the material while eliminating air from the process, in an attempt to reduce oxidation of free radicals. To capitalize on increased cross-linking and decreased free radical production, gamma irradiation in an inert gas has been utilized with success by many manufacturers. Electron beam (E-beam) radiation, in lieu of gamma irradiation, allows for the same amount of cross-linking at lower doses of radiation. The maximum amount of radiation provided by commercial E-beam accelerators is two orders of magnitude larger than that of a commercial gamma source [24, 32]. In addition to the high temperatures available with E-beam radiation, thermal stabilization with annealing or remelting continues to complement the sterilization and cross-linking methods mentioned above. However, with concerns about the mechanical properties after melting and annealing, a process that stabilizes the UHMWPE without the pitfalls of thermal processing was sought. In 2007, UHMWPE stabilization with the antioxidant Vitamin E was commercially available. Vitamin E diffusion into the polyethylene limits the need for post-irradiation thermal processing and therefore maintains the crystallinity of the product [33]. It was found to increase both wear resistance and oxidation resistance of the polyethylene [32, 33], (Fig. 2). The two-step diffusion process with Vitamin E at higher

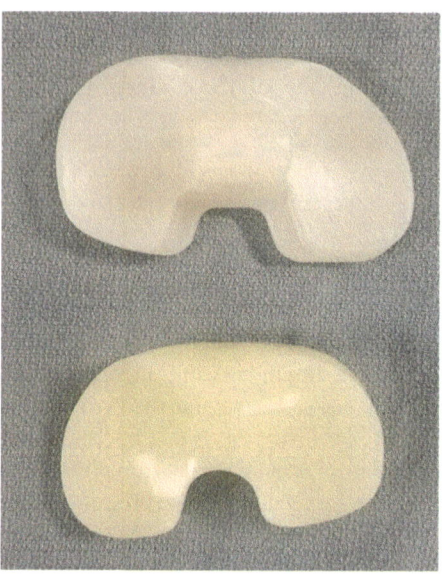

Fig. 2 The differences in a standard highly cross-linked polyethylene (top) and a Vitamin E infused highly cross-linked polyethylene (bottom) total knee inserts, showing the yellow coloration caused by the Vitamin E infusion. The goal of vitamin E is to stabilize the material to prevent oxidative degradation of the polyethylene. This also provides increased strength as additional steps of remelting are not required thereby reducing crystallinity

temperatures has improved distribution throughout the polyethylene. Currently the UHMWPE is heated to 120 °C (20 °C below its melting point of 140 ° C) to allow for evenly distributed integration of the antioxidant throughout the material while preventing temperatures that will cause changes in the molecular structures further reducing the crystallinity [33, 34]. There have been attempts at infusing the polyethylene resin with Vitamin E prior to consolidation; however, this process has provided obstacles to obtaining high enough levels of cross-linking to make it competitive with non-infused products [35]. Short-term studies suggest an improved wear rate of Vitamin E polyethylene to high cross-linked polyethylene without Vitamin E [36]. In addition, packaging has become more sophisticated with particle barrier material preventing oxygen exposure in irradiated implants; packaging has also been designed to allow particular gas particle penetration when chemically sterilized. The balance of these factors has led to the great successes realized in the field of arthroplasty. Comparisons of the different commercially available UHMWPEs are seen in Table 1.

Table 1 Commercially available UHMWPE liners in the US, with documentation of their name, initial release, raw material, cross-linking process, thermal stabilization, terminal sterilization [37]

Manufacturer	Biomet	DePuy		Smith and Nephew	Stryker		Wright	Zimmer		
Implant name	ArComXL	Marathon	AltrX	XLPE	Crossfire	X3	Lineage A class	Longevity	Durasul	Prolong
Initial release	2005	1998	2007	2005	1999	2005	2005	1999	1999	2001
Raw material	GUR 1050 isostatic rod	GUR 1050 extruded rod	GUR 1020 extruded rod	GUR 1050 extruded rod	GUR 1050 extruded rod	GUR 1020 extruded rod	GUR 1020 extruded rod	GUR 1050 molded sheet bar stock	GUR 1050 molded sheet preforms	GUR 1050 molded bar stock
Cross-linking process	Gamma (5 Mrad)	Gamma (5 Mrad)	Gamma (7.5 Mrad)	Gamma (10 Mrad)	Gamma (7.5 Mrad)	Gamma (3 Mrad x 3 sequences)	Gamma (7.5 Mrad)	E-beam (10 Mrad)	E-beam (9.5 Mrad)	E-beam (6.5 Mrad)
Thermal stabilization	Annealing	Remelting	Remelting	Remelting	Annealing	Annealing (3 sequences)	Remelting	Remelting	Remelting	Remelting
Terminal sterilization	Gas plasma	Gas plasma	Gas plasma	Ethylene oxide	Gamma irradiation in nitrogen (3 Mrad)	Gas plasma	Ethylene oxide	Gas plasma or ethylene oxide	Ethylene oxide	Gas plasma

5.3   Polyethylene: Case Reports 1–4 (Figs. 3, 4, 5 and 6)

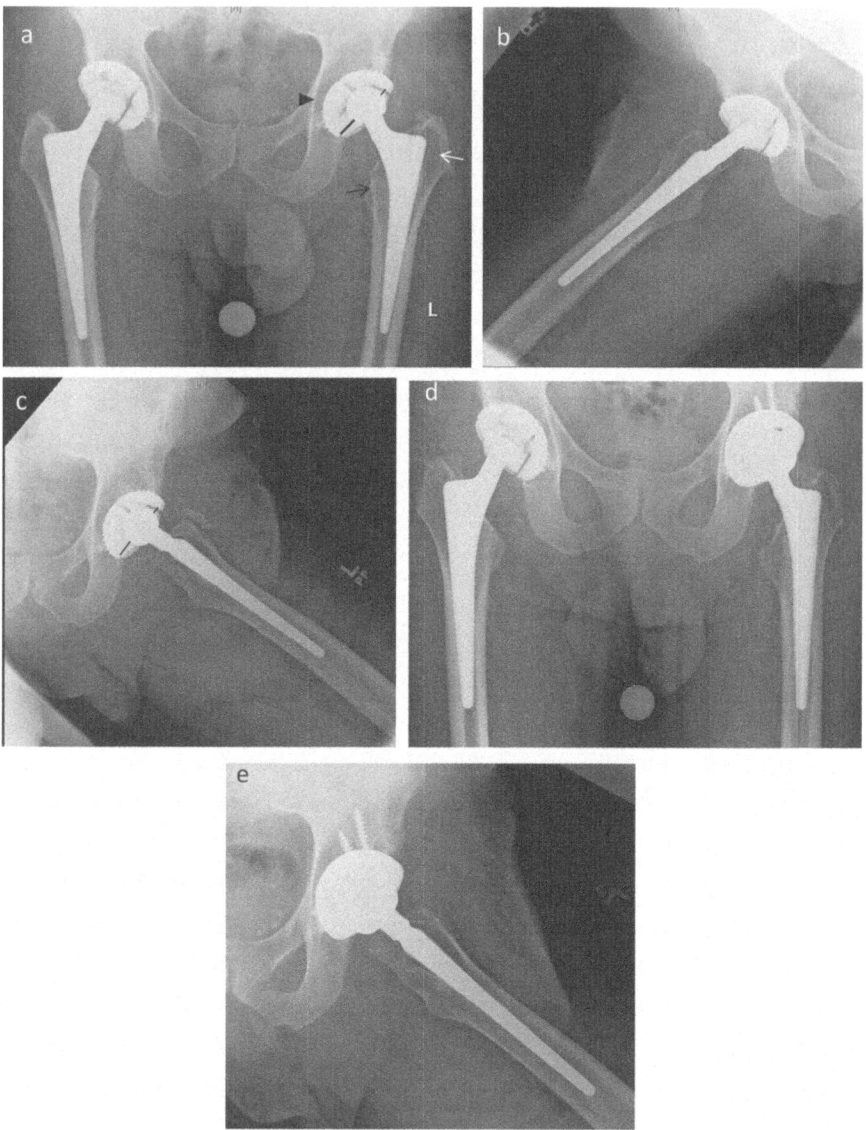

Fig. 3 70-year-old male who underwent staged bilateral total hip arthroplasties 18 and 20 years ago. (**a**) and (**b**) show the right hip at 20 years postoperative follow-up while (**a**) and (**c**) show the left hip at 18 years postoperative follow-up. The patient has left hip pain with sensation of his "hip sliding around in the socket" but denies right hip pain. Radiographs show eccentric polyethylene wear (black lines) with extensive osteolysis in Charnley acetabular zone one (black arrow head) and in Gruen femoral zones one (white arrow) and seven (black arrow). (**d**) and (**e**) show postoperative follow-up 1 year from revision of only the acetabular component

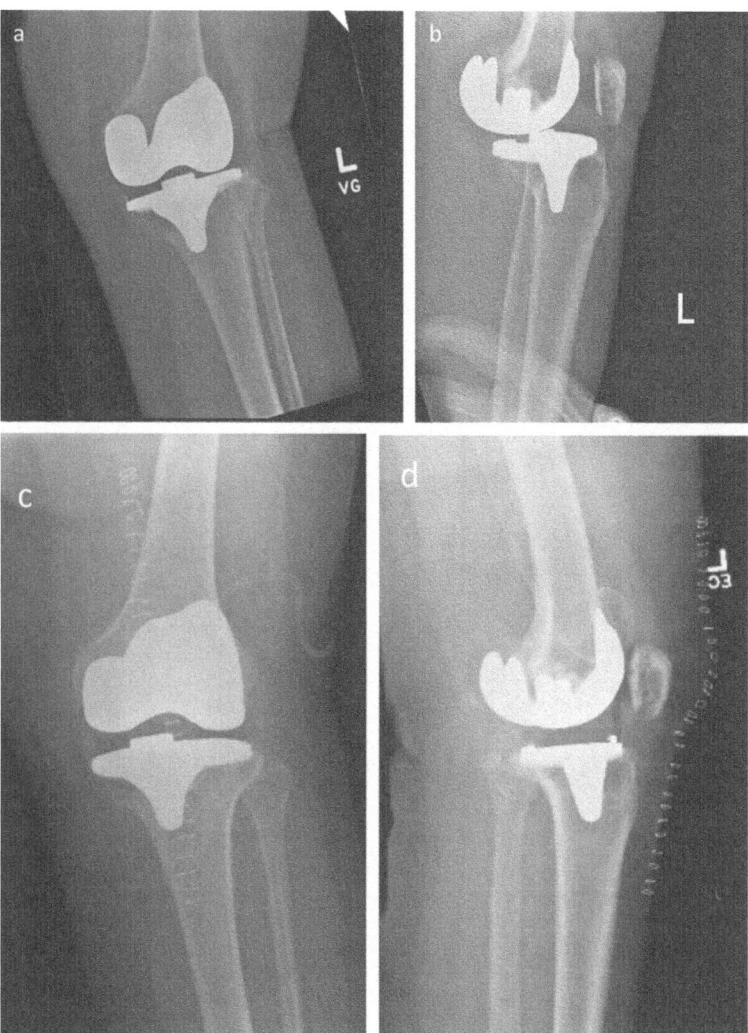

Fig. 4 89-year-old female presented with a total knee arthroplasty that had been performed 20 years ago with a Stryker Duracon implant (Kalamazoo, MI). She dealt with pain and increasing instability for 5 years. Radiographs at initial presentation are shown in (**a**) and (**b**) and display subluxation of the tibiofemoral joint secondary to significant polyethylene wear with limited osteolysis in the distal femur and proximal tibia. Despite the relative position of the implants and their stability along with the patient's age, it was elected to perform a revision of the polyethylene liner only. Postoperative imaging after undergoing revision of the polyethylene insert to a Stryker Duracon 16 mm polyethylene insert are shown in (**c**) and (**d**)

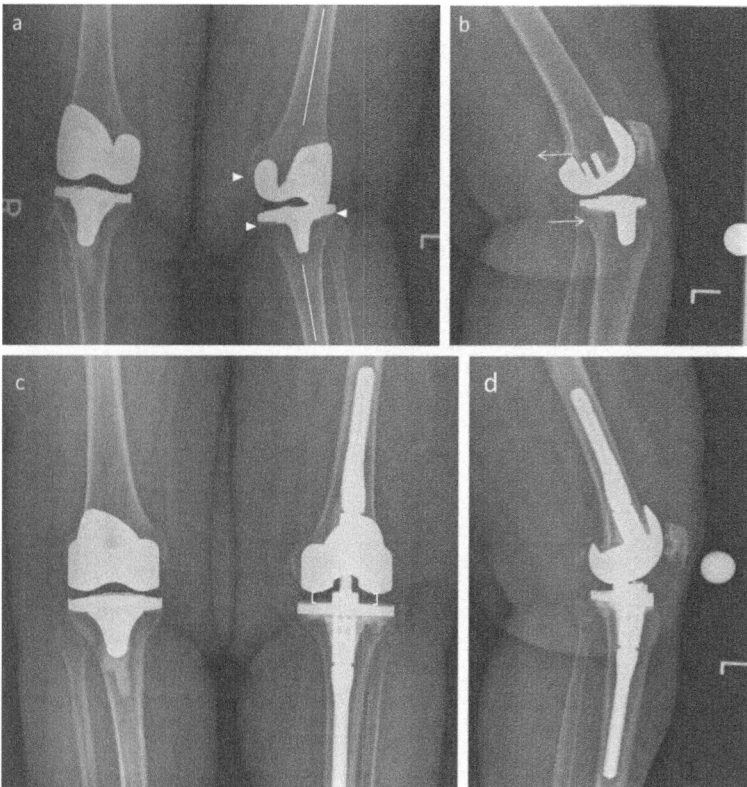

Fig. 5 74-year-old female underwent bilateral total knee arthroplasties 18 years [left knee, (**a**) and (**b**)] and 13 years [right knee, (**a**)] ago. The patient presented with one-year history of left knee pain and decreased activity. Radiographs show extensive tibial and femoral osteolysis (arrow heads) with eccentric polyethylene wear, valgus angulation (white lines) and tibiofemoral sub-luxation (white arrow). The patient underwent full revision knee arthroplasty with a Zimmer LCCK component (Zimmer Biomet, Warsaw, IN). Follow-up radiographs demonstrate a stable knee with correction of the subluxation and angulation with symmetric polyethylene (white brackets) (**c**) and (**d**)

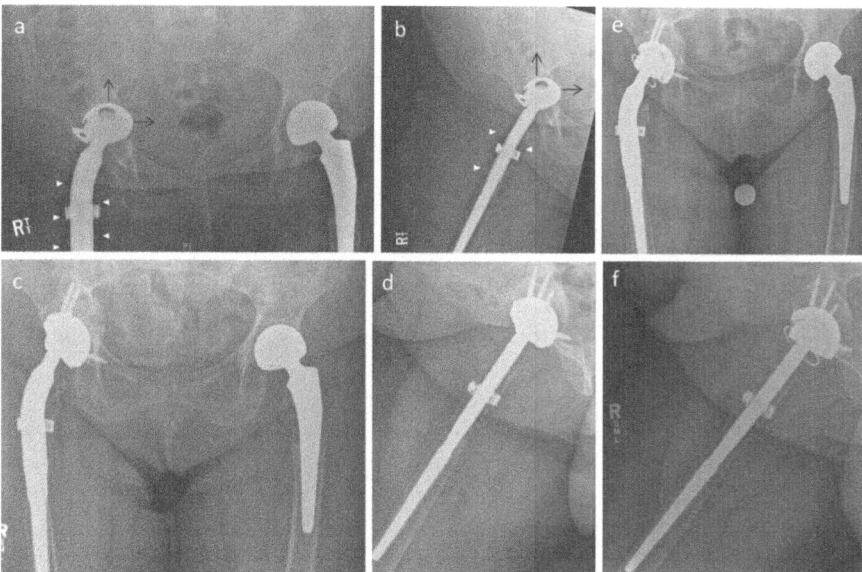

Fig. 6 70-year-old female underwent right total hip arthroplasty 20 years ago. The patient presented with progressive right hip pain resulting in eventual inability to bear weight. Anteroposterior (AP) and lateral radiographs (**a**) and (**b**) show eccentric polyethylene wear with proximal femoral bone deficiency (white arrow heads) due to osteolysis and acetabular osteolysis resulting in proximal migration of the acetabular component into the pelvis (black arrows). The patient underwent revision total hip arthroplasty, acetabular component only. (**c**) and (**d**) show the revised acetabular component to a titanium cup and a Modular Dual Mobility head (Stryker, Kalamazoo, MI). The patient presented with a periprosthetic hip dislocation 2 months postoperatively. She was further revised to a constrained liner for stability (**e**) and (**f**)

6 Bearing Surfaces: Metal

6.1 Metal Then

Using metal for orthopedic implants dates back to 1804 with fracture plating [38]. In the 1900s, dentists were using cobalt alloy (Co-Cr-Mo) which was soon adopted by the orthopedic surgeon in 1920s–30s for use in interpositional arthroplasty to prevent arthrodesis [32]. Stainless steel was introduced in the implant composition in 1926. The first joint arthroplasty meant to recreate and simulate the normal function of a joint did not appear until 1938 and was performed using stainless steel by Dr. Philip Wiles [38]. Stainless steel was replaced due to its lack of long-term corrosion resistance. Drs. McKee and Farrar used it in the 1950s–70s initially, for the first metal on metal hip arthroplasties but soon started using cobalt alloy for better long-term outcomes (Co-Cr-Mo) [16]. Cobalt chromium alloys were found to be subject to mechanically assisted corrosion. Titanium alloys were introduced as a beneficial material for use in orthopedic implants due to their high strength, similar modulus of elasticity to bone and self-passivation characteristics that assist with corrosion resistance. It has been found that the addition of 2–3% molybdenum to stainless steel decreases pitting and crevice corrosion [32].

Research on the different combinations of polycrystalline substances have led to the belief that metal alloys decrease the defects that define a substance's fatigue strength, modulus of elasticity and corrosion resistance. Lowering the point and lines defects as well as decreasing the size of the grain boundaries and volume defects within a particular metal allows for longer survivability when implanted into the host environment of the human body. There are several ways to address these defects including solid solution strengthening, cold and hot working of the material, thermomechanical processing (wrought and forged material), grain size reduction (powder metallurgy techniques) and precipitation hardening [32, 39, 40]. These methods are used in isolation or in combination with different metals to achieve the best functional material possible for its intended use. For instance, the decreased time required for work hardening of cobalt alloys and the increased number of carbides within the substance after being carbon forged makes this material very strong with an increased wear resistance profile compared to its unmolested counterpart [40]. Titanium alloys, on the other hand, have a bimodal structure that provides them with a lower modulus of elasticity than cobalt alloys, but they have high fatigue strength and are highly biocompatible. Zirconium improves its wear and corrosion resistance when oxidized to ZrO_2. Multiple properties can be affected by the different manufacturing techniques. Even after optimization of the crystalline material, surface modifications are implemented to further stabilize an implant for its function and survivability. Ion implantation, chemical and physical vapor deposition, nitriding and oxygen diffusion hardening are a few of the techniques used to increase wear and corrosion resistance while maintaining the necessary oxide layer that prevents adhesive wear [41–43].

Balancing the chemical, mechanical, electrical and biological factors of metal alloys when manufacturing these implants is a juggling act. Even though there are factors such as cyclic loading related to activity level of the patient, body mass index of the patient, requirement of bone graft for bony defects during implantation and the progressive bone loss associated with increased age, there are complications directly related to the implant biomaterial itself. Mechanically assisted corrosion (MAC) such as fretting or crevice corrosion can be seen at the junction between modular implants; this corrosion was initially recognized at the head-neck taper junction of the femoral prosthesis. Just as polyethylene wear does not end with production of small micro molecules, the debris formed by MAC can cause clinically relevant complications requiring lengthy revision surgeries. These complications include osteolysis, pseudotumor formation and hydrogen embrittlement [44–46]. Individual metals also have specific complications according to their biomaterial profile such as penetrating intergranular corrosion of Co-Cr-Mo, selective dissolution of beta phase titanium and large-scale pitting attack of Ti-6Al-4V [47, 48]. Fatigue crack initiation can occur due to MAC which, when combined with high cycle fatigue processes, will result in fatigue failure.

The use of these materials in fracture care versus total joint arthroplasty precludes a different biomaterial profile and desired function. Metal on metal articulations were popular and comprised approximately 30–50% of total hip arthroplasties in 2007–2008. Long-term outcomes of these articulations show that MAC plays a role in decreasing an implant's ability to prevent corrosion by disrupting its passive film layer. Once the film layer is disrupted, particles can interpose themselves between the two articular surfaces and cause abrasive (stripe) wear, tribochemical reactions and the dominant wear mechanism of metal on metal articulations, surface fatigue. Implant size and the adjacent articulating surfaces (whether a taper junction or the joint surfaces themselves) have been optimized with machining of the implants and digital technology to provide precise three-dimensional measurements of the implant. Previously there were polar, congruent and equatorial bearings based on the imperfections of human error during manufacturing and machining individual implants. Our current knowledge of implant articulations support boundary lubrication provided with polar bearings, meaning the inner bearing (femoral head) has a slightly smaller radius of curvature than the outer bearing (acetabular cup). Boundary lubrication decreases the presence of both abrasive and adhesive wear [49].

Position of an implant to distribute forces symmetrically along the bearing surface is important for long-term success as well. Malalignment of a bearing surface, such as vertical acetabular cup placement or impingement of an implant, can lead to edge loading. This type of loading results in asymmetrical implant wear, propagation of metal debris, pseudotumor formation, fatigue failure and ultimately necessitates revision procedures [50].

6.2 Metal Now

The knowledge concerning manufacturing, machining and combining different biomaterials has brought about the success and popularity of the metal alloys used today, including iron based (stainless steel), cobalt based, titanium based, zirconium alloys and tantalum alloys. Cobalt alloys (Co-Cr-Mo) and titanium alloys (Ti-6Al-4 V) are the primary metal biomaterials currently used in total joint arthroplasty. Metal on metal articulations can be successful when implanted in a desirable position and in the appropriate patient population. High carbide containing cobalt chromium alloy implants with a polar bearing design have the best long-term outcome data of all metal on metal articulations [32].

6.3   **Metal on Metal: Case Report 5 and 6 (Figs. 7 and 8)**

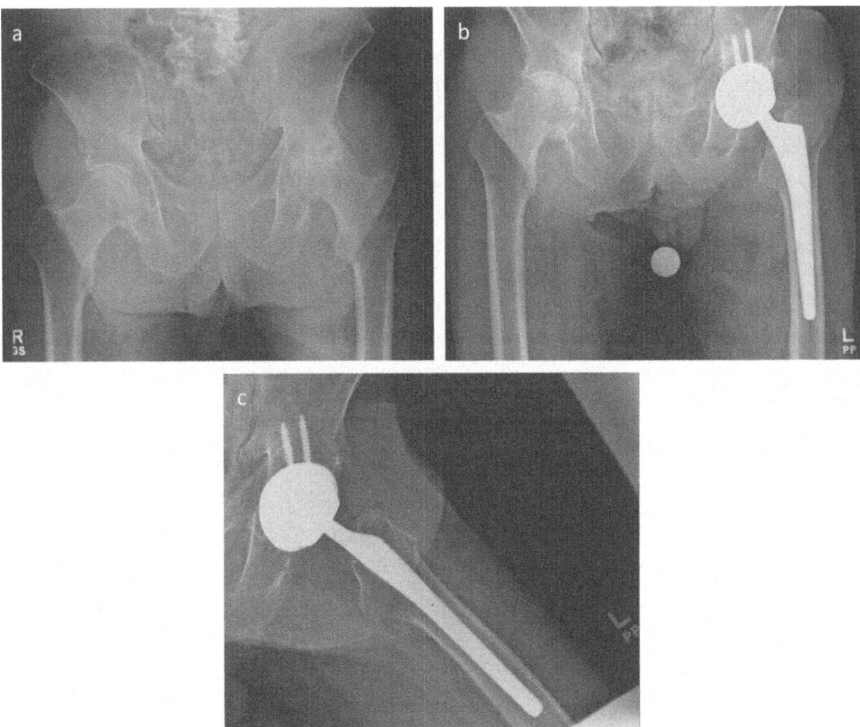

Fig. 7 78-year-old male underwent THA of the left hip for significant osteoarthritis (**a**). Total hip arthroplasty was performed with a metal on metal articulation utilizing a DePuy Synthes (Raynham, MA) Pinnacle 56 mm sector cup and a 40 mm large femoral head (**b**) and (**c**). The patient had significant relief of his pain and was able to continue participating in tennis and weight lifting activities. Ten years postoperatively, the patient has no pain, episodes of instability or limitations due to his left hip. Routine monitoring of his cobalt and chromium levels have remained normal with values of 0.7 ng/mL and 0.6 ng/mL, respectively. He will continue with yearly labs to ensure no progression or accumulation of systemic metal ions or concern for adverse local reactions

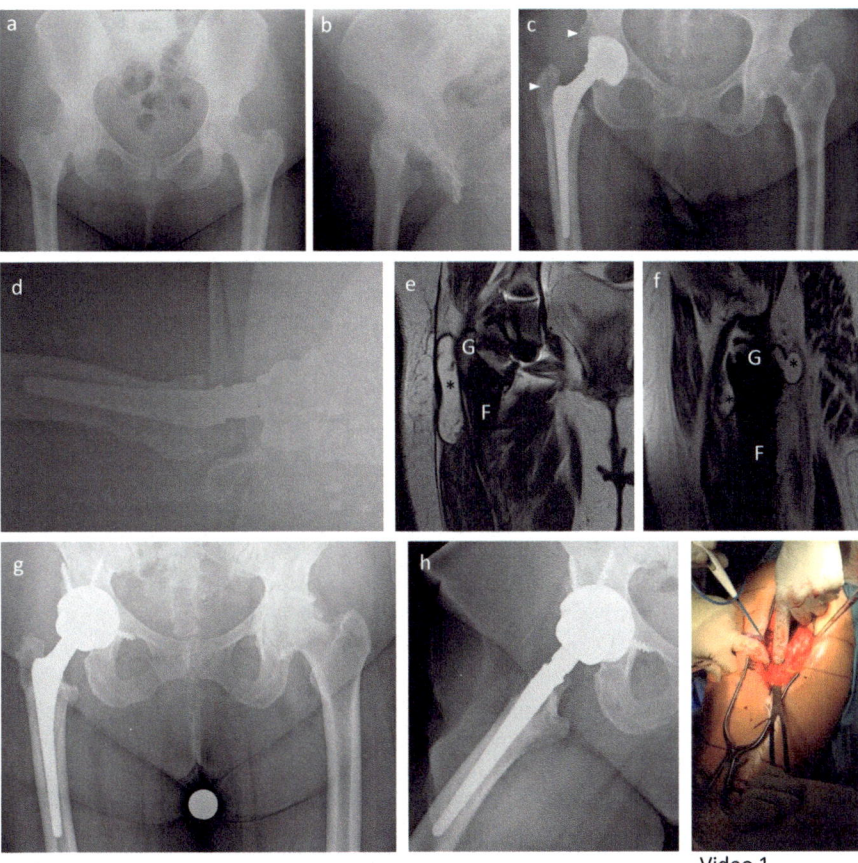

Video 1

Fig. 8 65-year-old female underwent THA of the right hip for severe osteoarthritis (**a**) and (**b**) with good relief of her pain for 7 years followed by progressive pain and instability. The patient was monitored for infection with a normal erythrocyte sedimentation rate of 15 mm/h and a C-reactive protein slightly elevated at 1.5 mg/dL. She had negative aspirates; her cobalt and chromium levels were elevated to 20 and 18.4 ng/mL, respectively. (**c**) and (**d**) show her right hip 10 years postoperatively from her primary THA showing mild osteolysis in both the femur and the acetabulum (arrowheads). (**e**) and (**f**) are the coronal and sagittal T2 cuts of the right hip magnetic resonance imaging, respectively. The greater trochanter (G) and femoral shaft (F) are marked and the metal on metal pseudotumor is visible on both images and identified with an asterisk (*). (**g**) and (**h**) are the AP and lateral radiographs of the patient's right hip after she was revised to a Biomet McLaughlin 50 mm diameter + 5 mm shell with a Biomet polyethylene liner with retention of her femoral stem and placement of a Wright Medical Group (Memphis, TN) 36 mm, 3.5 cobalt-chromium head. She also had a resection of the pseudotumor and a video of the metal on metal fluid collection encountered during the surgery is available in the supplemental material of this chapter (Video 1). At last follow-up, the patient was doing well with respect to her right hip but complained of progressive left hip pain due to severe osteoarthritis

7   Bearing Surfaces: Ceramic

7.1   Ceramic Then

Alumina was the first ceramic material to be used in total joint arthroplasty in the 1970s. The first ceramic zirconium femoral heads were introduced in Europe in 1985, followed by use in the United States in 1989 [51]. Both alumina and zirconia were the basic ceramic materials yielding increased strength and hardness to overcome increased cyclic loads and longevity of implants for a more youthful population. Ceramic provided the implant industry with an inorganic combination of both metal and non-metal material with a low surface roughness and a substantially higher hardness (approximately five-fold) than what was had with cobalt chromium [52, 53]. When oxidized, there is also a protective layer that prevents ceramic materials from undergoing an oxidative process (13-2). Unfortunately, as has been seen with the other biomaterials discussed in this chapter, for every benefit of a material's composition, there is a tradeoff. The tradeoff for ceramic's attractive hardness is its brittleness and chances of catastrophic failure when used as an orthopedic implant [54, 55].

Alumina was the initial material used for ceramic bearings and has a much lower wear rate of 0.025 mm/year compared to that of metal bearings [56]; even lower wear rates (around 0.022 mm/year) were observed when the first couple years of bedding in were ignored [57–59]. Alumina was not successful, however, due to its high incidence of catastrophic failure which led to a transition to zirconia, a material with higher bending strength and increased fracture toughness compared to alumina [51, 60]. Yttrium oxidized zirconium implants made an oxide stabilized material by forming a metastable microstructure. The balance between this microstructure and its stable monoclinical phase is temperature dependent; at lower temperatures, the material reverted to the monoclinical phase resulting in decreased strength and subsequent increased fracture incidence. In addition, this state of the material also resulted in increased surface roughness leading to polyethylene wear and eventual osteolysis. As most scientists would have entertained, combining the two materials in what is called zirconia toughened alumina would bring about an implant with the increased hardness and decreased surface roughness of alumina, while providing the properties of zirconia particles that assist in resisting fracture propagation [61, 62]. Current investigations are finding the different compositions of ceramic alloys that give the greatest benefit with the fewest complications. With little long-term evidence, a new zirconia alloy (Zr2.5Nb) was brought into use in 2003. This alloy has a great material profile and promising success rates; it has a 4–5 µm black oxide layer that becomes protective for the underlying ceramic material if the implant experiences normal anatomic, expected loads. If the protective oxide layer is scratched or flawed by means of dislocation or implant malposition, the exposure of the underlying material produces a gateway to complications of wear and local inflammatory responses [63]. In the ideal environment, the material has an increased hardness and decreased surface roughness due to its primary composition of zirconia, with increased fracture toughness and fatigue strength due to the addition of metal substrates [29, 64, 65]. Again, this material is new enough to have promising results but no long-term data exist on its success.

Material makeup of a bearing surface is only half of the battle. Manufacturing techniques have also improved the effectiveness of ceramics as total joint implants. Decreasing the grain size and increasing a material's purity aids in reducing catastrophic failures of ceramic components [29, 55, 61]. In addition, HIPing is used with ceramics, just as it is with polyethylene, to decrease the material's porosity, while tempering is used to increase toughness [29]. With these advanced techniques, the incidence of catastrophic failure has reduced from the 1970s to 2011 [55].

Furthermore, catastrophic failures are not the only complication with ceramic bearing surfaces. Incongruent articulation of the bearing surfaces can lead to edge loading that, in turn, results in stripe wear [66]. As discussed previously, stripe wear affects the oxide layer serving as the protective layer that increases fracture toughness and decreases surface roughness. Incongruent articulation can be a result of implant positioning, but it may also be related to differences in articular pressure during the swing and strike phases of gait [66]. Edge loading is also thought to contribute to audible squeaking of the hip prosthesis. This phenomenon is only seen with ceramic bearing surfaces with a 1–20% incidence of an audible squeak with range of motion of ceramic bearings [51, 58, 59, 67–70]. It has been argued that position of the implants does not cause this noisy provocation [71, 72], although there are studies attributing the squeaky bearing surface to shortened head length, lateralized cups, young age, obesity and the natural resonant frequencies associated with stem design [70, 72]. *In vitro* simulations have attributed the noise to a lack of lubrication and the material itself [73–75].

7.2   Ceramic Now

Every brand within the total joint industry has a line of ceramic implants; however, Ceram Tec (Laurens, SC) is the sole ceramic producer in North America and supplies industry. Implant companies differ according to the varying combinations of metal or non-metal as well as the presence and type of surface oxide layer they possess. The transition to ceramic on ceramic or ceramic on polyethylene has advantages over metal on metal for two reasons. The first is an excellent wear resistance profile giving it greater longevity for a younger population. The second is the reduction in femoral taper corrosion and the amount of metal ions released systemically and within the tissues [76–79]. In addition, research is continuing to expose other benefits of ceramic bearings such as decreased dislocation events and possibly decreased incidence of periprosthetic joint infection [76]. Current literature to support these suppositions is minimal, but enough significance has been shown for continued research on the subject.

The increased use of ceramic heads also followed the recall of multiple metal on polyethylene implants (Accolade I, TMZF, Stryker, Kalamazoo, MI) for trunnion corrosion and gross trunnion failure [80, 81]. There has been an increase in literature focused on trunnion corrosion and options for revision cases. In revision settings for taper corrosion at the head-neck junction with a well-fixed stem, ceramic has become a good alternative to decrease the metal-metal reaction causing fretting corrosion. Although ceramic on metal has a good track record, the surface changes on the tapered neck after head removal for trunnion corrosion has been shown to cause early failure of the ceramic head [82, 83]. To prevent femoral stem revision, titanium taper sleeves have provided a solution to prevent failure of the ceramic femoral head; however, there are few studies to support or refute the success of the metal-metal interaction of the original damaged stem taper with the titanium sleeve [83–85]. One study with an average follow-up of 2.1 years did not have any evidence of ceramic head failure or trunnion wear after revision surgery using ceramic heads with titanium sleeves [86].

7.3   Ceramic: Case Report 7 (Fig. 9)

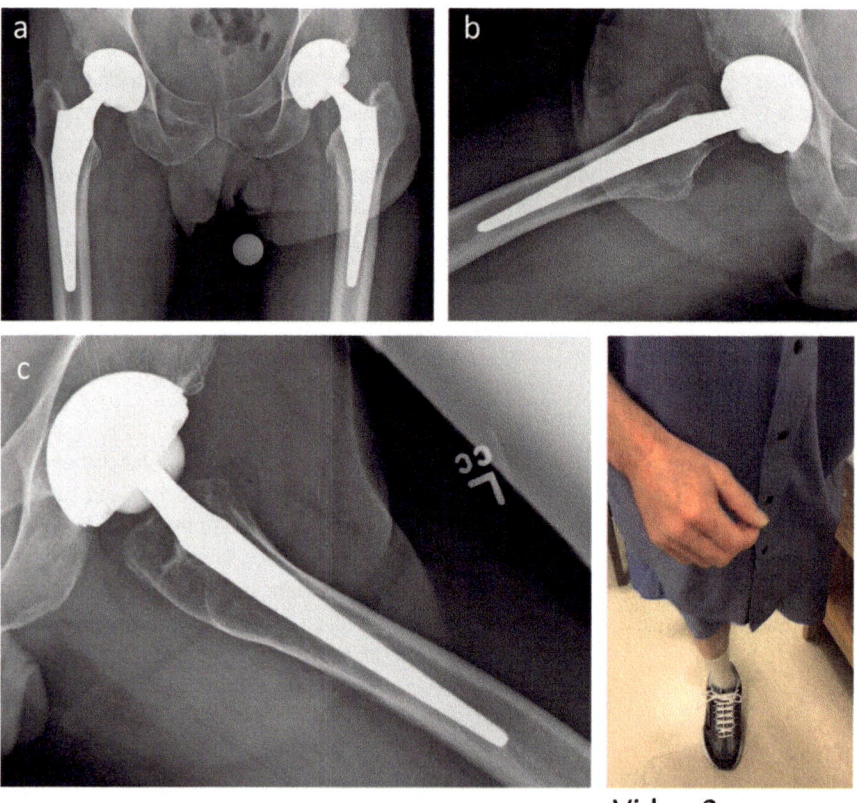

Video 2

Fig. 9 52-year-old male who underwent bilateral total hip arthroplasty for primary osteoarthritis 14 years ago with ceramic on ceramic hip prostheses. The patient presented 14 years postoperatively with right worse than left hip pain that was eventually determined to be psoas tendonitis. (**a**–**c**) show the patient's current radiographs including AP pelvis, right and left lateral images, respectively. The images show appropriate alignment and positioning of the implants with concentric femoral head placement in the acetabular shell and no signs of peri-implant osteolysis. The patient also has complaints of hip squeaking with ambulation and weight bearing range of motion demonstrated in the supplemental material of this chapter (Video 2)

8 Conclusion

The amazing advances within biomaterials have dramatically improved the lives of millions of patients around the world. With these advances have come new challenges that will require the marrying of biomaterial science and clinical experience to most benefit the patients. The perfect solution does not yet exist; however, the continued need to develop and create new surfaces and implants remains.

Acknowledgements The authors would like to acknowledge Dr. Adam Klein and Dr. Ryan Murphy for sharing their expertise and experiences. The authors would also like to thank Suzanne Danley for her critical review and editing. The work reported in this publication was supported by the National Institute of General Medical Sciences of the National Institutes of Health under Award Number 2U54GM104942-02. The Content is solely the responsibility of the authors and does not necessarily represent the official views of the National Institutes of Health.

References

1. Koenig L, et al. Estimating the societal benefits of THA after accounting for work status and productivity: a Markov model approach. Clin Orthop Relat Res. 2016;474(12):2645–54.
2. Ruiz D Jr, et al. The direct and indirect costs to society of treatment for end-stage knee osteoarthritis. J Bone Joint Surg Am. 2013;95(16):1473–80.
3. Elmallah RK, et al. Determining cost-effectiveness of total hip and knee arthroplasty using the short form-6D utility measure. J Arthroplast. 2017;32(2):351–4.
4. Kurtz S, et al. Projections of primary and revision hip and knee arthroplasty in the United States from 2005 to 2030. J Bone Joint Surg Am. 2007;89(4):780–5.
5. Kurtz S, et al. Prevalence of primary and revision total hip and knee arthroplasty in the United States from 1990 through 2002. J Bone Joint Surg Am. 2005;87(7):1487–97.
6. Kurtz SM, et al. International survey of primary and revision total knee replacement. Int Orthop. 2011;35(12):1783–9.
7. Neuprez A, et al. Patients' expectations impact their satisfaction following total hip or knee arthroplasty. PLoS One. 2016;11(12):e0167911.
8. Lim JB, et al. Comparison of patient quality of life scores and satisfaction after common orthopedic surgical interventions. Eur J Orthop Surg Traumatol. 2015;25(6):1007–12.
9. Naal FD, et al. Clinical improvement and satisfaction after total joint replacement: a prospective 12-month evaluation on the patients' perspective. Qual Life Res. 2015;24(12):2917–25.
10. Flego A, et al. Addressing obesity in the management of knee and hip osteoarthritis—weighing in from an economic perspective. BMC Musculoskelet Disord. 2016;17:233.
11. Gold HT, et al. Association of depression with 90-day hospital readmission after total joint arthroplasty. J Arthroplast. 2016;31(11):2385–8.
12. Bauer TW, Schils J. The pathology of total joint arthroplasty. II. Mechanisms of implant failure. Skeletal Radiol. 1999;28(9):483–97.
13. Blumenfeld TJ, Bargar WL. Early aseptic loosening of a modern acetabular component secondary to a change in manufacturing. J Arthroplast. 2006;21(5):689–95.
14. Bonsignore LA, Goldberg VM, Greenfield EM. Machine oil inhibits the osseointegration of orthopaedic implants by impairing osteoblast attachment and spreading. J Orthop Res. 2015;33(7):979–87.
15. Suhardi VJ, et al. A fully functional drug-eluting joint implant. Nat Biomed Eng. 2017;1.

16. McKee GK, Watson-Farrar J. Replacement of arthritic hips by the McKee-Farrar prosthesis. J Bone Joint Surg Br. 1966;48(2):245–59.
17. Charnley KEA. The classic: arthroplasty of hip: a new operation. Reprinted from Lancet pp. 1129–32, 1961. Clin Orthop Relat Res. 1973;95:4–8.
18. Bankston AB, et al. Comparison of polyethylene wear in machined versus molded polyethylene. Clin Orthop Relat Res. 1995;(317):37–43.
19. Faris PM, et al. Polyethylene sterilization and production affects wear in total hip arthroplasties. Clin Orthop Relat Res. 2006;453:305–8.
20. Rose RM, et al. On the origins of high in vivo wear rates in polyethylene components of total joint prostheses. Clin Orthop Relat Res. 1979;(145):277–86.
21. Landy MM, Walker PS. Wear of ultra-high-molecular-weight polyethylene components of 90 retrieved knee prostheses. J Arthroplast. 1988;3(Suppl):S73–85.
22. Oonishi H, et al. Wear of highly cross-linked polyethylene acetabular cup in Japan. J Arthroplast. 2006;21(7):944–9.
23. Muratoglu OK, et al. A novel method of cross-linking ultra-high-molecular-weight polyethylene to improve wear, reduce oxidation, and retain mechanical properties. Recipient of the 1999 HAP Paul award. J Arthroplast. 2001;16(2):149–60.
24. Muratoglu OK, et al. Effect of radiation, heat, and aging on in vitro wear resistance of polyethylene. Clin Orthop Relat Res. 2003;(417):253–62.
25. Tower SS, et al. Rim cracking of the cross-linked longevity polyethylene acetabular liner after total hip arthroplasty. J Bone Joint Surg Am. 2007;89(10):2212–7.
26. Birman MV, et al. Cracking and impingement in ultra-high-molecular-weight polyethylene acetabular liners. J Arthroplast. 2005;20(7 Suppl 3):87–92.
27. Oparaugo PC, et al. Correlation of wear debris-induced osteolysis and revision with volumetric wear-rates of polyethylene: a survey of 8 reports in the literature. Acta Orthop Scand. 2001;72(1):22–8.
28. Sochart DH. Relationship of acetabular wear to osteolysis and loosening in total hip arthroplasty. Clin Orthop Relat Res. 1999;363:135–50.
29. Kurtz SM. UHMWPE biomaterials handbook : ultra-high molecular weight polyethylene in total joint replacement and medical devices. 3rd ed. Amsterdam; Boston: Elsevier/WA, William Andrew is an Imprint of Elsevier; 2016. xxiv, 815 pages.
30. Kurtz SM, et al. Advances in the processing, sterilization, and crosslinking of ultra-high molecular weight polyethylene for total joint arthroplasty. Biomaterials. 1999;20(18):1659–88.
31. Peacock AJ. Handbook of polyethylene. New York, NY: Marcel Dekker; 2000.
32. Callaghan JJ. The adult hip: hip arthroplasty surgery. 3rd ed. Philadelphia: Wolters Kluwer; 2016. 2 volumes (xxxi, 1490, 30 pages).
33. Oral E, et al. Diffusion of vitamin E in ultra-high molecular weight polyethylene. Biomaterials. 2007;28(35):5225–37.
34. Oral E, et al. Wear resistance and mechanical properties of highly cross-linked, ultrahigh-molecular weight polyethylene doped with vitamin E. J Arthroplast. 2006;21(4):580–91.
35. Oral E, et al. The effects of high dose irradiation on the cross-linking of vitamin E-blended ultrahigh molecular weight polyethylene. Biomaterials. 2008;29(26):3557–60.
36. Shareghi B, Johanson PE, Karrholm J. Wear of vitamin E-infused highly cross-linked polyethylene at five years. J Bone Joint Surg Am. 2017;99(17):1447–52.
37. Currently Available HXLPE Implants Available in the US, in Zimmer Longevity Highly Cross-linked Polyethylene: Clinical Value Dossier. 2012.
38. WILES P. The surgery of the osteoarthritic hip. Br J Surg. 1958;45(193):488–97.
39. Reed-Hill R. Physical metallury principles. 2nd ed. New York: Van Nostrand; 1973.
40. Kuhn AT. Corrosion of co-Cr alloys in aqueous environments. Biomaterials. 1981;2(2):68–77.
41. Fenske G. Ion implantation, in ASM handbook: friction, lubrication and wear technology, vol. 1992. Metals Park, OH: ASM International. p. 850.
42. Bunshah R. PVD and CVD coatings, in ASM handbook: friction, lubrication and wear technology, vol. 1992. Metals Park, OH: ASM International. p. 840.

43. Streicher RM, et al. New surface modification for Ti-6Al-7Nb alloy: oxygen diffusion hardening (ODH). Biomaterials. 1991;12(2):125–9.
44. Jones DM, et al. Focal osteolysis at the junctions of a modular stainless-steel femoral intramedullary nail. J Bone Joint Surg Am. 2001;83A(4):537–48.
45. Urban RM, et al. Migration of corrosion products from modular hip prostheses. Particle microanalysis and histopathological findings. J Bone Joint Surg Am. 1994;76(9):1345–59.
46. Jacobs JJ, et al. Local and distant products from modularity. Clin Orthop Relat Res. 1995;(319):94–105.
47. Gilbert J. Mechanically assisted corrosion of biomedical alloys. In: American Society for Materials Handbook, Corrosion 2006. Materials Park, OH: ASM International. p. 826–36.
48. Gilbert JL, et al. In vivo oxide-induced stress corrosion cracking of Ti-6Al-4V in a neck-stem modular taper: emergent behavior in a new mechanism of in vivo corrosion. J Biomed Mater Res B Appl Biomater. 2012;100((2):584–94.
49. ASTM Annual Book of Standards. vol. 13. Philadelphia, PA: American Society for Testing and Materials; 1995.
50. Kwon YM, et al. Analysis of wear of retrieved metal-on-metal hip resurfacing implants revised due to pseudotumours. J Bone Joint Surg Br. 2010;92(3):356–61.
51. Masonis JL, et al. Zirconia femoral head fractures: a clinical and retrieval analysis. J Arthroplasty. 2004;19(7):898–905.
52. Huot JC, et al. The effect of radiation dose on the tensile and impact toughness of highly cross-linked and remelted ultrahigh-molecular weight polyethylenes. J Biomed Mater Res B Appl Biomater. 2011;97((2):327–33.
53. Atwood SA, et al. Tradeoffs amongst fatigue, wear, and oxidation resistance of cross-linked ultra-high molecular weight polyethylene. J Mech Behav Biomed Mater. 2011;4(7):1033–45.
54. Hannouche D, et al. Ceramics in total hip replacement. Clin Orthop Relat Res. 2005;(430):62–71.
55. Jeffers JR, Walter WL. Ceramic-on-ceramic bearings in hip arthroplasty: state of the art and the future. J Bone Joint Surg Br. 2012;94(6):735–45.
56. Hamadouche M, et al. Alumina-on-alumina total hip arthroplasty: a minimum 18.5-year follow-up study. J Bone Joint Surg Am. 2002;84-A(1):69–77.
57. Wroblewski BM, et al. Prospective clinical and joint simulator studies of a new total hip arthroplasty using alumina ceramic heads and cross-linked polyethylene cups. J Bone Joint Surg Br. 1996;78(2):280–5.
58. Wroblewski BM, Siney PD, Fleming PA. Low-friction arthroplasty of the hip using alumina ceramic and cross-linked polyethylene. A ten-year follow-up report. J Bone Joint Surg Br. 1999;81(1):54–5.
59. Wroblewski BM, Siney PD, Fleming PA. Low-friction arthroplasty of the hip using alumina ceramic and cross-linked polyethylene. A 17-year follow-up report. J Bone Joint Surg Br. 2005;87(9):1220–1.
60. Allain J, et al. Poor eight-year survival of cemented zirconia-polyethylene total hip replacements. J Bone Joint Surg Br. 1999;81(5):835–42.
61. De Aza AH, et al. Crack growth resistance of alumina, zirconia and zirconia toughened alumina ceramics for joint prostheses. Biomaterials. 2002;23(3):937–45.
62. Chevalier J. What future for zirconia as a biomaterial? Biomaterials. 2006;27(4):535–43.
63. Tribe H, et al. Advanced wear of an Oxinium™ femoral head implant following polyethylene liner dislocation. Ann R Coll Surg Engl. 2013;95(8):e133–5.
64. Hernigou P, Mathieu G, Poignard A. Oxinium, a new alternative femoral bearing surface option for hip replacement. Eur J Orthop Surg Traumatol. 2007;17(3):243–6.
65. Kop AM, Whitewood C, Johnston DJ. Damage of oxinium femoral heads subsequent to hip arthroplasty dislocation three retrieval case studies. J Arthroplasty. 2007;22(5):775–9.
66. Walter WL, et al. Edge loading in third generation alumina ceramic-on-ceramic bearings: stripe wear. J Arthroplasty. 2004;19(4):402–13.
67. Jarrett CA, et al. The squeaking hip: a phenomenon of ceramic-on-ceramic total hip arthroplasty. J Bone Joint Surg Am. 2009;91(6):1344–9.

68. Mai K, et al. Incidence of 'squeaking' after ceramic-on-ceramic total hip arthroplasty. Clin Orthop Relat Res. 2010;468(2):413–7.
69. Swanson TV, et al. Influence of prosthetic design on squeaking after ceramic-on-ceramic total hip arthroplasty. J Arthroplast. 2010;25(6 Suppl):36–42.
70. Kiyama T, Kinsey TL, Mahoney OM. Can squeaking with ceramic-on-ceramic hip articulations in total hip arthroplasty be avoided? J Arthroplast. 2013;28(6):1015–20.
71. Restrepo C, et al. The noisy ceramic hip: is component malpositioning the cause? J Arthroplast. 2008;23(5):643–9.
72. Stanat SJ, Capozzi JD. Squeaking in third- and fourth-generation ceramic-on-ceramic total hip arthroplasty: meta-analysis and systematic review. J Arthroplasty. 2012;27(3):445–53.
73. Currier SF, Mautner HG. On the mechanism of action of choline acetyltransferase. Proc Natl Acad Sci U S A. 1974;71(9):3355–8.
74. Chevillotte C, et al. The 2009 frank Stinchfield award: "hip squeaking": a biomechanical study of ceramic-on-ceramic bearing surfaces. Clin Orthop Relat Res. 2010;468(2):345–50.
75. Chevillotte C, et al. Hip squeaking: a 10-year follow-up study. J Arthroplast. 2012;27(6):1008–13.
76. Kurtz SM, et al. Do ceramic femoral heads reduce taper fretting corrosion in hip arthroplasty? A retrieval study. Clin Orthop Relat Res. 2013;471(10):3270–82.
77. Gill IP, et al. Corrosion at the neck-stem junction as a cause of metal ion release and pseudo-tumour formation. J Bone Joint Surg Br. 2012;94(7):895–900.
78. Chana R, et al. Mixing and matching causing taper wear: corrosion associated with pseudotumour formation. J Bone Joint Surg Br. 2012;94(2):281–6.
79. Hallab NJ, et al. Differences in the fretting corrosion of metal-metal and ceramic-metal modular junctions of total hip replacements. J Orthop Res. 2004;22(2):250–9.
80. Urish KL, et al. Trunnion failure of the recalled low friction ion treatment cobalt chromium alloy femoral head. J Arthroplast. 2017;32(9):2857–63.
81. Patel S, Talmo CT, Nandi S. Head-neck taper corrosion following total hip arthroplasty with Stryker meridian stem. Hip Int. 2016;26(6):e49–51.
82. Gührs J, et al. The influence of stem taper re-use upon the failure load of ceramic heads. Med Eng Phys. 2015;37(6):545–52.
83. MacDonald DW, et al. Fretting and corrosion damage in taper adapter sleeves for ceramic heads: a retrieval study. J Arthroplast. 2017;32(9):2887–91.
84. Esposito CI, et al. What is the trouble with trunnions? Clin Orthop Relat Res. 2014;472(12): 3652–8.
85. Jack CM, et al. The use of ceramic-on-ceramic bearings in isolated revision of the acetabular component. Bone Joint J. 2013;95-B(3):333–8.
86. Thorey F, et al. Early results of revision hip arthroplasty using a ceramic revision ball head. Semin Arthroplast. 2011;22:284–9.

Modulating Innate Inflammatory Reactions in the Application of Orthopedic Biomaterials

Tzuhua Lin, Eemeli Jämsen, Laura Lu, Karthik Nathan, Jukka Pajarinen, and Stuart B. Goodman

Keywords Orthopedic biomaterials · Inflammation · Macrophage polarization · Immunomodulation · Wear particles · Osteolysis · Bone remodeling · Tissue engineering · Mesenchymal stem cell · Osteoblast · Osteoclast · NF-kB

Abbreviations

CCL2	C-C motif chemokine ligand 2
FBGC	Foreign-body giant cell
GM-CSF	Granulocyte macrophage colony-stimulating factor
IFN-γ	Interferon gamma
IL	Interleukin
LPS	Lipopolysaccharide
M-CSF	Macrophage colony-stimulating factor
MSC	Mesenchymal stem cell
ODN	Oligodeoxynucleotide
OPG	Osteoprotegerin
PAMP	Pathogen-associated molecular pattern
PDGF	Platelet-derived growth factor
PGA	Poly-glycolic-acid
PLA	Poly-lactic-acid
PLGA	Poly-lactic-glycolic-acid

T. Lin · E. Jämsen · L. Lu · K. Nathan · J. Pajarinen
Departments of Orthopedic Surgery, Stanford University School of Medicine, Redwood City, CA, USA

S. B. Goodman (✉)
Departments of Orthopedic Surgery, Stanford University School of Medicine, Redwood City, CA, USA

Department of Bioengineering, Stanford University, Stanford, CA, USA
e-mail: goodbone@stanford.edu

© Springer International Publishing AG, part of Springer Nature 2018   199
B. Li, T. Webster (eds.), *Orthopedic Biomaterials*,
https://doi.org/10.1007/978-3-319-89542-0_10

PRR   Pattern-recognition receptor
RANKL   Receptor-activator of NF-κB ligand
RNAi   RNA interference
TGF-β   Transforming growth factor-β
TLR   Toll-like receptor
TNF-α   Tumor necrosis factor-α
VEGF   Vascular endothelial growth factor

1 Introduction

Musculoskeletal conditions affect millions of people and are the second leading cause of disability worldwide [1]. Through orthopedic care, patients can find pain relief, increased mobility, and improved quality of life, thus bettering the lives of individuals and improving society as a whole. Many orthopedic procedures require implants, and advances over the past half-century have allowed for the development of biomaterials with both enhanced mechanical and biological function [2]. Despite these improvements, many implants do not last forever: up to 15% of total joint implants require operative revision within 15 years of initial surgery [3, 4]. With over one million Americans undergoing total joint replacement annually, there is a need to improve the biological function and longevity of orthopedic biomaterials [5].

It is well known that inflammation is induced by implants and their resulting wear particles. On the other hand, early, transient inflammation is essential for proper bone formation and osseointegration of the implant [6]. This process is mediated through prostaglandins and macrophage-related inflammation, mimicking the natural fracture healing response beginning with acute inflammation and resolving into repair and regeneration of peri-implant tissues [7, 8]. However, if inflammation continues, the body may mount a foreign body chronic inflammatory reaction, leading to enhanced and persisting inflammation, bone resorption, osteolysis, and ultimately implant failure [3]. Wear particles produced from the bearing surface, modular connections, and motion at the interface of the implant and bone can activate the NALP inflammasome, the NF-κB pathway, and toll-like receptors (TLR)-2 and TLR-4 depending on the material, size of the particles, and surface topology, which are reviewed thoroughly by Cobelli et al., Gibon et al., and Lin et al. [4, 9, 10].

Chronic inflammation can arise from aberrant activity of the immune system to clear wear particles. If wear debris are too large for macrophages to remove, macrophages fuse to form multinucleated foreign body giant cells (FBGCs) [11]. Normally, these FBGCs are able to degrade or sequester wear particles, allowing for short-lived inflammation, and eventual resolution and repair; however, if the immune system is overwhelmed by excessive particles of appropriate size, inflammation persists [10]. This inflammation in conjunction with continued micromotion propagates wear debris formation, macrophage and T cell infiltration, and eventual osteolysis [10, 12]. As such, chronic inflammation is a vicious cycle of persistent inflammation, implant wear, and bone loss.

The longevity of an orthopedic implant is dependent upon the implant material, operative procedure, and patient-related factors [3]. Precise modulation of inflammation post-operatively has the potential to increase the lifespan of orthopedic implants. By integrating our understanding of the cellular and molecular mechanisms underlying prolonged inflammation, it is possible to develop optimized biomaterials that not only mitigate chronic inflammation, but also enhance osseointegration, vascularization, mechanical strength, and long-term healing.

2 Inflammation and Immunomodulating Strategy

2.1 Innate Immune Response and Macrophages

Cells of the innate immune system, particularly macrophages, are the main inflammatory mediators that drive successful implant integration or rejection [13, 14]. Macrophages also play a crucial role in tissue maintenance and regulation of inflammation [15, 16]. All tissues contain a specific set of macrophages known as tissue resident macrophages that remove damaged, senescent, and infected cells to maintain tissue homeostasis. These tissue resident macrophages originate from circulating, myeloid-derived, monocytes or, in some cases, are distributed among tissues during embryonic development and are maintained by local pools of precursor cells [17]. Macrophages are specialized to effectively phagocytize and remove cellular debris, microbes, and foreign substances identified as potentially dangerous or pathogenic. These foreign agents are recognized by various families of pattern recognition receptors (PRRs) that are expressed on the cell surface, endosomes, and cytoplasm of macrophages [18, 19]. Activation of PRRs initiates intracellular downstream signaling pathways that activate phagocytosis and can lead to the secretion of various cytokines and chemokines. The degree of this innate immune response depends on the nature and amount of the activating stimuli: phagocytosis of apoptotic cells induces an anti-inflammatory response to maintain tissue homeostasis and immune tolerance, while recognition of necrotic cells, damaged extracellular matrix, or pathogens initiates a pro-inflammatory response [19, 20]. Secreted inflammatory mediators then stimulate further cytokine secretion from other resident cells and recruit more immune cells to the site of inflammation. During acute or chronic inflammation, a large number of bone marrow-derived monocytes are recruited from the circulation to the inflamed tissues and undergo differentiation to inflammatory or tissue regenerative macrophages as guided by the local microenvironment [21].

Following the recognition of threatening agents, macrophages engulf these agents to restrict any deleterious effects on adjacent tissues. This phagocytosis process triggers enzymatic degradation of the engulfed material inside macrophages in some cases and, in the case of foreign biological structures, leads to antigen presentation to activate the adaptive immune response. As regulated by macrophages, the immune system thus attempts to efficiently dispose of the harmful substance, kill the microbes encountered, and prevent further tissue injury. As an inflammatory reaction proceeds and the danger becomes resolved, macrophages begin to promote tissue healing and regeneration by secreting extra-cellular matrix precursors, a range of growth factors such as vascular endothelial growth factor (VEGF) and platelet derived growth factor (PDGF), and anti-inflammatory cytokines [22]. The cytokine signaling in the local microenvironment thus coordinates the development of an immune response and macrophage function [23].

2.2 Macrophage Polarization

Macrophages are able to undergo functional changes as instructed by the surrounding cytokine milieu in tissues [16]. These phagocytes assume a distinct phenotype with divergent inflammatory, fibrotic, and regenerative properties necessary for different phases of inflammation and tissue regeneration. Initially, macrophages become activated to a pro-inflammatory phenotype following the recognition of a stimulating agent by PRRs. The specific factors promoting this classical macrophage activation, known as M1 polarization, is comprised of endogenous danger signals released from necrotic cells and pathogen-associated molecular patterns (PAMPs), such as lipopolysaccharide (LPS), released from various invading microorganisms. Activators with a similar effect also include cytokines, especially, interferon gamma (IFN-γ), granulocyte-monocyte colony stimulating factor (GM-CSF), and tumor necrosis factor alpha (TNF-α) [24]. Macrophages with a pro-inflammatory phenotype are essential for the early phase of repair, but prolonged inflammation by these M1 macrophages exacerbate tissue injury by actively phagocytizing potential pathogens, killing intracellular microbes by producing oxygen and nitrogen radicals, and vigorously secreting more inflammatory cytokines and chemokines.

In contrast, alternatively activated macrophages, also known as M2 polarization, have tissue regenerative, pro-fibrotic, and anti-inflammatory characteristics. Various subsets of this phenotype are induced by a combination of cytokines such as interleukin-4 (IL-4), IL-10, IL-13, transforming growth factor beta (TGF-β), glucocorticoids or macrophage colony-stimulating factor (M-CSF) [25]. For example, macrophages treated with IL-4 and IL-13 produce minimal amounts of proinflammatory cytokines, and their other secretory products stimulate cell growth and proliferation as well as collagen formation. Hence, M1 and M2 polarization are considered to represent the opposite ends of the continuum of macrophage phenotypes [16, 26]. Whereas M1 polarized macrophages predominate in a strong proinflammatory phase at an early stage of an inflammation, M2 polarization gradually takes over when the intrusive agents become cleared. The tissue under inflammatory signaling likely contains macrophages with mixed phenotypes, and crosstalk between these cells enables a proper healing process and the resolution of the inflammation.

2.3   Interaction Between Macrophages and Orthopedic Biomaterials

Tissue injury caused by surgical insertion of an orthopedic implant initially activates the immune system, but with time, the implant itself mediates inflammation as a foreign body [27]. The initial recognition of a biomaterial and the tissue trauma caused by the implantation is primarily performed by resident macrophages, which subsequently become activated to a pro-inflammatory phenotype and initiate an inflammatory response. Since many orthopedic implants, such as joint replacements, are generally designed for permanent tissue and bone integration, they are biologically non-degradable, and might provide a constant stimulus for macrophage activation. In particular, the release of particulate materials of a phagocytosable size (<10 μm in diameter) has proven to provide a constant stimulus for inflammatory macrophage activation; these phagocytosable particles have been shown to induce endosomal damage with subsequent activation of intracellular danger sensing mechanisms [28, 29]. A prolonged presence of M1 macrophages leads to an increased inflammatory status, fibrosis, and granulomatous tissue around the implant—a condition called the chronic foreign body reaction [11, 30].

The long-lasting inflammatory events at the bone-implant interface have been observed to significantly affect the bone repair process and cause implant failures via osteolysis [31, 32]. At the cellular level, pro-inflammatory mediators such as TNF-α favor bone resorption over bone formation by increasing the production of Receptor Activator of NF-κB Ligand (RANKL) and decreasing the production of osteoprotegerin (OPG) resulting in an altered RANKL/OPG ratio [33]. RANKL efficiently stimulates the activation and proliferation of osteoclast precursors whereas OPG acts as a decoy receptor inhibiting RANKL signaling. Sustained inflammation thus drives osteoclast formation and ultimately failure of the implant.

As implant-mediated inflammation closely involves adverse tissue reactions and bone regeneration, novel approaches for biomaterial engineering are being developed: a new generation of orthopedic biomaterials should be able to modulate the immune environment in order to favor osseointegration of the implant [34]. This immunomodulating strategy aims to extend the lifespan of the implant by minimizing the destructive and maximizing the regenerative effects of the immune response induced by the implant.

2.4   Modulation of Macrophage-Mediated Pro-Inflammatory Response

Macrophages are a prime target for immunomodulation in the application of orthopedic biomaterials. This is not only because these cells play an essential role in initiating and regulating the implant-mediated immune responses, but also because of their considerable heterogeneity and plasticity enable the modulation of their

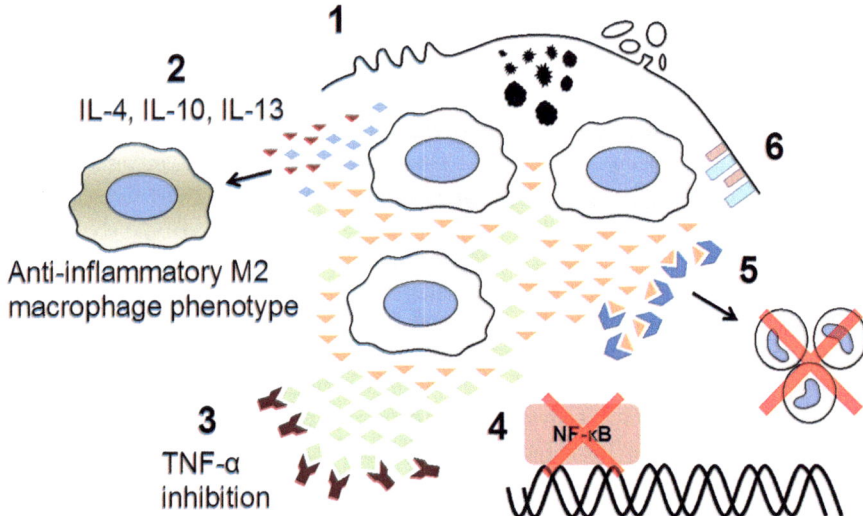

Fig. 1 Strategies to modulate the innate inflammatory reactions against orthopedic biomaterials. *1*. Optimize biomaterial characteristics, e.g. surface roughness, porosity, and generation of wear particles, *2*. Delivery of macrophage polarizing cytokines to drive the anti-inflammatory M2 macrophage polarization, *3*. Inhibition of pro-inflammatory cytokines such as TNF-α, *4*. Blockade of the transcription factor NF-κB, *5*. Inhibiting chemokines such as CCL2 to suppress monocyte recruitment, and *6*. Coupling biomaterials with anti-inflammatory and bioactive molecules

function [16, 30]. Methods controlling macrophage activation could discourage increases in the pro-inflammatory signaling, avoid excess fibrosis, and prevent bone loss around the implants. Thus, new therapeutic interventions are being pursued with the purpose of controlling chronic inflammation associated with implant materials by modulating macrophage polarization and thus their secretory products. Whereas continuous M1 activation impairs integration of the implant, signals that suppress the pro-inflammatory effects and support M2 polarization have emerged as an attractive means to facilitate implant integration [35, 36].

Several different strategies for macrophage-targeted immunomodulation around implants have been developed (Fig. 1) [37]. Since the degree of an implant-mediated immune reaction depends on the biomaterial characteristics, beneficial effects on macrophage function and implant integration may be achieved by modifying the physical and chemical properties of the biomaterial. For instance, the specific surface structure of the implant material and the amount of wear products accumulating in the surrounding tissues are important variables that determine the type of macrophage activation. Surface topography of the implant can be optimized for porosity, roughness, hydrophilicity, and the ability to produce wear particles in order to decrease initial monocyte adhesion and activation [34]. These micro- and nanoscale material characteristics largely determine the folding of absorbed proteins on the implant and consequent presentation of bioactive sites for macrophages. Moreover, TGF-β and PDGF directly modulate macrophage function and chemotaxis during wound healing without a foreign body and may play a similar role in peri-implant biology [38]. Implants loaded with these molecules could thus theoretically promote tissue and bone regeneration both directly and indirectly.

Incorporation of immunomodulatory agents into the implant constitutes another major strategy to modulate innate immune reactions. For example, macrophage-mediated inflammation could be controlled by the local release of M2 polarizing cytokines IL-4, IL-10, or IL-13 [35]. In particular, IL-4 has shown great potential to increase implant integration to bone and mitigate wear particle-induced inflammation in animal models [39–41]. Delivery of IL-4 alters the function of local M1 activated macrophages towards an anti-inflammatory M2 phenotype and dramatically reduces the production of pro-inflammatory cytokines. In addition, IL-4 has anti-osteoclastogenic effects that might promote osseointegration of the implant. IL-10 and IL-13 possess similar immune-regulatory properties. These cytokines have been reported to inhibit the expression of pro-inflammatory cytokines in macrophages and drive M2 activation [25]. Several studies that used a murine subcutaneous implantation model rather than delivery of wear particles demonstrated that the release of IL-4 attracts M2 macrophages and modulates the inflammatory response to improve the implant integration [42–44]. Interestedly, sequential delivery of M1 and M2 polarizing factors mimicking the natural course of tissue regeneration enhanced implant vascularization. The anti-inflammatory phenotypic switch promoted implant integration also by diminishing the formation of a fibrous capsule and increasing the quality of remodeled collagen around the implant. Further studies are needed to investigate the full potential of M2 polarizing cytokines in orthopedic applications.

In addition to favoring M2 polarization, improved tissue healing around an implant could potentially be achieved by directly inhibiting pro-inflammatory signals. For example, TNF-α, one of the most potent pro-inflammatory cytokines, promotes M1 macrophage polarization, enhances fibrosis, inhibits osteoblast differentiation, induces osteoclast formation, and thus mediates osteolysis around orthopedic implants [45, 46]. Blocking these effects by anti-TNF-α therapy provides a means to modulate implant-induced immune responses. Etanercept, a decoy receptor for TNF-α, was shown to mitigate wear particle-induced cytokine production from macrophages and reduce bone resorption in animals but was not effective in a small clinical trial [47, 48]. Similar results were obtained using an antisense oligonucleotide targeting to mouse TNF-α mRNA in a murine calvarial model [49]. However, blocking the effect of only one pro-inflammatory mediator among the complicated signal network may not be enough to prevent osteolysis in the long term. The compensatory actions of other pro-inflammatory cytokines, such as IL-1β and IL-6, could maintain an inflammatory status in the peri-implant tissue in the absence of TNF-α signaling. A combination of locally delivered cytokine inhibitors might thus prove to be more effective.

Transcription factor NF-κB serves as another target for immunomodulation in the context of implant-mediated immune response [35]. This transcription factor functions as a key regulator of multiple inflammatory cytokines and chemokines in macrophages, and becomes active as a result of a relevant PRR stimulus [10, 50].

Moreover, NF-κB mediates the RANKL signaling, which is integral for osteoclast differentiation and activation. Thus, inhibition of this transcription factor offers an intriguing possibility to attenuate the biomaterial-induced inflammation and osteolysis. This inhibitory effect has been demonstrated using a decoy oligodeoxynucleotide (ODN) that competitively binds NF-κB *in vitro* and *in vivo* with polyethylene particles as the adverse inflammatory stimulus; the suppression of intracellular signaling in macrophages resulted in less cytokine expression and osteoclast activation [51, 52]. NF-κB decoy may also suppress the production of chemokines essential for monocyte recruitment.

Preventing the continued recruitment of immune cells to the bone-implant interface could mitigate the inflammatory reaction and periprosthetic bone loss and constitutes another strategy for immunomodulation. Indeed, a chemokine directed immunomodulatory method was recently established using a mutant C-C motif chemokine ligand 2 (CCL2) protein to inhibit CCL2 signaling [53, 54]. Anti-CCL2 therapy suppressed macrophage recruitment to the implant in a murine model and prevented wear particle induced inflammation and bone loss.

Lastly, orthopedic biomaterials can be coupled with anti-inflammatory drugs such as glucocorticoids. These drugs elicit an alternative macrophage phenotype with an increased ability to recognize and scavenge dying cells. These macrophages suppress the production of numerous inflammatory mediators such as pro-inflammatory cytokines, chemokines, prostaglandins, leukotrienes, and proteolytic enzymes, whereas an enhanced expression of anti-inflammatory cytokine IL-10 has been reported. Other bioactive molecules that can also be considered as immunomodulatory, include TGF-β, VEGF, and PDGF [27]. These growth factors tightly regulate the healing process by targeting fibroblasts and endothelial cells, rather than macrophages. Moreover, TGF-β and PDGF directly modulate macrophage function and chemotaxis at least during wound healing without a foreign body [38]. Implants loaded with these molecules could thus potentially also promote bone regeneration and implant integration either directly or indirectly.

3 Sequential Modulation of Inflammatory Response for Optimal Bone Regeneration/Osseointegration

3.1 Essential Role of Acute Inflammation in Bone Regeneration

Determination of the appropriate timeframe of immunomodulation is critical for optimizing their application as orthopedic biomaterials. Acute phase inflammation is crucial for proper bone repair after trauma. Impairing early inflammatory conditions in a murine fracture model resulted in diminished stem cell recruitment and differentiation, fracture callus formation, and overall bone growth [55–58]. The inflammatory phase sparks the repair cascade by initiating angiogenesis, recruiting and stimulating the differentiation of mesenchymal stem cells (MSCs), and encouraging extracellular matrix synthesis [59–61].

Specific cytokines appear to be tied to the inflammatory phase of bone repair, namely TNF-α and IL-1 [62]. TNF-α and IL-1 are more commonly known for mediating foreign body reactions that can result in impaired tissue function and rejection of prosthetic implants [63]. Gerstenfeld et al. showed that a reduction in TNF-α signaling results in improper formation of fracture callus and delayed endochondral and intramembranous bone formation [56]. The key difference between pathological and therapeutic inflammation is that the latter is highly regulated, both in intensity, duration, and timing to provide a foundation for healing [62].

3.2 Transition of Macrophage Polarization Status for Optimal Bone Formation

Macrophage polarization status also plays a critical role in bone regeneration. M1 macrophages, despite releasing inflammatory cytokines, are highly angiogenic, stimulate early mineralization by MSCs, and support overall bone healing [64–67]. M2 macrophages, on the other hand, secrete anti-inflammatory cytokines such as IL-10 and IL-1Ra and have been associated with enhanced bone formation [68–70]. This proves to be a delicate balance that can result in failed bone regeneration if tipped too far one way or another. As such, the interplay between M1- and M2-dominated microenvironments is one that provides interesting avenues through which to pursue new immune-modulatory therapies.

After an injury, the acute inflammatory phase has been shown to last from 3–7 days before the anti-inflammatory phase begins to exert its longer-lasting influence [43, 71, 72]. Proper timing of the transition between the two phases is crucial to optimal bone regeneration. Indeed, Loi et al. showed that transition from M1 to M2-like macrophages at 72 hours resulted in significantly increased osteogenesis by MC3T3 osteoprogenitors in a co-culture model. Further studies exploring the mechanisms and temporal modulation of the M1 to M2 transition are warranted, as this could provide a prime early target for improved bone repair and implant integration.

The task of stimulating M1 macrophages to transition to M2 macrophages to enhance bone regeneration is one that is currently under investigation. One possible method is to utilize a controlled release system to maintain a short period of M1, followed by a transition to M2 polarization via cytokines such as IFN-γ, IL-4, and IL-10. Kumar et al. reported the development of a multi-domain peptide hydrogel that delivered IL-4 and CCL2 in a biphasic manner. This biphasic, sustained delivery was able to modulate both non-polarized (M0) and M1 macrophages towards an M2-like phenotype [73]. Finally, Spiller et al. utilized a decellularized bone scaffold to release IFN-γ over the first 3 days of repair, along with release of IL-4 over the first 6 days. The bone scaffolds were able to spur polarization towards an M2 phenotype *in vitro* and led to enhanced angiogenesis in an *in vivo* subcutaneous murine model [43]. The modulation of M1 to M2 is not limited to cytokine release systems; Rostam et al. has shown that physical and chemical modifications to biomaterial surfaces alone can shift the macrophage polarization towards M1 or M2 [74, 75].

4 Application of Immunomodulating Reagents on Orthopedic Biomaterials

The delivery method of various immunomodulating reagents to enhance the performance of orthopedic biomaterials is dependent on the physical and biological characteristics of the agent. The therapeutic molecules with different biological features including molecular size, hydrophilic/hydrophobic, stability (degradation rate), effective dose, and the optimal administration time points determine the optimal strategy for drug delivery. Different materials used for orthopedic implants can also influence the drug delivery efficiency. For example, the absorption of small peptides on the metal surface is ineffective compared to the application on a polymeric surface.

Surface coating and drug releasing materials are an interesting strategy to modulate the tissue environment surrounding orthopedic implants. The various strategies to apply these bioactive coating on orthopedic implants including hydrogel, layer-by-layer, and immobilization have been summarized comprehensively in other reviews [76, 77]. Agarwal et al. summarized strategies to enhance osseointegration of orthopedic implants by biomolecules such as growth factors, with similar delivery methods being potentially applicable for the delivery of immunomodulating reagents [78]. In the following section, the immunomodulating candidates are classified into four categories including: (1) protein, (2) nucleic acid, (3) small molecule, and (4) cell-based therapies (Table 1). The current development of administration strategies and the therapeutic effects in the application of orthopedic biomaterials are discussed.

Table 1 Delivery strategies for immunomodulating biomolecules

Biomolecules	Size	Delivery strategies	Reference
Protein	~150 kDa (large) 15-21 kDa (small) <5 kDa (peptide)	Hydrogel, layer-by-layer coating, immobilization, controlled release scaffold	[44, 47, 53, 82]
Nucleic acid	10-15 kDa (RNAi, ODN) >3000 kDa (plasmid)	Viral (lentivirus, adenovirus, adeno-associated virus, etc.) and non-viral (polymer, liposome, chitosan, etc.) vector (can be combined with other scaffold such as hydrogels)	[51, 89–94]
Small molecules	<0.9 kDa	Conjugation with polymeric carrier, controlled release scaffold	[10, 97–99]
Cell therapy (MSC)	~25 μm in diameter	Natural or synthetic scaffold, bone/inflammation-targeting vehicles	[113, 115, 116]

4.1   Protein-Based Biomolecules

The size of the immunomodulating protein determines the biomaterial coating strategy and release pattern. Large proteins such as antibodies or fusion protein inhibitors (~150 kDa) targeting pro-inflammatory cytokines or the associated receptors can be coated with hydrogels with larger pore sizes. Anti-TNFα antibodies conjugated with a hyaluronic acid hydrogel was applied to a burn wound and demonstrated an inhibitory effect on inflammation [79]. Direct treatment of a soluble TNFα inhibitor (Etanercept) mitigated wear particle-induced osteolysis [47]. However, no significant difference was observed between Etanercept and placebo-treated patients with acetabular loosening [48]. The results may be due to the limited number of patients, or the compensatory effects of other pro-inflammatory cytokines.

Small proteins including anti-inflammatory cytokines IL-4, IL-10, and IL-13 (ranging from 15 to 21 kDa) can be applied via a hydrogel with smaller pore size or layer-by-layer coating. A nanometer thickness IL-4 eluting layer-by-layer coated polypropylene mesh showed improved implant integration and enhanced M2 macrophage polarization in a subcutaneous implantation murine model [44]. Further validation is required to demonstrate the potential to improve osseointegration of IL-4 eluting bone implants. Another example of this protein delivery approach demonstrated that titanium rods coated with mutant CCL2 protein (7ND) with a layer-by-layer technique mitigated polyethylene wear particle-induced osteolysis in a murine femoral infusion model (See Sect. 2.4 for details) [53].

Small peptides with anti-microbial and immunomodulating activity have recently been identified [80, 81]. Compared to whole protein biomolecules, a higher concentration of small peptides could be potentially applied onto or within biomaterials and thus increase the immunomodulating efficiency [76]. Inhibition of NF-κB activation by a small peptide termed NEMO-binding domain peptide suppressed poly(methyl methacrylate) (PMMA) induced osteoclastogenesis and osteolysis in a murine calvarial model, yet the modulation of an inflammatory response was not characterized [82].

4.2   Nucleic Acid

Gene therapy is mediated through the delivery of nucleic acid-based biomolecules, including plasmid DNA, RNA interference (RNAi), micro-RNA, and ODN, to express proteins or modulate gene expression in the target cells. Delivery of naked nucleic acid is inefficient due to low cell attachment/uptake and rapid nuclease-mediated degradation. Therefore, viral and non-viral vectors are utilized to mediate the delivery of anti-inflammatory genes or silence pro-inflammatory gene expression *in vivo*. Viral vectors are efficient in transducing target gene expression *ex vivo* and thus are effective tools to induce gene expression in cell-based therapy (see

Sect. 4.4). In contrast, the immunogenicity and potential cytotoxicity effects of viral vectors may limit their direct translational use *in vivo*. Non-viral vectors, including calcium phosphate, liposomes, nano-hydroxyapatite [83], chitosan [84, 85], poly-ethyleneimine [86], and dendrimer [87], have lower immunogenicity and cytotoxicity but also lower transfection efficiency *in vivo*. Raftery et al. summarized the current development of delivering nucleic acid-based biomolecules in orthopedic biomaterials [88].

The combination of scaffolds and gene delivery vectors is a highly promising strategy for prolonged immunomodulation and controlled released of nucleic acid-based biomolecules. Previous studies showed that a combination of collagen or poly-lactic-co-glycolic acid (PLGA) scaffolds with viral or non-viral vectors delivered plasmid DNA or RNAi enhances tissue regeneration [89–94]. The therapeutic potential of immunomodulation using this strategy remains to be investigated in inflammatory bone disorders. Decoy ODN can be taken up by the cellular receptor in a sequence-specific manner [95]. The administration of decoy ODN without delivery vectors via local infusion was shown to mitigate orthopedic wear particle-induced osteolysis [51].

4.3   Small Molecules

Small molecule drugs have several advantages in clinical applications including the efficient administration and relative low cost for large-scale production. Steroids and molecular kinase inhibitors are potent anti-inflammatory small molecules that could be applied to orthopedic biomaterials. Signal transduction pathways including NF-κB and MAP kinase are crucial for the regulation of inflammatory responses [50, 96] and periprosthetic osteolysis [10, 97]. Titanium particles have induced VEGF expression and increased macrophage chemotactic activity in primary human macrophages, which was inhibited by MAPK kinase inhibitor PD98059 [98].

A daily injection of *N*- (2-hydroxypropyl) methacrylamide copolymer-dexamethasone conjugate mitigated osteolysis in the murine femur infused with PMMA particles [99]. Systemic bone loss was not observed in the conjugated dexamethasone injected mice. Several advanced drug-delivery strategies have been developed to apply dexamethasone in pre-clinical inflammatory disease models [100–102]. An inflammation-targeting hydrogel generated from ascorbyl palmitate was developed to deliver dexamethasone in an inflammatory bowel disease model [103]. While these drug delivery strategies have shown promise for the treatment of inflammatory disorders, the application in orthopedic biomaterials remains to be examined.

4.4 Cell-Based Therapy

MSCs-based therapy has been applied to bone tissue engineering and inflammatory disorders. The ability to modulate innate [104, 105] and adaptive immune responses [106] further underscored its translational potential to modulate inflammation associated with orthopedic biomaterials. Moreover, MSCs can serve as gene expression carriers to secrete immunomodulating cytokines such as IL-4 or IL-10 [107, 108]. The applications of MSC-based therapy in bone regeneration and immunomodulation are discussed in other reviews [109–111]. The following section focuses on scaffold and delivery strategies in MSC-based therapy.

MSC based therapy can be administered through local implantation or systemic delivery. Natural and synthetic scaffolds are crucial for the local administration of MSC-based therapy by providing the appropriate mechanical strength and cell viability [112]. Commonly used natural scaffolds in bone tissue engineering include collagen hyaluronic acid fibrin and poly(ε-caprolactone)/poly(vinyl alcohol)/chitosan-associated hybrid scaffolds. However purity issues and poor mechanical properties limit the application of natural scaffolds. Synthetic scaffolds including PLGA polyglycolic acid (PGA) and poly-L-lactic acid (PLA) enable the precise control of mechanical properties and stability of the scaffold to further enhance therapeutic efficiency. For example a macroporous and highly flexible gelatin-based scaffold with a microribbon-like structure has recently been demonstrated to increase MSC proliferation and bone regeneration [113]. However the biocompatibility of the synthetic scaffold remains a concern since degradation products could initiate inflammatory responses [114]. The systemic delivery of MSCs provides an alternative strategy of minimally invasive procedures to patients with orthopedic implants. Though MSCs can naturally migrate into inflammatory sites conjugating with antibodies targeting bone or inflammation-associated molecules can further enhance their homing efficiency [115, 116].

5 Conclusion

Transient acute inflammation is closely associated with successful osseointegration and bone regeneration in orthopedic biomaterial implantation. The transition between the pro-inflammatory M1 and anti-inflammatory M2 macrophage phenotypes has been shown to be a key step in bone regeneration. Alternatively, chronic inflammatory bone diseases associated with implants often exhibit excessive pro-inflammatory macrophage infiltration and the generation of wear particles. The combination of bone regenerating scaffolds and controlled drug releasing systems has great potential for advancing clinical applications of orthopedic biomaterials for a variety of conditions including aseptic loosening, osteonecrosis, and fracture non-union. Taken together, optimizing the timing and efficacy of the innate immune reaction provide a promising approach to harness the inflammatory response for therapeutic applications of orthopedic biomaterials.

Acknowledgement This work was supported by NIH grants 2R01AR055650, 1R01AR063717 and the Ellenburg Chair in Surgery at Stanford University. J. P. was supported by a grant from the Jane and Aatos Erkko foundation.

The authors have no conflicts of interest to declare.

References

1. Global Alliance for Musculoskeletal Health. Key facts from the global burden of disease. 2012, http://bjdonline.org/key-facts-and-figures/.
2. Navarro M, Michiardi A, Castano O, Planell JA. Biomaterials in orthopaedics. J R Soc Interface. 2008;5(27):1137–58.
3. Drees P, Eckardt A, Gay RE, Gay S, Huber LC. Mechanisms of disease: molecular insights into aseptic loosening of orthopedic implants. Nat Clin Pract Rheumatol. 2007;3(3):165–71.
4. Cobelli N, Scharf B, Crisi GM, Hardin J, Santambrogio L. Mediators of the inflammatory response to joint replacement devices. Nat Rev Rheumatol. 2011;7(10):600–8.
5. LBHN Service. Joint replacements in U.S. exceed 1 million a year, Pittsburgh Post-Gazette, Pittburgh Post-Gazette. 2013. http://www.post-gazette.com/news/health/2013/03/04/Joint-replacements-in-U-S-exceed.
6. Lewallen EA, Riester SM, Bonin CA, Kremers HM, Dudakovic A, Kakar S, Cohen RC, Westendorf JJ, Lewallen DG, van Wijnen AJ. Biological strategies for improved osseointegration and osteoinduction of porous metal orthopedic implants. Tissue Eng Part B Rev. 2015;21(2):218–30.
7. Vitkov L, Hartl D, Hannig M. Is osseointegration inflammation-triggered? Med Hypotheses. 2016;93:1–4.
8. Ma QL, Zhao LZ, Liu RR, Jin BQ, Song W, Wang Y, Zhang YS, Chen LH, Zhang YM. Improved implant osseointegration of a nanostructured titanium surface via mediation of macrophage polarization. Biomaterials. 2014;35(37):9853–67.
9. Gibon E, Amanatullah DF, Loi F, Pajarinen J, Nabeshima A, Yao Z, Hamadouche M, Goodman SB. The biological response to orthopaedic implants for joint replacement: part I: metals. J Biomed Mater Res B Appl Biomater. 2017;105(7):2162–73.
10. Lin TH, Tamaki Y, Pajarinen J, Waters HA, Woo DK, Yao Z, Goodman SB. Chronic inflammation in biomaterial-induced periprosthetic osteolysis: NF-kappaB as a therapeutic target. Acta Biomater. 2014;10(1):1–10.
11. Anderson JM, Rodriguez A, Chang DT. Foreign body reaction to biomaterials. Semin Immunol. 2008;20(2):86–100.
12. Goodman SB. Wear particles, periprosthetic osteolysis and the immune system. Biomaterials. 2007;28(34):5044–8.
13. Nich C, Takakubo Y, Pajarinen J, Ainola M, Salem A, Sillat T, Rao AJ, Raska M, Tamaki Y, Takagi M, Konttinen YT, Goodman SB, Gallo J. Macrophages-key cells in the response to wear debris from joint replacements. J Biomed Mater Res A. 2013;101:3033.
14. Ingham E, Fisher J. The role of macrophages in osteolysis of total joint replacement. Biomaterials. 2005;26(11):1271–86.
15. Fujiwara N, Kobayashi K. Macrophages in inflammation. Curr Drug Targets Inflamm Allergy. 2005;4(3):281–6.
16. Mosser DM, Edwards JP. Exploring the full spectrum of macrophage activation. Nat Rev Immunol. 2008;8(12):958–69.
17. Davies LC, Jenkins SJ, Allen JE, Taylor PR. Tissue-resident macrophages. Nat Immunol. 2013;14(10):986–95.
18. Mogensen TH. Pathogen recognition and inflammatory signaling in innate immune defenses. Clin Microbiol Rev. 2009;22(2):240–73.

19. Kawai T, Akira S. The role of pattern-recognition receptors in innate immunity: update on toll-like receptors. Nat Immunol. 2010;11(5):373–84.
20. Kono H, Rock KL. How dying cells alert the immune system to danger. Nat Rev Immunol. 2008;8(4):279–89.
21. Shi C, Pamer EG. Monocyte recruitment during infection and inflammation. Nat Rev Immunol. 2011;11(11):762–74.
22. Murray PJ, Wynn TA. Protective and pathogenic functions of macrophage subsets. Nat Rev Immunol. 2011;11(11):723–37.
23. Wynn TA, Vannella KM. Macrophages in tissue repair, regeneration, and fibrosis. Immunity. 2016;44(3):450–62.
24. Lech M, Anders HJ. Macrophages and fibrosis: how resident and infiltrating mononuclear phagocytes orchestrate all phases of tissue injury and repair. Biochim Biophys Acta. 2013;1832(7):989–97.
25. Martinez FO, Sica A, Mantovani A, Locati M. Macrophage activation and polarization. Front Biosci. 2008;13:453–61.
26. Wang N, Liang H, Zen K. Molecular mechanisms that influence the macrophage M1–M2 polarization balance. Front Immunol. 2014;5:614.
27. Franz S, Rammelt S, Scharnweber D, Simon JC. Immune responses to implants—a review of the implications for the design of immunomodulatory biomaterials. Biomaterials. 2011;32(28):6692–709.
28. Maitra R, Clement CC, Scharf B, Crisi GM, Chitta S, Paget D, Purdue PE, Cobelli N, Santambrogio L. Endosomal damage and TLR2 mediated inflammasome activation by alkane particles in the generation of aseptic osteolysis. Mol Immunol. 2009;47(2–3):175–84.
29. Caicedo MS, Desai R, McAllister K, Reddy A, Jacobs JJ, Hallab NJ. Soluble and particulate co-Cr-Mo alloy implant metals activate the inflammasome danger signaling pathway in human macrophages: a novel mechanism for implant debris reactivity. J Orthop Res. 2009;27(7):847–54.
30. Sridharan R, Cameron AR, Kelly DJ, Kearney CJ, O'Brien FJ. Biomaterial based modulation of macrophage polarization: a review and suggested design principles. Mater Today. 2015;18(6):313–25.
31. Goodman SB, Gibon E, Yao Z. The basic science of periprosthetic osteolysis. Instr Course Lect. 2013;62:201–6.
32. Purdue PE, Koulouvaris P, Potter HG, Nestor BJ, Sculco TP. The cellular and molecular biology of periprosthetic osteolysis. Clin Orthop Relat Res. 2007;454:251–61.
33. Hofbauer LC, Schoppet M. Clinical implications of the osteoprotegerin/RANKL/RANK system for bone and vascular diseases. JAMA. 2004;292(4):490–5.
34. Chen Z, Klein T, Murray RZ, Crawford R, Chang J, Wu C, Xiao Y. Osteoimmunomodulation for the development of advanced bone biomaterials. Mater Today. 2016;19(6):304–21.
35. Goodman SB, Gibon E, Pajarinen J, Lin TH, Keeney M, Ren PG, Nich C, Yao Z, Egashira K, Yang F, Konttinen YT. Novel biological strategies for treatment of wear particle-induced periprosthetic osteolysis of orthopaedic implants for joint replacement. J R Soc Interface. 2014;11(93):20130962.
36. Brown BN, Ratner BD, Goodman SB, Amar S, Badylak SF. Macrophage polarization: an opportunity for improved outcomes in biomaterials and regenerative medicine. Biomaterials. 2012;33(15):3792–802.
37. Morais JM, Papadimitrakopoulos F, Burgess DJ. Biomaterials/tissue interactions: possible solutions to overcome foreign body response. AAPS J. 2010;12(2):188–96.
38. Hosgood G. Wound healing. The role of platelet-derived growth factor and transforming growth factor beta. Vet Surg. 1993;22(6):490–5.
39. Wang Y, Wu NN, Mou YQ, Chen L, Deng ZL. Inhibitory effects of recombinant IL-4 and recombinant IL-13 on UHMWPE-induced bone destruction in the murine air pouch model. J Surg Res. 2013;180(2):e73–81.

40. Rao AJ, Nich C, Dhulipala LS, Gibon E, Valladares R, Zwingenberger S, Smith RL, Goodman SB. Local effect of IL-4 delivery on polyethylene particle induced osteolysis in the murine calvarium. J Biomed Mater Res A. 2013;101((7):1926–34.

41. Sato T, Pajarinen J, Behn A, Jiang X, Lin TH, Loi F, Yao Z, Egashira K, Yang F, Goodman SB. The effect of local IL-4 delivery or CCL2 blockade on implant fixation and bone structural properties in a mouse model of wear particle induced osteolysis. J Biomed Mater Res A. 2016;104(9):2255–62.

42. Minardi S, Corradetti B, Taraballi F, Byun JH, Cabrera F, Liu X, Ferrari M, Weiner BK, Tasciotti E. IL-4 release from a biomimetic scaffold for the temporally controlled modulation of macrophage response. Ann Biomed Eng. 2016;44(6):2008–19.

43. Spiller KL, Nassiri S, Witherel CE, Anfang RR, Ng J, Nakazawa KR, Yu T, Vunjak-Novakovic G. Sequential delivery of immunomodulatory cytokines to facilitate the M1-to-M2 transition of macrophages and enhance vascularization of bone scaffolds. Biomaterials. 2015;37:194–207.

44. Hachim D, LoPresti ST, Yates CC, Brown BN. Shifts in macrophage phenotype at the biomaterial interface via IL-4 eluting coatings are associated with improved implant integration. Biomaterials. 2017;112:95–107.

45. Hess K, Ushmorov A, Fiedler J, Brenner RE, Wirth T. TNFalpha promotes osteogenic differentiation of human mesenchymal stem cells by triggering the NF-kappaB signaling pathway. Bone. 2009;45(2):367–76.

46. Gilbert L, He X, Farmer P, Boden S, Kozlowski M, Rubin J, Nanes MS. Inhibition of osteoblast differentiation by tumor necrosis factor-alpha. Endocrinology. 2000;141(11):3956–64.

47. Childs LM, Goater JJ, O'Keefe RJ, Schwarz EM. Efficacy of etanercept for wear debris-induced osteolysis. J Bone Miner Res. 2001;16(2):338–47.

48. Schwarz EM, Campbell D, Totterman S, Boyd A, O'Keefe RJ, Looney RJ. Use of volumetric computerized tomography as a primary outcome measure to evaluate drug efficacy in the prevention of peri-prosthetic osteolysis: a 1-year clinical pilot of etanercept vs. placebo. J Orthop Res. 2003;21(6):1049–55.

49. Dong L, Wang R, Zhu YA, Wang C, Diao H, Zhang C, Zhao J, Zhang J. Antisense oligonucleotide targeting TNF-alpha can suppress Co-Cr-Mo particle-induced osteolysis. J Orthop Res. 2008;26(8):1114–20.

50. Lin TH, Pajarinen J, Lu L, Nabeshima A, Cordova LA, Yao Z, Goodman SB. NF-kappaB as a therapeutic target in inflammatory-associated bone diseases. Adv Protein Chem Struct Biol. 2017;107:117–54.

51. Lin TH, Pajarinen J, Sato T, Loi F, Fan C, Cordova LA, Nabeshima A, Gibon E, Zhang R, Yao Z, Goodman SB. NF-kappaB decoy oligodeoxynucleotide mitigates wear particle-associated bone loss in the murine continuous infusion model. Acta Biomater. 2016;41:273–81.

52. Sato T, Pajarinen J, Lin TH, Tamaki Y, Loi F, Egashira K, Yao Z, Goodman SB. NF-kappaB decoy oligodeoxynucleotide inhibits wear particle-induced inflammation in a murine calvarial model. J Biomed Mater Res A. 2015;103(12):3872–8.

53. Nabeshima A, Pajarinen J, Lin TH, Jiang X, Gibon E, Cordova LA, Loi F, Lu L, Jamsen E, Egashira K, Yang F, Yao Z, Goodman SB. Mutant CCL2 protein coating mitigates wear particle-induced bone loss in a murine continuous polyethylene infusion model. Biomaterials. 2017;117:1–9.

54. Keeney M, Waters H, Barcay K, Jiang X, Yao Z, Pajarinen J, Egashira K, Goodman SB, Yang F. Mutant MCP-1 protein delivery from layer-by-layer coatings on orthopedic implants to modulate inflammatory response. Biomaterials. 2013;34(38):10287–95.

55. Gerstenfeld L, Cho TJ, Kon T, Aizawa T, Tsay A, Fitch J, Barnes G, Graves D, Einhorn T. Impaired fracture healing in the absence of TNF-α signaling: the role of TNF-α in endochondral cartilage resorption. J Bone Miner Res. 2003;18(9):1584–92.

56. Gerstenfeld L, Cho T-J, Kon T, Aizawa T, Cruceta J, Graves B, Einhorn T. Impaired intramembranous bone formation during bone repair in the absence of tumor necrosis factor-alpha signaling. Cells Tissues Organs. 2001;169(3):285–94.

57. Yang X, Ricciardi BF, Hernandez-Soria A, Shi Y, Camacho NP, Bostrom MP. Callus mineralization and maturation are delayed during fracture healing in interleukin-6 knockout mice. Bone. 2007;41(6):928–36.
58. Glass GE, Chan JK, Freidin A, Feldmann M, Horwood NJ, Nanchahal J. TNF-α promotes fracture repair by augmenting the recruitment and differentiation of muscle-derived stromal cells. Proc Natl Acad Sci. 2011;108(4):1585–90.
59. Xing Z, Lu C, Hu D, Yu Y-y, Wang X, Colnot C, Nakamura M, Wu Y, Miclau T, Marcucio RS. Multiple roles for CCR2 during fracture healing. Dis Model Mech. 2010;3(7–8):451–8.
60. Gerstenfeld LC, Cullinane DM, Barnes GL, Graves DT, Einhorn TA. Fracture healing as a post-natal developmental process: molecular, spatial, and temporal aspects of its regulation. J Cell Biochem. 2003;88(5):873–84.
61. Kon T, Cho TJ, Aizawa T, Yamazaki M, Nooh N, Graves D, Gerstenfeld LC, Einhorn TA. Expression of osteoprotegerin, receptor activator of NF-κB ligand (osteoprotegerin ligand) and related proinflammatory cytokines during fracture healing. J Bone Miner Res. 2001;16(6):1004–14.
62. Mountziaris PM, Mikos AG. Modulation of the inflammatory response for enhanced bone tissue regeneration. Tissue Eng Part B Rev. 2008;14(2):179–86.
63. Luttikhuizen DT, Harmsen MC, Luyn MJV. Cellular and molecular dynamics in the foreign body reaction. Tissue Eng. 2006;12(7):1955–70.
64. Nicolaidou V, Wong MM, Redpath AN, Ersek A, Baban DF, Williams LM, Cope AP, Horwood NJ. Monocytes induce STAT3 activation in human mesenchymal stem cells to promote osteoblast formation. PLoS One. 2012;7(7):e39871.
65. Guihard P, Danger Y, Brounais B, David E, Brion R, Delecrin J, Richards CD, Chevalier S, Rédini F, Heymann D. Induction of osteogenesis in mesenchymal stem cells by activated monocytes/macrophages depends on oncostatin M signaling. Stem Cells. 2012;30(4):762–72.
66. Spiller KL, Anfang RR, Spiller KJ, Ng J, Nakazawa KR, Daulton JW, Vunjak-Novakovic G. The role of macrophage phenotype in vascularization of tissue engineering scaffolds. Biomaterials. 2014;35(15):4477–88.
67. Lu LY, Loi F, Nathan K, Lin Th, Pajarinen J, Gibon E, Nabeshima A, Cordova L, Jämsen E, Yao Z. Pro-inflammatory M1 macrophages promote Osteogenesis by mesenchymal stem cells via the COX-2-prostaglandin E2 pathway. J Orthop Res. 2017;35:2378.
68. Raggatt LJ, Wullschleger ME, Alexander KA, Wu ACK, Millard SM, Kaur S, Maugham ML, Gregory LS, Steck R, Pettit AR. Fracture healing via periosteal callus formation requires macrophages for both initiation and progression of early Endochondral ossification. Am J Pathol. 2014;184(12):3192–204.
69. Schlundt C, El Khassawna T, Serra A, Dienelt A, Wendler S, Schell H, van Rooijen N, Radbruch A, Lucius R, Hartmann S, Duda GN, Schmidt-Bleek K. Macrophages in bone fracture healing: their essential role in endochondral ossification. Bone. 2018;106:78–89.
70. Mantovani A, Sozzani S, Locati M, Allavena P, Sica A. Macrophage polarization: tumor-associated macrophages as a paradigm for polarized M2 mononuclear phagocytes. Trends Immunol. 2002;23(11):549–55.
71. Marsell R, Einhorn TA. The biology of fracture healing. Injury. 2011;42(6):551–5.
72. Claes L, Recknagel S, Ignatius A. Fracture healing under healthy and inflammatory conditions. Nat Rev Rheumatol. 2012;8(3):133–43.
73. Kumar VA, Taylor NL, Shi S, Wickremasinghe NC, D'Souza RN, Hartgerink JD. Self-assembling multidomain peptides tailor biological responses through biphasic release. Biomaterials. 2015;52:71–8.
74. Rostam H, Singh S, Vrana N, Alexander M, Ghaemmaghami A. Impact of surface chemistry and topography on the function of antigen presenting cells. Biomater Sci. 2015;3(3):424–41.
75. Rostam HM, Singh S, Salazar F, Magennis P, Hook A, Singh T, Vrana NE, Alexander MR, Ghaemmaghami AM. The impact of surface chemistry modification on macrophage polarisation. Immunobiology. 2016;221(11):1237–46.

76. Goodman SB, Yao Z, Keeney M, Yang F. The future of biologic coatings for orthopaedic implants. Biomaterials. 2013;34(13):3174–83.
77. Tobin EJ. Recent coating developments for combination devices in orthopedic and dental applications: a literature review. Adv Drug Deliv Rev. 2017;112:88.
78. Agarwal R, Garcia AJ. Biomaterial strategies for engineering implants for enhanced osseointegration and bone repair. Adv Drug Deliv Rev. 2015;94:53–62.
79. Friedrich EE, Sun LT, Natesan S, Zamora DO, Christy RJ, Washburn NR. Effects of hyaluronic acid conjugation on anti-TNF-alpha inhibition of inflammation in burns. J Biomed Mater Res A. 2014;102(5):1527–36.
80. Haney EF, Hancock RE. Peptide design for antimicrobial and immunomodulatory applications. Biopolymers. 2013;100(6):572–83.
81. Hilchie AL, Wuerth K, Hancock RE. Immune modulation by multifaceted cationic host defense (antimicrobial) peptides. Nat Chem Biol. 2013;9(12):761–8.
82. Clohisy JC, Hirayama T, Frazier E, Han SK, Abu-Amer Y. NF-kB signaling blockade abolishes implant particle-induced osteoclastogenesis. J Orthop Res. 2004;22(1):13–20.
83. Matsiko A, Levingstone TJ, O'Brien FJ, Gleeson JP. Addition of hyaluronic acid improves cellular infiltration and promotes early-stage chondrogenesis in a collagen-based scaffold for cartilage tissue engineering. J Mech Behav Biomed Mater. 2012;11:41–52.
84. Raftery R, O'Brien FJ, Cryan SA. Chitosan for gene delivery and orthopedic tissue engineering applications. Molecules. 2013;18(5):5611–47.
85. Raftery RM, Tierney EG, Curtin CM, Cryan SA, O'Brien FJ. Development of a gene-activated scaffold platform for tissue engineering applications using chitosan-pDNA nanoparticles on collagen-based scaffolds. J Control Release. 2015;210:84–94.
86. Tierney EG, Duffy GP, Hibbitts AJ, Cryan SA, O'Brien FJ. The development of non-viral gene-activated matrices for bone regeneration using polyethyleneimine (PEI) and collagen-based scaffolds. J Control Release. 2012;158(2):304–11.
87. Schatzlein AG, Zinselmeyer BH, Elouzi A, Dufes C, Chim YT, Roberts CJ, Davies MC, Munro A, Gray AI, Uchegbu IF. Preferential liver gene expression with polypropylenimine dendrimers. J Control Release. 2005;101(1–3):247–58.
88. Raftery RM, Walsh DP, Castano IM, Heise A, Duffy GP, Cryan SA, O'Brien FJ. Delivering nucleic-acid based Nanomedicines on biomaterial scaffolds for orthopedic tissue repair: challenges, progress and future perspectives. Adv Mater. 2016;28(27):5447–69.
89. Huang CL, Leblond AL, Turner EC, Kumar AH, Martin K, Whelan D, O'Sullivan DM, Caplice NM. Synthetic chemically modified mrna-based delivery of cytoprotective factor promotes early cardiomyocyte survival post-acute myocardial infarction. Mol Pharm. 2015;12(3):991–6.
90. Deng Y, Bi X, Zhou H, You Z, Wang Y, Gu P, Fan X. Repair of critical-sized bone defects with anti-miR-31-expressing bone marrow stromal stem cells and poly(glycerol sebacate) scaffolds. Eur Cell Mater. 2014;27:13–24. discussion 24-5.
91. Zhang M, Gao Y, Caja K, Zhao B, Kim JA. Non-viral nanoparticle delivers small interfering RNA to macrophages in vitro and in vivo. PLoS One. 2015;10(3):e0118472.
92. Sheedy FJ. Turning 21: induction of miR-21 as a key switch in the inflammatory response. Front Immunol. 2015;6:19.
93. Boehler R, Kuo R, Shin S, Goodman A, Pilecki M, Leonard J, Shea L. Lentivirus delivery of IL-10 to promote and sustain macrophage polarization towards an anti-inflammatory phenotype. Biotechnol Bioeng. 2014;111(6):1210–21.
94. Li Y, Fan L, Liu S, Liu W, Zhang H, Zhou T, Wu D, Yang P, Shen L, Chen J, Jin Y. The promotion of bone regeneration through positive regulation of angiogenic-osteogenic coupling using microRNA-26a. Biomaterials. 2013;34(21):5048–58.
95. Benimetskaya L, Loike JD, Khaled Z, Loike G, Silverstein SC, Cao L, el Khoury J, Cai TQ, Stein CA. Mac-1 (CD11b/CD18) is an oligodeoxynucleotide-binding protein. Nat Med. 1997;3(4):414–20.

96. Kaminska B. MAPK signalling pathways as molecular targets for anti-inflammatory therapy--from molecular mechanisms to therapeutic benefits. Biochim Biophys Acta. 2005;1754(1–2):253–62.
97. Purdue PE, Koulouvaris P, Nestor BJ, Sculco TP. The central role of wear debris in periprosthetic osteolysis. HSS J. 2006;2(2):102–13.
98. Miyanishi K, Trindade MC, Ma T, Goodman SB, Schurman DJ, Smith RL. Periprosthetic osteolysis: induction of vascular endothelial growth factor from human monocyte/macrophages by orthopaedic biomaterial particles. J Bone Miner Res. 2003;18(9):1573–83.
99. Ren K, Dusad A, Yuan F, Yuan H, Purdue PE, Fehringer EV, Garvin KL, Goldring SR, Wang D. Macromolecular prodrug of dexamethasone prevents particle-induced peri-implant osteolysis with reduced systemic side effects. J Control Release. 2014;175:1–9.
100. Urbanska J, Karewicz A, Nowakowska M. Polymeric delivery systems for dexamethasone. Life Sci. 2014;96(1–2):1–6.
101. Webber MJ, Matson JB, Tamboli VK, Stupp SI. Controlled release of dexamethasone from peptide nanofiber gels to modulate inflammatory response. Biomaterials. 2012;33(28):6823–32.
102. Wadhwa R, Lagenaur CF, Cui XT. Electrochemically controlled release of dexamethasone from conducting polymer polypyrrole coated electrode. J Control Release. 2006;110(3):531–41.
103. Zhang S, Ermann J, Succi MD, Zhou A, Hamilton MJ, Cao B, Korzenik JR, Glickman JN, Vemula PK, Glimcher LH, Traverso G, Langer R, Karp JM. An inflammation-targeting hydrogel for local drug delivery in inflammatory bowel disease. Sci Transl Med. 2015;7(300):300ra128.
104. Nemeth K, Leelahavanichkul A, Yuen PS, Mayer B, Parmelee A, Doi K, Robey PG, Leelahavanichkul K, Koller BH, Brown JM, Hu X, Jelinek I, Star RA, Mezey E. Bone marrow stromal cells attenuate sepsis via prostaglandin E(2)-dependent reprogramming of host macrophages to increase their interleukin-10 production. Nat Med. 2009;15(1):42–9.
105. Francois M, Romieu-Mourez R, Li M, Galipeau J. Human MSC suppression correlates with cytokine induction of indoleamine 2,3-dioxygenase and bystander M2 macrophage differentiation. Mol Ther. 2012;20(1):187–95.
106. Ren G, Zhang L, Zhao X, Xu G, Zhang Y, Roberts AI, Zhao RC, Shi Y. Mesenchymal stem cell-mediated immunosuppression occurs via concerted action of chemokines and nitric oxide. Cell Stem Cell. 2008;2(2):141–50.
107. Choi JJ, Yoo SA, Park SJ, Kang YJ, Kim WU, Oh IH, Cho CS. Mesenchymal stem cells overexpressing interleukin-10 attenuate collagen-induced arthritis in mice. Clin Exp Immunol. 2008;153(2):269–76.
108. Tan CQ, Gao X, Guo L, Huang H. Exogenous IL-4-expressing bone marrow mesenchymal stem cells for the treatment of autoimmune sensorineural hearing loss in a Guinea pig model. Biomed Res Int. 2014;2014:856019.
109. Pajarinen J, Lin TH, Nabeshima A, Jamsen E, Lu L, Nathan K, Yao Z, Goodman SB. Mesenchymal stem cells in the aseptic loosening of total joint replacements. J Biomed Mater Res A. 2017;105(4):1195–207.
110. Bernardo ME, Fibbe WE. Mesenchymal stromal cells: sensors and switchers of inflammation. Cell Stem Cell. 2013;13(4):392–402.
111. Le Blanc K, Mougiakakos D. Multipotent mesenchymal stromal cells and the innate immune system. Nat Rev Immunol. 2012;12(5):383–96.
112. Rosenbaum AJ, Grande DA, Dines JS. The use of mesenchymal stem cells in tissue engineering: a global assessment. Organogenesis. 2008;4(1):23–7.
113. Han LH, Conrad B, Chung MT, Deveza L, Jiang X, Wang A, Butte MJ, Longaker MT, Wan D, Yang F. Winner of the young investigator award of the Society for Biomaterials at the 10th world biomaterials congress, may 17-22, 2016, Montreal QC, Canada: microribbon-based hydrogels accelerate stem cell-based bone regeneration in a mouse critical-size cranial defect model. J Biomed Mater Res A. 2016;104(6):1321–31.

114. Uebersax L, Hagenmuller H, Hofmann S, Gruenblatt E, Muller R, Vunjak-Novakovic G, Kaplan DL, Merkle HP, Meinel L. Effect of scaffold design on bone morphology in vitro. Tissue Eng. 2006;12(12):3417–29.
115. Karp JM, Leng Teo GS. Mesenchymal stem cell homing: the devil is in the details. Cell Stem Cell. 2009;4(3):206–16.
116. Guan M, Yao W, Liu R, Lam KS, Nolta J, Jia J, Panganiban B, Meng L, Zhou P, Shahnazari M, Ritchie RO, Lane NE. Directing mesenchymal stem cells to bone to augment bone formation and increase bone mass. Nat Med. 2012;18(3):456–62.

Anti-Infection Technologies for Orthopedic Implants: Materials and Considerations for Commercial Development

David Armbruster

Keywords Infection · Biofilm · Orthopedic · Coating · Staphylococcus · Bacteria · Surface modification · Antibiotic · Antimicrobial · Silver · PMMA · Regulatory · FDA

1 Introduction

The development of modern orthopedic implants for internal fixation in the 1940s, and for arthroplasty in the 1960s revolutionized orthopedic surgery and led to significant improvements in functional outcomes, however these new technologies also brought with them new complications. Orthopedic implant related infection is a difficult clinical challenge, and for some orthopedic procedures is the most common cause of implant failure [1]. Surgery always carries the risk of infection, but the presence of a foreign body in the surgical site provides a non-vital surface where bacteria can adhere, form a biofilm, and avoid clearance by the immune system and systemic antibiotics. This significantly reduces the number of contaminating organisms required to cause a clinical infection [2].

Despite advances in operating room design and strict hospital infection protocols, implant related infection is still a significant clinical problem. The use of systemic antibiotic prophylaxis at the time of surgery to reduce infection rates has become the standard of care [3] but does not eliminate all implant related infection. Estimates of infection rates in primary knee arthroplasty range from 0.39 to 2.5% [4], and in internal fracture fixation can be 15% or higher in cases of significant comorbidities or high energy trauma. Deep implant related infection carries high costs in patient morbidity and hospital expenses. Treatment cost of an infected knee or hip implant can be five times the cost of a non-infected implant [5] and full treatment can take many months. In the case of orthopedic trauma implants, infection rates are highest in the lower extremities, where soft tissue coverage and blood flow are lower.

D. Armbruster (✉)
DePuy Synthes Biomaterials R&D, West Chester, PA, USA
e-mail: darmbrus@its.jnj.com

© Springer International Publishing AG, part of Springer Nature 2018   219
B. Li, T. Webster (eds.), *Orthopedic Biomaterials*,
https://doi.org/10.1007/978-3-319-89542-0_11

This unmet clinical need presents an opportunity for the development of new material technologies to make orthopedic implant surfaces less hospitable to bacterial colonization, while still performing well in the complex bone healing environment. The challenges facing development of next generation infection-resistant materials are not only technical however, regulatory requirements and commercial constraints are increasing, and must be overcome for promising material technologies to have any impact clinically. The history of anti-infection biomaterials R&D is filled with sophisticated technologies which showed great promise in the lab, only to be thwarted by manufacturing complexity or regulatory requirements. The variety of anti-infection technologies that have been evaluated and published in the scientific and clinical literature is too great to summarize exhaustively here. This chapter will instead focus on those technologies which are of greatest interest from an industry perspective or have been demonstrated to be effective in the clinic, and review the opportunities and obstacles for their commercial development.

2 Working Theories of Implant Related Infection

In assessing anti-infection technologies, it is important to have an operational theory of the etiology and life-cycle of bacterial infection. Most research in the field assumes the biofilm theory of infection is the most explanatory and predictive of clinical success, as captured by the term "the race for the surface" coined in the late 1980s by Gristina et al. [6, 7]. If bacteria can avoid the patient's immune system and systemic antibiotics long enough to attach to and proliferate on an implant surface, they will undergo a phenotypic shift away from the planktonic to a biofilm phenotype characterized by excretion of extracellular matrix and reduced metabolic activity. The biofilm thus provides a protected niche where the concentration of antibiotics needed to kill bacteria can be increased 1000-fold or more [8] and phagocytic cells of the immune system are thwarted. Detailed reviews of the role of bacterial biofilms in implant related infection have been published by Aricola, Costerton and others [9–12]. If biofilms are the primary mechanism for implant related infection, then successful anti-infection technologies should interfere with one or more steps in the process of biofilm formation, or kill contaminating bacteria before biofilm formation can begin.

The bacteria responsible for most infections of orthopedic trauma implants are *Staphylococcus aureus* and Gram negative rods. For prosthetic joint infections, the most common pathogens are coagulase negative staphylococci such as *Staphylococcus epidermidis* [13, 14]. Because some biofilm forming bacteria are not readily detected using traditional culture techniques, infections are at times difficult to accurately diagnose in the clinic, however more modern genetic methods promise to improve the accuracy and timeliness of infection diagnosis [15]. Techniques such as fluorescent *in situ* hybridization (FISH) staining and confocal microscopy have been used to evaluate biofilms *in situ* on retrieved orthopedic implants (see Fig. 1) [16].

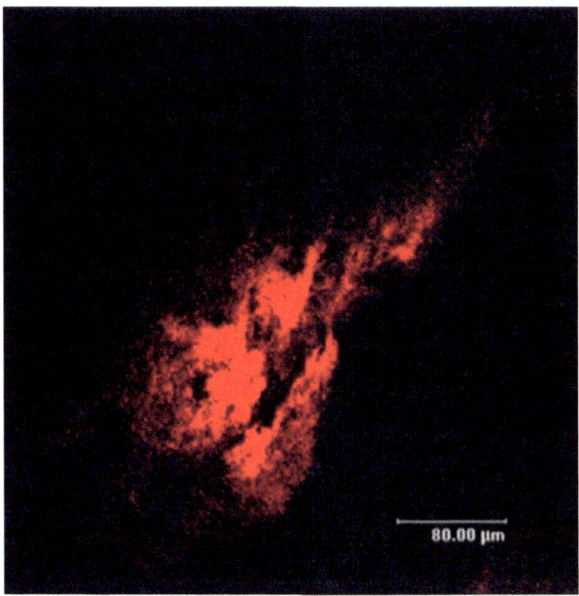

Fig. 1 Confocal image of biofilm membrane from retrieved titanium tibial nail, removed 22 months after implantation for fixation of a Gustilo Anderson Type IIIA open diaphyseal tibia fracture, due to nonunion. Sample was stained with a genus level FISH probe for Staphylococci, revealing very well developed biofilm occupying hundreds of cubic microns of tissue. Reprinted by permission from Springer; Clin Orthop Relat Res.; Molecular Techniques to Detect Biofilm Bacteria in Long Bone Nonunion: A Case Report., Palmer M, Costerton JW, Sewecke J, et al. 2011 [16]

Bacteria are remarkably resourceful in their virulence, and possess mechanisms other than biofilm formation that may potentiate implant related infection. *S. aureus* can become internalized into host fibroblasts and osteoblasts and remain in this protected niche for an extended period, providing a reservoir of bacteria for delayed infection [17]. Similarly, *S. epidermidis* can become internalized in phagocytic cells [18]. Small colony variants (SCVs) of *Staphylococci* are a slow-growing subpopulation caused by a genetic mutation which alters cell metabolism, making them resistant to common antibiotics. SCVs are associated with persistent bone infections and can increase intracellular uptake [19]. *S. aureus* can also migrate into and sequester itself in the canaliculi of cortical bone by a process of asymmetric binary fission and proliferation [20]. Infection of orthopedic implants can occur even years after the surgical procedure, due to hematogenous spread of bacteria from another part of the body. For any material technology to completely prevent orthopedic implant related infection it must contend with these various pathogenic processes.

The great variety of implant geometry, materials, and functional requirements within orthopedics also will play a role in the development of anti-infection technologies. Implants for total joint arthroplasty, spinal fusion, external fixation, or internal fixation vary greatly in requirements such as bone ingrowth, bearing surfaces, percutaneous exposure, and intraoperative contouring. It may not be possible to develop a single anti-infection material strategy that would be effective for all types of orthopedic implants.

3 Current Clinical Options

Orthopedic surgeons have adopted standard clinical practices to reduce the risk of surgical site infection, such as skin and nasal decolonization, parenteral antimicrobial prophylaxis, and glycemic control [21]. These measures have all resulted in improvements in infection rates, but have not been sufficient to prevent all implant related infections.

Surgeons often will use locally delivered antibiotics to help prevent infection in patients at high risk due to severe trauma or comorbidities. Beads formed from poly (methyl methacrylate) (PMMA) bone cement mixed with antibiotics are placed in the wound site to elute high levels of drug locally. These are implanted during an initial surgical procedure, then are removed prior to later definitive fixation and soft tissue closure. This practice was initially developed for treating bone infections, but has been adopted for prevention of infection in orthopedic trauma cases with severe soft tissue injury [22].

More recently, researchers have described the practice of sprinkling powdered antibiotics directly into the wound site before skin closure. Clinical case series have shown that this may reduce infection rates, however the practice may be more effective in spine surgery than in trauma applications, and questions remain about local toxicity risks [23]. Some orthopedic implants which are specifically modified to reduce bacterial colonization have been introduced commercially, such as gentamicin coated tibia nails or silver coated megaprostheses in use in Europe, however these are still niche technologies without wide application. There is a clear opportunity for development of new and broadly applicable technologies to aid surgeons in preventing implant related infection.

4 Biomaterial Strategies for Infection Prevention

Most material technologies to prevent implant related infection can be divided generally into three categories, as suggested by Romanò et al.; passive surface modifications, active surface modifications, and perioperative local antibiotic carriers or coatings [24]. Passive surface modification refers to surfaces that do not release an active agent into the surrounding tissue after implantation. Active surface modifications release an active agent locally, and therefore have an effect against bacteria at some distance from the implant surface. Perioperative technologies are anti-infective agents that are introduced to the implant or the surgical site only at the time of surgery, and can be either passive or active.

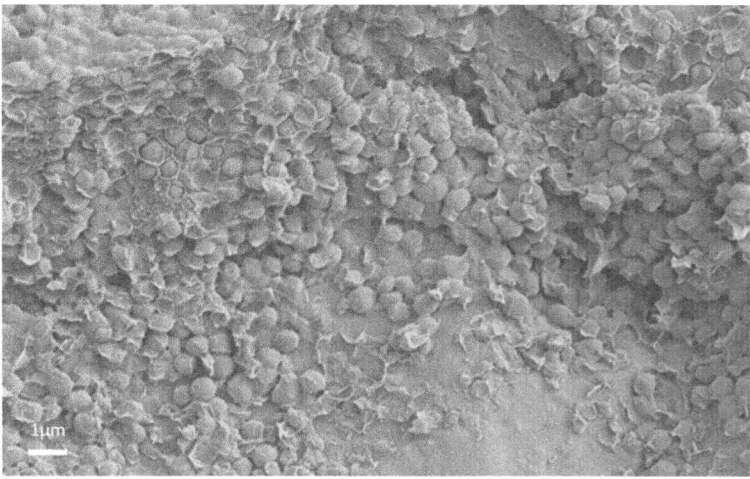

Fig. 2 *Staphylococcus aureus* biofilm grown *in vitro* on 1.0 mm titanium alloy (Ti6Al4V) wire. Wire was cultured in TSB medium for 48 h. Micrograph courtesy of Malavosklish Bikram-Liles and Peter Schaut, DePuy Synthes Biomaterials R&D

4.1 Passive Surface Modification

Bacterial colonization of an orthopedic implant may begin before the implant reaches the patient, as bacteria transported through the air of the operating room settle onto the exposed surgical instruments or implants during the surgical procedure. Partially for this reason, prolonged surgical time has been correlated with increased infection rate in both joint arthroplasty and orthopedic trauma surgery [25]. Bacteria from the skin, including commensal skin flora such as *S. epidermidis*, may be transmitted from the edges of an incision into the deeper layers of a surgical site to contaminate the implant. As bacteria encounter the implant surface, they contact either the clean implant surface, or a conditioning layer of proteins adsorbed to the implant from the host blood or wound fluids. A portion of the bacteria will reversibly adhere to the surface due to local conditions including surface energy, van der Waals forces and local electrostatic interactions. Many of these cells are only transiently attached and will desorb from the surface but those that remain will irreversibly adhere to the bulk of the conditioning layer on the implant surface, and begin multiplying to form microcolonies, leading to biofilm formation [26]. A mature biofilm develops a three-dimensional multi-layered structure that can be seen by confocal microscopy or scanning electron microscopy (SEM) as shown in Fig. 2. Numerous surface modification strategies have been evaluated with the intent of killing bacteria on contact with the implant surface, or using surface features to arrest bacterial growth and colonization.

4.1.1 Nanotopography

The most basic feature of an implant surface to modify may be surface roughness, which has the potential to affect the adhesion of bacterial cells at the sub-micron scale. A recent *in vivo* evaluation has shown that polished versus micro-roughened surfaces of titanium alloy or stainless steel plates do not affect infection rates by *S. aureus* in a rabbit contaminated plate model [27]. This implies that micron-scale features did not affect bacterial growth *in vivo*. However, introduction of nano-scale roughness to a metal surface by shot peening has been shown to reduce adhesion of Gram positive bacteria [28].

Titanium oxide (TiO_2) nanotubes on anodized titanium surfaces can reduce adhesion of *S. aureus* and *P. aeruginosa* in *in vitro* studies by over 90% in a manner that is a function of nanotube dimensions. As an additional benefit, TiO_2 nanotube surfaces can enhance osteoblast proliferation as well [29]. Nanoscale surface modification can also affect surface energy in a way that alters protein adsorption, allowing further control of bacteria-biomaterial interaction [30]. These technologies are very attractive commercially, because they result in durable surfaces and could potentially be implemented with relatively minor modifications of the standard manufacturing process.

Surface roughness modification without chemical modification is also desirable from a regulatory point of view, as it represents no increase in risk to the patient, and no change to the device regulatory pathway. Additional *in vivo* research is needed to evaluate whether the antibacterial effect of nano-textured surfaces might translate to reduction in clinical infection, and modification of surface nanotopography could potentially be combined with other technologies to give an additive effect toward building an antimicrobial implant.

4.1.2 Photocatalytic Titanium Oxide

One intriguing antibacterial surface modification is the photocatalytic property of titanium dioxide. TiO_2 is an N type semiconductor, and the anatase and rutile polymorphs of TiO_2 exhibit a photocatalytic reaction when exposed to UV and visible light respectively, with anatase considered more photochemically active [31]. Under aerobic conditions electrons and holes are photo-generated, then combine with molecular oxygen or hydroxyl groups at the material surface to generate superoxide and then hydroxyl radicals [32]. The antimicrobial properties of TiO_2 surfaces is likely due to the capacity of these free radicals to degrade organic molecules, damaging bacterial cell membranes and degrading bacterial toxins, and are therefore broadly effective against both Gram negative and Gram positive bacteria [33].

The photocatalytic effect is clearly attractive for the case of orthopedic implants made from anodized titanium, where creation of antimicrobial surfaces may be achieved by modifying anodizing bath chemistry and conditions [34]. TiO_2 surfaces have been generated by other standard methods such as sol-gel coating, electrophoretic deposition, chemical vapor deposition, or cathodic arc deposition. Because it is a surface phenomenon, coating thickness and microstructure play a role in the degree of photocatalytic effect [35]. UV-activated TiO_2 surfaces have been demonstrated to kill clinically relevant bacteria in both planktonic and biofilm state *in vitro* [36, 37].

Yue et al. demonstrated that activation of TiO_2 by 5 min of UV exposure was stable for up to 30 days in air, and effective to significantly reduce colonization by *S. aureus* and *S. epidermidis*. However, this effect dissipated after 30–60 min of exposure to an aqueous environment, and therefore would only provide antimicrobial coverage for implants prior to and shortly after implantation [37]. This limitation makes technologies based on photocatalytic TiO_2 less attractive for orthopedic implants than for other applications where the surface would remain dry. Thorough reviews of the science behind photocatalytic titanium oxide surfaces for biomedical implants have been published by Visai et al. and Foster et al. [31, 33].

4.1.3 Covalently Bound Antimicrobials

In 2005 the group of Noreen Hickock et al. at Thomas Jefferson University began publishing a series of studies demonstrating the ability of covalently bound vancomycin on titanium surfaces to kill *S. aureus* and *S. epidermidis in vitro* and *in vivo* [38, 39]. Vancomycin is linked to the passivated TiO_2 layer via a self-assembled monolayer formed by reaction with aminopropyltriethoxysilane (APTES). This is covalently bound to a flexible linker molecule (aminoethoxyethoxy-acetate, AEEA) to space the vancomycin away from the metal surface, then finally to vancomycin [40]. *In vitro* studies showed that the antimicrobial activity was maintained after three repeat challenges with *S. aureus*, and after 6 weeks of incubation in buffered saline [41]. In a sheep osteotomy infection model, this covalently tethered vancomycin demonstrated significantly less colonization by *S. aureus* and improved bone healing after 3 months relative to controls [42].

Although the concept is elegant, there are some limitations to this technology. The coating was not stable to mechanical shear forces such as seen during bone screw insertion, was only effective against Gram positive bacteria by nature of the antibiotic choice, and was only applicable to titanium implants due to the specific surface chemistry. More recently, a similar covalent chemistry utilizing polyethylene glycol (PEG) linker molecules or dopamine has been used to bind the broad-spectrum antibiotics bacitracin or enoxacin to oxidized titanium surfaces, and have been shown to prevent colonization by *S. aureus* in a rat femoral IM infection model [43]. The APTES linking technology has also been used to bind antimicrobial peptides derived from lactoferrin or melimine to anodized titanium surfaces, demonstrating strong *in vitro* anti-biofilm activity [44, 45].

These surface modification chemistries take advantage of the ability to covalently bind to TiO_2. In the orthopedic market this limits their use to implants made from titanium, which is most common in intramedullary nails, spinal rods and pedicle screws, maxillofacial plates and screws and some hip and knee prostheses. These technologies are less applicable for orthopedic trauma plate and screw systems, which are predominantly stainless steel in the US, for cobalt chrome in bearing surfaces for artificial hips and knees, and for PEEK spine implants. For an orthopedic manufacturer to invest in development of an antimicrobial technology, it is preferable to find a technology that can be used on all implant types, regardless of material.

One non-eluting surface treatment of interest that is material independent is a polycationic paint-on coating developed at MIT that has been demonstrated to prevent bacterial colonization *in vivo*. The hydrophobic polycation N,N-dodecyl,methyl-PEI, which contains a 750 kDa branched polyethylenimine, was painted onto stainless steel and titanium orthopedic trauma plates. Plate sections were inoculated *in vitro* with 10^4 CFU *S. aureus*, and showed a greater than 4-log reduction in bacterial colonization relative to uncoated controls. Coated and uncoated plates were implanted in a sheep tibia osteotomy model, and challenged with 2.5×10^6 CFU of *S. aureus*. After 1 month *in vivo*, the coated plates showed significantly better clinical, radiographic and histological outcomes [46].

4.2 Active Surface Modification

4.2.1 Antibiotic Bone Cement

Since the early 1970s, surgeons have been incorporating antibiotics into the poly(methyl methacrylate) (PMMA) cement used in hip and knee arthroplasty, to provide high local antibacterial concentrations for prolonged periods [46]. PMMA cements incorporating gentamicin, tobramycin, clindamycin or vancomycin have been commercialized and represent the standard of care for revision hip and knee arthroplasty procedures. Pre-molded PMMA spacers containing antibiotics are commonly used in 2-stage revisions of hip or knee prostheses to maintain the joint space prior to implant replacement [47–50]. A similar procedure has been developed for revision of infected intramedullary nails as well, with surgeons constructing custom intramedullary implants from PMMA and antibiotics using re-purposed disposable medical tubing as intraoperative casting molds [51–53].

PMMA beads incorporating antibiotics, mixed and formed intraoperatively, have become a standard method of treating osteomyelitis, and have also been adapted for delivering antibiotics locally in staged trauma surgery to prevent bacterial infection prior to definitive fixation and wound closure [22, 54]. Intraoperative mixing of antibiotics with PMMA cement allows antibiotic selection based on sensitivity of the infecting pathogen, however the antibiotic elution rate varies significantly based on cement manufacturer, mixing method, antibiotic choice, and antibiotic loading level [55–57]. Significant research has gone into optimizing antibiotic release rate from PMMA cements, however typically only a small fraction of the total antibiotic loading is released from the cured PMMA. Residual antibiotic release at levels below the minimum inhibitory concentration (MIC) of bacteria may contribute to the selection of antibiotic resistant strains, and antibiotic resistant bacteria have been isolated from PMMA beads retrieved from patients [58, 59].

Since antibiotic-loaded PMMA implants used solely for antibiotic drug delivery must eventually be surgically removed, bioabsorbable matrices for antibiotic local delivery are preferable in some situations. A fully bioabsorbable orthopedic trauma screw containing ciprofloxacin to prevent implant related infection has been developed, but has seen limited clinical use [60].

4.2.2 Antibiotic Coated Implants

Orthopedic implants coated with antibiotics in a bioabsorbable drug-delivery coating were described as early as 1996 by Price et al. [61–63]. This approach has been applied commercially to orthopedic trauma implants by DePuy Synthes with the development of the ETN PROtect intramedullary tibial nail, which contains gentamicin sulfate incorporated into a biodegradable poly(D,L-lactic acid) (PDLLA) coating matrix [64]. The coating is approximately 10 μm thick, and applied to all surfaces of the implant by a dip coating process. Gentamicin is released rapidly from the coating during the initial hour after implantation, then more slowly for approximately two weeks, and the polymer is completely degraded in approximately 6 months [65, 66]. The strong adherence of the polymer matrix coating to the titanium surface provides abrasion resistance as the nail is inserted into the intramedullary space of the tibia.

This CE (European Conformity) marked product has been shown to be effective at reducing infection rate in severe trauma in a clinical case series, but larger controlled trials have not been performed [67, 68]. Other research groups have published preclinical data on similar antibiotic coatings in absorbable polymer carriers [69, 70] however these have not yet been commercialized. Polylactide and polyglycolide based polymers have been used extensively for drug delivery applications in the pharmaceutical industry, as well as in bioabsorbable medical devices such as sutures and orthopedic implants. Because of this, the safety profile of these polymers is well established, and there are high quality sources of commercial raw materials. These are important considerations which help simplify translation to commercial production.

Sol-gel derived silica coatings, which are processed at room temperature from solutions of tetraethoxysilane (TEOS) alcohol and water, can be used to incorporate various antimicrobial molecules including triclosan and vancomycin, and give sustained release for 6 weeks or more. A TEOS sol-gel coating with 20% by weight triclosan has been shown to prevent infection of percutaneous and intramedullary implants in animal models [71, 72]. The ability to tune drug release parameters by varying processing conditions, and the ability to encapsulate a wide variety of organic molecules make sol-gel derived ceramic coatings a very versatile matrix for drug release applications.

Anti-infection implant coatings containing antibiotic drugs benefit from the proven efficacy and safety of the drug component, however the regulatory pathway to market is complicated because drug-device combination products must meet both drug and device standards. This typically means increased clinical data requirements for approval as well as more stringent manufacturing and quality standards. In the United States, the regulatory pathway for this type of technology is addressed in the FDA guidance document "Premarket Notification [510(k)] Submissions for Medical Devices that Include Antimicrobial Agents", published in 2007 [73]. This document lays out the requirements for predicate devices and clinical data to support 510(k) submissions, and states that a predicate device must have "the same device design and the same antimicrobial agent for the same indication for use".

In the United States, manufacturers of drug-device combination products must meet both device and drug requirements for Good Manufacturing Practices (GMP's), which include increased requirements for raw material testing, lot release testing, and shelf life retains [74]. The addition of a drug substance and degradable delivery matrix to an otherwise very stable metal implant may significantly reduce shelf life stability. For these reasons, although adding an antibiotic coating to an implant may be clinically very effective and technically feasible, few antibiotic coated products have seen commercialization in orthopedics.

4.2.3   Bone Graft Substitutes with Antibiotics

Calcium sulfate ($CaSO_4$) and calcium phosphate (βTCP or hydroxyapatite) based bone void fillers are used to fill defects in bones following trauma or tumor resection. These materials are supplied as preformed granules, or can be mixed as a paste or putty intraoperatively. These materials are replaced by native bone as they degrade, but prior to complete remodeling the ceramics are foreign bodies and are subject to biofilm mediated infection. Addition of antibiotics to the ceramics during mixing can reduce the risk of infection, and several products with pre-mixed antibiotics have been commercialized in Europe, including Herafill-G and Cerement G with gentamicin, Cerement V with vancomycin, and Osteoset-T with tobramycin. Antibiotic loaded bone void fillers are most commonly used to treat active bone infection, or in revision arthroplasty, with little clinical literature on their use in primary surgeries. In one clinical case series of 26 patients, calcium sulfate beads loaded with vancomycin effectively prevented infection in open long bone fractures, with no inhibition of bone healing. However, two patients experienced wound drainage related to the calcium sulfate resorption, which is the most commonly reported adverse event [75].

A composite TCP bone void filler with absorbable polymer drug release matrix has been developed by the specialty pharmaceutical company PolyPid. TCP granules are coated with a solution of 75:25 poly (DL-lactide-co-glycolide), cholesterol and phospholipid containing the antibiotic doxycycline hyclate. The unique formulation provides a linear release of antibiotic for up to 28 days in 5% FBS solution *in vitro*. The antimicrobial efficacy was evaluated *in vivo* in a rabbit tibia infection model; 0.3 g of composite granules were implanted in a 6 mm drill hole in the proximal metaphysis and inoculated with 1.5×10^5 CFU of *S. aureus*. Radiographs at 7,

14, and 21 days showed significantly less bone resorption for the doxycycline composite granules than inoculated controls with plain TCP. Microbiological analysis of retrieved bone samples at 21 days showed a 2-log reduction in bacterial burden relative to controls [76].

4.2.4 Antimicrobial Silver Coatings

The antimicrobial properties of the element silver have been known for centuries. Silver nitrate formulations have appeared in pharmacopeia dating to ancient Rome, and the surgeon Dr. J. Marion Sims pioneered the use of antibacterial silver wire sutures in 1852 [77]. More recently, antimicrobial silver coatings have been commercialized for urinary catheters, central venous catheters, endotracheal tubes and wound dressings. Silver coated urinary catheters and central venous catheters have been shown in some clinical studies to reduce infection rates, however the effects are dependent on clinical setting and patient population [78–80]. There are numerous different silver coating technologies varying in chemistry, silver loading, silver release rate, and even mechanism of action. The goal of new coating technology development has been to control silver ion release rate and achieve therapeutic effect without toxic side effects.

Silver Antimicrobial Mechanism of Action

The antimicrobial mechanism of silver ions is to bind sulfydryl (thiol) groups in bacterial membrane proteins and metabolic enzymes [81, 82]. Disruption of bacterial respiratory enzymes leads to production of reactive oxygen species (ROS) and oxidative damage to cell membranes and proteins, which is also the mechanism of toxicity to human cells. *In vivo*, silver is released from metallic sliver as Ag^+ ions in solution. Silver ions are bound by physiological anions (such as Cl^-, $S2^-$, or $-SH$), removing free Ag^+ from solution, and reducing both toxicity and antimicrobial efficacy [83]. Precipitation of metallic silver or silver chloride or selenide salts produces deposits in tissues which remain stable and relatively inert but can lead to argyria, or bluish discoloration of skin. Alan Lansdown of Imperial College, London, has published a very thorough review of the use of silver in medical applications [84].

Current Commercial Products with Antimicrobial Silver

Silver coated external fixation pins were cleared for clinical use in the United States by EBI Medical in 1996 (EBI Medical K961433), and in Europe by Orthofix [85]. These were coated with elemental silver using the SPI-Argent ion beam assisted deposition (IBAD) process from Spire Corp. *In vitro* studies of these pins showed some effect to reduce bacterial adhesion, reducing colonization of *S. aureus* by

80–95%, but complete killing was not observed [86]. In an *in-vivo* study performed in sheep, percutaneous silver coated ex-fix pins and conventional stainless steel pins were inoculated with *S. aureus*. After 2½ weeks, the silver coating reduced pin tip infection rate from 84 to 62%, as well as improving the stability of pins in bone [87]. In clinical use this relatively small effect did not translate to a clear benefit. In one small study of 19 patients, the infection rate with silver coated ex-fix pins was 21% compared with 30% for stainless steel pins, however there was no effect on average clinical scoring of pin sites [88]. In a similar study of 24 patients including 106 ex-fix pins (50 silver coated, 56 controls), the infection rate was reduced from 43 to 30%, which was not statistically significant. Concerns over a significant increase in silver serum concentrations in patients led to the early termination of this study [85].

Recent clinical publications indicate a greater benefit for silver antimicrobial coatings on megaprostheses typically used for joint reconstruction in orthopedic oncology. Infection rates associated with these implants can be over 20%, and are related to patient risk factors and prolonged operating times [89]. Implantcast GmbH commercialized a version of their MUTARS® megaprosthesis implant system in the EU in 2002 with a galvanically deposited elemental silver coating approximately 10–15 μm thick. Multiple clinical studies have shown that this implant system is effective to reduce deep infection rates by up to 75% [90–92]. However, several publications have reported cases of argyria, or grey-blue discoloration of overlying soft tissue due to silver deposits [93, 94].

Newer silver coating technologies designed to reduce the total amount of silver released and avoid local toxicity issues have been commercialized on megaprostheses by Stanmore (Stryker) and Waldemar Link in the EU. Stanmore's Agluna® technology (from Accentus Medical Ltd.) is a silver modification of a titanium surface by anodization to give micron scale silver-containing domains, and has been used clinically on their METS megaprosthesis implants since 2006. This coating results in low milligram levels of silver on each implant, compared with one gram or more of silver per implant for the Implantcast product [95]. The Stanmore implant has been shown in one clinical case control study of 170 patients to reduce infection rates by 47% [96]. Waldemar Link® has modified their Megasystem-C® prosthesis system in the EU with the PorAg® coating, which releases silver ions as well as electrons through an electrochemical reaction. In a clinical case series, this implant resulted in infection rates of 0% for primary implants, and 9.5% for revision surgeries [97]. It is important to note that for each of these megaprosthesis implants the silver coating is applied only to the soft-tissue contacting portion of the implant, not the intramedullary components.

Silver Coating Technologies in Development

The clinical success of these silver based coatings has correlated with an increase in silver based technologies evaluated in *in vitro* and *in vivo* preclinical studies. A coating technology developed by Bio-Gate AG, HyProtect™, has been commercialized on tibial plateau leveling osteotomy (TPLO) plates for the canine veterinary market in the US, and evaluated on hip implants in a canine preclinical study [98]. This technology consists of silver nanoparticles, 5–50 nm in diameter formed by a physical vapor deposition process. These nanoparticles are embedded in a SiO_xC_y plasma polymer layer formed by chemical vapor deposition. Multiple layers of silver nanoparticles may be embedded to increase the silver loading. The plasma polymer layer is slowly degraded while the silver is released by oxidation and dissolution from the embedded nanoparticles. A single layer composite coating containing approximately 1.5 µg/cm^2 silver released 73% of the silver content after 28 days *in vivo* in a rabbit soft tissue model [99].

Numerous other researchers have evaluated the antimicrobial activity of silver nanoparticles in various matrices, and there are too many preclinical publications to review in detail here. The ability to enhance silver ion release rate via the high surface area of nano-scale particles is an attractive capability, however the size dependent cytotoxicity of silver nanoparticles raises concerns for clinical use in the bone healing environment [100, 101]. Technologies such as binding silver nanoparticles to TiO_2 nanotubes on implant surfaces may harness the nano-scale benefits for ion release rate while reducing particle related toxicity [102, 103].

Potential for Toxicity of Silver in Orthopedics

The adoption of antimicrobial silver in orthopedics has so far been limited, in part due to concerns over both efficacy and safety. Multiple cytotoxicity studies of silver nanoparticles and silver salts show cytotoxicity to human mesenchymal stem cells (MSCs) above 1 µg/mL silver ion concentration. Cytotoxic effects of silver to MSCs include reduced differentiation to osteoblasts [104, 105], decreased cell viability, cytokine release and impaired chemotaxis [106, 107]. The *in vitro* toxicity of silver salts to L929 fibroblasts has been demonstrated at 1 ppm, with antimicrobial effectiveness against *S. aureus* demonstrated at 0.1 ppm [108].

Studies of local toxicity *in vivo* have also raised concerns. In a unique study of soft tissue implant toxicity using a hamster dorsal skinfold chamber, a commercially pure silver implant led to a significant local effect on microvasculature, with leukocyte activation and extravasation. This resulted in severe inflammation and edema associated with the implant [109]. A canine hip prosthesis model has been used to evaluate the effect of coatings on direct bone contact healing. One study was performed to evaluate the effect on bone integration of a galvanic metallic silver coating on the intramedullary component of the hip implant. After one year of implantation, direct bone healing was poor compared to titanium controls, with pullout forces significantly reduced relative to controls (average of 24 N vs. 3764 N) [110]. A more recent study of a nanoparticulate silver coating on the same canine hip stem, but with much lower total silver content, also demonstrated a reduced pullout strength at 12 months (average of 440 N), but with greater implant stability observed [98]. These studies suggest that silver coatings should be used with caution on surfaces where direct bone healing is required, such as the intramedullary portions of hip or knee implants.

Several clinical reports have identified the risk of local argyria, or skin discoloration, in patients receiving the Implantcast Mutars® silver coated megaprostheses. In a single institution case series of 32 patients, 23% developed local argyria after a mean of 26 months, but showed no renal, hepatic, or neurological toxicity [94]. However, several other clinical case series with the same implant showed no sign of argyria. In summary, the preclinical and clinical data indicate that silver can be a safe and effective antimicrobial agent in orthopedics, if the application is chosen appropriately. Silver ions have been shown *in vitro* to negatively affect stem cell recruitment and differentiation, which are critical events in bone healing, so it's use in locations where direct bone integration is required should be carefully evaluated.

4.2.5 Antimicrobial Iodine Coatings

Iodine has also been evaluated as an antimicrobial coating and has demonstrated some clinical efficacy. Professor Hiroyuki Tsuchiya from Kanazawa University and his colleagues have published extensively on the use of iodine coated titanium implants for infection prevention in orthopedic surgery. Titanium or Ti6Al4V alloy implants were anodized in a solution containing povidone-iodine, forming an oxide film 5–10 μm thick with approximately 10 $\mu g/cm^2$ iodine. This coating was applied to a variety of orthopedic implants, including hip prostheses, spinal implants, megaprostheses, knee prostheses, and trauma implants [111–114]. These implants were used in revision surgeries where an active implant related infection was treated, or in high risk patients with significant risk factors for infection. The iodine coated implants were associated with very low infection rates and good implant biocompatibility and stability, however, larger randomized controlled clinical trials are needed to fully demonstrate the effectiveness of this iodine coating. Like some other surface coatings, this technology is limited in application to titanium or titanium alloy surfaces.

4.3 Perioperative Local Antibiotics

4.3.1 Direct Local Application of Antibiotics

Due to local disruption of blood flow and poor vascularity in bone, systemic antibiotics administered during orthopedic surgery may not be available at the implant surface at sufficient concentration to kill contaminating bacteria. By locally administering antibiotics, much higher concentrations can be achieved. Direct sprinkling of powdered antibiotics such as vancomycin into the surgical site prior to skin closure has been described clinically in spine and trauma surgery [23, 115, 116].

Fig. 3 DAC® hydrogel coating is spread onto a plate and a screw for osteosynthesis in an ankle fracture. Reprinted by permission from Springer; J Orthop Traumatol; Fast-resorbable antibiotic-loaded hydrogel coating to reduce post-surgical infection after internal osteosynthesis: a multicenter randomized controlled trial., Malizos K, Blauth M, Danita A, et al., 2017

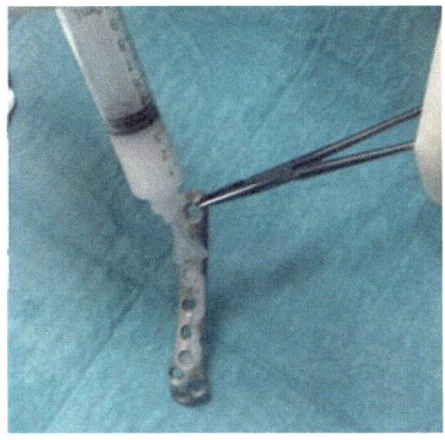

Meta-analyses of retrospective cohort clinical studies indicate the technique is effective to reduce infection rates in spinal surgery, although the studies are typically small [117–119]. The limited number of randomized controlled trials do not show as strong an effect for locally applied antibiotic powder, indicating the need for more high quality clinical trials in this area [120]. Local infusion of vancomycin or gentamicin solution to the joint space following single stage revision total hip arthroplasty (THA) has shown good efficacy to prevent infection [121] as has local administration of liquid aminoglycoside solutions to open fractures during surgical fixation [122].

4.3.2 Local Antibiotic Carriers

In order to improve on direct antibiotic application, a number of local drug delivery technologies have been developed that are device-independent. The biotechnology company Novagenit has commercialized their DAC® gel product in Europe, which is a hydrogel coating that can be mixed with the surgeon's choice of antibiotics in an application syringe, and applied to an implant intraoperatively as shown in Fig. 3. The gel is composed of low molecular weight hyaluronic acid derivatized with poly-D,L-lactic acid. A syringe containing 300 mg of powder is reconstituted with 5 mL of sterile water for injection and up to 50 mg/mL of antibiotic. The gel is coated onto the orthopedic implant prior to implantation as well as injected into the surgical site. *In vitro* studies have shown that 80% of the antibiotic is released at 24 h [123]. DAC® gel was used with gentamicin or vancomycin in a randomized controlled study of closed fractures treated with metal plates and screws or with an IM nail. In 126 patients, there were no infections in the treatment group, as compared to 6 infections in 127 patients in the control group [124].

The generic pharmaceutical company Dr. Reddy's Laboratories has developed a bioabsorbable phospholipid based gel, designated DFA-02 containing 1.88% vancomycin and 1.68% gentamicin. The gel is formed by combining a water solution of the antibiotics with lecithin and sesame oil, then emulsifying with high shear mixing. The DFA-02 gel has been tested in a rat femur segmental defect plating model, challenged with *S. aureus*, and compared with PMMA beads containing gentamicin and tobramycin. After two weeks *in vivo* the antibiotic gel resulted in significantly fewer animals with culturable bacteria in their wounds [125]. The DFA-02 has been evaluated clinically in the US for efficacy in a non-orthopedic indication; in a study of 445 patients undergoing abdominal surgery there was no significant difference between the treatment group and placebo control [126].

A different novel point of care antibiotic coating formulation also based on phosphatidylcholine has been evaluated at the University of Arkansas. This formulation is made by kneading antibiotics into pure phosphatidylcholine at 37 °C at concentrations up to 25%, and then spreading on an implant surface. An *in vivo* study of this formulation in a rabbit radius intramedullary infection model showed greater than 4-log reduction in contaminating *S. aureus* bacteria seven days after implantation, as well as improved clinical signs of infection [127, 128].

5   Regulatory and Commercial Considerations

5.1   Preclinical Data

Development of an orthopedic implant-related infection is a battle between the human immune system and a bacterial biofilm community, which are both extremely complex systems down to the molecular scale. Preclinical evaluation of anti-infection technologies for orthopedic implants is complicated, and neither *in vitro* nor *in vivo* models perfectly predict clinical success. Initial *in vitro* studies of anti-microbial efficacy often rely on time tested microbiological assays such as zone of inhibition (ZOI) studies or colonization assays performed in bacterial growth agar. But bacterial agar was developed in the 1880s to rapidly grow tuberculosis colonies in clinical microbiology labs, and in many ways does not represent conditions in the surgical wound site or bone healing environment [129]. The use of more complex growth media including serum proteins in microbiological culture will significantly change bacterial adhesion and biofilm formation on metal or plastic surfaces [130]. More advanced *in vitro* culture conditions such as drip flow biofilm reactors or minimum biofilm eradication concentration (MBEC) assays may give more relevant information on the effectiveness of a technology to prevent biofilm formation, rather than just kill planktonic bacteria [131, 132].

There is a need for improved animal models of implant related infection as well. Much-published models such as rabbit intramedullary k-wire implants with bacterial inoculum were developed in the early 1990s [133, 134] and have been extremely valuable as a screening tool for anti-infection technologies. However high dose inoculum of planktonic bacteria, often at levels of 10^6 CFU or higher, does not represent the clinical situation. The large majority of bacteria in our bodies and in the environment exist as biofilms, and bacteria contaminating a surgical site are likely in the biofilm phenotype when they arrive at the orthopedic implant, having originated from commensal skin flora at the surgical incision site or airborne particles from the environment. New animal models of orthopedic implant-related infection which employ biofilm inoculum or delayed initiation of antibacterial treatment may give a more accurate assessment of efficacy for anti-infection technologies [115, 135, 136].

5.2 Regulatory and Market Hurdles

The biggest hurdles to commercialization of new anti-infection technologies are often regulatory approval and the cost of assembling the required preclinical and clinical data. Regulatory oversight is of course essential and scrutiny of anti-infection technologies, especially drug-device combination products, has increased as these technologies have become more common [73]. Unfortunately in orthopedics there are two factors working against the commercialization of new technologies; the fragmentation of the market into different anatomies and implant technologies, and the regulatory requirement for indication-specific clearances. In other words, if a clinical trial is performed on a new anti-infection coating on a tibial nail to support regulatory approval, a second clinical trial will likely be required for regulatory approval of the same coating applied to a knee prosthesis or to an ankle plating system.

Compounding this dilemma, a clinical trial to demonstrate prevention of infection requires many more patients than a clinical trial to demonstrate treatment of infection, because the lower the underlying infection rate the more patients are needed to demonstrate statistical significance. A clinical trial powered to demonstrate statistically significant infection reduction in primary hip arthroplasty for example, where infection rates are 1–2%, would require thousands of patients. For these reasons, commercialization of anti-infection technologies in orthopedics may be limited to high-volume, high-infection rate indications where the economics can support the required investment, or to technologies that are flexible enough to be used in multiple different indications or on different implant types.

Finally, the value of a new technology must be estimated relative to the available alternatives. For example, a hydrogel carrier for local delivery of antibiotics may give better clinical results than direct antibiotic sprinkling, but how much better? Does the incremental improvement in outcomes justify the cost of a large randomized controlled multi-center clinical trial? Despite the large number of anti-infection technologies that have been demonstrated in preclinical testing in recent decades, relatively few have been successfully commercialized. The non-technical hurdles to clinical development are in many cases more significant than the technical hurdles. However, future advances in anti-infection technologies must rely on the expectation that we can find technological solutions to the non-technical challenges facing industry, if these challenges are well understood and factored into the assessment of new technologies from the start.

6 Summary

Implant related bacterial infection is a significant clinical problem in orthopedic surgery, and there is clear opportunity for development of new material technologies to help reduce infection rates. Many anti-infection technologies have been evaluated preclinically, however few have been commercialized. Commercially successful technologies include implant coatings that release antibiotics or silver ions locally in the wound site, to kill bacteria before they can form a biofilm on the implant surface. These technologies have been applied only in very limited indications however, and there is still a significant unmet clinical need. The next generation of anti-infection technologies must be selected and designed to overcome the regulatory and commercial challenges as well as technical challenges if they are to make a significant clinical impact.

References

1. Bozic KJ, Kurtz SM, Lau E, et al. The epidemiology of revision total knee arthroplasty in the United States. Clin Orthop Relat Res. 2010;468(1):45–51.
2. Southwood RT, Rice JL, McDonald PJ, et al. Infection in experimental hip arthroplasties. J Bone Joint Surg Br. 1985;67(2):229–31.
3. Rezapoor M, Parvizi J. Prevention of periprosthetic joint infection. J Arthroplasty. 2015;30(6):902–7.
4. Voigt J, Mosier M, Darouiche R. Systematic review and meta-analysis of randomized controlled trials of antibiotics and antiseptics for preventing infection in people receiving primary total hip and knee prostheses. Antimicrob Agents Chemother. 2015;59(11):6696–707.
5. Stambough JB, Nam D, Warren DK, et al. Decreased hospital costs and surgical site infection incidence with a universal decolonization protocol in primary total joint arthroplasty. J Arthroplasty. 2017;32(3):728–34.
6. Gristina AG. Biomaterial-centered infection: microbial adhesion versus tissue integration. Science. 1987;237(4822):1588–95.
7. Busscher HJ, van der Mei HC, Subbiahdoss G, et al. Biomaterial-associated infection: locating the finish line in the race for the surface. Sci Transl Med. 2012;4(153):153rv10.
8. Zaborowska M, Tillander J, Brånemark R, et al. Biofilm formation and antimicrobial susceptibility of staphylococci and enterococci from osteomyelitis associated with percutaneous orthopaedic implants. J Biomed Mater Res B Appl Biomater. 2016. https://doi.org/10.1002/jbm.b.33803.
9. Arciola CR, Campoccia D, Speziale P, et al. Biofilm formation in Staphylococcus implant infections. A review of molecular mechanisms and implications for biofilm-resistant materials. Biomaterials. 2012;33(26):5967–82.
10. Costerton JW, Montanaro L, Arciola CR. Biofilm in implant infections: its production and regulation. Int J Artif Organs. 2005;28(11):1062–8.
11. Archer NK, Mazaitis MJ, Costerton JW, et al. Staphylococcus aureus biofilms: properties, regulation, and roles in human disease. Virulence. 2011;2(5):445–59.
12. Stoodley P, Ehrlich GD, Sedghizadeh PP, et al. Orthopaedic biofilm infections. Curr Orthop Pract. 2011;22(6):558–63.
13. Torbert JT, Joshi M, Moraff A, et al. Current bacterial speciation and antibiotic resistance in deep infections after operative fixation of fractures. J Orthop Trauma. 2015;29(1):7–17.

14. Drago L, De Vecchi E, Bortolin M, et al. Epidemiology and antibiotic resistance of late prosthetic knee and hip infections. J Arthroplasty. 2017 Mar;32(8):2496–500.
15. Palmer MP, Altman DT, Altman GT, et al. Can we trust intraoperative culture results in nonunions? J Orthop Trauma. 2014;28(7):384–90.
16. Palmer M, Costerton JW, Sewecke J, et al. Molecular techniques to detect biofilm bacteria in long bone nonunion: a case report. Clin Orthop Relat Res. 2011;469(11):3037–42. https://doi.org/10.1007/s11999-011-1843-9.
17. Fraunholz M, Sinha B. Intracellular staphylococcus aureus: live-in and let die. Front Cell Infect Microbiol. 2012;2:43.
18. Broekhuizen CA, de Boer L, Schipper K, et al. Staphylococcus epidermidis is cleared from biomaterial implants but persists in peri-implant tissue in mice despite rifampicin/vancomycin treatment. J Biomed Mater Res A. 2008;85(2):498–505.
19. Sendi P, Proctor RA. Staphylococcus aureus as an intracellular pathogen: the role of small colony variants. Trends Microbiol. 2009;17(2):54–8. https://doi.org/10.1016/j.tim.2008.11.004.
20. de Mesy Bentley KL, Trombetta R, Nishitani K, et al. Evidence of staphylococcus aureus deformation, proliferation, and migration in canaliculi of live cortical bone in murine models of osteomyelitis. J Bone Miner Res. 2017;32(5):985–90.
21. Berríos-Torres SI, Umscheid CA, Bratzler DW, Healthcare Infection Control Practices Advisory Committee, et al. Centers for disease control and prevention guideline for the prevention of surgical site infection. JAMA Surg. 2017;152(8):784–91.
22. Hake ME, Young H, Hak DJ, et al. Local antibiotic therapy strategies in orthopaedic trauma: practical tips and tricks and review of the literature. Injury. 2015;46(8):1447–56.
23. Singh K, Bauer JM, LaChaud GY, et al. Surgical site infection in high-energy peri-articular tibia fractures with intra-wound vancomycin powder: a retrospective pilot study. J Orthop Traumatol. 2015;16(4):287–91.
24. Romanò CL, Scarponi S, Gallazzi E, et al. Antibacterial coating of implants in orthopaedics and trauma: a classification proposal in an evolving panorama. J Orthop Surg Res. 2015;10:157.
25. Willis-Owen CA, Konyves A, Martin DK. Factors affecting the incidence of infection in hip and knee replacement: an analysis of 5277 cases. J Bone Joint Surg Br. 2010;92(8):1128–33. https://doi.org/10.1302/0301-620X.92B8.24333.
26. Garrett TR, Bhakoo M, Zhang Z. Bacterial adhesion and biofilms on surfaces. Progress in Natural Science. 2008;18(9):1049–56.
27. Metsemakers WJ, Schmid T, Zeiter S, et al. Titanium and steel fracture fixation plates with different surface topographies: influence on infection rate in a rabbit fracture model. Injury. 2016;47(3):633–9.
28. Bagherifard S, Hickey DJ, de Luca AC, et al. The influence of nanostructured features on bacterial adhesion and bone cell functions on severely shot peened 316L stainless steel. Biomaterials. 2015;73:185–97.
29. Kummer KM, Taylor EN, Durmas NG, et al. Effects of different sterilization techniques and varying anodized TiO_2 nanotube dimensions on bacteria growth. J Biomed Mater Res B Appl Biomater. 2013;101(5):677–88. https://doi.org/10.1002/jbm.b.32870. Epub 2013 Jan 29.
30. Bhardwaj G, Webster TJ. Reduced bacterial growth and increased osteoblast proliferation on titanium with a nanophase TiO_2 surface treatment. Int J Nanomedicine. 2017;12:363–9.
31. Visai L, De Nardo L, Punta C, et al. Titanium oxide antibacterial surfaces in biomedical devices. Int J Artif Organs. 2011;34(9):929–46.
32. Fujishima A, Zhangb X, Tryk DA. TiO_2 photocatalysis and related surface phenomena. Surf Sci Rep. 2008;63:515–82.
33. Foster HA, Ditta IB, Varghese S, et al. Photocatalytic disinfection using titanium dioxide: spectrum and mechanism of antimicrobial activity. Appl Microbiol Biotechnol. 2011;90(6):1847–68.
34. Roach MD, Williamson RS, Blakely IP, et al. Tuning anatase and rutile phase ratios and nanoscale surface features by anodization processing onto titanium substrate surfaces. Mater Sci Eng C Mater Biol Appl. 2016;58:213–23.

35. Lilja M, Welch K, Astrand M, et al. Effect of deposition parameters on the photocatalytic activity and bioactivity of TiO₂ thin films deposited by vacuum arc on Ti-6Al-4V substrates. J Biomed Mater Res B Appl Biomater. 2012;100(4):1078–85.

36. Dunlop PS, Sheeran CP, Byrne JA, et al. Inactivation of clinically relevant pathogens by photocatalytic coatings. J Photochem Photobiol A. 2010;216(2):303–10.

37. Yue C, Kuijer R, Kaper HJ, et al. Simultaneous interaction of bacteria and tissue cells with photocatalytically activated, anodized titanium surfaces. Biomaterials. 2014;35(9):2580–7.

38. Jose B, Antoci V Jr, Zeiger AR, et al. Vancomycin covalently bonded to titanium beads kills Staphylococcus aureus. Chem Biol. 2005;12(9):1041-1048.

39. Antoci V Jr, Adams CS, Parvizi J, et al. The inhibition of Staphylococcus epidermidis biofilm formation by vancomycin-modified titanium alloy and implications for the treatment of periprosthetic infection. Biomaterials. 2008;29(35):4684–90.

40. Shapiro IM, Hickok NJ, Parvizi J, et al. Molecular engineering of an orthopaedic implant: from bench to bedside. Eur Cell Mater. 2012;23:362–70.

41. Antoci V Jr, Adams CS, Hickok NJ, et al. Vancomycin bound to Ti rods reduces periprosthetic infection: preliminary study. Clin Orthop Relat Res. 2007;461:88–95.

42. Stewart S, Barr S, Engiles J, et al. Vancomycin-modified implant surface inhibits biofilm formation and supports bone-healing in an infected osteotomy model in sheep: a proof-of-concept study. J Bone Joint Surg Am. 2012;94(15):1406–15.

43. Nie B, Ao H, Long T, et al. Immobilizing bacitracin on titanium for prophylaxis of infections and for improving osteoinductivity: An *in vivo* study. Colloids Surf B Biointerfaces. 2017;150:183–91.

44. Chen R, Willcox MD, Ho KK, et al. Antimicrobial peptide melimine coating for titanium and its *in vivo* antibacterial activity in rodent subcutaneous infection models. Biomaterials. 2016;85:142–51.

45. Godoy-Gallardo M, Mas-Moruno C, Fernández-Calderón MC, et al. Covalent immobilization of hLf1-11 peptide on a titanium surface reduces bacterial adhesion and biofilm formation. Acta Biomater. 2014;10(8):3522–34.

46. Schaer TP, Stewart S, Hsu BB, et al. Hydrophobic polycationic coatings that inhibit biofilms and support bone healing during infection. Biomaterials. 2012;33(5):1245–54. https://doi.org/10.1016/j.biomaterials.2011.10.038.

47. Levin PD. The effectiveness of various antibiotics in methyl methacrylate. J Bone Joint Surg Br. 1975;57(2):234–7.

48. Buchholz HW, Elson RA, Engelbrecht E, et al. Management of deep infection of total hip replacement. J Bone Joint Surg Br. 1981;63-B(3):342–53.

49. Cui Q, Mihalko WM, Shields JS, et al. Antibiotic-impregnated cement spacers for the treatment of infection associated with total hip or knee arthroplasty. J Bone Joint Surg Am. 2007;89(4):871–82.

50. Bertazzoni Minelli E, Benini A, et al. Antimicrobial activity of gentamicin and vancomycin combination in joint fluids after antibiotic-loaded cement spacer implantation in two-stage revision surgery. J Chemother. 2015;27(1):17–24. https://doi.org/10.1179/19739478 13Y.0000000157.

51. Conway J, Mansour J, Kotze K, et al. Antibiotic cement-coated rods: an effective treatment for infected long bones and prosthetic joint nonunions. Bone Joint J. 2014;96-B(10):1349–54.

52. Thonse R, Conway J. Antibiotic cement-coated interlocking nail for the treatment of infected nonunions and segmental bone defects. J Orthop Trauma. 2007;21(4):258–68.

53. Ohtsuka H, Yokoyama K, Higashi K, et al. Use of antibiotic-impregnated bone cement nail to treat septic nonunion after open tibial fracture. J Trauma. 2002;52(2):364–6.

54. Eckman JB Jr, Henry SL, Mangino PD, et al. Wound and serum levels of tobramycin with the prophylactic use of tobramycin-impregnated polymethylmethacrylate beads in compound fractures. Clin Orthop Relat Res. 1988;237:213–5.

55. Lewis G. Properties of antibiotic-loaded acrylic bone cements for use in cemented arthroplasties: a state-of-the-art review. J Biomed Mater Res B Appl Biomater. 2009;89(2):558–74.

56. Meyer J, Piller G, Spiegel CA, et al. Vacuum-mixing significantly changes antibiotic elution characteristics of commercially available antibiotic-impregnated bone cements. J Bone Joint Surg Am. 2011;93(22):2049–56.
57. Chang Y, Tai CL, Hsieh PH, et al. Gentamicin in bone cement: a potentially more effective prophylactic measure of infectionin joint arthroplasty. Bone Joint Res. 2013;2(10):220–6.
58. Neut D, van de Belt H, Stokroos I, et al. Biomaterial-associated infection of gentamicin-loaded PMMA beads in orthopaedic revision surgery. J Antimicrob Chemother. 2001;47(6):885–91.
59. Anagnostakos K, Hitzler P, Pape D, et al. Persistence of bacterial growth on antibiotic-loaded beads: is it actually a problem? Acta Orthop. 2008;79(2):302–7. https://doi.org/10.1080/17453670710015120.
60. Mäkinen TJ, Veiranto M, Knuuti J, et al. Efficacy of bioabsorbable antibiotic containing bone screw in the prevention of biomaterial-related infection due to Staphylococcus aureus. Bone. 2005;36(2):292–9.
61. Price JS, Tencer AF, Arm DM, et al. Controlled release of antibiotics from coated orthopedic implants. J Biomed Mater Res. 1996;30(3):281–6.
62. Gollwitzer H, Ibrahim K, Meyer H, et al. Antibacterial poly(D,L-lactic acid) coating of medical implants using a biodegradable drug delivery technology. J Antimicrob Chemother. 2003;51(3):585–91.
63. Kälicke T, Schierholz J, Schlegel U, et al. Effect on infection resistance of a local antiseptic and antibiotic coating on osteosynthesis implants: an *in vitro* and *in vivo* study. J Orthop Res. 2006;24(8):1622–40.
64. Lucke M, Schmidmaier G, Sadoni S, et al. Gentamicin coating of metallic implants reduces implant-related osteomyelitis in rats. Bone. 2003;32(5):521–31.
65. Nast S, Fassbender M, Bormann N, et al. *In vivo* quantification of gentamicin released from an implant coating. J Biomater Appl. 2016;31(1):45–54.
66. Vester H, Wildemann B, Schmidmaier G, et al. Gentamycin delivered from a PDLLA coating of metallic implants: *In vivo* and *in vitro* characterisation for local prophylaxis of implant-related osteomyelitis. Injury. 2010;41(10):1053–9.
67. Fuchs T, Stange R, Schmidmaier G, et al. The use of gentamicin-coated nails in the tibia: preliminary results of a prospective study. Arch Orthop Trauma Surg. 2011;131(10):1419–25.
68. Metsemakers WJ, Reul M, Nijs S. The use of gentamicin-coated nails in complex open tibia fracture and revision cases: a retrospective analysis of a single centre case series and review of the literature. Injury. 2015;46(12):2433–7.
69. Neut D, Dijkstra RJ, Thompson JI, et al. A biodegradable gentamicin-hydroxyapatite-coating for infection prophylaxis in cementless hip prostheses. Eur Cell Mater. 2015;29:42–55.
70. Metsemakers WJ, Emanuel N, Cohen O, et al. A doxycycline-loaded polymer-lipid encapsulation matrix coating for the prevention of implant-related osteomyelitis due to doxycycline-resistant methicillin-resistant Staphylococcus aureus. J Control Release. 2015;209:47–56.
71. Radin S, Ducheyne P, Kamplain T, et al. Silica sol-gel for the controlled release of antibiotics. I. Synthesis, characterization, and *in vitro* release. J Biomed Mater Res. 2001;57(2):313–20.
72. Qu H, Knabe C, Radin S, et al. Percutaneous external fixator pins with bactericidal micron-thin sol-gel films for the prevention of pin tract infection. Biomaterials. 2015;62:95–105.
73. FDA. Draft guidance for industry and FDA staff: premarket notification [510(k)] submissions for medical devices that include antimicrobial agents 2007. https://www.fda.gov/downloads/medicaldevices/deviceregulationandguidance/guidancedocuments/ucm071396.pdf. Accessed 29 May 2017.
74. FDA. Guidance for industry and FDA staff: current good manufacturing practice requirements for combination products. 2015. https://www.fda.gov/downloads/regulatoryinformation/guidances/ucm429304.pdf. Accessed 29 May 2017.
75. Cai X, Han K, Cong X, et al. The use of calcium sulfate impregnated with vancomycin in the treatment of open fractures of long bones: a preliminary study. Orthopedics. 2010;33(3). https://doi.org/10.3928/01477447-20100129-17.

76. Emanuel N, Rosenfeld Y, Cohen O, et al. A lipid-and-polymer-based novel local drug delivery system-BonyPid™: from physicochemical aspects to therapy of bacterially infected bones. J Control Release. 2012;160(2):353–61. https://doi.org/10.1016/j.jconrel.2012.03.027.
77. Alexander JW. History of the medical use of silver. Surg Infect (Larchmt). 2009;10(3):289–92.
78. Lai NM, Chaiyakunapruk N, Lai NA, et al. Catheter impregnation, coating or bonding for reducing central venous catheter-related infections in adults. Cochrane Database of Syst Rev. 2013; (6): CD007878.
79. Lam TB, Omar MI, Fisher E, et al. Types of indwelling urethral catheters for short-term catheterisation in hospitalised adults. Cochrane Database Syst Rev. 2014; (9):CD004013.
80. Karchmer TB, Giannetta ET, Muto CA, et al. A randomized crossover study of silver-coated urinary catheters in hospitalized patients. Arch Intern Med. 2000;160(21):3294–8.
81. Feng QL, Wu J, Chen GQ, et al. A mechanistic study of the antibacterial effect of silver ions on Escherichia coli and Staphylococcus aureus. J Biomed Mater Res. 2000;52(4):662–8.
82. Lansdown AB. A pharmacological and toxicological profile of silver as an antimicrobial agent in medical devices. Adv Pharmacol Sci. 2010;2010:910686.
83. Mulley G, Jenkins AT, Waterfield NR. Inactivation of the antibacterial and cytotoxic properties of silver ions by biologically relevant compounds. PLoS One. 2014;9(4):e94409.
84. Lansdown AB. Silver in healthcare: its antimicrobial efficacy and safety in use. Cambridge: RSC Publishing; 2010. https://doi.org/10.1039/9781849731799.
85. Massè A, Bruno A, Bosetti M, et al. Prevention of pin track infection in external fixation with silver coated pins: clinical and microbiological results. J Biomed Mater Res. 2000;53(5):600–4.
86. Wassall MA, Santin M, Isalberti C, et al. Adhesion of bacteria to stainless steel and silver-coated orthopedic external fixation pins. J Biomed Mater Res. 1997;36(3):325–30.
87. Collinge CA, Goll G, Seligson D, et al. Pin tract infections: silver vs uncoated pins. Orthopedics. 1994;17(5):445–8.
88. Coester LM, Nepola JV, Allen J, et al. The effects of silver coated external fixation pins. Iowa Orthop J. 2006;26:48–53.
89. Schmidt-Braekling T, Streitbuerger A, Gosheger G, et al. Silver-coated megaprostheses: review of the literature. Eur J Orthop Surg Traumatol. 2017;27(4):483–9.
90. Donati F, Di Giacomo G, D'Adamio S, et al. Silver-coated hip megaprosthesis in oncological limb savage surgery. Biomed Res Int. 2016;2016:9079041.
91. Piccioli A, Donati F, Giacomo GD, et al. Infective complications in tumour endoprostheses implanted after pathological fracture of the limbs. Injury. 2016;47(Suppl 4):S22–8.
92. Hussmann B, Johann I, Kauther MD, et al. Measurement of the silver ion concentration in wound fluids after implantation of silver-coated megaprostheses: correlation with the clinical outcome. Biomed Res Int. 2013;2013:763096.
93. Hardes J, von Eiff C, Streitbuerger A, et al. Reduction of periprosthetic infection with silver-coated megaprostheses in patients with bone sarcoma. J Surg Oncol. 2010;101(5):389–95.
94. Glehr M, Leithner A, Friesenbichler J, et al. Argyria following the use of silver-coated megaprostheses: no association between the development of local argyria and elevated silver levels. Bone Joint J. 2013;95-B(7):988–92.
95. Hardes J, Ahrens H, Gebert C, et al. Lack of toxicological side-effects in silver-coated megaprostheses in humans. Biomaterials. 2007;28(18):2869–75.
96. Wafa H, Grimer RJ, Reddy K, et al. Retrospective evaluation of the incidence of early periprosthetic infection with silver-treated endoprostheses in high-risk patients: case-control study. Bone Joint J. 2015;97-B(2):252–7.
97. Scoccianti G, Frenos F, Beltrami G, et al. Levels of silver ions in body fluids and clinical results in silver-coated megaprostheses after tumour, trauma or failed arthroplasty. Injury. 2016;47(Suppl 4):S11–6.
98. Hauschild G, Hardes J, Gosheger G, et al. Evaluation of osseous integration of PVD-silver-coated hip prostheses in a canine model. Biomed Res Int. 2015;2015:292406.
99. Khalilpour P, Lampe K, Wagener M, et al. Ag/SiO(x)C(y) plasma polymer coating for antimicrobial protection of fracture fixation devices. J Biomed Mater Res B Appl Biomater. 2010;94(1):196–202.

100. Zielinska E, Tukaj C, Radomski MW, et al. Molecular mechanism of silver nanoparticles-induced human osteoblast cell death: protective effect of inducible nitric oxide synthase inhibitor. PLoS One. 2016;11(10):e0164137.
101. Rosário F, Hoet P, Santos C, et al. Death and cell cycle progression are differently conditioned by the AgNP size in osteoblast-like cells. Toxicology. 2016;368-369:103–15.
102. Esfandiari N, Simchi A, Bagheri R. Size tuning of Ag-decorated TiO_2 nanotube arrays for improved bactericidal capacity of orthopedic implants. J Biomed Mater Res A. 2014;102(8):2625–35.
103. Zhao Y, Xing Q, Janjanam J, et al. Facile electrochemical synthesis of antimicrobial TiO_2 nanotube arrays. Int J Nanomedicine. 2014;9:5177–87.
104. Pauksch L, Hartmann S, Rohnke M, et al. Biocompatibility of silver nanoparticles and silver ions in primary human mesenchymal stem cells and osteoblasts. Acta Biomater. 2014;10(1):439–49.
105. Sengstock C, Diendorf J, Epple M, et al. Effect of silver nanoparticles on human mesenchymal stem cell differentiation. Beilstein J Nanotechnol. 2014;5:2058–69.
106. Necula BS, van Leeuwen JP, Fratila-Apachitei LE, et al. *In vitro* cytotoxicity evaluation of porous TiO_2-Ag antibacterial coatings for human fetal osteoblasts. Acta Biomater. 2012;8(11):4191–7.
107. Greulich C, Kittler S, Epple M, et al. Studies on the biocompatibility and the interaction of silver nanoparticles with human mesenchymal stem cells (hMSCs). Langenbecks Arch Surg. 2009;394(3):495–502.
108. Ning C, Wang X, Li L, et al. Concentration ranges of antibacterial cations for showing the highest antibacterial efficacy but the least cytotoxicity against mammalian cells: implications for a new antibacterial mechanism. Chem Res Toxicol. 2015;28(9):1815–22.
109. Kraft CN, Hansis M, Arens S, et al. Striated muscle microvascular response to silver implants: a comparative *in vivo* study with titanium and stainless steel. J Biomed Mater Res. 2000;49(2):192–9.
110. Welz C Osteointegrative eigenschaften einer silberbeschichtung von hüftendoprothesen bei hunden. Dissertation. Westfälische Wilhelms Universität Münster; 2008.
111. Kabata T, Maeda T, Kajino Y, et al. Iodine-supported hip implants: short term clinical results. Biomed Res Int. 2015;2015:368124.
112. Demura S, Murakami H, Shirai T, et al. Surgical treatment for pyogenic vertebral osteomyelitis using iodine-supported spinal instruments: initial case series of 14 patients. Eur J Clin Microbiol Infect Dis. 2015;34(2):261–6.
113. Shirai T, Tsuchiya H, Nishida H, et al. Antimicrobial megaprostheses supported with iodine. J Biomater Appl. 2014;29(4):617–23.
114. Tsuchiya H, Shirai T, Nishida H, et al. Innovative antimicrobial coating of titanium implants with iodine. J Orthop Sci. 2012;17(5):595–604.
115. Tennent DJ, Shiels SM, Sanchez CJ Jr, et al. Time-dependent effectiveness of locally applied vancomycin powder in a contaminated traumatic orthopaedic wound model. J Orthop Trauma. 2016;30(10):531–7.
116. Armaghani SJ, Menge TJ, Lovejoy SA, et al. Safety of topical vancomycin for pediatric spinal deformity: nontoxic serum levels with supratherapeutic drain levels. Spine (Phila Pa 1976). 2014;39(20):1683–7.
117. Khan NR, Thompson CJ, DeCuypere M, et al. A meta-analysis of spinal surgical site infection and vancomycin powder. J Neurosurg Spine. 2014;21(6):974–83.
118. Kang DG, Holekamp TF, Wagner SC, et al. Intrasite vancomycin powder for the prevention of surgical site infection in spine surgery: a systematic literature review. Spine J. 2015;15(4):762–70.
119. Chiang HY, Herwaldt LA, Blevins AE, et al. Effectiveness of local vancomycin powder to decrease surgical site infections: a meta-analysis. Spine J. 2014;14(3):397–407.
120. Tubaki VR, Rajasekaran S, Shetty AP. Effects of using intravenous antibiotic only versus local intrawound vancomycin antibiotic powder application in addition to intravenous antibiotics on postoperative infection in spine surgery in 907 patients. Spine (Phila Pa 1976). 2013;38(25):2149–55.

121. Whiteside LA, Roy ME. One-stage revision with catheter infusion of intraarticular antibiotics successfully treats infected THA. Clin Orthop Relat Res. 2017;475(2):419–29.
122. Lawing CR, Lin FC, Dahners LE. Local injection of aminoglycosides for prophylaxis against infection in open fractures. J Bone Joint Surg Am. 2015;97(22):1844–51.
123. Giavaresi G, Meani E, Sartori M, et al. Efficacy of antibacterial-loaded coating in an *in vivo* model of acutely highly contaminated implant. Int Orthop. 2014;38(7):1505–12.
124. Malizos K, Blauth M, Danita A, et al. Fast-resorbable antibiotic-loaded hydrogel coating to reduce post-surgical infection after internal osteosynthesis: a multicenter randomized controlled trial. J Orthop Traumatol. 2017;18(2):159–69.
125. Penn-Barwell JG, Murray CK, Wenke JC. Local antibiotic delivery by a bioabsorbable gel is superior to PMMA bead depot in reducing infection in an open fracture model. J Orthop Trauma. 2014;28(6):370–5.
126. Bennett-Guerrero E, Berry SM, Bergese SD, et al. A randomized, blinded, multicenter trial of a gentamicin vancomycin gel (DFA-02) in patients undergoing abdominal surgery. Am J Surg. 2017;213(6):1003–9.
127. Jennings JA, Carpenter DP, Troxel KS, et al. Novel antibiotic-loaded point-of-care implant coating inhibits biofilm. Clin Orthop Relat Res. 2015;473(7):2270–82.
128. Jennings JA, Beenken KE, Skinner RA, et al. Antibiotic-loaded phosphatidylcholine inhibits staphylococcal bone infection. World J Orthop. 2016;7(8):467–74.
129. Hesse W. Walther, Angelina Hesse-Early Contributors to Bacteriology. ASM News. 1992;58(8):425–8.
130. Kinnari TJ, Peltonen LI, Kuusela P, et al. Bacterial adherence to titanium surface coated with human serum albumin. Otol Neurotol. 2005;26(3):380–4.
131. Incani V, Omar A, Prosperi-Porta G, et al. Ag5IO6: novel antibiofilm activity of a silver compound with application to medical devices. Int J Antimicrob Agents. 2015;45(6):586–93.
132. ASTM International. Standard test method for testing disinfectant efficacy against Pseudomonas aeruginosa biofilm using the MBEC assay. E2799-11. West Conshohocken, PA; 2011.
133. Melcher GA, Claudi B, Schlegel U, et al. Influence of type of medullary nail on the development of local infection. An experimental study of solid and slotted nails in rabbits. J Bone Joint Surg Br. 1994;76(6):955–9.
134. Lambe DW Jr, Ferguson KP, Mayberry-Carson KJ, et al. Foreign-body-associated experimental osteomyelitis induced with Bacteroides fragilis and Staphylococcus epidermidis in rabbits. Clin Orthop Relat Res. 1991;266:285–94.
135. Williams DL, Haymond BS, Woodbury KL, et al. Experimental model of biofilm implant-related osteomyelitis to test combination biomaterials using biofilms as initial inocula. J Biomed Mater Res A. 2012;100(7):1888–900.
136. Williams DL, Costerton JW. Using biofilms as initial inocula in animal models of biofilm-related infections. J Biomed Mater Res B Appl Biomater. 2012;100(4):1163–9.

Platelet Rich Plasma: Biology and Clinical Usage in Orthopedics

Dukens LaBaze and Hongshuai Li

Keywords Platelet · Platelet rich plasma · Orthopedics surgery · Sports medicine · Tendon · Tendinopathy · Ligament · Cartilage · Osteoarthritis · Muscle · Growth factor · Cell proliferation · Cell differentiation · Fibrosis

1 Introduction

Biological research in the areas of skeletal, cartilaginous, tendinous, and muscular tissues has led to the advancement of various products designed to augment healing. Platelet-rich plasma (PRP) has been in clinical use and researched since the 1970s because of its regenerative properties. In this chapter, we will first define PRP and its components and discuss various methods of preparation and isolation. The second section will focus on the clinical applications of PRP on tissue specific pathologies including tendinopathy, ligamentous injuries, osteoarthritis, and muscle injuries in the field of orthopedic surgery and sports medicine. Lastly, the latest progress in PRP research will be briefly listed and some promising future directions will be discussed.

D. LaBaze · H. Li (✉)
Musculoskeletal Growth & Regeneration Laboratory, Department of Orthopedics Surgery, School of Medicine, University of Pittsburgh, Pittsburgh, PA, USA
e-mail: Hongshuai.li@pitt.edu

© Springer International Publishing AG, part of Springer Nature 2018   243
B. Li, T. Webster (eds.), *Orthopedic Biomaterials*,
https://doi.org/10.1007/978-3-319-89542-0_12

2 Biology of Platelet Rich Plasma

2.1 What is PRP (PRP Definition)?

Despite any solid consensus as to its definition, platelet-rich plasma (PRP) is accepted by major scientific communities as the plasma fraction of autologous blood which has a platelet concentrate above baseline. The normal range of platelets in whole blood in a healthy individual is 150,000–450,000 per μL. The reported platelet concentration of PRP in the literature ranges from as low as 200,000 platelets/μL to equal or greater than 1000,000 platelets/μL. Other names by which PRP has been known include platelet-enriched plasma (PeRP), platelet-rich concentrate (PRC), plasma rich in platelets, preparation rich in growth factors (PRGF), platelet-rich fibrin (PRF), autologous platelet concentrate (APC), and autogenous platelet gel (APG) [1, 2]. It has been constantly reported that PRP has a greater concentration of growth factors, cytokines, and other useful substances which have been demonstrated to stimulate and/or aid various healing processes, which gained its popularity as to be a great source of autologous growth factors. For preparation of PRP, various protocols are used, with an underlying principle of concentrating platelets to higher than its physiological level, then injecting this concentrate in the tissue where healing is desired. However, a clear consensus over the best method of preparation and usage of the PRP is still lacking. Much of the confusion is due to the variability in results and interpretations in current basic and clinical studies [3, 4].

It is worth mentioning here another autologous blood product named autologous conditioned serum (ACS), which could cause confusion sometimes with PRP, as both are derived from whole blood and can be used for direct injections to treat musculoskeletal conditions in clinic. ACS is generated by incubating venous blood in the presence of medical grade glass beads. Peripheral blood leukocytes produce elevated amounts of endogenous anti-inflammatory cytokines, including the inhibitor of IL-1, the IL-1 receptor antagonist (IL-ra), which accumulates in the serum [5]. Following centrifugation and extraction, ACS can be injected into the affected region, especially in osteoarthritic joints [6]. Although also derived from autologous blood, the concept and the indications for application are different from PRP [7].

2.2 Principles for PRP Isolation and Classification

The use of PRP to improve healing has been explored considerably during the last decade after firstly been described by Whitman et al. [8]. Since then, many different PRP isolation techniques have been developed, and many different PRP preparation systems were commercially marketed. Table 1 summarizes commonly used PRP preparation kits and key characterizations of their final products. It is worth mentioning that the design of PRP isolation kits is constantly evolving due to the rapid

Table 1 Major PRP isolation kits, protocols, and characteristics

System	Blood volume	Anticoagulant	Single or double spin	Final volume of PRP	Final platelet concentration	Leukocytes	Activator
ACP-DS (Arthrex)	9 ml	ACD-A	Single spin 350 g, 5 min	3 ml	X2-X3	NO	N/A
Fibrinet (Cascade; Musculoskeletal Tissue Foundation)	9–18 ml	N/A	Single spin (for PRP) 1100 g, 6 min; double spin (for PRFM) 3600-4500 g, 25 min	4–9 ml	X1-X1.5	NO	CaCl₂
GPS III (Biomet)	27–110 ml	ACD-A	Single spin, 3200 rpm, 15 min	3–12 ml	X3-X8	Yes	AT/ CaCl₂
Magellan (Medtronic, Minneapolis)	30–60 ml	ACD-A	Double spin	6 ml	X3-X7	Yes	CaCl₂
Endoret (PRGF) (BTI Biotechnology Institute)	9 ml	Sodium citrate	Single spin 580 g, 8 min	4 ml	X2-X3	NO	CaCl₂
Smart prep (Harvest Technologies, Plymouth)	20–120 ml	ACD-A	Double spin	3–20 ml	X4-X6	Yes	BT/ CaCl₂
Selphyl system (Selphyl, Bethlehem)	8 ml	Sodium citrate	Single spin 1100 g, 6 min	4–5 ml	X2-X3	No	CaCl₂
RegenKit-A-PRP (ReganLab, Le Mont-sur-Lausanne)	8 ml	ACD-A	Single spin 1500 g, 8 min	4–5 ml	X1.6	NO	N/A
Pure PRP (Emcyte Genesis CS)	50 ml	Sodium citrate	Double spin 1st spin at 3800 rpm, 1.5 min; 2nd spin at 3.800 rpm, 5 min	5–7 ml	X3-X5	Customizable	N/A

CaCl₂ Calcium Chloride, *AT* autologous thrombin, *BT* bovine thrombin, *ACD-A* anticoagulant citrate dextrose solution A progress of PRP basic science research. Readers are advised to refer to the latest updated information from the manufacture for the latest development.

Although these systems differ widely in their general procedures, choices of anti-coagulants, abilities to collect and concentrate platelets, centrifugation forces and time durations, and even the cellular composition of their final products, they all share some common principles. All protocols follow a **generic sequence** that consists of whole blood collection, and an initial centrifugation to separate red blood cells (RBCs) from the rest of the blood, with or without the second centrifugation to further concentrate platelets and other components.

2.2.1 Principle for PRP Isolation

Generally, PRP is prepared by a process known as "differential centrifugation". In differential centrifugation, acceleration force and centrifugation duration are adjusted to sediment certain cellular constituents in whole blood based on their different specific gravities. At $37°$, the specific gravity of whole blood is 1.050–1.060; plasma's specific gravity is 1.025–1.029; RBC's specific gravity is approximately 1.095, which is higher than white blood cells (WBCs) (1.063–1.085) and platelets (approximately 1.030). Platelets in normal human have a skewed distribution of densities, which overlapped with other cellular components (mainly with WBCs). Due to the overlapping of specific gravities, double centrifugation process is usually used to obtain high concentration of platelets. In the double centrifugation method, an initial centrifugation to separate RBC from the rest of the blood is followed by a second centrifugation to further pellet platelets, which is then re-suspended in the smaller plasma volume to obtain a relatively higher concentrated platelets [9]. Fig. 1 describes a double centrifugation process of PRP isolation. Whole blood is collected with anticoagulant and is immediately processed for centrifugation. The first centrifugation step is designed to separate RBCs from the remaining whole blood volume. After centrifugation, the whole blood separates into three layers: a bottom layer that consists mostly of RBCs; an upper layer that contains mostly platelets and plasma; and a thin intermediate layer that is known as the buffy coat which is rich in WBCs and platelets. The part that will be collected and processed to the second centrifugation depends on the desired final product. To produce pure PRP (P-PRP), the upper layer and superficial buffy coat are transferred to an empty sterile tube leaving behind the rest of the buffy coat which contains concentrated WBCs. To produce leukocyte rich PRP (L-PRP), the entire layer of buffy coat and few RBCs are transferred. To further concentrate the final PRP product, the second centrifugation step can be applied. This centrifugation step is to soft pellet platelets without rupture the cell membrane. After centrifuge, the upper portion of the volume, which is composed mostly of PPP (platelet-poor plasma), is removed. The soft platelet pellets are then re-suspended in the remaining plasma (usually 1/3–½ of the total plasma volume) to create more concentrated PRP. Finally, the obtained platelet concentrate is applied to the surgical site with a syringe, with or without platelet activators to trigger platelet activation and fibrin polymerization.

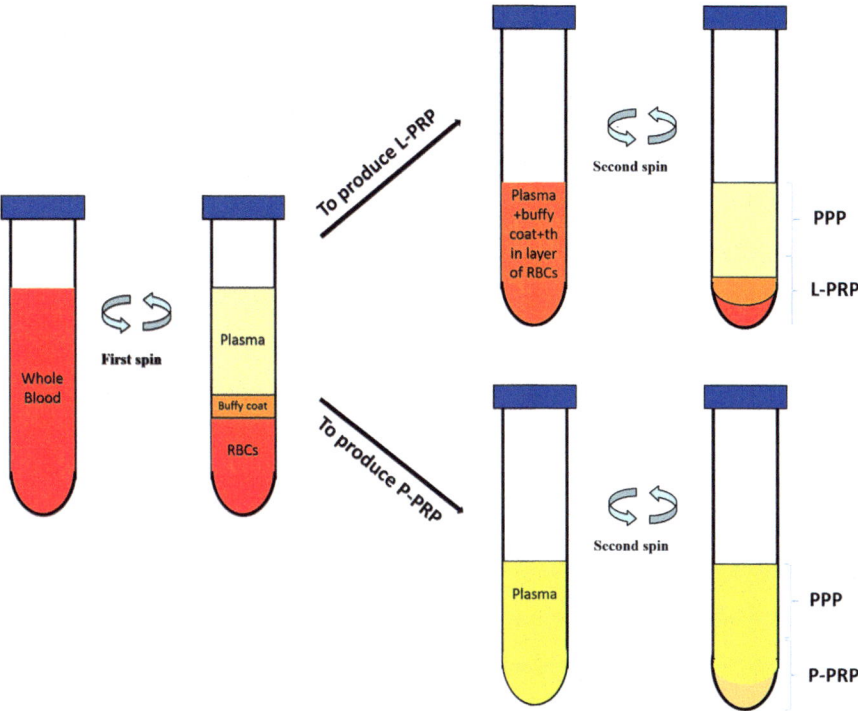

Fig. 1 Schematic drawing of PRP isolation process by differential centrifugation. Abbreviations: RBC: red blood cells; L-PRP: Leukocyte rich platelet-rich plasma; P-PRP: Pure platelet-rich plasma

As demonstrated in Table 1 and discussed in many reviews [10, 11], PRP is prepared by centrifugation varying the relative centrifugal forces, temperature and time. Although there are numerous protocols in the current literature that described the optimal conditions for centrifugation [10], no solid consensus has been reached due to the lack of a clear golden standard for PRP compositions, such as the inclusion of leukocytes, the optimal concentration of platelets, the optimal volume for injection etc. Nevertheless, the readers and researchers should be familiar with the basic principles of PRP isolation and understand the variations among different protocols in the current literature.

2.2.2 PRP Classification

As evidenced and discussed, there are many forms of PRP; and there is much to learn about its preparation, application, concentration, and timing for application, to name a few variables. With many variations in PRP preparation, classification methods have been created to assist in categorizing different PRPs and comparing

Table 2 Platelet-rich plasma classification proposed by Dohan et al. [12]

Categories	Description
Pure platelet-rich plasma (P-PRP)	Also known as Leukocyte-Poor Platelet-Rich Plasma. Products are prepared without leukocytes and with a low-density fibrin network after activation.
Pure platelet-rich fibrin (P-PRF):	Also known as Leukocyte-Poor Platelet-Rich Fibrin. Products are prepared without leukocytes and with a high-density fibrin network. These products presented in a strongly activated gel form, and cannot be injected as a solution or used like traditional fibrin glues. However, because of their strong fibrin matrix, they can be handled like a real solid material which can hold sutures.
Leukocyte and platelet-rich plasma (L-PRP)	These products are prepared with leukocytes and with a low-density fibrin network after activation. It is in this category that the majority of commercial systems belongs.
Leukocyte- and platelet-rich fibrin (L-PRF)	Products are prepared with leukocytes and with a high-density fibrin network.

Table 3 Platelet-rich plasma classification proposed by Mishra et al. [13]

	White blood cells	Activation?	Platelet concentration
Type 1	Increased over baseline	No activation	A: 5× or>; B: <5×
Type 2	Increased over baseline	Activated	A: 5× or>; B: <5×
Type 3	Minimal or no WBCs	No activation	A: 5× or>; B: <5×
Type 4	Minimal or no WBCs	Activated	A: 5× or>; B: <5×

different studies. Just as the rapid progress of our understanding about PRP, the classification methods also keep evolving as our understanding deepens.

The first classification system was proposed by Dohan et al. in [12], and is now widely accepted as a milestone in the unification of the terminology for different PRP products. This classification system separated the PRP products by two key parameters: the presence of leukocytes and the fibrin architecture. Four main categories can be generally defined (Table 2):

The first PRP classification specific for sports medicine was proposed by Mishra et al. at 2012 (Table 3) [13]. This classification system separated PRP products into four types based on leukocytes and platelets concentrations, and activation. Two subtypes were proposed based on platelet concentrations. Sub-type A PRP is 5× or more the blood concentration of platelets, and sub-type B PRP being less than 5 times the blood concentration of platelets. The 5 times cutoff is debatable, as the effective concentration of platelets in the final PRP product were largely unclear; and there was no clear scientific evidence to show the optimal platelet concentration for multiple clinical conditions. Also, the relative concentration to the baseline is more meaningful to evaluate the ability of concentrating platelets by different PRP preparation protocols rather than to predict its effectiveness. Absolute platelet concentration and final volume are more accurate parameters to predict the amount of growth factors that could be delivered based on the fact that the concentration of

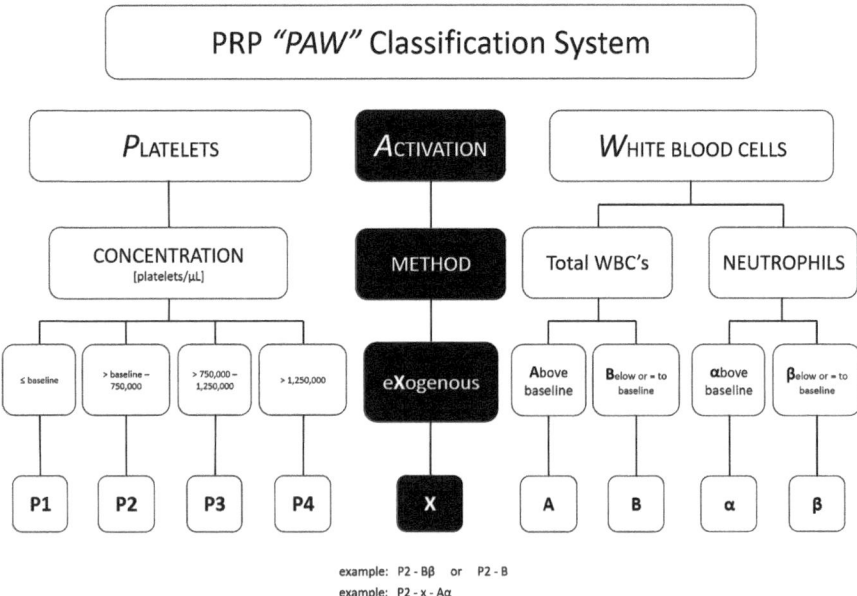

Fig. 2 The PAW classification system proposed by DeLong et al. [14]

growth factors are positively related to the platelet concentration. In 2012, DeLong et al. published the PAW (Platelets, Activation, White blood cells) classification system which is based on those three major components, and is very similar to the proposal of Mishra et al. [13], but focused more on the platelet quantity (absolute number) (Fig. 2) [14].

The 2009 terminology and classification system is a crucial step [12], which laid down foundation for PRP classification. The following developed classification systems added more detailed parameters to fine tune this very complicated product. Without a doubt, PRP classification systems will keep evolving with new developments in PRP preparation and application.

2.3   Biologics of PRP

PRP is a part of concentrated autologous blood, which has a higher concentration of platelets than whole blood. While platelets have been the primary focus when studying PRP in the past, other important constituents have been identified that can contribute to the overall function of PRP to the healing process, which include WBCs, RBCs, and other components in the plasma (Table 4). In this part, we will focus on platelets and its released factors, leukocyte poll, red blood cells, and extracellular vesicles; the current understanding of their roles in tissue healing will be discussed as well.

Table 4 Summary of cellular and molecular components of PRP relevant to musculoskeletal tissue healing

	Component
Plasma	
Proteins	Albumin, globulins, fibrinogen, complement, and clotting factors
Electrolytes	Chloride, sodium, potassium, and calcium
Hormones	IGF1, estrogens, progesterone, androgens, ACTH, and HGH
Biomarkers	COMP, CD11b, protein C, microRNA, osteocalcin, and osteonectin
Platelets	
Alpha granules	Adhesive proteins, clotting factors, and GFs (PDGF, TGF-β, VEGF, FGF, EGF, and HGF)
Dense granules	Calcium, magnesium, ADP, ATP, histamine and neurotransmitters (serotonin)
Lysosomes	Lysosomal enzymes
Leukocytes	
Neutrophils	
Primary granules	Myeloperoxidase, acid hydrolases, defensins, and serine proteases
Secondary granules	Collagenase, lactoferrin, cathelicidin, bactericidal phagocytins, and lysoyme
Tertiary granules	Gelatinase and proteases
Monocytes	Platelet-activating factor, TGF- β, VEGF, FGF, and EGF
Erythrocytes	ATP, S-nitrosothiols, nitric oxide, hydrogen sulfide, hemoglobin, and free radicals.

Abbreviations: *ACTH* adrenocorticotrophic hormone, *ATP* adenosine triphosphate, *COMP* cartilage oligomeric matrix protein, *EGF* epidermal growth factor, *FGF* fibroblastic growth factor, *HGF* hepatocyte growth factor, *HGH* human growth hormone, *VEGF* vascular endothelial growth factor.

2.3.1 Platelet and Platelet Released Factors

Platelets are the major component of PRP. Platelets are small enucleate cytoplasmic fragments of multinucleated megakaryocytes in bone marrow; it ranges from 2 to 3 μm in size. The concentration of platelets in healthy human is 150,000–450,000/μL. Inactive platelets have a discoid shape with an open canalicular system. Many native and exogenous molecules can activate platelets, including collagen, platelet-activating factor, serotonin, calcium, thromboxane A2 (TXA2), and thrombin etc. [15]. The activated platelet undergoes cytoskeleton restructuring to develop multiple filopodia from the location of the canaliculi to initiate exocytosis and degranulation process [16]. Numerous active substances can be released from its intracellular granules. The secretome of platelets consist the core for the clinical application of PRP on tissue healing and regeneration. Proteomic studies have shown that platelets contain over 800 proteins with numerous post-translational modifications, such as phosphorylation, resulting in over 1500 protein-based bioactive factors [17, 18]. Only a portion of these proteins' physiologic actions have been studied so far, including growth factors (GFs) and peptide hormones. In this section, we will go

Table 5 Key regenerative growth factors stored in alpha-granules and their functions

Growth factors	Function
PDGF	Stimulates cell proliferation, chemotaxis, and differentiation. Stimulates angiogenesis
TGF-β	Stimulates production of collagen type I and type III, angiogenesis, re-epithelialization, and synthesis of protease inhibitors to inhibit collagen breakdown.
VEGF	Stimulates angiogenesis by regulating endothelial cell proliferation and migration.
EGF	Influences cell proliferation and cyto-protection Accelerates re-epithelialization Increases tensile strength in wounds Facilitates organization of granulation tissue
bFGF	Stimulates angiogenesis Promotes stem cell proliferation and differentiation Promotes collagen production and tissue repair
IGF-1	Regulates cell proliferation and differentiation Influences matrix secretion from osteoblasts and production of proteoglycan, collagen, and other non-collagen proteins.

through the basics of platelet secretion and discuss mainly the substances that are relevant or important for tissue healing.

Platelet Alpha Granules

α-granules are 300- to 500-nm micro-vesicles with a proteome count of approximately 284 [19]. These include bioactive molecules such as adhesive proteins (fibrinogen, von Willebrand factor) and receptors (αIIbB3 and GPVI receptors), clotting factors (V, XI, XIII, and prothrombin), anticoagulation cargos (tissue factor pathway inhibitor-TFPI, protein S, protease nexin-2), fibrinolytic factors (anti-thrombin, plasmin, and plasminogen), other basic proteins, membrane glycoproteins, and many GFs [20]. The GFs stored in α-granules drawn most of the attentions as they formed the scientific basis of the clinical applications of PRP to healing tissue injuries [21, 22]. Table 5 summarizes the key growth factors that are released from platelet and their biological functions, including platelet-derived growth factor (PDGF), transforming growth factor-β1 (TGF-β1), vascular endothelial growth factor (VEGF), epidermal growth factor (EGF), basic fibroblast growth factor (bFGF), and insulin-like growth factor-1 (IGF-1).

Dense Granules

Dense granules, also known as delta granules, are vesicles 250–300 nm in size. They contain primarily substances that promote blood clotting, such as calcium (clotting factor IV), magnesium, polyphosphate, ADP, ATP, GDP, adenosine, and histamine [23]. A deficiency of delta granules results in mild bleeding disorders. Serotonin, a neurotransmitter, was also found abundant in dense granules [23]. When released, serotonin promotes hemostasis by constricting vascular tone and permeability. Besides its role in hemostasis, it is known that serotonin acts as either a hormone or neurotransmitter and can both positively and negatively regulate bone mass [24]. The release of substances from dense granules was described as "fast" after platelet activation. It has been shown that the release of [^{3}H]-serotonin occurred more rapidly than PF-4 from alpha-granules or beta-hexosaminidase from lysosomes (lambda granules), regardless of the agonist used to stimulate platelets [25].

The Lambda Granules

Lambda granules are lysosomal-type organelles, which contains several hydrolytic enzymes to aid clot resorption. As healing progresses, tissue plasminogen activator secreted by the endothelium activates lysosomal enzymes, which convert plasminogen to plasmin and lyse the clot. Besides, it was reported that plasminogen activators also play a role in homeostasis of muscle fibers and the adjoining extracellular matrix, including fracture repair [26, 27].

Regulation of Platelet Secretion

As evidenced by accumulating evidences, the roles of platelets extend far beyond initial platelet adhesion and aggregation at the site of injury. In addition to coagulation, platelets are essential to many other biological/pathological processes such as inflammation, antimicrobial responses, and wound healing [28]. As the secretome studies have shown, platelets secrete more than 800 molecules during its degranulation process. It is still not clear how platelets keep balance among all the potent mitogenic, pro-angiogenic, anti-angiogenic, pro-inflammatory, anti-inflammatory and adhesive factors released from granules under certain activation conditions. It was believed that the degranulation process is finely controlled [28], simply because if all the contents of the intracellular stores were released in an uncontrolled manner, the thrombus formation and all the following events would be similarly uncoordinated. To explain this, two hypotheses were proposed. One is that platelet granules are not uniform, especially alpha-granule cargos, and that certain factors may be differentially packaged and thus released "selectively". Ma et al. observed that platelet stimulation with specific protease-activated receptor (PAR) 1 or PAR4 agonist resulted in the preferential release of VEGF (a pro-angiogenic factor) or endostatin (anti-angiogenic factor) [29]. Indeed, distinct localization of fibrinogen and vWF (two of the main alpha-granule cargos) have been clearly documented by spinning-disk confocal microscopy [30]. Immuno-gold labeling also showed alpha-granule populations containing either endostatin or VEGF but not both [31].

As discussed, it is clear that the release of factors from platelets is somehow regulated depends on the agonist used. In the current clinical practices, an activation step before PRP injection is applied. In most protocols, thrombin and/or calcium chloride ($CaCl_2$) is used to activate the platelets, but some physicians prefer to inject PRP in its resting form, relying on the spontaneous platelet activation occurring after exposure to the native collagen present in the injured tissues. Currently, there is a lack of evidence on the most suitable method for PRP activation, and the choice of strategy for activation is mainly empirically based on clinical practice rather than supported by well-designed studies on the final platelet releasate and under the various clinical indications. How to control the release of certain growth factors that are desired for the targeted tissue is still a holy grail for the PRP research. Studies have demonstrated that $CaCl_2$, thrombin, and collagen type I all can stimulate immediate initial release of GFs which is sustained over 10 days from a PRP clot [32, 33]. Collagen type I as an activator produces an overall less GF release and more sustained manner than $CaCl_2$, thrombin or combined [32]. Further studies should aim at further investigating the effect of the different activation strategy on platelet concentrates according to the targeted tissue and injury mechanisms, in order to optimize the *in vivo* effect of the released bioactive molecules and therefore increase PRP healing potentials.

2.3.2 Leukocytes

Depends on different isolation protocol, PRP products are available as PRP containing Leukocytes (L-PRP) or without (P-PRP). Leukocytes include neutrophils, monocytes, macrophages, and lymphocytes which play a central role in immune response to pathogens, inflammations, and wound healing. The presence of leukocytes in PRP is very controversial and continues to be a focus of investigation to establish indications or contraindications for its use. Several studies have supported the use of leukocyte-rich PRP, arguing that the leukocytes potentiate the release of cytokines from platelets to improve healing, and confers antimicrobial properties to reduce infection rates, as demonstrated *in vitro* [34, 35]. Others argue that the release of these cytokines causes a highly inflammatory reaction, predisposing to fibrosis and structurally weaker tissue, [36, 37]. As to intra-articular usage, leukocyte-rich PRP has been shown to cause increased post-injection pain, cell death, and excessive activation of synoviocyte than leukocyte-poor PRP [38]. At present, no clear consensus exists regarding the utility of leukocytes in PRP and further study is warranted.

2.3.3 Red Blood Cells

As described in PRP preparation part, a thin layer of RBCs might be included in some final PRP products to obtain more concentrated platelet. The role of RBCs in PRP injections is also largely unknown, with little data on the specific effects of this component within PRP. There has been evidence to suggest that RBCs might be deleterious to cartilage, as they may be harmful to synoviocytes [38], resulting in

Table 6 Characteristics of different extracellular vesicles

	Exosomes	Micro-vesicles	Apoptotic bodies
Origin	Endocytic pathway	Plasma membrane	Plasma membrane
Size	40–120 nm	50–1000 nm	500–2000 nm
Function	Intercellular communication	Intercellular communication	Facilitate phagocytosis
Markers	Alix, Tsg101, tetraspanins (CD81, CD63, CD9), flotillin	Integrins, selectins, CD40	Annexin V, phosphatidylserine
Contents	Proteins and nucleic acids (mRNA, miRNA and other non-coding RNAs)	Proteins and nucleic acids (mRNA, miRNA and other non-coding RNAs)	Nuclear fractions, cell organelles.

the release of catabolic mediators that may worsen cartilage damage and contribute to joint degeneration. Caution should be exercised when utilizing PRP preparations that contain considerable amount of RBC. Further study is necessary to determine the clinical effect of RBC within PRP for the treatment of other musculoskeletal tissue pathologies.

2.3.4 Extracellular Vehicles (EVs)

Extracellular vesicles (EVs) are plasma membrane-derived vesicles, released from cells during stress conditions, including activation and apoptosis, which contains specific proteins, RNA, miRNA, and lncRNA [39]. EVs are present in peripheral blood and body fluids and constitute a heterogeneous population of particles highly variable in size, composition, concentration, cellular origin, and functional properties [40]. EVs are released by various cells and play a vital role in cell communication by transferring their contents from the host cells to the recipient cells [41–43]. Many diverse names have been used to refer to these vesicles released by healthy cells including ectosomes, microparticles, and shedding micro-vesicles, just to name a few. Researchers are now encouraged to use the term EVs as a generic term for all secreted vesicles. EVs may be broadly classified into exosomes, micro-vesicles (MVs, also known as microparticles or ectosomes), and apoptotic bodies according to their size and to cellular origin as summarized in Table 6. Exosomes and MVs are both released by healthy cells, although they differ in several aspects. Exosomes are nanometer-sized (40–120 nm) vesicles of endocytic origin that form by inward budding of the limiting membrane of multi-vesicular endosomes (MVEs). However, MVs bud from the cell surface and their size may vary between 50 nm and 1000 nm. Due to their difference origins, they also differed from their contents and corresponding surface markers (Table 6). For more detailed and in-depth info on this part, please refer the reviews [44, 45] .

In the plasma, EVs are continuously released from variety of cell types, including RBCs, WBCs, endothelial cells, and platelets [46]. Seventy to ninety percent of circulating EVs are derived from platelets [47, 48]. Platelets generate EVs in response to agonists, complement activation, shear forces, senescence, and cytoskeletal abnormalities [49]. Recent studies demonstrated that platelet-derived EVs participate in a variety of important biological and pathological processes via intracellular communication, including but not limited to clotting, angiogenesis, inflammation, immunoregulation, and tumor progression. As a result, platelet-derived EVs represent another important regulating pathway other than traditional communicating pathways such as growth factors, cytokines etc. EVs, especially exosome, have been extensively studied for their role in stimulating tissue regeneration, in many *in vitro* and *in vivo* models, demonstrating that they can confer proangiogenic, proliferative, anti-apoptotic and anti-inflammatory actions through transporting RNA and protein cargos. However, the platelet-derived EVs contents and their functions in the context of regenerative properties of PRP remains largely unknown. The possibility that an important or even a major effect of PRP may derive from EVs should be considered for future investigations. Moreover, platelet-derived EVs may also contribute to the overall variations observed in PRP. Indeed, the MVs in blood from normal healthy transfusion donors is highly variable and affected by diet and exercise. This observation raises the question that if PRPs prepared from different donors (healthy or disease/injury status) also differ from each other regarding EVs. Clearly, more studies are needed to understand or elucidate potential functions or usages of EVs from PRP in the context of regenerative medicine.

3   Clinical Applications of Platelet-Rich Plasma in Orthopedics Surgery

PRP is increasingly used in a variety of tissue injuries, and the usage of PRP in orthopedics field mainly focused on the soft tissue injuries and sports medicine related indications. However, a large number of variables, including method of preparation, composition, medical conditions, injury mechanisms etc. influence the clinical outcomes. The wild-spread variations preclude interpretation of the effectiveness of PRP, and prevents comparison between studies and makes replication by others impossible. The effectiveness and potential adverse effects of this treatment require high quality studies prior to widespread clinical applications. In this part, instead of discussing the variations and the inconsistency or controversial clinical results, we will summarize the current scientific evidence that supports the usage of PRP and present only the results of high level (mainly level-one) randomized clinical trials (RCTs) with the focus on tendon, ligament, cartilage, and muscle injuries.

3.1 Tendons

Tendons are connective tissues that transmit forces from muscle to bone. Physiology permits tendons to have unique properties; tendons have tensile strength that is equal to bone and per unit area stronger than muscle while still being flexible. The properties of tendons are a result of the parallel arrangement of collagen which allow the preservation of contractile energy transfer from muscle to bone. The collagen in tendons are made from fibroblast-like cells called Tenocytes, these cells are the predominant cell type of this tissue. The other roughly 5% of cells are tendon stem cells (TSCs). Injury can occur when tendons approach, reach, or exceed maximum transmittable forces [50].

PRP has demonstrated benefit to tendon healing *in Vitro*. PRP has a dose-dependent benefit to the proliferation of tenocytes [51] and TSCs [52]. Though dose dependent effects are observed, higher concentrations can be deleterious. Benefits to proliferation are observed up to a concentration of 0.5×10^6 plt/μL, a concentration exceeding 3.0×10^6 plt/μL in media started to cause poor proliferation [53]. A rabbit *in vitro* study had similar findings with benefit observed up to a concentration of 10% culture media being PRP with waning effects when PRP exceeded 20% [54].

Different PRP preparation methods may differently affect tendon healing. A study by Zhou et al. evaluated effects L-PRP and P-PRP on the proliferation and differentiation of TSCs [54]. L-PRP was more effective at promoting proliferation and significantly increased gene expression of collagen type III. P-PRP had higher collagen type I/ collagen type III ratio, catabolic markers such as matrix metalloproteinase-1 (MMP-1), MMP-13 and increased inflammatory markers [54]. PRP gel prepared with calcium and thrombin had better results than non-activated PRP and PRP activated with calcium only [55]. Those findings still need to be tested and confirmed in clinical trials or at least in larger clinically relevant animal models.

The clinical trials of PRP on tendon are mainly focused on tendinopathies. Only level-one RCTs that use PRP as a treatment for tendinopathies focusing on the patellar tendon, Achilles tendon, rotator cuff tendons, and lateral elbow tendons were presented and analyzed.

For the Patellar tendon, 6 level one RCTs were published (Table 7). Two studies described the intra-operative PRP application to heal patellar tendon gap after graft harvesting during ACL reconstruction [56, 57]. Both studies reported supercity of PRP groups over controls in terms of pain control and tissue healing. Four other studies evaluated the efficacy of PRP as a conservative treatment for tendinopathies [58–61]. The literature showed overall superior results when compared to controls, but heterogeneous preparation and therapeutic protocols, differing in terms of number of injections performed and time interval between administrations. In most of the published studies, the injective treatment was followed by a standard rehabilitation program.

For the Achilles tendon, five RCTs were published (Table 8). Four of them dealt with chronic Achilles tendinopathy [62–66] and one with acute tendon rupture [67].

Table 7 Summary of the clinical trials dealing with PRP application in the patellar tendon

Publication	N of patients	Pathology	Protocol	PRP characteristics	Follow up	Main findings
[57]	44 patients in total: n = 23 PRGF; n = 21 surgery alone	ACL injury	ACL reconstruction using patellar tendon graft, and application of PRGF at the donor site after graft harvesting. Control group: ACL reconstruction alone	PLT: N/A L: N/A ACT: $CaCl_2$	24 months	PRGF application provided faster improvement in pain but no long-term difference in clinical scores.
[59]	20 patients in total: n = 10 one PRP injection; n = 10 two PRP injections	Patellar tendinopathy	1 vs 2 ultrasound-guided injections of PRP (at one week interval)	PLT: $8.5 - 9 \times 10^5$ per mm^3 L: No ACT: $CaCl_2$	12 months	No statistical inter-group differences between one or two injections.
[58]	40 patients in total: n = 20 one injection; n = 20 two injections	Patellar tendinopathy	1 vs 2 injections of PRP (at two weeks interval)	PLT: $2\times$ basal value. L: No ACT: No	2-years (minimum)	Two injections of PRP provided better results when compared to a single injection.
[60]	23 patients in total: n = 13 ultrasound-guided dry needling; n = 10 PRP injection	Patellar tendinopathy	1 ultrasound-guided dry needling 1 ultrasound-guided injection of PRP	PLT: N/A L: yes ACT: N/A	6 months	PRP administration provided faster recovery at 12 weeks evaluation, whereas no clinical difference was reported at the final 26 weeks follow-up.
[61]	46 patients in total: n = 23 PRP; n = 23 ESWT	Patellar tendinopathy	Two ultrasound-guided injections of PRP at one week interval Three focused extracorporeal shock wave therapy	PLT: $0.89 - 1.1 \times 10^6$ per mm^3 L: N/A ACT: No	12 months	PRP administration provided significantly better improvement than ESWT in VISA-P and VAS scores at 6- and 12-month follow-up.

(continued)

Table 7 (continued)

Publication	N of patients	Pathology	Protocol	PRP characteristics	Follow up	Main findings
[56]	27 patients in total: n = 12 PRP; n = 15 surgery alone	ACL injury	PRP added to the site of patellar tendon harvest after ACL reconstructive surgery Control group: ACL reconstruction alone	PLT: $1185 \pm 404 \times 10^3$ per mm^3 L: $0.91 \pm 0.81 \times 10^3$ per mm^3 ACT: Thrombin and CaCl$_2$	6 months	PRP injection reduced pain in the immediate post-operative period. PRP did not improve patients' functional scores after ACL reconstruction with a patellar tendon graft.

Table 8 Summary of the clinical trials dealing with PRP application in the Achilles tendon

Publication	N patients	Pathology	Protocol	PRP characteristics	F-up	Main findings
[66]	60 in total: n = 20 high -volume injection (HVI); n = 20 PRP; n = 20 saline control	Achilles tendinopathy	Eccentric training combined with either 1) one HVI; 2) Four PRP injections every 2 weeks; 3) placebo (saline)	PLT: n/a L: yes ACT: no	6 months	Treatment with HVI or PRP in combination with eccentric training seems more effective in reducing pain, improving activity level, and reducing tendon thickness and intratendinous vascularity than eccentric training alone. HVI showed more effectiveness in improving all outcomes than PRP in the short term.
[64]	24 in total: n = 12 PRP n = 12 saline	Achilles tendinopathy	1 ultrasound-guided injection or PRP	PLT: 8× basal value L: n/a ACT: n/a	12 months	PRP injection did not provide better outcome with respect to placebo after 3 months.
[65]	20 in total: n = 10 PRP; n = 10 eccentric exercise	Achilles tendinopathy	1 injection of PRP	PLT: n/a L: n/a ACT: n/a	6 months	No inter-group statistically significant difference.
[66]	30 in total: n = 16 surgery + PRP n = 14 surgery alone	Achilles tendon rupture	1 injection of PRP	PLT: $3.673 \pm 1.051 \times 10^6$ per mm^3 L: n/a ACT: CaCl$_2$	12 months	No significant inter-group differences regarding healing of Achilles tendon. Lower biomechanical performance in PRP-augmented group.
[62] [63]	54 in total: n = 27 PRP n = 27 saline	Achilles tendinopathy	1 ultrasound-guided injection of PRP 1 ultrasound-guided injection of saline	PLT: n/a L: n/a ACT: no	52 weeks	No significant inter-group difference regarding pain and activity at final follow-up. No differences in tendon structure between groups.

Looking at PRP application to manage chronic Achilles pathology, three RCTs documented no-better clinical outcomes when PRP was used alone as a treatment when compared with traditional therapies, which showed a huge discrepancy with other lower level case series studies. Most recent RCT published by Boesen et al. showed effectiveness when PRP was used as an augmentation along with eccentric training; however, no better than high volume injection (HVI) [66]. For what regards PRP application for acute Achilles tendon rupture, the study authored by Schepull et al. [67] even revealed that PRP addition could be detrimental in tissue healing since no biomechanical advantages were found, and lower performance were reported in PRP groups when compared to the "suture-alone" group.

For lateral elbow tendons, 14 level one RCTs were published (Table 9), and all of them focused on lateral elbow tendinopathy. By far, a single PRP injection is the most preferred protocol, with some authors performing a second or more injections in case of poor clinical response. In addition to the sham or saline controls, PRP therapy was also compared with other therapeutic approaches including autologous whole blood, corticosteroids, local anesthetic, or laser therapy. The comparison between PRP and corticosteroids injections demonstrated overall superior results for PRP treatment: Perbooms et al. [68] were the first group documented superior clinical outcomes after PRP treatment at 1-year follow-up, and Gosens et al. [69] confirmed the better outcome at 2-years follow-up. The authors pointed out that comparing to corticosteroid therapy, PRP therapy is more durable and requires much less re-intervention. Also the RCTs authored by Yadav et al. [70], Lebiedzinski et al. [71] and Khaliq et al. [72] confirmed superior results for a single injection of PRP compared to corticosteroids. On the other hand, there are also three well designed level one trials which failed to reveal beneficial effects of PRP injections when compared with corticosteroids [73–75]. Two trials compared PRP with local anesthetic injections. Mishra et al. [76] reported a better clinical outcome of PRP injection both at 12 and 24 weeks compared to bupivacaine injection based on a large multicenter RCT including 230 patients. Behera et al. [77] in a RCT on a smaller group of patients confirmed the beneficial effects of PRP. Interestingly, controversial results have also been found in 4 RCT trials compared with autologous whole-blood injections [78–81]. The overall response after intra-tendinous injection of whole blood was consistently satisfactory in all the published studies; and PRP did not show a clear advantage over autologous whole blood in terms of pain relief and functional recovery. These findings raised the debate about the necessity of further PRP isolation from blood as autologous whole blood itself seems to provide satisfactory and comparable clinical benefit in the treatment of lateral elbow tendinopathy.

For the rotator cuff injuries (Table 10), PRP was used either as conservative treatment for the management of chronic tendinopathy that was not responsive to previous therapeutic attempts, or used as an augmentation during or immediately after surgical arthroscopic cuff repair. In the case of PRP application as a conservative option, the 2 level-one RCTs [83, 84] showed no better outcomes when compared with placebo controls in terms of clinical scores and other objective measurements, which is quite contrasting with other lower level clinical trials. Conversely, Rha

Table 9 Summary of the clinical trials dealing with PRP application in the lateral epicondylitis

Publication	N of patients	Pathology	Protocol	PRP characteristics	F-up	Main findings
[75]	60 in total: n = 20 neocaine; n = 20 dexamethasone; n = 20 PRP	Lateral epicondylitis	1 injection for each treatment group	PLT: n/a L: n/a ACT: n/a	6 months	PRP injection did not provide better results than corticosteroids or local anesthetic.
[72]	102 in total: n = 51 corticosteroid; n = 51 PRP	Lateral epicondylitis	1 injection of corticosteroid 1 injection of PRP	PLT: n/a L: n/a ACT: n/a	3 weeks	A single PRP injection demonstrated better improvement in term of pain control than corticosteroid injection.
[70]	60 in total: n = 30 PRP; n = 30 corticosteroid	Lateral epicondylitis	1 injection of PRP 1 injection of corticosteroid	PLT: 1×10^6 per mm^3 L: n/a ACT: n/a	3 months	PRP treatment demonstrated better results at long term evaluation with respect to corticosteroids.
[74]	50 in total: n = 25 PRP; n = 25 saline	Lateral epicondylitis	2 ultrasound-guided injections of PRP at one-month interval 2 ultrasound-guided injections of saline at one-month interval	PLT: 1.6× basal value L: no ACT: n/a	12 months	Two PRP injection did not show better results compared to saline injections.
[71]	120 in total: n = 64 PRP; n = 56 steroid+lidocaine	Lateral epicondylitis	1 injection of PRP 1 injection of corticosteroid	PLT: n/a L: n/a ACT: n/a	12 months	ACP treatment provided long lasting beneficial effects with respect to corticosteroid injection.
[82]	30 in total: n = 15 PRP; n = 15 corticosteroid	Lateral epicondylitis	1 injection of PRP 1 injection of corticosteroid	PLT: n/a L: n/a ACT: n/a	6 months	PRP provided superior biological healing of the lesion and longer lasting beneficial effects.
[77]	25 in total: n = 15 PRP; n = 10 bupivacaine	Lateral epicondylitis	1 ultrasound guided injection of PRP 1 ultrasound guided injection of bupivacaine	PLT: $6 - 8 \times 10^5$ per mm^3 L: no ACT: $CaCl_2$	12 months	PRP injection provided significantly superior improvement in pain and function.

(continued)

Table 9 (continued)

Publication	N of patients	Pathology	Protocol	PRP characteristics	F-up	Main findings
[80]	40 in total: n = 20 PRP; n = 20 autologous whole blood	Lateral epicondylitis	1 injection of PRP 1 injection of autologous whole blood	PLT: 220000 ± 23,000 per mm³ L: yes ACT: n/a	8 weeks	PRP injection was more effective than autologous blood in pain control at 8 weeks evaluation.
[79]	64 in total: n = 33 PRP; n = 31 autologous whole blood	Lateral epicondylitis	1 injection of PRP 1 injection of autologous whole blood	PLT: 4.8× basal value L: 6740 ± 1396 per mm³ in PRP group	12 months	PRP was not better than whole bold at long term follow-up in any parameter considered.
[73]	60 in total: n = 20 PRP; n = 20 glucocorticoid; n = 20 placebo saline solution	Lateral epicondylitis	1 ultrasound-guided injection for each group	PLT: n/a L: n/a ACT: n/a	3 months	No inter-group differences in pain reduction or disability at 3 months; PRP injection was more painful.
[76]	225 in total: n = 112 PRP; n = 113 bupivacaine	Lateral epicondylitis	1 injection of PRP, and 1 injection of bupivacaine	PLT: 5× base value L: yes ACT: n/a	6 months	Significantly better performance of PRP compared with control group (bupivacaine).
[78]	150 in total: n = 80 PRP; n = 70 autologous whole blood	Lateral epicondylitis	2 ultrasound-guided injections of PRP or Whole blood at one-month interval	PLT: 652 × 10³ per mm³ L: n/a ACT: n/a	6 months	Autologous whole blood and PRP can be used as effective second-line therapy. No intergroup difference reported.
[68] [69]	100 in total: n = 51 PRP; n = 49 corticosteroid	Lateral epicondylitis	1 injection for each group	PLT: n/a L: n/a ACT: n/a	2 years	Significantly better performance in PRP group in terms of pain relief and functional improvement.
[81]	28 in total: n = 14 autologous whole blood n = 14 PRP	Lateral epicondylitis	1 ultrasound-guided injections of PRP or Whole blood	PLT: 235 – 1292 × 10³ per mm³ L: no ACT: no	6 months	Statistically significant difference in pain control only at 6 weeks follow-up. No other inter-group difference.

Table 10 Summary of the clinical trials dealing with PRP application in the rotator cuff injuries

Publication	N of patients	Pathology	Protocol	PRP characteristics	F-up	Main findings
[98]	62 in total: n = 32 PRP; n = 30 control	Rotator cuff tear	Arthroscopic double-row repair + intra-op injection of PRP Arthroscopic double-row repair	PLT: n/a L: n/a ACT: $CaCl_2$	12 months	PRP injection provided a significantly higher rate of tendon healing with respect to surgery alone.
[89]	120 in total: n = 60 PRP; n = 60 control	Rotator cuff tear	Arthroscopic double-row repair + intra-op injection of PRP Arthroscopic double-row repair	PLT: n/a L: n/a ACT: n/a	24 months	The intraoperative injection of PRP showed no significant effect on the clinical and patient-reported outcomes up to 24 months.
[99]	110 in total: n = 56 PRP; n = 54 control	Rotator cuff tear	Arthroscopic single-row repair+1 injection of PRP Arthroscopic single-row repair	PLT: $4.74 \pm 0.3 \times 10^5$ per mm^3 L: no ACT: $CaCl_2$	24 months	PRP injection demonstrated superior structural healing in a large rotator cuff tear, with higher vascularization of rotator cuff and surrounding tissues in the early phases.
[92]	40 in total: n = 20 PRP; n = 20 control	Rotator cuff calcification	Arthroscopic debridement of the calcification+ intra-op PRP injection Arthroscopic debridement of the calcification	PLT: n/a L: yes ACT:n/a	12 months	PRP injection did not demonstrate beneficial effect on rotator cuff healing.
[95]	35 in total: n = 17 PRP; n = 18 control	Rotator cuff tear	Arthroscopic double-row repair with addition of PRP clots Arthroscopic double-row cuff repair	PLT: n/a L: yes ACT: n/a	12 months	PRP augmentation during double-row repair did not produce better clinical or structural outcomes.

(continued)

Table 10 (continued)

Publication	N of patients	Pathology	Protocol	PRP characteristics	F-up	Main findings
[87]	48 in total: n = 25 PRP; n = 23 control	Rotator cuff tendinopathy	Arthroscopic acromioplasty+1 intra-op sub-acromial PRP injection Arthroscopic acromioplasty	PLT: n/a L: Yes ACT: autologous thrombin	24 months	PRP injection did not provide beneficial effect on clinical outcomes. Potentially detrimental effects of PRP to the long-term structural properties of the tendon.
[84]	70 in total: n = 35 PRP; n = 35 physical therapy	Rotator cuff tear	3 intra-articular injections of PRP at one- week interval; standard physical therapy for 15 session	PLT: 2.1–2.5× basal value L: 1.1–1.3 × basal value ACT: $CaCl_2$	12 months	Comparable results were obtained between physical therapy and PRP.
[93]	60 in total: n = 30 PRP; n = 30 control	Rotator cuff tear	Arthroscopic double-row cuff repair followed by 1 post-op PRP injection Arthroscopic double-row cuff repair	PLT: 470,000 per mm^3 L: no ACT: $CaCl_2$	16 weeks	Delivery of PRP post-op did not improve early rotator cuff healing or functional recovery.
[101]	25 in total: n = 12 PRP; n = 13 saline	Rotator cuff tear	Arthroscopic single-row repair+PRP during surgery and after 4 weeks Arthroscopic single-rwo repair+saline during surgery and after 4 weeks	PLT: n/a L: n/a ACT:	6 weeks	PRP injections did not provide superior pain relief. No statistical inter-group differences in functional outcomes.
[102]	65 in total: n = 33 PRP; n = 32 control	Rotator cuff tear	Arthroscopic double-row suture bridge repair+intra-op PRP injection Arthroscopic double-row suture bridge repair	PLT: n/a L: yes ACT: n/a	12 months	The injection of PRP did not demonstrate beneficial effects on tendon healing and functional outcome. Only a possible analgesic effect has been documented.

[97]	55 in total: n = 28 PRP; n = 27 control	Rotator cuff tear	Single-row repair Single-row repair+intra-op PRP injected at the tendon-bone interface	PLT: n/a L: n/a ACT: autologous thrombin + CaCl$_2$	24 months	PRP treatment did not provide better functional results at 24-month follow-up for small- and medium-sized tears.
[103]	20 in total: n = 10 PRP; n = 10 control	Rotator cuff tear	Arthroscopic double-row repair+ PRP clots placed at the tendon-bone interface Arthroscopic double-row repair	PLT: n/a L: n/a ACT: n/a	3 months	Application of PRP determined higher early vascularization that might potential predispose to an increased cellular response and healing potential.
[86]	28 in total: n = 14 PRP; n = 14 control	Rotator cuff tear	Arthroscopic repair+intra-op PRF injection Arthroscopic repair	PLT: n/a L: n/a ACT: n/a	24 months	PRP application did not improve the clinical outcome and the healing rate compared with standard repair.
[96]	48 in total: n = 24 PRP; n = 24 control	Rotator cuff tear	Arthroscopic double-row repair+PRP gel at the tendon-bone interface Arthroscopic double-row repair	PLT: 1000×10^3 per mm^3 L: n/a ACT: Ca-gluconate	15.9 months for PRP; 17.3 months for control	The re-tear rate in the PRP group was significantly lower than control group. Clinical outcomes showed no statistical difference between groups.
[83]	40 in total: n = 20 PRP; n = 20 saline	Rotator cuff tendinopathy	1 ultrasound-guided injection of PRP into the sub-acromial space 1 ultrasound-guided injection of saline solution	PLT: n/a L: n/a ACT:no	1 year	No significant inter-group differences regarding quality of life, pain, disability, and shoulder range of motion at 1-year follow-up.

(continued)

Table 10 (continued)

Publication	N of patients	Pathology	Protocol	PRP characteristics	F-up	Main findings
[91]	63 in total: n = 32 PRGF; n = 31 control	Rotator cuff tear	Arthroscopic double-row repair+intra-op PRP application Arthroscopic double-row repair	PLT: 600×10^3 per mm^3 L: yes ACT: $CaCl_2$	1 year	No significant inter-group differences in cuff healing and in functional scores.
[94]	60 in total: n = 30 PRP; n = 30 control	Partial rotator cuff tear	Arthroscopic single-row repair+PRP membrane clot placed onto the repair site Arthroscopic single-row repair	PLT: n/a L: n/a ACT: $CaCl_2$	1 year	PRP did not show significant improvement in perioperative morbidity, clinical outcomes, VAS and structural integrity. MRI showed comparable re-tear rate and tendon healing.
[85]	39 in total: n = 20 PRP; n = 19 control	Tendinosis or partial tear of supraspinatus tendon	1 ultrasound-guided injection of PRP 1 ultrasound-guided dry needling procedure	PLT: n/a L: n/a ACT: n/a	6 months	PRP administration determined significant reduction in pain and disability when compared to dry needling.
[104]	80 in total: n = 40 PRP; n = 40 control	Rotator cuff tear	Arthroscopic single-row repair+PRP membrane placed at the tendon-bone interface Arthroscopic single-row repair	PLT: $>400 \times 10^3$ per mm^3 L: 7×10^3 per mm^3 ACT: Ca-Gluconate	13 months	PRP group showed better tendon repair but not better functional outcome.
[90]	79 in total: n = 40 PRFM n = 39 control	Rotator cuff tear	Arthroscopic repair+PRP membrane placed at the tendon-bone interface Arthroscopic repair	PLT: n/a L: n/a ACT: $CaCl_2$	12 months	No differences in tendon-to-bone healing between the PRFM and control groups; no significant differences in healing by ultrasound at 6 and 12 weeks.

| [100] | 53 in total:
n = 26 PRP;
n = 27
control | Rotator cuff tear | Arthroscopic single-row repair+ intra-op PRP injection
Arthroscopic single-row repair | PLT: n/a
L: n/a
ACT: autologous thrombin + $CaCl_2$ | 24 months | PRP treatment significantly reduced pain in the first post-op month and provide better clinical score than controls at 3 months. For grade I-II tears, PRP was able to provide better results even at 24 months. |
| [88] | 88 in total:
n = 43
PRFM;
n = 45
control | Rotator cuff tear | Arthroscopic double-row repair+PRP membrane placed at the tendon-bone interface
Arthroscopic double-row repair | PLT: n/a
L: n/a
ACT: $CaCl_2$ | 20.2 months | PRP did not improve the healing of the rotator cuff. |

et al. [85] reported that 2 PRP injections could yield improved symptoms and restored shoulder motility at 6 months follow-up when compared to dry needling alone.

With respect to the application as an augmentation during surgery, 10 studies [86–95] failed to show any beneficial effects of PRP augmentation compared to the surgical procedure alone. Contrarily, 5 RCTs clearly demonstrated beneficial effects after PRP administration. In particular, Jo [96], Malavolta [97], Zhang [98], and Pandey [99] reported significant lower re-tear rate when PRP was used as an augmentation, while Randelli [100] and Pandey [99] documented better pain relief in PRP group and also superior clinical scores at various follow-up evaluations.

3.2   ligament

The application of PRP on ligament healing is mainly focused on anterior cruciate ligament (ACL) ruptures or tears, due to the fact that ACL injuries are among the most common sport-related injuries and ACL reconstruction is one of the most frequently performed procedures in sports medicine. Despite overall "good" clinical outcomes reported at mid/long-term follow-up, ACL reconstruction is still not a "100%-success" procedure; and the rate for full recovery (back to sports) is <60%, with a high re-injury rate in professional sport players. A clear demanding exist especially from professionals who need to return to the field as soon as possible [105]. Researchers is therefore investigating novel strategies to enhance ACL healing with the intension to further accelerate recovery time, and to reduce the failure rate. PRP administration is one of the few options that aim for enhancing ACL reconstruction. So far, *In vitro* and pre-clinical animal studies have demonstrated overall promising results: PRP administration increases the gene and protein expressions of collagen, and also contributes to the reduced apoptosis and stimulated fibroblast metabolic activity; it was observed in animal models that PRP application was able to produce stronger graft which has superior biomechanical properties. Despite the encouraging results from animal studies, the results from clinical trials are still inconclusive. Figueroa et al. [105] systematically reviewed PRP usage in ACL surgery. Among 11 prospective comparative or randomized controlled trials included, only 4 were level-one RCTs [106–109] which were summarized in Table 11. In 2009, Nin et al. [108] published a controlled study of 100 patients undergoing ACL bone-tendon-bone reconstruction with or without PRP. The PRP was placed on the graft and in the tibial tunnel. This study presented blinded MRI assessment and clinical evaluation at 24 months as endpoints. Their results did not show any statistically significant differences between the groups for inflammatory parameters (perimeters of the knee and C-reactive protein level), MRI appearance of the graft (although there was an improvement consistent with 20%–25% based on platelet use), and clinical evaluation scores (visual analog scale, IKDC, and KT-1000 arthrometer). In 2010, Vogrin et al. [107] compared ACL hamstring reconstruction without PRP and ACL reconstruction with PRP on the graft and in both

Table 11 Summary of the clinical trials dealing with PRP application in the ACL injury

Publication	N of patients	Protocol	PRP characteristics	F-up	Main findings
[106]	46 in total: n = 23 PRP; n = 23 control	Quadrupled hamstring graft	PLT: 7× L: n/a ACT: n/a	3 months	Graft maturation: N/A Tunnel healing/widening: despite slightly less tunnel widening in the PRP group, there was no significant difference between the groups at the femoral opening and mid tunnel or t the tibial opening and mid tunnel. Clinical evaluation: arthrometric results had improved significantly in both groups. No inter-group differences.
[109]	50 in total: n = 25 PRP; n = 25 control (standard surgical procedure without PRP application)	Double-looped hamstring	PLT:12× L: n/a ACT: autologous human thrombin Magellan autologous platelet separator was used	3 months	Patients treated with PRP showed significantly better anteroposterior knee stability than patients in the control group.
[107]	50 in total: n = 25 PRP; n = 25 control (standard surgical procedure without PRP application)	Double-looped hamstring	PLT:12× L: n/a ACT: autologous human thrombin. Magellan autologous platelet separator was used	3 months	Graft maturation: after 4–6 weeks, in the intra-articular part of the graft, there was no evidence of revascularization in either group. At 10–12 weeks, the results suggested initiation of revascularization in the PRP group, but without a statistical significance between the groups. Tunnel Healing/widening: After 4–6 weeks, the PRP-treated group showed a significantly higher level of vascularization at the osteoligamentous interface than the control group. Clinical evaluation: N/A

(continued)

Table 11 (continued)

Publication	N of patients	Protocol	PRP characteristics	F-up	Main findings
[108]	100 in total	Bone-patellar tendon-Bone	PLT: 5× L: ACT:	24 months	Graft maturation: the results did not show any statistically significant differences between the groups for MRI appearance of the graft. Tunnel healing/widening: N/A Clinical evaluation: the results did not show any statistically significant differences between the groups for inflammatory parameters and clinical evaluation scores.

tunnels in a randomized controlled, double-blind study. Outcome was assessed by MRI at 3 months postoperatively. In the same year, Vogrin et al. published another article [109] in which anteroposterior knee stability with the KT-2000 arthrometer (MEDmetric) was evaluated before surgery and at 3 and 6 months after surgery in ACL reconstructed patients with and without PRP. Their study focused on vascularization in their MRI analysis but found no differences in the intra-articular portion of ACL grafts with PRP treatment and without such treatment at 12 weeks. They also assessed vascularization at the bone-ligament interface and found a significantly better score for the group treated with PRP ($P < .001$). In addition, Vogrin et al. noted significantly better anteroposterior knee stability in patients treated with PRP than in patients in the control group. In 2013, Mirzatolooei et al. [106] reported a study comparing ACL reconstructions by hamstring graft with or without PRP administration into the bone tunnels. The width of the bone tunnels was measured by CT scans of the knees (right after surgery and at 3 months postoperatively). Their data revealed that despite slightly less tunnel widening in the PRP group, there was no significant difference between the groups at the femoral opening or the mid-femoral tunnel or at the tibial opening or mid-tibial tunnel. Clinical results were also similar between groups.

Based on the limited clinical evidence available so far, it seems there is some promising evidence that the addition of PRP to the graft or tunnels could be a synergic factor in acquiring maturity more quickly than grafts with no PRP, which the clinical implication of this remaining unclear. Considering tunnel healing and tunnel widening, it seems that PRP offers little or no benefit. Besides the common variations for PRP research, such as different PRP preparations, the dose, interval, and methods of application etc., clinically, there are also an infinite number of confounding variables.

3.3 Cartilage

The articular cartilage has very poor regenerative capacity due to its avascular nature. Clinical and laboratory attempts using biological approaches including growth factors provides promise for the treatment of disabling articular cartilage injuries and diseases. Many anabolic growth factors have been intensively tested to promote articular cartilage growth including TGF-β1, IGF-1, FGF-2, VEGF, and PDGF. Those factors can also be found in high concentration in PRP products, which set up a scientific promise that PRP could be beneficial to the cartilage injuries. Besides abundant anabolic growth factors, anti-inflammatory properties of platelet concentrates may help lower the inflammation in the synovial tissue which would lead to a reduction of matrix-metalloproteinases, known as cartilage-matrix degrading enzymes, which is a well-known pathological and progressive factor for OA.

In vitro studies using porcine, bovine, and human chondrocytes demonstrate that PRP can promote the proliferation of chondrocytes. Culture of porcine chondrocytes for 72 h in 10% PRP had significantly more DNA content than cells cultured in 10% PPP and 10% FBS [110]. PRP promotes the proliferation of primary chondrocytes isolated from patients whom had a total joint replacement for up to 12 days in both a monolayer and three-dimensional setting [111]. The proliferative benefit to human chondrocytes can be observed for up to 20 days in both monolayer and 3D when co-cultured with Platelet releasates (PRPr) [112]. Though there are clear benefits to proliferation, there is no consensus on the effects regarding differentiation. Porcine and human chondrocyte in both monolayer and 3D environments had increased mRNA transcription and protein synthesis of aggrecan and type II collagen when cultured with PRP and PRPr [110–112]. One study observed reduced expression levels of aggrecan, type II collagen, and bone morphogenetic protein-2 were in human chondrocytes co-cultured with PRP. Though they also reported increased proliferation in PRP, chondrocytes dedifferentiated towards a fibroblast-like phenotype [113]. Another study also demonstrated increased proliferation in PRP however primary chondrocytes in Fetal Calf serum were clearly differentiated with matrix deposition and was not observed in the PRP groups [114]. This lack of consistency among the published reports may be attributable to the heterogeneity of study designs, variations in PRP preparations, and differences in PRP delivery.

Apart from its beneficial effects on the chondrocytes proliferation, PRP has also been demonstrated to have anti-inflammation potential in an osteoarthritic milieu. It was observed that PRP releasate completely reversed IL-1β induced inflammatory response of chrondrocytes isolated from patients with OA [115]. Further study suggested that PRP exerts its anti-inflammatory effects via inhibition of NF-κB activation through increasing gene expression of hepatocyte growth factor (HGF) and TNF-α [116]. These findings provide preclinical support for the concept that PRP application could reduce chondrocyte inflammatory responses and restoring anabolic activity via deactivation of NF-κB. However, due to the presence of concentrated leukocytes, PRP was originally thought to be pro-inflammatory [117]. Further investigation revealed that the most represented pro-inflammatory cytokines, IL-1β, only showed a slight increase after platelet activation, whereas the anti-inflammatory molecules, including IL-4 and IL-10, increased more than 5 times [116]. With more refined PRP isolation protocol, the anti-inflammatory effect is likely to predominate in PRP formulations in which the presence of leukocytes is substantially reduced.

For clinical data (Table 12), PRP was mainly tested on OA patients with pure-PRP as the main formula. Overall, more positive effects of PRP injections were reported. The first positive level one RCT compared PRP with placebo for the treatment of knee OA was reported by Patel et al. [118], who compared the efficacy of PRP injections (single injection and two injections of P-PRP 3 eeks apart) with single saline injection. Their data demonstrated significantly improved WOMAC scores in all PRP groups, but worsened in saline group. A more recent FDA-sanctioned RCT study confirmed that P-PRP is effective for the pain relief and functional improvement with regards to knee OA [119]. The efficacy of PRP in the treatment of OA was also compared with HA administration. Cerza et al. [120] reported that the improvement of WOMAC scores was more significant in the P-PRP-treated group than the HA group at each time point and in all grades

Table 12 Summary of the clinical trials dealing with PRP application for the treatment of OA

Publication	N of patients	Protocol	PRP characteristics	F-up	Main findings
[125]	102 patients in total: n = 33 PRP group; n = 34 HA group; n = 35 Ozon group	Three groups: PRP group: 2 injections of PRP (1-month interval); HA group: single injection of HA; Ozone group: 4 injections of Ozone	PLT: 9–13× baseline. L: n/a ACT: No	12 months	PRP was more effective than HA and ozone injections. Two applications can provide at least 1-year improvement than HA injection.
[126]	162 patients in total: n = 39 three PRP injection group; n = 44 single PRP injection; n = 39 HA group; n = 40 saline control	Four groups receiving 3 intra-articular injections of PRP, one injection of PRP, one injection of HA, and a saline control	PLT: around 5× baseline L: n/a ACT: $CaCl_2$ and PRP were freeze-thawed once before injection	6 months	Significant improvement were observed in PRP and HA groups when compared with control. Three PRP injections showed significantly better results can single injection and HA groups. No difference between single PRP injection and HA groups. The PRP effects is clearer in the early OA subgroup but not in advanced OA group.
[119]	30 patients in total: n = 15 PRP group; n = 15 saline group	3 injections of PRP or HA in a weekly basis	PLT: 2–3× baseline L: No ACT: CaCl2	12 months	P-PRP showed beneficial effects in terms of pain relief and functional improvement with regards to knee OA. No adverse events were reported.

(continued)

Table 12 (continued)

Publication	N of patients	Protocol	PRP characteristics	F-up	Main findings
[124]	192 patients in total: n = 94 PRP group; n = 89 HA group	3 injections of PRP or HA in a weekly basis	PLT: 4.6 ± 1.4 × baseline. L: 1.1 ± 0.5 × baseline. ACT: $CaCl_2$, and PRP were freeze-thawed once before injection	12 months	Both treatments proved to be effective in improving knee functional status and reducing symptoms, but no significant intergroup difference at any follow-up in any of the clinical scores.
[118]	78 patients (156 knees) in total: n = 52 single injection of PRP; n = 50 two injections of PRP; n = 46 single saline injection control	Three groups: Single or twice PRP injection (3 weeks apart), or single saline injection	PLT: 310.14 × 10^3/ul (3× baseline) L: No ACT: $CaCl_2$	6 months	WOMAC parameters improved after PRP injections (both single and twice injections), whereas worsened after saline infiltration. However, no significantly differences were found between PRP groups; and the improvement of PRP groups can only last for 6 months.
[122]	96 patients in total: n = 48 PRP group; n = 48 HA group	3 injections of CaCl2-activated P-PRP every two weeks, or single HA injection	PLT: n/a L: No ACT: $CaCl_2$ (activated in glass container before injection) PRGF-Endoret (BTI Biotechnology Institute, Spain)	48 weeks	PRP injection is significantly more efficient in reducing pain, stiffness and improving physical function than HA.

(continued)

Table 12 (continued)

Publication	N of patients	Protocol	PRP characteristics	F-up	Main findings
[120]	120 patients in total: n = 60 PRP group; n = 60 HA group	4 intra-articular injections of PRP (5.5 ml) or HA (20 mg/2 ml) in a one-week interval	PLT: n/a L: No ACT: n/a	6 months	Treatment with PRP showed a significantly better clinical outcome than did with HA. Treatment with HA did not seem to be effective in the patients with grade III knee OA.
[121]	176 patients in total: n = 80 PRP group; n = 87 HA group	3 intra-articular injections of PRP or HA in a one-week interval	PLT: n/a L: No ACT: CaCl₂ (activated in glass container before injection) PRGF-Endoret (BTI Biotechnology Institute, Spain)	6 months	PRP injection showed superior short-term beneficial results when compared with HA.
[123]	109 patients in total: n = 54 PRP group; n = 55 HA group	3 injections of PRP or HA in a one-week interval	PLT: 5× baseline. L: 1.2 x baseline ACT: No. the PRPs were freeze-thawed once before injection	12 months	No significant difference in all scores. Only a trend favoring PRP in patients with early OA.

of OA. Comparable results were also observed in a multicenter, double-blind RCT [121]. Besides WOMAC scores, PRP also yielded better results than HA in terms of the secondary outcome measures assessing pain, stiffness and physical function [122].

On the other hand, a single-center, double blind RCT including 109 matched patients demonstrated that PRP treatment did not lead to statistically significant differences in all scores evaluated with respect to HA injections at 12-month follow-up [123]. The same group published another similarly designed RCT with a higher n number and concluded the same [124]. However, unlike the aforementioned RCTs, which used fresh P-PRP, these trials prepared PRP by double-spinning technique which include high concentration of leukocytes, and the PRP was freeze-thawed once before injection. Although no definitive comparative study was performed to address the adverse effects of leucocytes in PRP, the current clinical data suggest P-PRP should be the better choice for the treatment of OA.

In summary, PRP intra-articular knee injections may be an effective alternative treatment for knee OA for patients who do not adequately symptomatically response to more traditional treatments. However, current studies are, at best, inconclusive regarding the efficacy of PRP treatment. Significant variations in administration schedule likely make it difficult to draw definitive conclusions about PRP in general. Future studies need to evaluate the impact of WBCs and RBCs for OA treatment.

3.4 Muscle

Every year recreational and professional sports cause thousands of lesions to muscle, ligament and tendons. Muscle strains are among the top 10 sports injuries, they comprise about 35% of all sports injuries. The goal of PRP treatment for various muscle injuries is to reduce the time needed to resume training and competition and to avoid relapses. Compare to other sports related injuries, only three high quality level-one RCTs were published (Table 13). The scarcity of clinical trials in muscle injuries compared with other injuries is attributable to the fact that this treatment on athletics was banned by the WADA [127] until 2011 because of the suspicion that PRP could enhance athletic performance in base of its content in ergogenic factors included in the WADA list, that is, IGF-1, IGF-BP3, growth hormone, VEGF, PDGF-BB, bFGF and HGF [127]. Thus, the possibility that PRP injection could alter the systemic levels of these substances has been examined [128–130]. Because PDGF-BB and TGF-β1 do not show circadian variations [128], any increase in the circulating levels could be attributed to systemic effects of PRP injections. However, intramuscular injections of pure PRP (2.5 ml) did not alter the circulating levels of IGF-1, PDGF-AB, PDGF-BB, TGF-β1, TGF-β2, VEGF, EGF and FGF up to 24 h after injection [129, 130].

Only three level-one RCTs (Table 13) were published [131–134], two of them reported negative outcomes, which demonstrated no beneficial effects of PRP injections on the acute hamstring injuries in terms of time until patients could resume sports activities. One recent RCT analyzed acute muscle injuries including hamstrings, gastrocnemius and quadriceps, with time to return to play as the primary outcome [134]. The authors reported significantly shortened time return to sports in PRP group when compared with control group. Similar as most of the other studies, the results from those RCTs on acute muscle injuries are contradictory. It is too early to draw a conclusion based on this extremely limited clinical information. Besides, there are also concerns about the fibrosis were posted by researchers regarding the use of PRP for muscle injury.

Skeletal muscle repair after injury includes a complex and well-coordinated regenerative response [135]. Muscle healing occurs in a series of overlapping phases, including inflammation, regeneration, and fibrosis [136]. It has been observed that the natural healing process is often inefficient and hindered by the formation of scar tissue (fibrosis), which causes the tendency for muscle injury to recur [136, 137]. The optimal strategy to improve muscle healing should be enhancing muscle regeneration while reducing fibrosis. TGF-β1 is a key factor in the devel-

Table 13 Summary of the clinical trials dealing with PRP application for muscle injuries

Publication	Medical conditions	N of patients	Protocol	PRP characteristics	F-up	Main findings
[132]	Acute hamstring injuries	80 competitive and recreational athletes in total: n = 41 PRP group; n = 39 placebo group (saline)	Two injections (5–7 days apart) of either PRP or saline followed by standardized rehabilitation program	PLT: n/a L: n/a ACT:n/a ACP double-syringe system (Arthrex) was used	6 months 1 year	No beneficial or intramuscular PRP injections, as compared with placebo injections, in patients with acute hamstring injuries.
[133]	Acute hamstring injuries	90 athletes in total: n = 30 PRP group; n = 30 PPP group; n = 30 no injection group	Single injection of PPR, PPP, or no injection. All patients undergone standardized rehabilitation program	PLT: n/a L: n/a ACT: no GPSIII system (Biomet Becover) was used	6 months	No benefit of a single PRP injection over intensive rehabilitation in athletes who have sustained acute. MRI positive hamstring injuries.
[134]	Grade 2 acute muscle injuries including hamstrings, gastrocnemius and quadriceps	72 athletes in total: n = 34 PRP group; n = 38 Control group	Single intra-lesion injection of PRP followed by standard rehabilitation program 2 days after injection. The volume of the PRP was equal to the volume of the muscle injury determined under ultrasound guidance	PLT: n/a L: n/a ACT: no	Time to return to play is the primary outcome. 2 years follow up for recurrence rate	The mean time to return to play was significantly shorter in PRP group than control group. Lower pain severity scores were observed in the PRP group. No difference in the recurrence rate after 2 year follow-up between groups.

opment of fibrosis in the kidneys, liver, lungs, and skeletal muscle [138]. TGF-β1 is an anti-apoptotic agent for myo-fibroblasts and may trigger trans-differentiation of fibroblasts to activated myo-fibroblasts [139–141]. Additionally, TGF-β1 stimulates M2-like macrophages, which have pro-fibrotic activity [142, 143]. Recent studies also describe phenotypic transitions of endothelial cells into pro-fibrotic mesenchymal cells under the effects of TGF-β1 [144, 145]. It was reasoned that TGF-β1 in PRP could induce muscle fibrosis because of its ability to stimulate collagen synthesis and fibroblast proliferation [146]. PRP contains high levels of TGF-β1 which is a key factor responsible for the development of fibrosis [138, 147–149]. Concerns remain that PRP injection may lead to elevated fibrosis, and therefore hinder optimal muscle healing [150].

PRPs actions on fibrosis are paradoxical because they also contain anti-fibrotic molecules, such as HGF and serum amyloid protein, which showed inhibition of fibrosis in different models, in part by regulating macrophage function [151]. Although it is still not clear, several *in vivo* studies indicate that PRP injection may promote the formation of fibrosis in muscle [152–154]. In a recently published rat study [153], the PRP treated group tended to have more fibrotic tissue compared with the control group. In a rat gastrocnemius contusion model, co-treatment with PRP and losartan, an indirect TGF-β1 inhibitor, significantly reduced fibrosis and improved function compared to PRP alone [154]. These initial studies suggest that blocking the function of pro-fibrotic factor(s) in PRP, such as TGF-β1, could improve the beneficial effects of PRP on muscle healing by reducing fibrosis. As a proof-of-concept, a neutralizing Ab against TGF-β1 was utilized with the idea of blocking the function of TGF-β1 within PRP while keeping the functions of beneficial factors for more optimized muscle recovery. Data demonstrated that neutralizing TGF-β1 within PRP significantly promotes muscle regeneration while significantly decreasing collagen deposition in a cardiotoxin induced skeletal muscle injury model [155]. Not only did the neutralization reduce fibrosis, it enhanced angiogenesis, prolonged satellite cell activation, and recruited a great number of M2 macrophages to the injury site, which also contributed to the efficacy that the TGF-β1 minus PRP had on muscle healing [155]. This work supports the notion that eliminating "negative/deleterious" factor(s) could improve the beneficial effect imparted by PRP on muscle injury, which will lead to optimized muscle healing.

3.5   Minimum Information for Studies Evaluating Biologics in Orthopedics (MIBO)

Despite the vast number of scientific literature on the PRP topic, many authors noticed that the published experimental results are difficult to sort and interpret. The clinical results are mixed or at least controversial and finally the relevance of use is debatable. The reason of this regrettable result was highlighted in nearly every debate and conference: (1) many different techniques for the production of platelet concentrates for surgical use are available, which leads to very different final products; (2) no proper or at least standard terminology to classify and describe the different platelet concentrates; (3) a lack of accurate or at least standardized characterization criteria of the tested platelet concentrate in most articles, leading to a huge literature of thousands of articles constituting a "blind library of knowledge" [12]. As an ongoing effort to reduce the variations and confusion in the expanding literatures, an expert consensus on a minimum reporting requirements for studies evaluating the efficacy of PRP were reached at the American Academy of Orthopedic Surgeons/Orthopedic Reserch Society (AAOS/ORS) Biologic Treatments for Orthopedic Injuries Symposium in 2015 and the American Orthopedic Society for Sports Medicine (AOSSM) Biologic Treatments for Sports Injuries II Think Tank in 2015 [156]. This minimum checklist (Table 14) could be used as a guide for authors, reviewers, and editors for sufficient report of experimental details.

3.6 In Summary

Although many animal and basic science studies assessed the use and the effectiveness of PRP, the beneficial effects of PRP for augmenting tissue healing are still theoretical. Evidence-based literature suggests that success varies depending on its preparation method and composition, medical condition of the patient, anatomic location and severity of the lesion, technique of administration, tissue type, and peri-procedural care. In response to a growing interest among patients and surgeons in PRP, recent studies report outcomes for a variety of conditions. Further critical review and rigorous clinical studies are required to formulate a cost effective, efficacious algorithm for the use of PRP in patients with varying musculoskeletal conditions. To quote Fu et al., clinical knowledge, clinical evidence, and clinical applications "remain in their infancy" [157].

References

1. Alsousou J, et al. The biology of platelet-rich plasma and its application in trauma and orthopaedic surgery: a review of the literature. J Bone Joint Surg Br. 2009;91(8):987–96.
2. Anitua E, et al. New insights into and novel applications for platelet-rich fibrin therapies. Trends Biotechnol. 2006;24(5):227–34.
3. Khan M, Bedi A. Cochrane in CORR ((R)): platelet-rich therapies for musculoskeletal soft tissue injuries (Review). Clin Orthop Relat Res. 2015;473(7):2207–13.
4. Sanchez M, et al. Platelet-rich therapies in the treatment of orthopaedic sport injuries. Sports Med. 2009;39(5):345–54.
5. Alvarez-Camino JC, Vazquez-Delgado E, Gay-Escoda C. Use of autologous conditioned serum (Orthokine) for the treatment of the degenerative osteoarthritis of the temporomandibular joint. Review of the literature. Med Oral Patol Oral Cir Bucal. 2013;18(3):e433–8.
6. Baltzer AW, et al. Autologous conditioned serum (Orthokine) is an effective treatment for knee osteoarthritis. Osteoarthr Cartil. 2009;17(2):152–60.

7. Wehling P, et al. Autologous conditioned serum in the treatment of orthopedic diseases: the orthokine therapy. BioDrugs. 2007;21(5):323–32.

8. Whitman DH, Berry RL, Green DM. Platelet gel: an autologous alternative to fibrin glue with applications in oral and maxillofacial surgery. J Oral Maxillofac Surg. 1997;55(11):1294–9.

9. Li H, Li B. PRP as a new approach to prevent infection: preparation and in vitro antimicrobial properties of PRP. J Vis Exp. 2013;74

10. Dhurat R, Sukesh M. Principles and methods of preparation of platelet-rich plasma: a review and author's perspective. J Cutan Aesthet Surg. 2014;7(4):189–97.

11. Degen RM, et al. Commercial separation systems designed for preparation of platelet-rich plasma yield differences in cellular composition. HSS J. 2017;13(1):75–80.

12. Dohan Ehrenfest DM, Rasmusson L, Albrektsson T. Classification of platelet concentrates: from pure platelet-rich plasma (P-PRP) to leucocyte- and platelet-rich fibrin (L-PRF). Trends Biotechnol. 2009;27(3):158–67.

13. Mishra A, et al. Sports medicine applications of platelet rich plasma. Curr Pharm Biotechnol. 2012;13(7):1185–95.

14. DeLong JM, Russell RP, Mazzocca AD. Platelet-rich plasma: the PAW classification system. Arthroscopy. 2012;28(7):998–1009.

15. Yun SH, et al. Platelet activation: the mechanisms and potential biomarkers. Biomed Res Int. 2016;2016:9060143.

16. Angiolillo DJ, Ueno M, Goto S. Basic principles of platelet biology and clinical implications. Circ J. 2010;74(4):597–607.

17. Qureshi AH, et al. Proteomic and phospho-proteomic profile of human platelets in basal, resting state: insights into integrin signaling. PLoS One. 2009;4(10):e7627.

18. Senzel L, Gnatenko DV, Bahou WF. The platelet proteome. Curr Opin Hematol. 2009;16(5):329–33.

19. Maynard DM, et al. Proteomic analysis of platelet alpha-granules using mass spectrometry. J Thromb Haemost. 2007;5(9):1945–55.

20. Blair P, Flaumenhaft R. Platelet alpha-granules: basic biology and clinical correlates. Blood Rev. 2009;23(4):177–89.

21. Boswell SG, et al. Platelet-rich plasma: a milieu of bioactive factors. Arthroscopy. 2012;28(3):429–39.

22. Freymiller EG. Platelet-rich plasma: evidence to support its use. J Oral Maxillofac Surg. 2004;62(8):1046. author reply 1047-8.

23. Meyers KM, Holmsen H, Seachord CL. Comparative study of platelet dense granule constituents. Am J Phys. 1982;243(3):R454–61.

24. Ducy P. 5-HT and bone biology. Curr Opin Pharmacol. 2011;11(1):34–8.

25. Jonnalagadda D, Izu LT, Whiteheart SW. Platelet secretion is kinetically heterogeneous in an agonist-responsive manner. Blood. 2012;120(26):5209–16.

26. Suelves M, et al. uPA deficiency exacerbates muscular dystrophy in MDX mice. J Cell Biol. 2007;178(6):1039–51.

27. Rundle CH, et al. Fracture healing in mice deficient in plasminogen activator inhibitor-1. Calcif Tissue Int. 2008;83(4):276–84.

28. Golebiewska EM, Poole AW. Platelet secretion: from haemostasis to wound healing and beyond. Blood Rev. 2015;29(3):153–62.

29. Ma L, et al. Proteinase-activated receptors 1 and 4 counter-regulate endostatin and VEGF release from human platelets. Proc Natl Acad Sci U S A. 2005;102(1):216–20.

30. Sehgal S, Storrie B. Evidence that differential packaging of the major platelet granule proteins von Willebrand factor and fibrinogen can support their differential release. J Thromb Haemost. 2007;5(10):2009–16.

31. Italiano JE Jr, et al. Angiogenesis is regulated by a novel mechanism: pro- and antiangiogenic proteins are organized into separate platelet alpha granules and differentially released. Blood. 2008;111(3):1227–33.

32. Cavallo C, et al. Platelet-rich plasma: the choice of activation method affects the release of bioactive molecules. Biomed Res Int. 2016;2016:6591717.

33. Fufa D, et al. Activation of platelet-rich plasma using soluble type I collagen. J Oral Maxillofac Surg. 2008;66(4):684–90.
34. Moojen DJ, et al. Antimicrobial activity of platelet-leukocyte gel against Staphylococcus aureus. J Orthop Res. 2008;26(3):404–10.
35. Li H, et al. Unique antimicrobial effects of platelet-rich plasma and its efficacy as a prophylaxis to prevent implant-associated spinal infection. Adv Healthc Mater. 2013;2(9):1277–84.
36. Portela GS, et al. L-PRP diminishes bone matrix formation around autogenous bone grafts associated with changes in osteocalcin and PPAR-gamma immunoexpression. Int J Oral Maxillofac Surg. 2014;43(2):261–8.
37. McCarrel TM, Minas T, Fortier LA. Optimization of leukocyte concentration in platelet-rich plasma for the treatment of tendinopathy. J Bone Joint Surg Am. 2012;94(19):e1431–8.
38. Braun HJ, et al. The effect of platelet-rich plasma formulations and blood products on human synoviocytes: implications for intra-articular injury and therapy. Am J Sports Med. 2014;42(5):1204–10.
39. Burnier L, et al. Cell-derived microparticles in haemostasis and vascular medicine. Thromb Haemost. 2009;101(3):439–51.
40. Mause SF, Weber C. Microparticles: protagonists of a novel communication network for intercellular information exchange. Circ Res. 2010;107(9):1047–57.
41. Simak J, Gelderman MP. Cell membrane microparticles in blood and blood products: potentially pathogenic agents and diagnostic markers. Transfus Med Rev. 2006;20(1):1–26.
42. Goubran HA, et al. Platelet microparticle: a sensitive physiological "fine tuning" balancing factor in health and disease. Transfus Apher Sci. 2015;52(1):12–8.
43. Burnouf T, et al. An overview of the role of microparticles/microvesicles in blood components: are they clinically beneficial or harmful? Transfus Apher Sci. 2015;53(2):137–45.
44. Antwi-Baffour S, et al. Understanding the biosynthesis of platelets-derived extracellular vesicles. Immun Inflamm Dis. 2015;3(3):133–40.
45. Lu M, et al. Recent advances on extracellular vesicles in therapeutic delivery: challenges, solutions, and opportunities. Eur J Pharm Biopharm. 2017;119:381–95.
46. Arraud N, et al. Extracellular vesicles from blood plasma: determination of their morphology, size, phenotype and concentration. J Thromb Haemost. 2014;12(5):614–27.
47. Flaumenhaft R. Formation and fate of platelet microparticles. Blood Cells Mol Dis. 2006;36(2):182–7.
48. Keuren JF, et al. Effects of storage-induced platelet microparticles on the initiation and propagation phase of blood coagulation. Br J Haematol. 2006;134(3):307–13.
49. Heijnen HF, et al. Activated platelets release two types of membrane vesicles: microvesicles by surface shedding and exosomes derived from exocytosis of multivesicular bodies and alpha-granules. Blood. 1999;94(11):3791–9.
50. James R, et al. Tendon: biology, biomechanics, repair, growth factors, and evolving treatment options. J Hand Surg Am. 2008;33(1):102–12.
51. Wang X, et al. Proliferation and differentiation of human tenocytes in response to platelet rich plasma: an in vitro and in vivo study. J Orthop Res. 2012;30(6):982–90.
52. Zhang J, Wang JH. Platelet-rich plasma releasate promotes differentiation of tendon stem cells into active tenocytes. Am J Sports Med. 2010;38(12):2477–86.
53. Giusti I, et al. Platelet concentration in platelet-rich plasma affects tenocyte behavior in vitro. Biomed Res Int. 2014;2014:630870.
54. Zhou Y, et al. The differential effects of leukocyte-containing and pure platelet-rich plasma (PRP) on tendon stem/progenitor cells - implications of PRP application for the clinical treatment of tendon injuries. Stem Cell Res Ther. 2015;6:173.
55. Jo CH, et al. Platelet-rich plasma stimulates cell proliferation and enhances matrix gene expression and synthesis in tenocytes from human rotator cuff tendons with degenerative tears. Am J Sports Med. 2012;40(5):1035–45.
56. de Almeida AM, et al. Patellar tendon healing with platelet-rich plasma: a prospective randomized controlled trial. Am J Sports Med. 2012;40(6):1282–8.

57. Seijas R, et al. Pain in donor site after BTB-ACL reconstruction with PRGF: a randomized trial. Arch Orthop Trauma Surg. 2016;136(6):829–35.
58. Zayni R, et al. Platelet-rich plasma as a treatment for chronic patellar tendinopathy: comparison of a single versus two consecutive injections. Muscles Ligaments Tendons J. 2015;5(2):92–8.
59. Kaux JF, et al. Using platelet-rich plasma to treat jumper's knees: exploring the effect of a second closely-timed infiltration. J Sci Med Sport. 2016;19(3):200–4.
60. Dragoo JL, et al. Platelet-rich plasma as a treatment for patellar tendinopathy: a double-blind, randomized controlled trial. Am J Sports Med. 2014;42(3):610–8.
61. Vetrano M, et al. Platelet-rich plasma versus focused shock waves in the treatment of jumper's knee in athletes. Am J Sports Med. 2013;41(4):795–803.
62. de Vos RJ, et al. Platelet-rich plasma injection for chronic Achilles tendinopathy: a randomized controlled trial. JAMA. 2010;303(2):144–9.
63. de Jonge S, et al. One-year follow-up of platelet-rich plasma treatment in chronic Achilles tendinopathy: a double-blind randomized placebo-controlled trial. Am J Sports Med. 2011;39(8):1623–9.
64. Krogh TP, et al. Ultrasound-guided injection therapy of achilles tendinopathy with platelet-rich plasma or saline: a randomized, blinded, placebo-controlled trial. Am J Sports Med. 2016;44(8):1990–7.
65. Kearney RS, Parsons N, Costa ML. Achilles tendinopathy management: a pilot randomised controlled trial comparing platelet-richplasma injection witperh an eccentric loading programme. Bone Joint Res. 2013;2(10):227–32.
66. Boesen AP, et al. Effect of high-volume injection, platelet-rich plasma, and sham treatment in chronic midportion achilles tendinopathy: a randomized double-blinded prospective study. Am J Sports Med. 2017;45(9):2034–43. https://doi.org/10.1177/0363546517702862.
67. Schepull T, et al. Autologous platelets have no effect on the healing of human achilles tendon ruptures: a randomized single-blind study. Am J Sports Med. 2011;39(1):38–47.
68. Peerbooms JC, et al. Positive effect of an autologous platelet concentrate in lateral epicondylitis in a double-blind randomized controlled trial: platelet-rich plasma versus corticosteroid injection with a 1-year follow-up. Am J Sports Med. 2010;38(2):255–62.
69. Gosens T, et al. Ongoing positive effect of platelet-rich plasma versus corticosteroid injection in lateral epicondylitis: a double-blind randomized controlled trial with 2-year follow-up. Am J Sports Med. 2011;39(6):1200–8.
70. Yadav R, Kothari SY, Borah D. Comparison of local injection of platelet rich plasma and corticosteroids in the treatment of lateral epicondylitis of humerus. J Clin Diagn Res. 2015;9(7):RC05–7.
71. Lebiedzinski R, et al. A randomized study of autologous conditioned plasma and steroid injections in the treatment of lateral epicondylitis. Int Orthop. 2015;39(11):2199–203.
72. Khaliq A, et al. Effectiveness of platelets rich plasma versus corticosteroids in lateral epicondylitis. J Pak Med Assoc. 2015;65(11 Suppl 3):S100–4.
73. Krogh TP, et al. Treatment of lateral epicondylitis with platelet-rich plasma, glucocorticoid, or saline: a randomized, double-blind, placebo-controlled trial. Am J Sports Med. 2013;41(3):625–35.
74. Montalvan B, et al. Inefficacy of ultrasound-guided local injections of autologous conditioned plasma for recent epicondylitis: results of a double-blind placebo-controlled randomized clinical trial with one-year follow-up. Rheumatology (Oxford). 2016;55(2):279–85.
75. Palacio EP, et al. Effects of platelet-rich plasma on lateral epicondylitis of the elbow: prospective randomized controlled trial. Rev Bras Ortop. 2016;51(1):90–5.
76. Mishra AK, et al. Efficacy of platelet-rich plasma for chronic tennis elbow: a double-blind, prospective, multicenter, randomized controlled trial of 230 patients. Am J Sports Med. 2014;42(2):463–71.
77. Behera P, et al. Leukocyte-poor platelet-rich plasma versus bupivacaine for recalcitrant lateral epicondylar tendinopathy. J Orthop Surg (Hong Kong). 2015;23(1):6–10.

78. Creaney L, et al. Growth factor-based therapies provide additional benefit beyond physical therapy in resistant elbow tendinopathy: a prospective, single-blind, randomised trial of autologous blood injections versus platelet-rich plasma injections. Br J Sports Med. 2011;45(12):966–71.

79. Raeissadat SA, et al. Is Platelet-rich plasma superior to whole blood in the management of chronic tennis elbow: one year randomized clinical trial. BMC Sports Sci Med Rehabil. 2014;6:12.

80. Raeissadat SA, et al. Effect of platelet-rich plasma (PRP) versus autologous whole blood on pain and function improvement in tennis elbow: a randomized clinical trial. Pain Res Treat. 2014;2014:191525.

81. Thanasas C, et al. Platelet-rich plasma versus autologous whole blood for the treatment of chronic lateral elbow epicondylitis: a randomized controlled clinical trial. Am J Sports Med. 2011;39(10):2130–4.

82. Gautam VK, et al. Platelet-rich plasma versus corticosteroid injection for recalcitrant lateral epicondylitis: clinical and ultrasonographic evaluation. J Orthop Surg (Hong Kong). 2015;23(1):1–5.

83. Kesikburun S, et al. Platelet-rich plasma injections in the treatment of chronic rotator cuff tendinopathy: a randomized controlled trial with 1-year follow-up. Am J Sports Med. 2013;41(11):2609–16.

84. Ilhanli I, Guder N, Gul M. Platelet-rich plasma treatment with physical therapy in chronic partial supraspinatus tears. Iran Red Crescent Med J. 2015;17(9):e23732.

85. Rha DW, et al. Comparison of the therapeutic effects of ultrasound-guided platelet-rich plasma injection and dry needling in rotator cuff disease: a randomized controlled trial. Clin Rehabil. 2013;27(2):113–22.

86. Antuna S, et al. Platelet-rich fibrin in arthroscopic repair of massive rotator cuff tears: a prospective randomized pilot clinical trial. Acta Orthop Belg. 2013;79(1):25–30.

87. Carr AJ, et al. Platelet-rich plasma injection with arthroscopic acromioplasty for chronic rotator cuff tendinopathy: a randomized controlled trial. Am J Sports Med. 2015;43(12):2891–7.

88. Castricini R, et al. Platelet-rich plasma augmentation for arthroscopic rotator cuff repair: a randomized controlled trial. Am J Sports Med. 2011;39(2):258–65.

89. Flury M, et al. Does pure platelet-rich plasma affect postoperative clinical outcomes after arthroscopic rotator cuff repair? A randomized controlled trial. Am J Sports Med. 2016;44(8):2136–46.

90. Rodeo SA, et al. The effect of platelet-rich fibrin matrix on rotator cuff tendon healing: a prospective, randomized clinical study. Am J Sports Med. 2012;40(6):1234–41.

91. Ruiz-Moneo P, et al. Plasma rich in growth factors in arthroscopic rotator cuff repair: a randomized, double-blind, controlled clinical trial. Arthroscopy. 2013;29(1):2–9.

92. Verhaegen F, Brys P, Debeer P. Rotator cuff healing after needling of a calcific deposit using platelet-rich plasma augmentation: a randomized, prospective clinical trial. J Shoulder Elb Surg. 2016;25(2):169–73.

93. Wang A, et al. Do postoperative platelet-rich plasma injections accelerate early tendon healing and functional recovery after arthroscopic supraspinatus repair? A randomized controlled trial. Am J Sports Med. 2015;43(6):1430–7.

94. Weber SC, et al. Platelet-rich fibrin matrix in the management of arthroscopic repair of the rotator cuff: a prospective, randomized, double-blinded study. Am J Sports Med. 2013;41(2):263–70.

95. Zumstein MA, et al. SECEC Research Grant 2008 II: Use of platelet- and leucocyte-rich fibrin (L-PRF) does not affect late rotator cuff tendon healing: a prospective randomized controlled study. J Shoulder Elb Surg. 2016;25(1):2–11.

96. Jo CH, et al. Platelet-rich plasma for arthroscopic repair of large to massive rotator cuff tears: a randomized, single-blind, parallel-group trial. Am J Sports Med. 2013;41(10):2240–8.

97. Malavolta EA, et al. Platelet-rich plasma in rotator cuff repair: a prospective randomized study. Am J Sports Med. 2014;42(10):2446–54.

98. Zhang Z, Wang Y, Sun J. The effect of platelet-rich plasma on arthroscopic double-row rotator cuff repair: a clinical study with 12-month follow-up. Acta Orthop Traumatol Turc. 2016;50(2):191–7.

99. Pandey V, et al. Does application of moderately concentrated platelet-rich plasma improve clinical and structural outcome after arthroscopic repair of medium-sized to large rotator cuff tear? A randomized controlled trial. J Shoulder Elb Surg. 2016;25(8):1312–22.

100. Randelli P, et al. Platelet rich plasma in arthroscopic rotator cuff repair: a prospective RCT study, 2-year follow-up. J Shoulder Elb Surg. 2011;20(4):518–28.

101. Hak A, et al. A double-blinded placebo randomized controlled trial evaluating short-term efficacy of platelet-rich plasma in reducing postoperative pain after arthroscopic rotator cuff repair: a pilot study. Sports Health. 2015;7(1):58–66.

102. Werthel JD, et al. Arthroscopic double row cuff repair with suture-bridging and autologous conditioned plasma injection: functional and structural results. Int J Shoulder Surg. 2014;8(4):101–6.

103. Zumstein MA, et al. Increased vascularization during early healing after biologic augmentation in repair of chronic rotator cuff tears using autologous leukocyte- and platelet-rich fibrin (L-PRF): a prospective randomized controlled pilot trial. J Shoulder Elb Surg. 2014;23(1):3–12.

104. Gumina S, et al. Use of platelet-leukocyte membrane in arthroscopic repair of large rotator cuff tears: a prospective randomized study. J Bone Joint Surg Am. 2012;94(15):1345–52.

105. Figueroa D, et al. Platelet-rich plasma use in anterior cruciate ligament surgery: systematic review of the literature. Arthroscopy. 2015;31(5):981–8.

106. Mirzatolooei F, Alamdari MT, Khalkhali HR. The impact of platelet-rich plasma on the prevention of tunnel widening in anterior cruciate ligament reconstruction using quadrupled autologous hamstring tendon: a randomised clinical trial. Bone Joint J. 2013;95-B(1):65–9.

107. Vogrin M, et al. Effects of a platelet gel on early graft revascularization after anterior cruciate ligament reconstruction: a prospective, randomized, double-blind, clinical trial. Eur Surg Res. 2010;45(2):77–85.

108. Nin JR, et al. Has platelet-rich plasma any role in anterior cruciate ligament allograft healing? Arthroscopy. 2009;25(11):1206–13.

109. Vogrin M, et al. The effect of platelet-derived growth factors on knee stability after anterior cruciate ligament reconstruction: a prospective randomized clinical study. Wien Klin Wochenschr. 2010;122(Suppl 2):91–5.

110. Akeda K, et al. Platelet-rich plasma stimulates porcine articular chondrocyte proliferation and matrix biosynthesis. Osteoarthr Cartil. 2006;14(12):1272–80.

111. Chien CS, et al. Incorporation of exudates of human platelet-rich fibrin gel in biodegradable fibrin scaffolds for tissue engineering of cartilage. J Biomed Mater Res B Appl Biomater. 2012;100(4):948–55.

112. Spreafico A, et al. Biochemical investigation of the effects of human platelet releasates on human articular chondrocytes. J Cell Biochem. 2009;108(5):1153–65.

113. Gaissmaier C, et al. Effect of human platelet supernatant on proliferation and matrix synthesis of human articular chondrocytes in monolayer and three-dimensional alginate cultures. Biomaterials. 2005;26(14):1953–60.

114. Kaps C, et al. Human platelet supernatant promotes proliferation but not differentiation of articular chondrocytes. Med Biol Eng Comput. 2002;40(4):485–90.

115. van Buul GM, et al. Platelet-rich plasma releasate inhibits inflammatory processes in osteoarthritic chondrocytes. Am J Sports Med. 2011;39(11):2362–70.

116. Bendinelli P, et al. Molecular basis of anti-inflammatory action of platelet-rich plasma on human chondrocytes: mechanisms of NF-kappaB inhibition via HGF. J Cell Physiol. 2010;225(3):757–66.

117. Sundman EA, Cole BJ, Fortier LA. Growth factor and catabolic cytokine concentrations are influenced by the cellular composition of platelet-rich plasma. Am J Sports Med. 2011;39(10):2135–40.

118. Patel S, et al. Treatment with platelet-rich plasma is more effective than placebo for knee osteoarthritis: a prospective, double-blind, randomized trial. Am J Sports Med. 2013;41(2):356–64.

119. Paterson KL, et al. Intra-articular injection of photo-activated platelet-rich plasma in patients with knee osteoarthritis: a double-blind, randomized controlled pilot study. BMC Musculoskelet Disord. 2016;17:67.
120. Cerza F, et al. Comparison between hyaluronic acid and platelet-rich plasma, intra-articular infiltration in the treatment of gonarthrosis. Am J Sports Med. 2012;40(12):2822–7.
121. Sanchez M, et al. A randomized clinical trial evaluating plasma rich in growth factors (PRGF-Endoret) versus hyaluronic acid in the short-term treatment of symptomatic knee osteoarthritis. Arthroscopy. 2012;28(8):1070–8.
122. Vaquerizo V, et al. Comparison of intra-articular injections of plasma rich in growth factors (PRGF-Endoret) versus Durolane hyaluronic acid in the treatment of patients with symptomatic osteoarthritis: a randomized controlled trial. Arthroscopy. 2013;29(10):1635–43.
123. Filardo G, et al. Platelet-rich plasma vs hyaluronic acid to treat knee degenerative pathology: study design and preliminary results of a randomized controlled trial. BMC Musculoskelet Disord. 2012;13:229.
124. Filardo G, et al. Platelet-rich plasma intra-articular knee injections show no superiority versus viscosupplementation: a randomized controlled trial. Am J Sports Med. 2015;43(7):1575–82.
125. Duymus TM, et al. Choice of intra-articular injection in treatment of knee osteoarthritis: platelet-rich plasma, hyaluronic acid or ozone options. Knee Surg Sports Traumatol Arthrosc. 2017;25(2):485–92.
126. Gormeli G, et al. Multiple PRP injections are more effective than single injections and hyaluronic acid in knees with early osteoarthritis: a randomized, double-blind, placebo-controlled trial. Knee Surg Sports Traumatol Arthrosc. 2017;25(3):958–65.
127. Engebretsen L, et al. IOC consensus paper on the use of platelet-rich plasma in sports medicine. Br J Sports Med. 2010;44(15):1072–81.
128. Aoto K, et al. Circadian variation of growth factor levels in platelet-rich plasma. Clin J Sport Med. 2014;24(6):509–12.
129. Schippinger G, et al. Does single intramuscular application of autologous conditioned plasma influence systemic circulating growth factors? J Sports Sci Med. 2012;11(3):551–6.
130. Schippinger G, et al. Influence of intramuscular application of autologous conditioned plasma on systemic circulating IGF-1. J Sports Sci Med. 2011;10(3):439–44.
131. Reurink G, et al. Platelet-rich plasma injections in acute muscle injury. N Engl J Med. 2014;370(26):2546–7.
132. Reurink G, et al. Rationale, secondary outcome scores and 1-year follow-up of a randomised trial of platelet-rich plasma injections in acute hamstring muscle injury: the Dutch Hamstring Injection Therapy study. Br J Sports Med. 2015;49(18):1206–12.
133. Hamilton B, et al. Platelet-rich plasma does not enhance return to play in hamstring injuries: a randomised controlled trial. Br J Sports Med. 2015;49(14):943–50.
134. Rossi LA, et al. Does platelet-rich plasma decrease time to return to sports in acute muscle tear? A randomized controlled trial. Knee Surg Sports Traumatol Arthrosc. 2017;25(10):3319–25.
135. Garg K, Corona BT, Walters TJ. Therapeutic strategies for preventing skeletal muscle fibrosis after injury. Front Pharmacol. 2015;6:87.
136. Huard J, Li Y, Fu FH. Muscle injuries and repair: current trends in research. J Bone Joint Surg Am. 2002;84-A(5):822–32.
137. de Jong S, et al. Fibrosis and cardiac arrhythmias. J Cardiovasc Pharmacol. 2011;57(6):630–8.
138. Border WA, Noble NA. Transforming growth factor beta in tissue fibrosis. N Engl J Med. 1994;331(19):1286–92.
139. Anitua E, Troya M, Orive G. Plasma rich in growth factors promote gingival tissue regeneration by stimulating fibroblast proliferation and migration and by blocking transforming growth factor-beta1-induced myodifferentiation. J Periodontol. 2012;83(8):1028–37.
140. Anitua E, et al. Plasma rich in growth factors (PRGF-Endoret) stimulates proliferation and migration of primary keratocytes and conjunctival fibroblasts and inhibits and reverts TGF-beta1-Induced myodifferentiation. Invest Ophthalmol Vis Sci. 2011;52(9):6066–73.
141. Koulikovska M, et al. Platelet-rich plasma prolongs myofibroblast accumulation in corneal stroma with incisional wound. Curr Eye Res. 2015;40(11):1102–10.

142. Vidal B, et al. Fibrinogen drives dystrophic muscle fibrosis via a TGF-beta/alternative macro-phage activation pathway. Genes Dev. 2008;22(13):1747–52.
143. Novak ML, Koh TJ. Phenotypic transitions of macrophages orchestrate tissue repair. Am J Pathol. 2013;183(5):1352–63.
144. van Meeteren LA, ten Dijke P. Regulation of endothelial cell plasticity by TGF-beta. Cell Tissue Res. 2012;347(1):177–86.
145. Piera-Velazquez S, Li Z, Jimenez SA. Role of endothelial-mesenchymal transition (EndoMT) in the pathogenesis of fibrotic disorders. Am J Pathol. 2011;179(3):1074–80.
146. Ignotz RA, Massague J. Transforming growth factor-beta stimulates the expression of fibronectin and collagen and their incorporation into the extracellular matrix. J Biol Chem. 1986;261(9):4337–45.
147. Bonner JC. Regulation of PDGF and its receptors in fibrotic diseases. Cytokine Growth Factor Rev. 2004;15(4):255–73.
148. Rhee S, Grinnell F. P21-activated kinase 1: convergence point in PDGF- and LPA-stimulated collagen matrix contraction by human fibroblasts. J Cell Biol. 2006;172(3):423–32.
149. Jinnin M, et al. Regulation of fibrogenic/fibrolytic genes by platelet-derived growth factor C, a novel growth factor, in human dermal fibroblasts. J Cell Physiol. 2005;202(2):510–7.
150. Kelc R, Vogrin M. Concerns about fibrosis development after scaffolded PRP therapy of muscle injuries: commentary on an article by Sanchez et al.: "Muscle repair: platelet-rich plasma derivates as a bridge from spontaneity to intervention.". Injury. 2015;46(2):428.
151. Sugiura T, et al. Increased HGF and c-Met in muscle tissues of polymyositis and der-matomyositis patients: beneficial roles of HGF in muscle regeneration. Clin Immunol. 2010;136(3):387–99.
152. Rothan HA, et al. Three-dimensional culture environment increases the efficacy of platelet rich plasma releasate in prompting skin fibroblast differentiation and extracellular matrix formation. Int J Med Sci. 2014;11(10):1029–38.
153. Delos D, et al. The effect of platelet-rich plasma on muscle contusion healing in a rat model. Am J Sports Med. 2014;42(9):2067–74.
154. Terada S, et al. Use of an antifibrotic agent improves the effect of platelet-rich plasma on muscle healing after injury. J Bone Joint Surg Am. 2013;95(11):980–8.
155. Li H, et al. Customized platelet-rich plasma with transforming growth factor beta1 neutral-ization antibody to reduce fibrosis in skeletal muscle. Biomaterials. 2016;87:147–56.
156. Murray IR, et al. Minimum information for studies evaluating biologics in orthopae-dics (MIBO): platelet-rich plasma and mesenchymal stem cells. J Bone Joint Surg Am. 2017;99(10):809–19.
157. Hogan MV, et al. Tissue engineering of ligaments for reconstructive surgery. Arthroscopy. 2015;31(5):971–9.

Bioresorbable Materials for Orthopedic Applications (Lactide and Glycolide Based)

Balaji Prabhu, Andreas Karau, Andrew Wood, Mahrokh Dadsetan, Harald Liedtke, and Todd DeWitt

Keywords Bioresorbable · Polymer · Lactide · Glycolide · Poly(lactic-co-glycolic acid) · Calcium phosphate · Mechanical performance · Bioactivity

1 Introduction

A material capable of serving a purpose and then disappearing within the human body is no magic folklore but based on years of rigorous scientific evidence, proven clinical data, and wide commercial use. With over five decades of clinical use as materials for orthopedic applications, these types of materials, known as bioresorbable materials, continue to find use in novel applications such as sutures, screws, stents, scaffolds, and even synthetic skin. Their continued development can be attributed to advancements in novel synthesis techniques, processing technologies, implant design development, and innovative surgical techniques. This has resulted in growing interest for use of these materials in regenerative medicine and patient-specific treatments in the orthopedic space. With increasing life expectancy, more active lifestyles, younger patient demographics, faster healing, advanced robotic surgical techniques, and reduced hospitalization costs, the need for 'biologically smart materials' is continually on the rise. Strategic selection and optimization of a bioresorbable material for a specific application meets this needs leading to the implant effectively serving its purpose *in vivo* and being able to smartly remove itself without incurring any extraneous effort from the native biological system.

B. Prabhu (✉) · A. Wood · M. Dadsetan
Evonik Industries AG, Birmingham, AL, USA
e-mail: Balaji.Prabhu@Evonik.com; Andrew.Wood@Evonik.com; Mahrokh.Dadsetan@Evonik.com

A. Karau · H. Liedtke
Evonik Industries AG, Darmstadt, Germany
e-mail: Andreas.Karau@Evonik.com; Harald.Liedtke@Evonik.com

T. DeWitt
Evonik Industries AG, Piscataway, NJ, USA
e-mail: Todd.DeWitt@Evonik.com

© Springer International Publishing AG, part of Springer Nature 2018
B. Li, T. Webster (eds.), *Orthopedic Biomaterials*,
https://doi.org/10.1007/978-3-319-89542-0_13

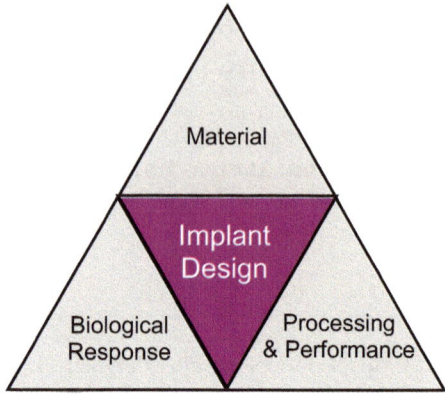

Fig. 1 Typical factors governing implant success

Among this class of materials, those that are based on lactide and glycolide have seen the most expansive development and subsequent clinical use.

Bioresorbable materials are able to be degraded in physiological environments into products that are either metabolized into non-toxic degradation products or thoroughly bio-absorbed. These materials are unique in this aspect compared to commonly used biostable implants, such as metallic devices, which require a secondary surgery for removal. Typical metallic implants have mechanical properties much higher than the native tissue system resulting in stress-shielding where the bone is unable to regenerate to a pre-implant state due to the lack of a mechanical load. A benefit of bioresorbable materials is their ability to gradually transfer increasing mechanical loads to the surrounding new tissue system as they degrade thereby allowing the body to heal naturally. In addition, bioresorbable materials offer clinicians the advantage of enhanced post-operative imaging diagnostics as they are transparent to X-rays and therefore allow for observation of tissue system regeneration more so than metallic implants. Further, these bioresorbable materials can be fabricated into implants through a number of processing methods which allow manufacturers to create complex geometries that can even be designed on a patient-to-patient basis. For an implant to be successful, a detailed understanding of the material properties, processing, part performance, and biological response is critical (Fig. 1). A basic outline of the considerations for a successful bioresorbable implant material was offered more than 20 years ago, that bioresorbable implant materials should [1]:

- Not provoke negative responses *in vivo*
- Fully resorb after serving its purpose
- Be processable into three-dimensional structures
- Degrade at the rate of local tissue regeneration
- Have a surface permitting cellular adhesion
- Have a design to allow cell growth & diffusion

Although a number of materials have been investigated for potential use as bioresorbable orthopedic implants, the lactide and glycolide based polymer materials

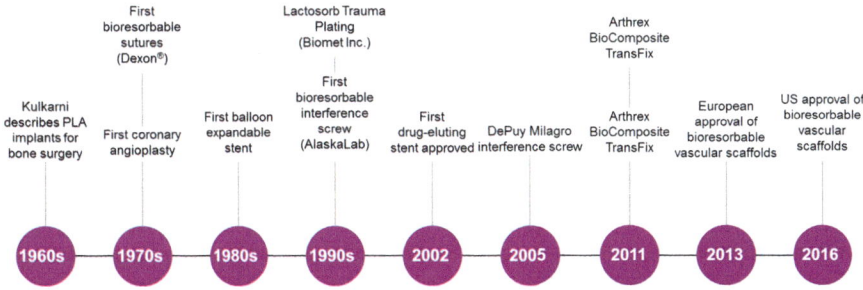

Fig. 2 A historical overview in the development of bioresorbable materials for orthopedic and cardiovascular applications

previously mentioned have a long history of proven biocompatibility and clinical efficacy. The first research on low-molecular weight poly(glycolic acid) (PGA) was conducted nearly 125 years ago by Bischoff and Walden [2]. From those beginnings, the first biodegradable suture was developed in the early 1960s by the American Cyanamid Company in an attempt to avoid batch variation from commonly used materials such as regenerated collagens [3]. Since this first use, the application of these glycolide and lactide based material has seen an increase in development and acceptance as biocompatible, bioresorbable materials. The widespread use of bioresorbable materials in surgery started in the 1970s with the first marketed suture (DEXON®) opening a new era in biomaterials in clinical settings. As the manufacturing technology developed, utilization of bioresorbable materials has expanded to various medical applications such as interference screws and cardiovascular scaffolds. A brief overview of this history is shown in Fig. 2.

By definition, bioresorbable materials degrade when implanted with by-products being eliminated from the body by biological processes under physiological conditions [4]. Obviously, both the implant in its initial state as well as the by-products produced during implant degradation must be biocompatible, that is, they must not cause any adverse biological or immunological reactions if the material is to be accepted as suitable for the application. Bioresorbable materials must also be able to provide sufficient mechanical properties comparable to the native bone at the implant site but also maintain their properties long enough to offer the advancing bone structure time to fully regrow. To this point, they are ideal for orthopedic implants due to their slowly changing mechanical properties which allow the implant to provide the initial mechanical strength required to carry the applied load after which it begins to degrade, slowly transferring the applied load to the surrounding bone. During this phase, the ideal degradation rate of the bioresorbable implant would match the rate at which the bone is able to regenerate. This matched rate will allow for an increasing load to be transferred from the implant to the bone thus transmitting mechanical signals to the resident cells encouraging tissue remodeling in accordance with Wolff's law [5, 6].

2 Bioresorbable Polymers

Polymers are large molecules that are composed of monomer repeating units which are connected with chemical covalent bonds. Polymers cover a broad range of materials in our daily life including natural polymers, such as cellulose and proteins, as well as the synthetic polymers, such as poly(lactic acid) (PLA), poly(glycolic acid) PGA, and poly(lactic-*co*-glycolic acid) (PLGA). According to the component heterogeneity, polymers can be divided into two major categories: homopolymers and copolymers. Homopolymers contain only one type of repeating monomer unit, such as PLA, whereas copolymers are composed of more than one species of monomers, such as PLGA. Both of these type of polymers are essential in material science as they provide a variety of choices for various applications. Homopolymers, such as PLA and PGA, exhibit different properties with altered molecular characteristics, including molecular weight and synthetic routines while copolymers, such as PLGA, can broaden the application window with advanced controlling techniques in the polymer composition and structure. The ratio of monomers as well as the organization order of the repeating units allow the modification of their performances to meet diverse requirements which offers tremendous benefit for bioresorbable materials as the capability to adjust the mechanical, biological, and degradation characteristics of a given material can be easily adjusted to meet the requirements of various applications. The ease in adjustement of these type of polymers is due in large part to them belonging to a larger group of polymers known as poly(α-hydroxy) acids (PAHAs). These PAHAs are repeating chains of hydroxyl acid monomers covalently linked by an ester group (-COO-) formed by the hydroxyl group (-OH) and the carboxylic acid group (-COOH) on adjacent carbon atoms [7, 8]. The repeating units of the previously mentioned PLA, PGA, and PLGA are most commonly used and their structures are shown in Fig. 3.

2.1 Poly(glycolic acid) (PGA)

PGA is the simplest linear PAHA made from the glycolide monomer and can be synthesized through several different processes such as polycondensation, ring-opening, and solid-state polycondensation. Since the first marketed PGA product, DEXON®, was produced by Davis and Geck in 1960s from ring opening polymerization, numerous efforts have been made to investigate ways of advancing the properties of PGA by modulating its synthetic routines. Factors including reaction temperature, catalyst content, and monomer and solvent amount are modulated in order to achieve PGA products with high monomer conversion, elevated reaction yield, and desired molecular weight.

Structurally, PGA is highly crystalline resulting in relatively high mechanical properties and poor solubility in most organic solvents. The glass transition temperature (T_g) of PGA, above which temperature polymers exhibit soft and rubbery

poly(glycolic acid) (PGA) 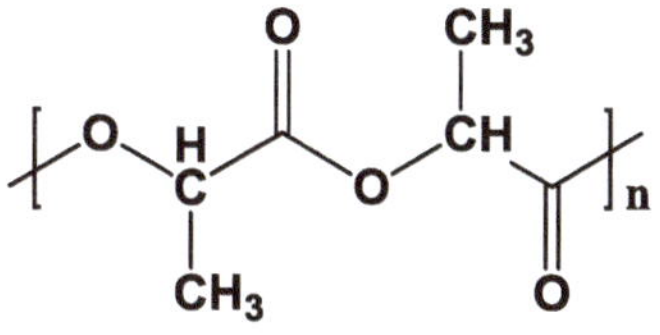   poly(lactic acid) (PLA)

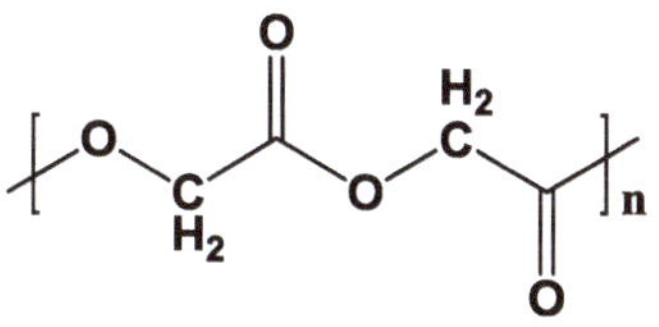

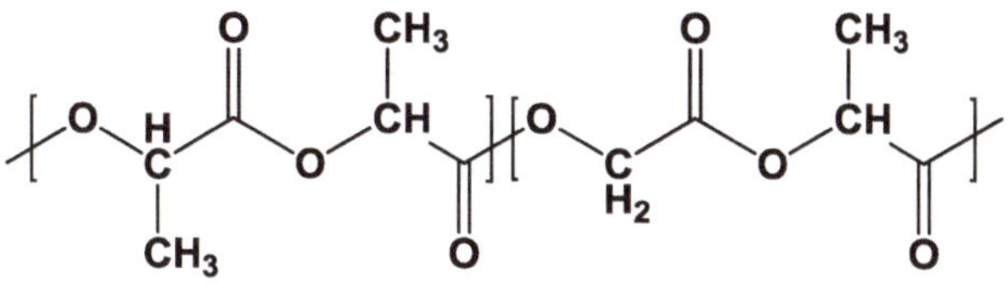

poly(lactic-co-glycolic acid) (PLGA)

Fig. 3 Structures of lactide and glycolide polymers

Fig. 4 The stereoisomers of lactide

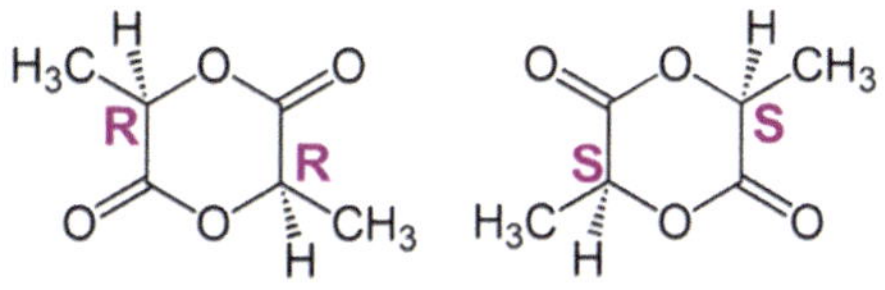

D-lactide   L-lactide

properties and below which polymers are hard and stiff, is highly dependent on molecular weight and can be modulated within the physiological window from 35 to 40°C with the melting temperature (T_m) being 220–225 °C. Due to the controllable mechanical and degradation properties along with the biocompatibility, PGA has been widely applied in biomedical applications such as suture and fibers [9].

2.2 Poly(lactic acid) (PLA)

PLA is polymerized from a lactic acid monomer (2-hydroxy propionic acid) which can be obtained via chemical methods or fermentation from natural renewable resources, such as corn, potato starch, and sugar. The asymmetric property of the lactic acid monomer results in the stereoisomerism for the PLA polymer. Lactic acid has two stereoisomeric configurations, dextrorotatory (D) and levorotatory (L), as shown in Fig. 4.

The stereoisomerism is important to note since the diversified space arrangements of lactide determine what stereoisomer the PLA homopolymers result in:

Table 1 Overview of commercially available RESOMER® bioresorbable material with varying compositions, degradation times, and general applications

Polymer name	Composition	Resorption	Crystallinity	Inherent viscosity	Application
RESOMER® L 206 S	Poly(L-lactide)	>3 years	Semi-Crystalline	0.8–1.2	Orthopedic, sports medicine, plates, screws, nails, stents
RESOMER® L 207 S	Poly(L-lactide)	>3 years	Semi-Crystalline	1.5–2.0	
RESOMER® L 209 S	Poly(L-lactide)	>3 years	Semi-Crystalline	2.6–3.2	
RESOMER® L 210 S	Poly(L-lactide)	>3 years	Semi-Crystalline	3.3–4.3	
RESOMER® LR 704 S	Poly(L-lactide-co-D,L-lactide) 70:30	2–3 years	Amorphous	2.0–2.8	Orthopedic, sports medicine, screws, nails fixation devices, screws, pins
RESOMER® LR 706 S	Poly(L-lactide-co-D,L-lactide) 70:30	2–3 years	Amorphous	3.3–4.2	
RESOMER® LR 708	Poly(L-lactide-co-D,L-lactide) 70:30	2–3 years	Amorphous	5.7–6.5	
RESOMER® LG 855 S	Poly(L-lactide-co-glycolide) 85:15	1–2 years	Amorphous	2.5–3.5	
RESOMER® R 207 S	Poly(D, L-lactide)	1–2 years	Amorphous	1.3–1.7	Soft tissue, meshes
RESOMER® X 206 S	Polydioxanone	<6 months	Semi-Crystalline	1.5–2.2	Sutures
RESOMER® C 209	Polycaprolactone	>2 years	Semi-Crystalline	0.8–1.2	Sutures, wound dressing, tissue engineering scaffold
RESOMER® C 212	Polycaprolactone	>2 years	Semi-Crystalline	1.13–1.38	

Data provided by Evonik Industries

PLLA, PDLA, or PDLLA [10]. The properties of these polymers are highly dependent on the crystallinity, molecular weight and tacticity. PLLA and PDLA are both optically active, semi-crystalline polymers and can be obtained from polymerization of L-lactide and D-lactide monomers, respectively. Particularly, PLLA is bioactive and exhibits a similar modulus and tensile strength to PGA due to its crystalline structure whereas PDLLA has lower tensile strength and higher elongation because of the more amorphous structure arising from the random distribution of lactide (LA) monomers [11]. Compared to PGA, the presence of pendant methyl group in LA monomers increases the hydrophobicity resulting in an extended degradation time as well as a higher T_g. PLA can be synthesized from similar methods used to produce PGA including ring opening polymerization and polycondensation methods. PLA is extensively used in the orthopedic devices with typical commercialized PLA polymers, such as RESOMER® polymers shown in Table 1, exhibiting various properties for biomedical purposes and are synthesized from both ring-opening polymerizationand polycondensation methods. For example, by controlling the molecular weight of PLLA, RESOMER® L 206 S, L 207 S, L 209 S, and L 210 S exhibit resorption period greater than 3 years whereas the inherent viscosity is

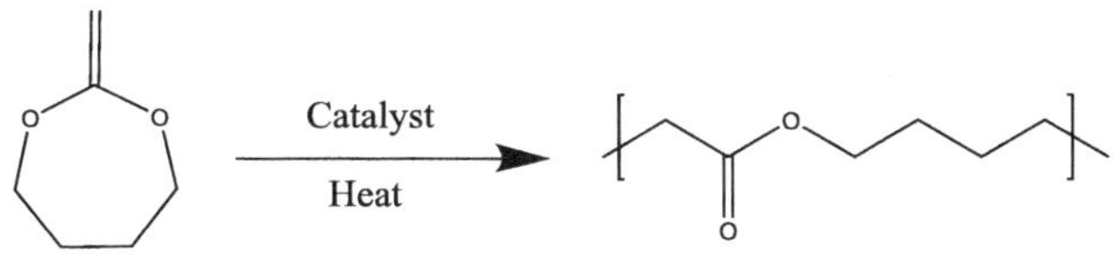

Fig. 5 Synthesis routine of polycaprolactone via ring-opening polymerization

different from each other providing various application options, such as orthopedic, sports medicine, screws, and nails. In addition, the amorphous PDLLA products, such as RESOMER® LR 704 S, LR 706 S, and LR 708 show faster degradation which is within 2–3 years due to the absence of crystallinity. Along with other biodegradable homopolymers, such as RESOMER® X 206 S (polydioxanone) which can degrade within 6 months or RESOMER® C 209 (polycaprolactone) which can survive *in vivo* for more than 2 years, bioresorbable materials provide a complete platform for biomedical applications.

2.3 Poly(lactic-co-glycolic acid) (PLGA)

A copolymer is a polymer that is formed by polymerizing more than one type of monomer and have properties that fall within the range of their individual constituent homopolymers. PLGA is such a copolymer composed of PGA and PLA. A benefit of copolymers like PLGA is that their properties can be easily adjusted by varying the ratios of the individual components. PLGA is normally preferred where precise control over degradability and necessary mechanical properties are required. It offers various parameters in adjusting the degradation rate, such as the ratio between components, chain length, hydrophobicity/hydrophilicity balance, and crystallinity. As the content of PGA increases, the hydrophilicity of the polymer increases and the crystallinity of PLA is increased which result in the faster degradation. RESOMER® LG 855 S has similar inherent viscosity with homopolymer RESOMER® L 209 S whereas the degradation rate is almost twofold faster.

2.4 Polycaprolactone (PCL)

As one of the earliest synthetic polymers for biomedical applications, polycaprolactone (PCL) has attracted attention since the 1930s. PCL can be synthesized by various methods such as anionic or cationic catalyst involved ring-opening polymerization from caprolactone monomer, as shown in Fig. 5, or through a free-radical polymerization method from 2-methylene-1-3-dioxepane monomer [12]. The thermal properties of PCL are dictated by the crystallinity of the polymer with the T_g occurring at -60 °C and melting point range of 59–64 °C which allows for

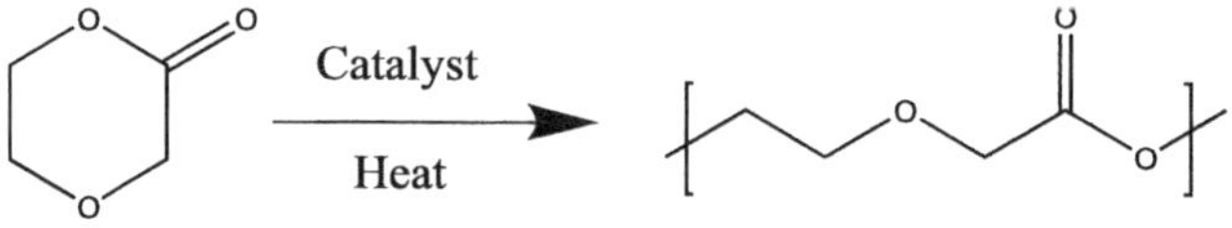

Fig. 6 Synthesis routine of polydioxanone via ring-opening mechanism

easy formability or processability into highly structured forms such as foams at relatively low temperatures. Due to these exceptional thermal properties, PCL is applicable for most of the currently used processing methods for materials fabrication, such as particulate leaching, electrospinning, and additive manufacturing [13]. During recent decades, PCL-based medical devices, such as sutures (Monocryl®, Ethicon, Inc), wound dressings, contraceptive devices, fixation devices, dental canal filling material (Resilon™, Parker), have been widely investigated [14–18].

Degradation of PCL is possible within the body due to the ester bonds chemical lability towards hydrolysis, but the rate of ester hydrolysis under physiological conditions declines sharply as the number of chain carbon atoms increases [13, 19, 20]. Studies have investigated two steps of degradation for PCL polymers, hydrolysis of ester group and intracellular degradation of small PCL fragments [21]. While PCL's slow rate of degradation in the body can take between 2 and 4 years for a total degradation depending on the starting molecular weight, it should be noted that the degradation time of PCL is also dependent on the compositional makeup of the polymer. By adding hydrophilic polymers the time can be adjusted to a shorter term. For instance, by copolymerization with lactides or glycolides, the degradation rate of PCL can be altered [18].

PCL has particular properties that are attractive for the design of tunable biomaterials such as slow crystallization kinetics and low melting temperatures in the physiological range. Slow degradation rates, with relatively minimal acid generation, can be valuable for prolonged drug release or implants which require longer-term stability [22]. Despite its long term degradation rate, PCL has been found extensively in use as drug delivery vehicles, cell cultivation, and in implants for regenerative medicine and drug release. In the fields of tissue engineering and drug delivery, PCL-based formulations as copolymers or as blends with synthetic or biopolymers have received particular attention, such as wound dressings [23–25], root canal filling material in dentistry [15], and tissue regenerative systems [26].

2.5 Polydioxanone (PDO)

Polydioxanone is a representative poly(ester ether) polymer that can be synthesized by ring-opening polymerization from monomer, *p*-dioxanone, as shown in Fig. 6, resulting in a semi-crystalline structure. The T_g of the material is dependent on the molecular weight and can be adjusted between −10 and 0 °C. Homopolymer PDO has been utilized for surgical sutures for 30 years due to the specific properties such as exceptional flexibility and low modulus attributable to the ether bond and additional methylene bonds in PDO repeating unit. Due to the ester-based backbone structure, PDO can be hydrolytically degraded in physiological environment [27, 28]. During the degradation process, the elastic modulus and morphology both change in a time-dependent manner. The degradation of PDO can be modulated within 6–12 months with different molecular weight. PDO was also the first bioresorbable polymer used for bioresorbable blood vessel ligating clips as alternatives to metallic clips with other applications including fixation devices and fasteners [29].

In contrast to other bioresorbable materials, such as PLA or PLGA, PDO can be processed into materials at a relatively lower temperature due to its lower T_g. For the production of sutures, PDO is generally extruded into fibers, however care should be taken to process the polymer to the lowest possible temperature in order to avoid its spontaneous depolymerization back to the monomer, paradioxanone [30]. Other processes such as electrospinning PDO have also been utilized to fabricate sutures and scaffold for tissue engineering as well as vascular grafts [31].

3 Bioresorbable Degradation

The degradation of lactide and glycolide based materials, such as RESOMER® polymers, follows the brief degradation process as described above with degradation initially eroding the polymer after which point hydrolytic scission of the hydrophilic moieties breaks the polymer into smaller section that are expelled through physiological pathways. At the initial stage, water hydrates the polymer from surface to the interior structure by diffusion. The diffusion and wetting rate are highly dependent on the polymer size, porosity, and hydrophilicity. The ester groups, which covalently link monomers, are hydrolyzed and cleaved when water is in the surrounding environment, resulting in the conversion of the long chains into short chains and even small segments the degradation process continues. It should be noted that the initial degradation occurs in amorphous regions with a reduction of local molecular weight and this process would not result in significant loss in mechanic properties as the material still holds integrity due to crystalline portions. Produced small molecules and monomers, such as lactic acid and glycolic acid from RESOMER®, are eliminated through biochemical degradative pathways. As shown in Fig. 7, lactic acid can be directly metabolized through the Krebs cycle/tricarboxylic acid cycle resulting in the carbon dioxide and water as the final product. Glycolic acid can be either directly cleared out as urine through kidney or participate in the Krebs cycle as pyruvate with the same excretion pathway as lactic acid.

In order to achieve optimal biomedical outcomes, the degradation and ultimate resorption of bioresorbable materials must be designed to survive implantation long enough to allow for healing at the implant site. The largest determining factor in the degradation of these implants is the rate and mechanism which erodes the polymer on a macro-scale, depicted in Fig. 8: surface erosion and bulk degradation [32, 33].

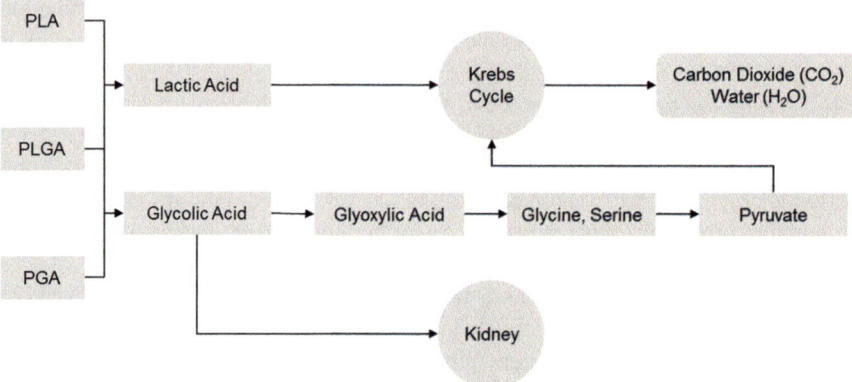

Fig. 7 Biological and physiological pathways for the degradation of PLA, PGA, and PLGA *in vivo*

Fig. 8 Effects of surface
and bulk erosion on the
bioresorbable implant

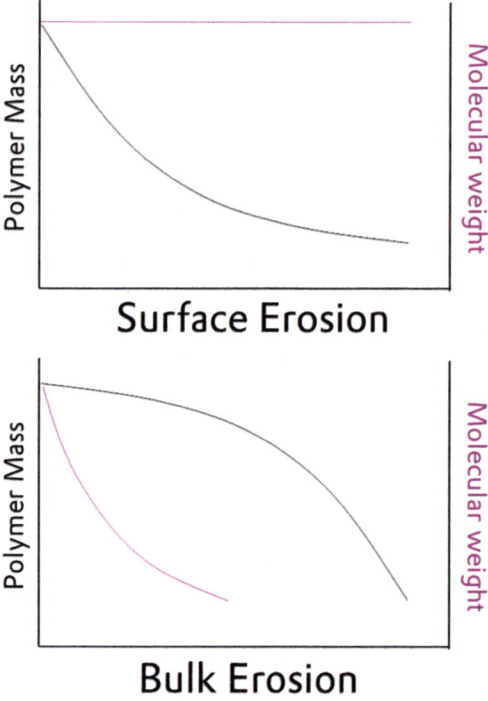

The type of erosion that may be desired is different for each application. For instance in a drug delivery application, the active therapeutics are not hydrolytically stable and therefore the drug delivery vehicle must be formulated to undergo surface erosion in order to protect the therapeutics from premature hydrolysis. This type of degradation results in an exponential decrease of polymer mass whereas the polymer molecular weight remains constant within the bulk of the material. In contrast, orthopedic implant materials should maintain their mechanical properties until the newly regenerating bone is formed. This requires that the implant lose integrity homogeneously through the bulk degradation mechanism in which the molecular weight and material's mass decrease simultaneously. In general, bulk degradation starts with the long polymeric chain being broken down into smaller segments through unspecific hydrolytic attack. During this reaction, the water contained in the surrounding biological fluid hydrolyzes the polymer and breaks the chain down into smaller segments. As the water attacks the hydrophilic regions of the polymer, the resulting smaller segments of the polymer now continue the degradation of the polymer into smaller and smaller chains through the autocatalysis process [34].

The ability to match the rate of degradation to the rate of new tissue regeneration is one of the most advantageous characteristics of bioresorbable materials because it not only facilitates a complete healing of the implant site but also eliminates the need for a removal surgery. As these implants degrade, they are also able to release any additives that have been introduced to the material that can help further facilitate tissue regeneration. Since the rate of degradation can be tightly controlled, the rate of this release of bioactive components can be adjusted so that the regenerating tissue system is continually promoted by biochemical cues in the surrounding environment.

3.1 Factors Affecting Degradation

Factors that can affect the rate of material degradation are numerous and can be divided into four major categories according to the trigger resources, including inherent polymer properties, (e.g., polymer composition, molecular weight, crystallinity, stereoisomerization) secondary ingredients (e.g., additives), material processing and manufacturing methods (e.g., extrusion, injection molding and compression molding), and post-processing techniques (e.g., sterilization or annealing) [35, 36] as shown in Fig. 9.

3.1.1 Inherent Polymer Factors

Naturally the material degradation rate is highly determined by the composition which regulates the hydrophilicity of the polymer [18]. As described above, PLGA is one type of copolymer which has a significant advantage of adjustable degradability resulting from the various compositions that can be achieved. For example, PLGA (50/50, LA/GA) exhibits faster chain scission compared with PLGA (85/15) since the more hydrophilic gycolic acid segment accelerates the degradation of the copolymer when its compositional ratio is increased.

The polymer composition also determines the crystallinity which can also play a role in affecting the degradation rate of the material. According to the degradation

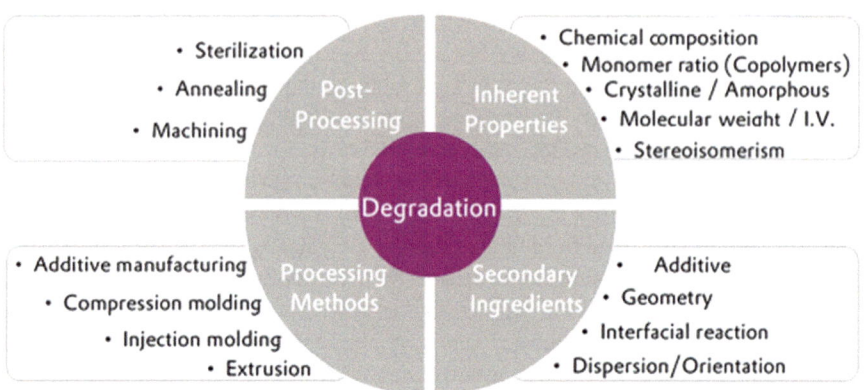

Fig. 9 Factors influencing the degradation of bioresorbable materials

dynamics, copolymer PLGA polymers initiate the degradation from the amorphous regions while the crystalline regions requires longer time for the ester bond cleavage. Similar phenomena have been reported for PLLA and PDLLA polymers where a faster degradation occurred in amorphous PDLLA polymer whereas the semicrystalline PLLA degraded in a slower manner. According to this concept, Evonik produced RESOMER® L series polymers (L 206 S, L 207 S, etc.) exhibiting longer duration *in vivo* compared with LR polymer (LR 704 S, LR 706 S, etc.).

Besides the influence from chemical factors, the degradation is also significantly dependent on the initial molecular weight. A higher polymer molecular weight typically results in a longer time required to degrade the materials. For example, studies have shown high molecular weight PDLLA particles exhibit much longer degradation time compared with spheres made from lower molecular weight. Similar studies have also reported that degradation rate constant of PLGA polymer with large molecular weight is larger than that of lower molecular weight [37]. A possible explanation is that large polymer chains have more free volume and repeating units that can be hydrolytically attacked. In addition, for PLLA polymers, reports have also demonstrated that decreasing the molecular weight of PLLA can expedite the degradation [35]. This is normally explained by the inverse proportion relationship between PLLA molecular weight and degree of crystallinity.

Considering the factor that the stereochemistry of PLA would affect crystallinity, the content of PLLA in PDLLA and PLGA copolymers has been demonstrated to be able to adjust the degradation rate. For example, studies have investigated the affect that the content of PLA has on the enzymatic biodegradation of PLDA copolymers and found that 92% PLA was the critical content amount that resulted in significant acceleration of the degradation [38, 39].

3.1.2 Secondary Ingredients

In order to enhance the functionality and mechanical properties of the material, it is not unusual to find that additives included in orthopedic composite materials are also biologically active [40]. Due to the mechanism of degradation, the pattern could also be altered by the existence of additives, such as therapeutics, salts, plasticizers, and catalyst residues. On one hand, chemical compounds, such as acids and bases, could catalyze the ester-bond degradation in hydrolysis and therefore elevate the degradation kinetics. On the other hand, these active ingredients can also neutralize the acidic groups produced in the degradation and prohibit the further degradation.

One specific additive category that has been widely used in bioresorbable orthopedic implants is calcium phosphate based salts, such as hydroxyapatite, tricalcium phosphate and biphasic calcium phosphate. These types of salts enhance the bioactivity and osteo-conductivity for implants by providing biocompatible scaffold and surface to adsorb proteins, construct bone matrix and support osteoblasts cells proliferation. Great benefits can be achieved from additive-included composites upon implantation and degradation. The duration of the implants can be modulated by adjusting the content of additives. More importantly, the degradation of calcium phosphate salts, such as hydroxyapatite, can trigger the release of calcium and phosphate ions in the biological environment and initiate bone formation. It should be noted that the presence of additives could not only modulate the properties of material but also provide a novel solution in adjusting the degradation profiles of orthopedic implant materials.

4 Mechanical Performance

The benefits of bioresorbable implant materials in their ability to serve a functional purpose for an allotted period of time after which point they disappear from the body is highly advantageous over typical biostable implants. The implant must offer adequate mechanical properties for the affected site and maintain those properties for a specific amount of time before degrading. With bioresorbable materials, once degradation begins to occur, a subsequent loss in mechanical properties is initiated. As the polymeric chains are hydrolytically attacked, the polymer chains become smaller thus the mechanical properties are reduced. This process continues until the implant is dissolved entirely at which point, if the implant has been designed appropriately, the native bone should have fully regenerated and overtaken the implant site. This rate of degradation and tissue growth should be synergistic, as shown in Fig. 10, in order to achieve the optimal success in treatment outcomes. These bioresorbable implants also promote regeneration of pre-implant condition tissue system due to their gradual degradation. For example, a bioresorbable orthopedic implant can be designed to remove the issue of stress-shielding commonly found with typical metallic implants. These metallic implants have mechanical properties much

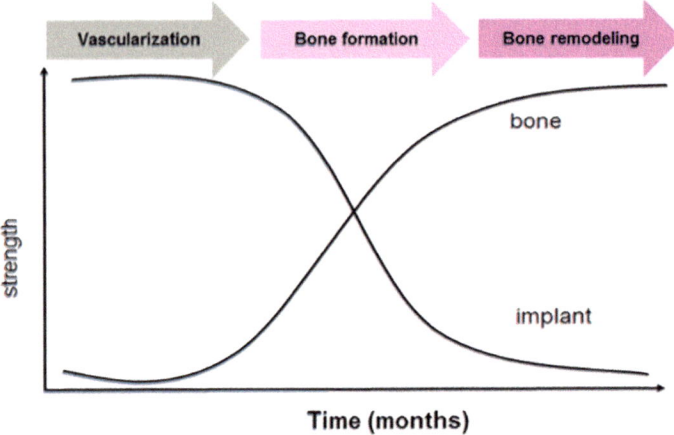

Fig. 10 An ideal bioresorbable material would match the rate of tissue regeneration with rate of implant degradation leading to the implant degrading completely once the tissue was fully regenerated

higher than those of the surrounding bone therefore the bone mineral density at the implant size decreases drastically over time. When a secondary procedure is carried out to remove the metallic implant, re-fracture of the bone at the implant site is a common occurrence as the bone itself has become more cancellous in morphology without having to carry a sustained load. This gradual load transfer provided by bioresorbable materials allows for the regenerating bone to increase in density thus promoting optimal healing with no need for revisionary procedures.

4.1 Factors Affecting Mechanical Performance

Similar to degradation, a number of factors can influence the resultant mechanical properties of the implant including material type and processing methods along with the implant design and environment. At the time of implantation, the bioresorbable implant, either as a virgin material or a composite, must be capable of carrying the initial loads that are placed on it through normal physiological occurrences. Compared to widely available metallic and biostable implant materials, bioresorbable materials have a wide range of mechanical properties depending on the polymer matrix and fillers. By selecting both the appropriate polymer species as well as the processing conditions and implant design, the mechanical properties can typically be selected so that they are capable of serving their structural purpose *in vivo*. Bioresorbable implant materials must also be evaluated for their mechanical properties as the material breaks down, a factor that does not need to be analyzed for non-bioresorbable implants [41]. *In vitro* studies of the effect of degradation on mechanical properties of bioresorbable composites have been carried out by a number of groups, with the consensus being that increasing the concentration of bioactive ceramic within the polymer matrix rapidly increased the rate of dissolution and resultant decline of mechanical properties due to the decreased amount of polymer material existing in the free space between ceramic particles [42]. Evaluation of the elastic modulus decline for a PLA/HA composite revealed that over the course of twenty days of submersion, the modulus decreased 8–17% depending on the concentration of HA within the composite with higher rates of mechanical decline being attributed to greater HA concentrations. Further studies corroborate with this decline in mechanical properties *in vivo* with one particular results showing that PLLA/HA composites decreased in bending strength and bending modulus over 36 weeks of 78.6% and 73.2%, respectively [43, 44].

The added consideration of implant site must be taken into account as the environment surrounding the implant will have a direct impact on the degradation properties of the device. In regions with high vascularity and a subsequent large volume of biological fluid diffusion, the rate of degradation is accelerated as the fluids both attack and break down the implant as well as rapidly carry away any fragmented sections of the polymer to be further broken down [45]. In a highly vascularized region, a porous implant design will facilitate more fluid flow throughout the implant and therefore expose an increased amount of surface area to the surrounding biological fluid flow and therefore the rate of degradation will accelerate proportionally. A non-porous implant on the other hand will not allow for this diffusion of biological fluid into the bulk of the device but these type of implants are more prone to self-catalyzed degradation where the acidic end groups of the broken polymer chain act to increase the localized pH level leading to increased degradation from the interior to the exterior of the implant [46–48].

4.2 Mechanical Enhancement via Additives.

While the mechanical properties of a virgin implant are dependent on the type of material and its method of processing, when a composite material is used, micromechanical effects must also be taken into consideration. As the concentration of additive increases, a saturation point is typically reached in which the mechanical properties rapidly decline as the additive goes from being a reinforcing member to the majority component in the composite, as depicted graphically in Fig. 11. Before this saturation point, which tends to occur near an additive concentration of 30–40%wt., the additives are able to act as hindrances to crack propagation within the material. Methods of reinforcement of bioresorbable material have been undertaken in an effort to increase the properties of the virgin materials by adding reinforcing materials or modifying the processing conditions and implant design. The inclusion of additive species for reinforcement must be understood fully as while the additives may increase the mechanical properties, they also have an effect on the biological response as well as the degradation profile of the composite.

Fig. 11 As the additive concentration increases, an increase in modulus can be expected while a proportional decrease in strength may also occur. This example is particularly true of the inorganic ceramic fillers commonly used in composite orthopedic implants

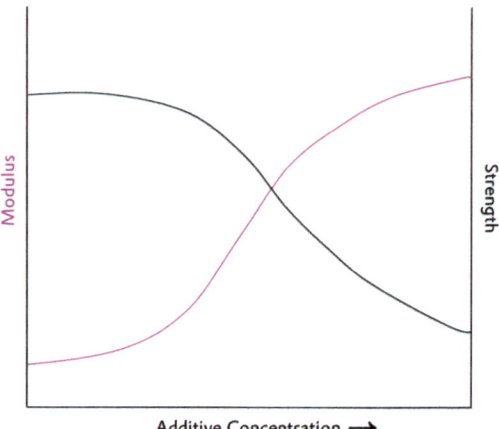

Additive Concentration ⟶

4.3   Effect of Implant Design on Mechanical Performance

As is the case for any implant, the mechanical factors of the material and the implant design need to be matched to those of the localized tissue near the treatment site. For instance, the design of interference screws has changed overtime with different variables in the design being changed and optimized to find a screw that can withstand the torque applied upon insertion as well as maintain stability within the bone by having appropriate pull-out resistance. The varied designs are typically tested through computational methods as an initial step after which point experimental validation of the design is undertaken to evaluate the potential advantages of each design iteration [49]. The thread profile as well as the geometry of an interference screw has a direct correlation to the load required to insert the screw and pull-out strength [50]. By increasing the thread pitch (the distance between individual threads), the amount of axial force required to insert the screw was dramatically decreased by approximately 83% along with an increase in fixation strength of nearly 50% for the most extreme cases. A similar study reported that the additional design factor of the screw head plays an important role in the failure of the screw upon insertion [51]. By changing the driver/interface design to a Torx design instead of a hex or a Tri-Lobe design, the amount of stress concentrations were decreased and the maximum torsional strength of the screw was increased. Once the screw design is optimized, the implementation of bioresorbable interference screws has been shown to also offer fixation strength comparable to that of titanium interference screws [52].

Recently, further developments in the design and implementation of various screw types and screw designs have sought to enhance the fixation strength through innovative engineering methods. SonicWeld Rx® is one such implant is a pin produced by KLS Martin and is able to achieve increased fixation strength and bone response through its ability to instantly shape its design to the porous nature of the bone at the implant site [53, 54]. As the pin is inserted into a pilot hole, it is exposed

Fig. 12 Pre-drilling, insertion, fusion, and the resultant fixture when the SonicWeld Rx® is used. Image courtesy of KLS Martin

to ultrasonic waves which cause the pin to vibrate. The vibration of the pin against the native bone structure along with the ultrasound waves produce a small amount of heat allowing the pin to form itself to the interior bone porosity as shown in Fig. 12.

5 Bioactivity

The furthering of bioresorbable materials has been focused on developing more biologically responsive materials with the capability to promote tissue system regeneration for materials in an effort to, for example, avoid the use of autografts and allografts in bone regeneration applications [55]. Specific to bone regeneration applications, autografts are bone grafts which are taken from a separate anatomic site of the patient and have consistently offered the most positive response at the implant site among all materials used. Allografts are bone grafts which are taken from a separate subject (such as a cadaver) and are the second most commonly used method for bone repair, substitution, or augmentation. Although these types of implants have been successful in the past, newly developed materials with the ability to promote positive responses *in vivo* are able to avoid the costs, donor-site morbidity, availability, and the potential viral transfer that can be encountered when using autografts and allografts.

Synthetic materials used for bone repair can be classified based on their response *in vivo* as being either bioinert or bioactive [56]. Bioinert materials are those which are unable to elicit a positive cellular response and tend to result in the formation of fibrous tissue on the surface of the materials. Bioactive materials, however, are capable of promoting bone tissue formation, for example, and therefore directly bone with the surrounding bone-material interface. Further, bioresorbable materials, such as lactide and glycolide based implants, have been consistently used in orthopedic applications due to the positive cellular response and clinical advantages with their implementation significantly increasing over the last four decades. They include cylinder rods, plates, screws, tacks, suture anchors, mesh, plugs, wires, arrows, and drug delivery devices. Lactide and glycolide based materials offer a lot of advantages over metallic implants such as an appropriate amount of mechanical strength when necessary, the ability to reach a degradation rate similar to the rate of new tissue formation, and the elimination of the need for an additional operation for implant removal.

5.1 Inorganic Additives

Calcium phosphate (CaP) based materials are widely used in development of bioactive materials for orthopedics. These materials promote bone formation due to their compositional similarity to bone mineral along with their bioresorption, bioactivity, and osteoconductivity [57]. Osteoconductivity is the ability of the material to serve as a scaffold or template to guide formation of the newly forming bone along their surfaces. CaP ceramics adsorb circulating proteins (from the biologic environment) on their surfaces followed by the attachment, proliferation, and differentiation of bone cells such as osteoblasts, osteoclasts, osteocytes, and bone-lining cells leading to matrix production. For the sake of bioactive implants, the ability for a material to promote the attachment, proliferation, and differentiation of the cells are critical for the success of the implant. In particular, the osteoblasts are the main targeted cell as they are those that are responsible for the production and mineralization of the bone matrix. Once these osteoblasts have attached, the further functions of bone matrix maintenance and bone resorption can be carried out by the osteocytes and osteoclasts, respectively. The addition of bioactive additives such as CaP ceramics can enhance this osteoblast adhesion and are therefore commonly used to promote bioactivity in typically biostable materials [58].

Following are CaP materials that are known to be osteoconductive and used in medical orthopedic implant applications.

5.1.1 Calcium Phosphate Based

In order to increase the bioactivity of a typically bioinert material, additives must match the mechanical and/or biochemical cues *in vivo* in order to promote an ideal response. Among the potential additives, three inorganic minerals have seen the most use in bioresorbable materials as they are synthetically manufactured minerals salient to those naturally found in hard tissue systems. Hydroxyapatite (HA), tricalcium phosphate (TCP), and biphasic calcium phosphate (BCP) are all various forms of calcium phosphate, a mineral constituting approximately 70% of calcified bone [57, 59, 60].

Hydroxyapatite (HA)

When compared to other calcium phosphate based minerals for bone regeneration, HA is the most stable form under physiological temperatures, pH, and chemical environments [61]. Upon implantation and degradation, HA, as well as other calcium phosphate minerals, release both calcium and phosphate ions into the surrounding biological fluid which initiate bone formation [62–65]. For example, combining HA with PLA has been shown to not only offer a method to adjust the rate of implant degradation, but also increase formation of a biological apatite surface and bone cell activity for the promotion of osteogenesis *in vivo*.

Tricalcium Phosphate (TCP)

Tricalcium phosphate (TCP), another type of calcium phosphate mineral, is both resorbable and bioactive [66]. Structurally, TCP exists in two forms known as α-TCP and β-TCP with the latter being most commonly used in orthopedic applications due to its strength in promoting bone regeneration [66–68]. Further, β-TCP has been shown in some instances to dissolve *in vivo* and therefore release bone formation promoting calcium and phosphate minerals more effectively than HA [69]. Studies show that after coating PLGA scaffolds with β-TCP, an increased rate of cellular adhesion, proliferation, and spreading can be achieved [67].

Biphasic Calcium Phosphate (BCP)

With the advantages of HA and β-TCP being well-studied and documented, the selection of biphasic calcium phosphate (BCP) was a natural evolution. By combining HA and β-TCP in various ratios, BCP is formed and offers the advantages of each component. For application in bioresorbable implants, tailoring of the ratio of HA/β-TCP in BCP allows control of biological response, degradation rate, and mechanical properties making it a highly desirable additive for a range of orthopedic implant applications [59, 68, 70]. Compared to only using HA as an additive, composite scaffolds consisting of PLGA and BCP (15%wt. HA: 85%wt. β-TCP), showed improved osteogenesis and bone formation [71]. This benefits of this additive have led to a number of market offerings of polymers containing it as a bioactivity enhancing ingredient such as the BioComposite™ material from Arthrex.

Calcium Sulfate

Being used as a bone graft substitute material, calcium sulfate provides the same osteoconductivity and biodegradability as common calcium phosphate based materials, but also has shown to have specific advantages for some applications [72]. While calcium phosphate additives such as β-TCP may take up to 18 months to be fully resorbed, the absorption of calcium sulfate occurs between 4 and 12 weeks allowing for enhanced bone growth during the early stages of post-treatment. Smith & Nephew, Inc.'s REGENESORB Interference Screw has taken the advantages of each of these types of additives and combined them in a PLGA matrix [73]. This mixture of a slowly degradation calcium phosphate additive along with the more quickly degrading calcium sulfate additive offers an implant which promotes rapid bone ingrowth as well as sustained bone regeneration through the addition of both a rapidly absorbed calcium sulfate additive and slowly absorbed β-TCP to a PLGA matrix.

5.2 Other Additives

Other non-typical additives such as graphene oxide, fullerene derivatives, fumaric acid monoethyl ester, and even bone growth factor have also been investigated with bioglass being widely researched for use in orthopedic medical devices [74–77]. Bioactive glass, commonly shortened to bioglass, is found in more recent studies as a potential additive for bioresorbable polymeric implants. Bioactive glasses are a family of amorphous ceramic materials discovered in 1969 and are capable of achieving remarkable high integration into surrounding bone upon implantation. More recently, the effects of bioactive glass additives to resorbable materials has been investigated to understand their impact on the degradation, osseointegration, and bone reconstruction for orthopedic implants. Study shows that a PLDLA matrix with a 30 wt.% 45S5 bioglass additive was recently found to display significantly accelerated degradation in *in vitro* studies when compared to virgin PLDLA. While the virgin PLDLA material remained stable for the entire 6 month immersion, the bioglass composite lost 35% of its initial weight during that time. The composite had already lost 15% of its weight in just the first 15 days. An *in vivo* study of this composite in rabbits revealed that the bioglass triggered osseointegration, most prominently within the first month after implantation [78]. This bioactive glass additive was shown to hold strong potential for orthopedic fixation applications.

Although promising in their *in vivo* response, further studies are required to determine how these materials function when used in conjunction with bioresorbable materials, as the addition of bioglass typically increases the rate of degradation of the matrix material in the composite system [79]. When comparing four PLGA matrices with $CaCO_3$, HA, 45S5 bioglass, and ICIE4 bioglass additives, both bioglass composites caused the matrix to degrade prematurely whereas the other two fillers possessed good degradation properties and also increased the thermomechanical properties of the polymer. A study of S53P4 bioglass added to a PCL/PDLLA matrix showed similar results of premature degradation through molar mass reduction [80]. Therefore, while rapid degradation may be beneficial for some applications, such as fixation devices, the mechanisms underlying this response must be understood fully and remedied if these compositions are to be used in long-term bone reconstruction implants. Study also showed that the bioactivity of another bioglass, coating electrospun PLGA with nanoparticles of the bioceramic willemite (Zn_2SiO_4) to improve bone reconstruction, as tested in a rat model. The results of *in vivo* experiments showed good bone reconstruction with no inflammation or other adverse effects, demonstrating this bioglass' potential for bone engineering implant applications [81]. Several reviews have been recently published which further characterize the benefits, challenges, and advances in using these additives in bioresorbable orthopedic implants [14, 82–85].

6 Biocompatibility

For bioresorbable materials, biocompatibility can be viewed as the ability for the material to serve its functional role in the regeneration of the tissue system being treated and not provoke negative cellular or system reactions such as infection, inflammation, or rejection. Specific to bioresorbable materials, the initial composition of the material is not the only factor that must be considered – the by-products released upon degradation must also be considered. Two types of biocompatibility evaluation methods are normally conducted for any material that is being screened as a potentially implantable composition, including *in vitro* and *in vivo* tests. For the *in vitro* evaluation, implanted materials and degradation by-products are examined by cellular toxicity assessment experiments, such as agar diffusion test and filter test [86]. These tests are generally based on the health performance and viability of cells after culturing them with either implanted materials or extracted degradation by-products for certain time. Due to the sensitivity of temperature, extracts of PLGA implanted materials are normally achieved after culturing in phosphate-buffered solution (PBS) at 37 °C and 70 °C which mimics the short-term and long-term toxicity evaluations. The *in vivo* evaluations of bioresorbable materials include tests such as physical examination, histology, radiography, and biomechanical testing in animal models. Specifically, radiography is utilized to locate the implants and the histology is utilized to evaluate the inflammatory response and osteolytic reactions of local tissues. Moreover, bone and soft tissues growth are also critical in determining the implants biocompatibility.

It is worth mentioning that the rate at which the implant degrades and the subsequent biological response occurs is highly dependent on the implant geometry. Fibrous scaffolds, for example, degrade more rapidly than the same material in its bulk form due to the increased surface area of the fiber which exposes more of the material to enzymatic or hydrolytic degradation. With this in mind, it is thus crucial to know the by-products of such degradation pathways and to understand their potential for acute and chronic affects *in vivo*. The lactide and glycolide based polymer implants have repeatedly shown that the initial composition as well as the degradation by-products result in no significant finding of negative effects with the average level of incidence being less than 5% for most cases of implantation [86, 87]. Great tissue receptivity of PLGA implants have been reported in femurs of rats and rabbits as well as mandibular of monkeys [88–90]. Moreover studies have demonstrated the accelerated healing of bones and tissues in rabbit and rat femoral osteotomies models when biodegradable, synthetic polymers are used [91]. Specifically, composite interference screws composed of PLGA/β-TCP have been shown to promote osteoconductivity values of 63% and a degradation of 88% original mass in a clinical review of nearly 700 patients who underwent knee or shoulder treatment procedures with less than 5% showing any adverse effects [92].

7 Processing and Fabrication

The need for improved implant devices with advanced designs and superior mechanical and biological properties has driven the development of novel techniques for processing these bioresorbable materials. Historically, the most common methods of processing these bioresorbable polymers employed melting followed by an extrusion or molding technique. However, recent years have seen an increase in novel techniques such as textile processing technologies and three-dimensional (3D) printing technologies which offer numerous advantages compared to typical processing. With the introduction of additive species to polymeric materials, re-optimization of conventional processing methods has been investigated as the introduction of additive species into the material offers increased mechanical and biological properties, but changes the processing parameters. The thermoplastic nature of lactide and glycolide based materials allow these bioresorbable polymers to be processed into various orthopedic implants, examples of which are shown in Fig. 13, through the use of a range of both conventional and novel techniques as will be outlined in the following sections.

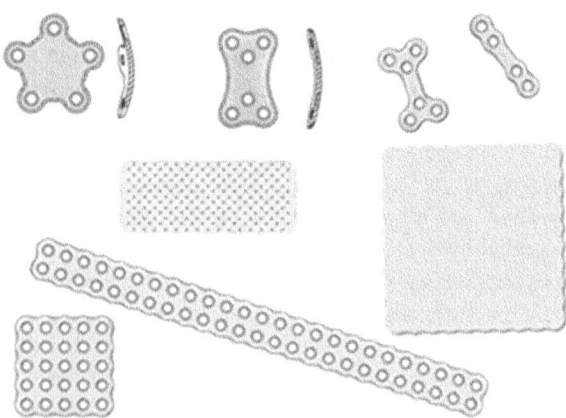

Fig. 13 Examples of RESOMER® based orthopedic implants fabricated through conventional processing methods. Image courtesy of KLS Martin

7.1   Material Effect on Pre-Processing and Processing

Besides the intrinsic properties of PLA-based polymer, environmental factors, such as temperature, humidity and pressure, are essential variables to consider in the degradation of the material. It is preferred to minimize the product degradation during the processing in order to achieve the superior mechanical properties and desired product degradation performance. According to the mechanism of PLA polymer degradation, moisture and humidity should be kept at optimal levels in all the processing procedures in order to attenuate the hydrolysis of the polymers. High moisture contents lead to vapor entrapment in the processed samples while low moisture contents result in high processing temperature, which is detrimental to material degradation. Semi-crystalline material's drying temperature is preferred to be above the T_g for a certain time. Drying methods can be vacuum drying, low temperature oven drying, nitrogen hopper drying, and compressed air hopper drying. On the other hand, amorphous biodegradable material's optimal drying temperature should be below the T_g for typically a longer time compared with semi-crystalline material using equipment such as vacuum oven drying, nitrogen hopper drying, and compressed air hopper drying. In order to achieve the optimal physical and mechanical properties of biodegradable materials, processing conditions of extrusion, injection molding, compression molding, and additive manufacturing are extremely important.

Due to different intrinsic properties of the polymer, the processing temperature needs to be adjusted accordingly. High processing temperature could lead to degradation of the polymer and generate monomers which play an important role in determining the material mechanical properties as well as material degradations [93]. It is important to have tight control of the process parameters. Monomers, such as lactic acid (LA) and glycolic acid (GA), can work as catalyst in the hydrolysis process and change the materials degradation performance.

The degradation rate of implants *in vivo* can also be altered by the material type, shape, size and porosity, to name a few. In theory, smaller particles with high pore density as well as large pore size would accelerate the degradation due to the enhanced surface-to-volume ratio and provides extensive surface for water to attack. However, while a highly porous implant would allow for more biological fluid to enter the inner bulk of the material, the segments of the polymer that have been attacked hydrolytically are more easily able to be expelled from the implant. With a low porosity implant, these chain segments reside within the implant for longer amounts of time before being removed and therefore the autocatalysis mechanism is accelerated within the implant.

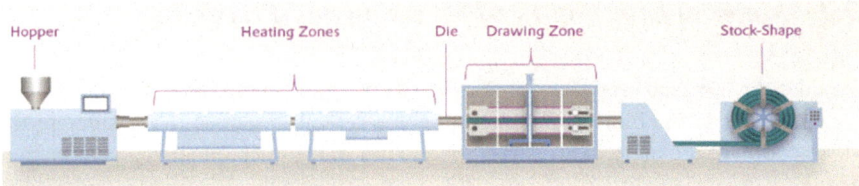

Fig. 14 An illustrative view of the extrusion process depicting the hopper for supplying material, the heating zones where the material melts and is moved forward by screws, the die which shapes the extrudate, a drawing zone which serves to pull the extrudate from the die, and the final stock-shape collection. Image courtesy of Evonik Industries

7.2   Conventional Processing Methods

Typical fabrication methods implemented in bioresorbable materials include extrusion, injection molding, and compression molding. These methods have a long past and have been optimized for most application needs. The recent need to enhance the biological activity as well as the mechanical properties of these materials however has required manufacturers to again optimize a process that now contains both a polymer matrix as well as the inorganic filler [79, 94]. Extrusion and molding process are typical conventional methods to make PLA-based implants/products and the temperature of these methods must be optimized as the temperature directly affect polymer degradation [46].

7.2.1   Extrusion

Extrusion is a continuous process, as shown in Fig. 14, which forces polymer melt through a die with an opening. Polymer materials in the form of pellets or flakes are fed into a hopper which serves to hold the material and can also be set up to keep the material dry. If additives are to be introduced to the material they can be added to the hopper at this point or they can be added downstream in the process once the polymer has melted. The materials are then conveyed forward from the hopper by a feeding screw and forced through a die which shapes the polymer melt into a continuous polymer product such as a rod, filament, or tube. A pelletizer can also be added to the extrusion line which segments the rod or filament extrudate into small granules able to be processed with subsequent fabrication methods. To form pelletized material, a single-screw extruder is typically used. If the desired material is a composite material, the use of a twin-screw extruder will commonly be chosen for compounding.

The process of extrusion is typically implemented as a pre-fabrication technique while the final parts will be processed through other methods. Some instances of extrusion occur where the final part is produced through this method, such as cardiovascular scaffolds (stents), balloons, catheters, or wires, but the majority of industrial use of extrusion is carried out to produce rods, filaments, or granules for secondary processing techniques. Orthopedic implants such as screw and pins can also be machined into desirable shapes or geometries from the extruded rods normally fabricated through single-screw extrusion. For composites, the effect of the additive on melt viscosity, thermal profiles, and solidifying times must be taken into account and optimized. Specifically, when particulate additives are introduced into a virgin material during compounding on a twin-screw extruder, the design of the screw as well as the melt profile, moisture content, and additive properties of the composite must be optimized [95]. Commonly, additive particles are prone to agglomerations within the polymer matrix which cause poor mechanical properties as well as non-uniform biological responses for the composite material [96, 97].

Neat resorbable polymers resins such as PLLA, PGA, or PLGA typically require the introduction of such additive species to enhance the mechanical and biological response of implants made from these materials. In these resorbable polymer composites, while rigid inorganic fillers, such as calcium phosphates, contribute to enhanced stiffness of the material, the strength and toughness of the composite material is largely dependent on filler type, filler morphology, particle size, filler dispersion quality, filler concentration, and the interfacial interaction between filler and the polymer phase. Excessive amounts of fillers may generate voids between the resin matrix and unwetted filler surface which causes a deterioration in the composite's properties. For example, the incorporation of bioglass spheres has been shown to have a direct effect on the mechanical properties of a PLLA composite with increasing concentration decreasing the mechanical properties [98]. Inorganic fillers with larger aspect ratios normally give composites improved tensile and flexural strength compared to composites which incorporate particulate fillers. The alignment of such high aspect ratio fillers through optimized processing techniques contributes to even higher mechanical properties along the alignment axis than in the transverse direction. By decreasing the size of the filler to the nano- or microscale, the fillers have larger surface area that is able to interact with more polymer chains thus further enhancing the composite's mechanical properties through improved interfacial interactions. This optimization of additive dispersion is particularly important in the application of resorbable materials for orthopedic implants as these additives are commonly introduced to the virgin polymer species in order to enhance the mechanical and biological properties. The reinforcement effect in resorbable polymer composites is largely dependent on filler dispersion throughout the polymer matrix, the size and geometry of the filler, and the interfacial interaction between filler and the polymer phase. Through optimization of processing conditions, better dispersion of filler particles, as seen in the SEM image in Fig. 15, can improve mechanical properties as well as enable bioactive functions, such as osteointegration. Optimization of extrusion when implementing additive species into the polymeric matrix is a process which must be tightly controlled in order to create materials with reproducible characteristics.

Through the implementation of additives, such as those listed above, the final implant can initiate positive cellular response leading to an effective implant. Incorporation of bioglass spheres has also been shown to have a direct effect on the

Fig. 15 A SEM image
showing the effects of
optimizing the processing
parameters of extrusion.
The result is a homogenous
microstructure without
defects, such as porosity
and agglomerations,
typically found in
composite materials. Image
courtesy of Evonik
Industries

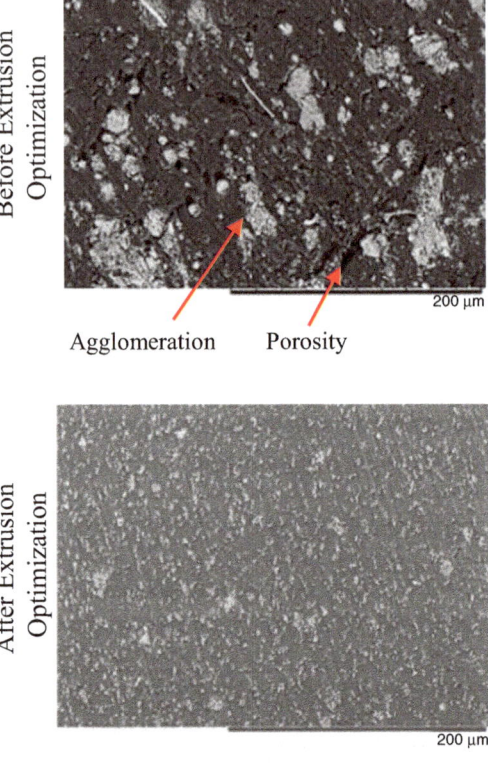

Fig. 16 The injection molding process consisting of a hopper, a single-screw, sequential heating zones, and the mold where the melt is shaped to the final part. Image courtesy of Evonik Industries

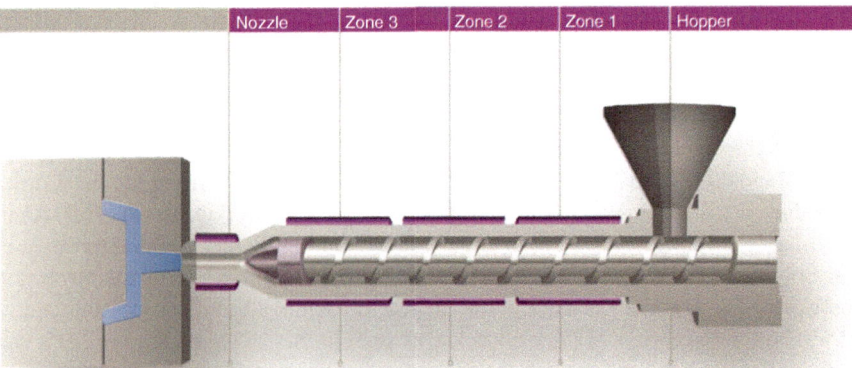

mechanical properties of a PLLA composite with increasing concentration decreasing the mechanical properties [98].

7.2.2 Injection Molding

Injection molding is one of the most widely used manufacturing methods to process thermoplastic polymers into a wide range of orthopedic implants. Products such as interference screws, pins, and suture anchors all require dimensional precision without sacrificing the mechanical and biological properties of the material therefore the processing must be optimized. A typical injection molding process, shown in Fig. 16, uses a reciprocating screw that functions like an extruder's screw for delivering the polymer melt to the mold. The screw action generates sufficient injection pressure to fill mold cavities and produce functional parts. During the injection molding cycle, a measured amount of polymer is gradually melted and metered to the end of the barrel through the action of the revolving screw. Once the melt flow is at the nozzle section of the barrel, it is at its optimal temperature and viscosity to be injected into the mold when the screw pushes forward. The screw must stay in the forward position long enough to maintain mold cavity packing pressure while the parts begin to solidify. Once packing pressure is released the screw retracts and plasticizes material for the next injection molding cycle. Most injection molding machines used for resorbable biopolymers and composites are reciprocating screw type machines. Temperature control of the mold is also critical. If the melted polymer cools quickly it can lead to the formation of a frozen, amorphous layer of skin on the part. The interior of the part cools more slowly and is therefore more crystalline than this frozen surface. Post-processing steps such as annealing can create homogenous crystallinity throughout the part.

The injection molding process is capable of producing parts, such as interference screws, from virgin bioresorbable polymeric materials as well as composite bioresorbable materials. When processing materials using this technique, a number of critical parameters must be considered. First, the material being used must have good flowability in order to adequately fill the mold cavity which may contain high-precision geometric features, such as thin walls. By first understanding the rheological properties of the material to be processed, molds can be designed that allow this mold filling repeatedly. Secondly, the thermal stability of the material must be understood. This consideration is different from the flowability in that thermal stability encompasses a material's ability to withstand processing temperatures without forming crosslinks (chemical or physical) as well as avoid thermal degradation. Third, the moisture level of the material must be closely monitored throughout the injection molding process. Not only can an increase in moisture content affect the process itself, bioresorbable materials may decrease in molecular weight upon exposure to increasing moisture content due to their potential for hydrolytic scission. Lastly, the injection, packing, and holding pressure of the process must be considered. Melt flow and pressure dependence of viscosity are dependent on polymer molecular structure, chain architecture, filler type, filler shape, filler concentra-

Fig. 17 An example of
injection molded screws
for orthopedic implants
using RESOMER®. Image
courtesy of Evonik
Industries

tion, interfacial adhesion between filler and polymer matrix, to name a few. Since
the flowability of the polymeric material as well as their composites are an exponential function of pressure, the pressure must be optimized to not only allow for infill
into the mold but also be high enough to hold the melt in the mold cavity. Degradation
of resorbable polymers and composites during injection molding mainly occurs in
the plasticizing zone where heat exposure is at its highest. Plasticizing time depen-
dents on the amount of polymer or composites to be injected to the mold. Analysis
of the injection molded parts for their molecular weight, mechanical properties, and
microstructure can help fine tune injection molding parameters to control part quality. An example of a common bioresorbable orthopedic medical implant, an interference screw, fabricated through injection molding are shown in Fig. 17.

Although injection molding is a complicated process requiring an good overall
understanding of the a number of variables, such as rheological properties, thermal
properties, structural properties, injection pressure, holding pressure, mold cooling
temperature and cooling rate, runner and gate design, it has been used widely to
produce orthopedic implant devices with complex designs. For instance, injection
molding a PLLA-based interference screw displayed the effect of cooling rate on
the resulting crystallinity of the implant, dropping to nearly half the pre-injection
value [99]. The degradation and biological characteristics of the screw remained
unchanged even with the change in crystallinity with cell viability remaining above
95% for the duration of the study.

Innovations in developing new materials suitable for the injection molding process have driven the development of new products based on resorbable biopolymers
and their composites [100]. Advances in injection molding technology play another
important role to produce innovative products. Some examples of these types of
technology are the use of gas (mainly nitrogen) assist injection molding which
reduces the overall weight of the fabricated product, micro-injection molding techniques that enable fabrication of miniaturized products with maximum accuracy
and precision, and multi-component injection molding that can combine various
materials together producing products that cannot be achieved by conventional
extrusion compounding [101].

7.2.3   Compression Molding

Compression molding is a fabrication method which can be used to create a wide range of geometries for implant devices such as trauma fracture repair plates and acetabular liners for hip replacement systems. This relatively inexpensive process, compared to injection molding, consists of two hot plates which serve to melt the material to be molded into final geometry of the part. The bottom plate holds the base of the mold, which typically contains a hole in the shape of the desired part. The top of the mold contains a plunger-like protrusion in the same shape, which presses the material into the mold during the compression process. The pellets or granules are poured into the bottom half being sure to add enough material to the mold so that it overflows with material once molten thus ensuring a fill of the mold. After adding the material, the top plate is pressed the top of the mold into the bottom of the mold and bottom plate, and both plates are heated to the same temperature to ensure even heat distribution. The continuous supply of pressure to the material melt promotes the removal of internal porosity as well as allows the appropriate melt time to remove residual stresses. Once the dwell time for the melt has been reached, the compression mold is generally slowly cooled if elevated mechanical properties are required as quenching the mold results in a lower degree of crystallinity and thus decreased mechanical properties. While the mechanical properties of an implant device fabricated through compression molding are generally higher than those found with injection molding, the long dwell time of the material within the heated mold is a cause for concern specific to bioresorbable materials. Materials such as those that are lactide and glycolide based are prone to premature degradation if they are exposed to levels of intense thermal processes for prolonged periods of time and must therefore be considered when selecting an appropriate processing method.

A particular advantage of compression molding compared to injection molding is that the requirement of the material to flow is not as critical of a consideration. While injection molding requires good flow properties of the polymer or composites in order for the material to pass through mold runners, gates, and fill cavity, compression molding does not require this level of material flow. Compression molding is therefore beneficial when processing composites materials which can become problematic with the incorporation of fillers or other reinforcing agents, such as fibers, which typically increase polymer melt viscosity dramatically. Further, compression molded parts usually do not contain residual stresses and therefore can have higher crystallinity and better uniformity. While this can be beneficial, it comes at the cost of longer processing time as the thermal cycling time for the mold to heat and cool is much longer than for injection molding. Further processing factors to be considered include compression pressure, holding pressure, mold temperature, preheating temperature and time, molding temperature, hold time, size of the polymer or composite pellets, cooling rate, and size of the part. Another contrast to injection molding is that compression molding is a slower process and is generally reserved for low-volume production or for proof-of-concept designs for which an investment in expensive injection mold tooling is not feasible [79, 102].

For application in orthopedic implants, compression molding is most commonly used to make fixation plates such as those used in craniomaxillofacial (CMF) procedures or the treatment of distal radius fractures. These compression molding technologies provide an alternative method for combining such resorbable polymers, composites or biologically active agents into products to enhance structural and mechanical properties as well as biological response functions [103]. Recently, high-pressure compression molding was implemented to fabricate a porous composite scaffold for cranial bone regeneration by combining HA with a PLLA polymer matrix. The mechanical properties were elevated and the interconnected porous structure of the plate enhanced cell proliferation resulting in the healing of a calvarial defect *in vivo* [104].

7.3   Novel Methods (Additive Manufacturing)

Additive manufacturing has recently been gaining interest as a method of processing and fabricating implants from bioresorbable materials. Additive manufacturing encompasses a range of manufacturing techniques that involve generating three-dimensional parts by distributing layers of material on top of each other and joining them through heat or other methods from a precursor powder or filament. Among techniques implemented under additive manufacturing, several specialized techniques, such as selective laser sintering (SLS) and fused deposition modeling (FDM), have seen the most use and implementation in the manufacturing of implants. For orthopedic implant applications, additive manufacturing allows for the fabrication of high precision parts, rapid production times, patient specificity, and is able to be used with a wide range of desirable materials. One of the largest advantages of additive manufacturing is its ability to tightly control the geometric specificities of the implant design. For example, by controlling the exterior and interior porosity of the fabricated implant, highly advantageous results in regards to the cellular interaction *in vivo* can be achieved as the porosity not only allows for diffusion of by-products of degradation but also serves to guide the distribution and proliferation of the native cells [105]. The considerations to take into account when using additive manufacturing techniques are different than those of the typical processing methods. Below is outlined an overview of these methods with their recent advancements and future directions in regards to bioresorbable materials for orthopedic applications [106].

7.3.1   Fused Deposition Modelling (FDM)

The process of fused deposition modeling (FDM) is similar to the above mentioned process of extrusion principle. The desired implant material is melted in a chamber (typically a syringe) and then forced through a die in the form of a rapidly cooling filament, as outlined in the schematic process in Fig. 18. The ability of the FDM to

Fig. 18 A schematic view of the FDM process where a heated head serves to melt the filament material and then deposit the melt onto the print platform as directed by the controlling computer. Image provided by BMC Musculoskeletal Disorders, 2012, S. Tefler et al.

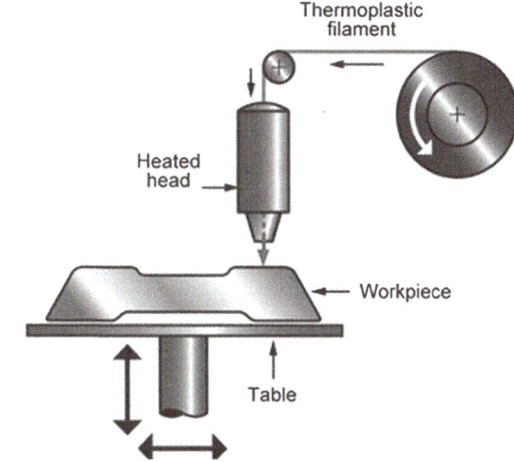

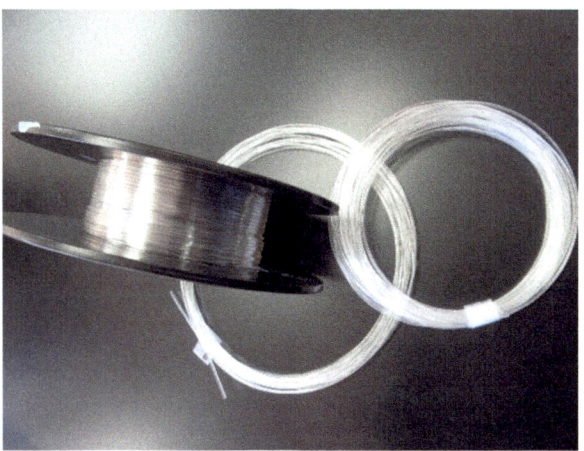

Fig. 19 Representative image of RESOMER® filaments used in the FDM process. Image courtesy of Evonik Industries

translate a computer assisted drawing (CAD) model into a complex, three-dimensional physical product advances the process of typical melt extrusion. The position of the nozzle where the extrudate exits is controlled by a computer that allows for the polymer to be placed at tightly controlled locations as a layer-by-layer product is formed. The starting material is a filament, shown in Fig. 19, and is generally a virgin material although a composite material can also be used as long as the additive species can exit the extrusion die port without causing adverse processing conditions, e.g., clogging. The low start-up cost and the ability to control the directional layering on the resulting parts and its effect on the resultant mechanical response are advantages provided by FDM) technologies that are not typically achievable with conventional methods.

Recent studies have shown cellular response to FDM implants is comparable and in some cases superior to those manufactured using typical processing methods. A PLGA scaffold fabricated via FDM was able to provide the interior porosity and mechanical properties to act as synthetic articular cartilage allowing for a high degree of distribution and growth of chondrocyte cells [107]. Further, the above-mentioned ability to control the directionality of the deposited filaments within the scaffold has shown the ability to guide cells along a proliferation path as they tend to grow along the central axis of the fiber [107, 108]. Additionally, the use of hydrogels laden with living cells has been advanced rapidly in recent years and shows potential in fabricating implants which allow for the rapid production of patient-specific implants as well as alleviate the potential for host-rejection [109].

The advantages offered by FDM show promise for the industrial manufacturing of implants but a more interesting insight is the potential for the fabrication of implantable devices in the operating room. With the relatively low-cost associated with FDM techniques, the possibility for clinicians to conduct a scan of the implant site, create a CAD model of the region of implant, and fabricate an implant design with parameters for a particular patient are becoming closer to a reality.

7.3.2 Selective Laser Sintering (SLS)

Selective laser sintering (SLS) is a newer technology introduced by Carl Deckard and DTM Corporation in 1992 and offers multiple advantages compared to FDM [110, 111]. Whereas FDM creates a product through a controlled geometry melt extrusion process, SLS fuses particulates of material together in the layer-by-layer fashion outlined in Fig. 20 [112]. Specifically, a thin layer of the material in powder

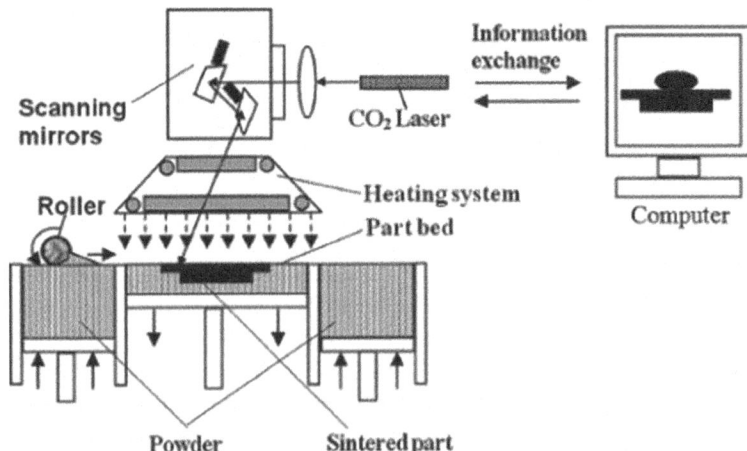

Fig. 20 A schematic view of the SLS (and SLSS) process. A computer-directed laser scans over a layer of particle material. After these particles have been sintered together, a new layer or particles is spread across the print bed and the process begins again. Image provided by the International Polymer Processing Journal of the Polymer Processing Society, **26**(4), 416–423 (2011), C.Y. Hao et al.

form is placed onto a print-bed at which point a laser, typically a CO_2 type, fuses select particles together following the geometry constraints of a CAD model. In contrast to FDM methods, SLS offers much higher resolution and dimensional control of the part geometry as well as a higher quality of the surface properties of the finished product. In terms of economic feasibility, when the SLS technique is used it is often for the production of a finished implant as it is much more expensive than FDM techniques [113].

When selecting to use the SLS technique to fabricate implants, some of the most dominant variables are those relating to the interaction of the laser and the particles. These variables are typically grouped into two umbrella categories: (1) laser related and (2) material related [114]. Since the thermal input in SLS-based techniques occurs through the interaction of the material with a laser, the laser power, as well as scanning size (a variable that can be grouped together with laser power in a term known as laser energy density), is a major controlling parameter. As the laser scans over a section of material, an increase in laser power has been shown to increase the resulting mechanical properties of the implant [114–117]. The same result has also been shown to be brought on by increasing the scan speed of the laser—that is, how long the laser rest on one position before moving to the next. Both of these occurances can be understood by analyzing the localized heating effect on the material as it interacts with the laser. With increasing the laser power, the localized melting of the material is increased. This induces a flowing effect on the material powder and results in a more efficient melt and subsequent fusion of the powders. By not allowing the laser to irradiate the powder for prolonged periods of time, the breakdown of the material structure is reduced allowing for, in the case of polymeric powders, the molecular weight to remain nearly constant and resultant mechanical property to remain largely unchanged. Control of the cooling rate of the powder bed allows the crystallinity of the final part to be controlled and therefore the mechanical and degradation properties adjusted.

The consideration for powder material, the second umbrella group, must also be taken into account. First and foremost, the material must be able to interact with the irradiation source (it must be capable of being melted by the laser). Further, the powder size must be tightly controlled as the larger the powder diameter becomes, the lower the resulting mechanical properties will be [114, 117–119]. Additionally, the material must not degrade prematurely when interacting with the laser—a consideration that has required advanced techniques to be developed for the processing of bioresorbable materials via SLS methods). Examples of SLS printed samples is shown in Fig. 21.

Since the localized heating encountered when using SLS methods results in an instant exposure to highly elevated temperatures, the most commonly used implant-grade materials processed through this method are those which have excellent thermal stability. Included in this group are metallic materials and high degradation temperature polymers such as polyamides and polyaryl ether ketones [103, 120–124]. Regarding bioresorbable polymers, the use of PLA has been investigated as a potential material for SLS fabricated implantable devices [125–128]. Of particular interest in the field of bone tissue engineering, the incorporation of calcium phosphate

Fig. 21 Example of printed samples via SLS using RESOMER® material. Image courtesy of Evonik Industries

particles into the SLS fabricated implants has shown the ability to provide for improved cell viability which leads to increased new bone growth as well as cell migration and proliferation [129, 130]. PLLA and PLLA/HA composites have been successfully printed using the SLS technique with promising results.

Although the thermal exposure time of the material used in SLS is minimal, the risk of premature degradation of bioresorbable polymers exists when any type of thermal processing method is used for these materials. In order to eliminate this risk, the development of an improved SLS method known as surface-selective laser sintering (SSLS) was developed. This process is identical to SLS with only the composition of the material to be processed varying. With SSLS, a thin, biocompatible coating is applied to the surface of the powders which absorbs irradiation. When the laser interacts with the material, only localized surface heating results allowing the part to be fused together without damaging the bioactive interior material [131]. Although the up-front costs associated with SSLS (and SLS) printers limits their potential use in clinical applications, manufacturers of medical implant devices can implement this technique in order to rapidly produce limitless designs and commercial volumes of implant from a wide range of materials in order to meet the growing needs of regenerative medicine applications.

7.4 Other Methods

Spinning polymer fibers is another approach to generate bioresorbable scaffolds. Spinning itself is not necessarily a novel bioresorbable processing technique. Sutures and textile reinforcement patches are often made from PLA, PGA, and PLGA copolymer fibers and make up one of the most widely-used and high-volume forms for these materials in clinical applications [100, 132]. Traditional spinning methods include solution-spinning (also called wet-spinning), dry-spinning, and melt-spinning can be used to create composites with unique mechanical, thermal, and degradation properties. Solution spinning involves dissolving a polymer in a solvent and spinning the solution in a chemical bath. The fiber precipitates and solidifies as it is lifted out of the bath. Dry-spinning follows a similar process, but once the polymer is dissolved in a solvent, the fiber forms through the evaporation of the solvent while spinning by subjection to an air stream or stream of inert gas. Finally, melt-spinning occurs when a polymer is melted and spun with the fibers solidifying by a cooling process after spinning. This method is most often used to create fibers used for sutures, as surgical reinforcement textiles, and to fabricate 3D scaffolds with their final properties depending largely on the initial molecular weight of the polymer [9, 100, 132, 133].

7.4.1 Electrospinning

Electrospinning is a novel method of creating bioresorbable polymers fibers. Although electrospinning was first developed over one-hundred years ago, its use has only gained popularity for biomedical applications within the last 15 years [100]. In the electrospinning process, a polymer solution is extruded from a needle into an electrostatic field. A positive electrical charge is applied to the needle tip from a high voltage power supply thus charging the exiting polymer solution. The solution is then drawn through the electrostatic field to a grounded collecting substrate. The resulting fibers can be micro- or nano-scale with controlled orientation based on the method of collection. If aligned fibers are desired, a rotating collection substrate is employed, otherwise a stationary collection substrate is used. The fiber diameter is controlled by multiple processing variables including solution viscosity, rate of the solution exiting the needle, the amount of electrical charge applied to the solution, and the distance to the collection substrate [100]. Electrospinning has quickly become a leading processing technique for developing bioresorbable polymer fibers for scaffolds largely due to the low start-up costs as well as its versatility in producing nano- and micro-sized fibers from various polymer materials.

PLGA scaffolds containing Santa Barbara Amorphous 15 (SBA15), a type of silica nanoparticle, were fabricated and subsequently crosslinked through the introduction of glutaraldehyde vapor [134]. Human MSCs were then seeded onto the membranes. The resultant membranes successfully achieved a porous structure with even SBA15 distribution facilitating good cell proliferation and adhesion with excellent biocompatibility. These electrospun membranes were determined to hold promise in bone repair and regeneration applications. A similar study seeded hMSCs onto PLGA scaffolds containing a graphene oxide (GO) additive [74]. They found that GO-PLGA scaffolds could be easily and repeatedly produced with electrospinning, which possessed a smooth surface and porous structures that promoted cell proliferation and adhesion.

As stated previously, the effect of the implant geometry has an appreciable effect on the biological response. In the electrospinning technique, the geometry is typically varied by changing the fiber diameter, amount of fiber deposited, or the orientation of the fibers. Variation of the solvent used during processing can also have an effect as a more volatile solvent will vaporize more rapidly and can result in the presence of pores. Electrospun meshes of aligned polycaprolactone (PCL) fibers and randomly-oriented PLGA fibers resulted in randomly-oriented cell proliferation on the PLGA regions and higher alignment on the PCL regions, demonstrating a typical trend in this type of scaffold fabrication. Cells will proliferate along the long axis of the fibers and can be controlled by varying the collecting apparatus [135]. 3D composite scaffolds with aligned PCL fibers and randomly oriented PLGA and PLGA/PCL fibers were also fabricated and showed promise in cell-proliferation but failed mechanically due to the stress concentrations in the aligned PCL regions. Additional studies have focused on electrospinning composite material scaffolds using a novel approach for combining the two polymers. In this case, PLLA scaffolds seeded with bone tissue were compared to PLLA/PEO scaffolds for their ability to enhance bone formation. The PLLA/PEO scaffolds were created by simultaneously electrospinning PLLA and PEO using a custom-made, dual-electrospinning apparatus. The final PLLA/PEO bone-tissue scaffolds possessed a mineralized surface with enhanced porosity and pore size, which demonstrated potential for this scaffold design in bone formation applications [136]. Further studies on fabricating composite scaffolds through electrospinning coated PLGA nanofibers with Willemite (Zn_2SiO_4) powder, seeded cells demonstrated the potential for using ceramic bioglass to support bone regeneration and reconstruction in orthopedic applications [81]. Another study investigated the biologic activity of electrospun composite scaffolds containing β-TCP, siloxane-containing vaterite (SiV), and PLLA. Histological studies on the composite materials, after seeding with rhBMP-2, indicated successful promotion of new bone growth and effective delivery of the rhBMP-2 protein [137]. Electrospinning has already demonstrated vast potential in the processing of bioresorbable polymers for orthopedic applications. Newer refinements to the electrospinning process show additional great potential for future development and commercialization of this process.

7.5 Effect of Post-Processing

After products have been fabricated from bioresorbable polymers, they often go through post-processing steps prior to commercial release. These post processing steps may include annealing and/or sterilization.

7.5.1 Annealing

An annealing treatment is frequently utilized for implant materials, such as PLGA polymers and HA/PLGA composites, in order to increase the crystallinity and therefore the mechanical [138–140]. Theoretically, energy is required as driving force to help crystalline region grow. Further, the growing rate of crystallinity is dependent on the addition and packing of other polymer chains and is governed by thermal energy. On one hand, annealing temperature should be above T_g which provides sufficient energy for polymer chains to form crystalline regions. On the other hand, neither the dwell time nor temperature should cause degradation of polymers in the annealing process. Significantly different degradation performances of PLGA materials after annealing at 115 °C for 60 min have been seen [138]. Annealing can increase the crystallinity and retard degradation, to a certain extent, which is relevant during radiation as well as the hydrolytic degradation processes [140].

7.5.2 Sterilization

Since bioresorbable implants are used for long-term implantation, it is required to sterilize them before application. Currently, there are three sterilization methods that are frequently used in biomedical field:

Radiation sterilization: Gamma (γ) radiation is preferred for sterilizing orthopedic implants since it can penetrate into the materials and eliminate bacterial species via local elevated temperatures without leaving any residues. Considering its cost effectiveness, this method is regarded as the standard sterilization method. However, due to the high energy from gamma sterilization, the risk of decreasing the molecular weight always rises concerns with further applications. Shoichet and colleagues found 50% decrease of molecular weight after treating PLGA materials at a dose of 2.5 Mrad [141]. Even though dry ice bags are usually applied couple with gamma irradiation to avoid the high temperature, the potential degradation is still addressed as a disadvantage for this method.

E-beam: Compared with gamma radiation, accelerated electron irradiation has also been investigated. As a novel method, e-beam appears to be more practical for bioresorbable materials due to its lower energy since the loss of molecular weight is not as significant as γ-irradiation. However, due to the high density of the materials, electron radiation can be easily absorbed before penetrating throughout the material and therefore may result in incomplete irradiation. Even though the energy of e-beam is lower than γ-irradiation, a slight decrease in the polymeric molecular weight can still be observed for the molecular weight leading to a slight drop in mechanical properties. For example, the utilization of e-beam to irradiate PLGA and PLLA polymers at the intensity of 50 Mrad exhibited thermal and morphological changes, such as a decrease in crystallinity [142]. In order to avoid these types of morphological changes, most irradiative treatments are done with doses ranging from 25 to 33 kGy.

Gas (ethylene oxide) sterilization: The use of ethylene oxide provides more mild conditions for sterilization and is widely used for lactide and glycolide based materials. However, this process requires long term degassing after sterilization to remove ethylene oxide residuals from the implant. In addition, the ethylene oxide can react as a plasticizer in the polymer which can affect the materials property and structure. Even though the molecular weight is not significantly affected in this method, the deformation of the material structure is influenced. Up to 60% shrinkage was reported earlier after ethylene oxide sterilization for 2 h followed by 15 h degassing [141].

8 Current Applications

Bioresorbable lactide and glycolide based implants have shown proven efficacy and performance over the years and have been widely used in orthopedic applications and sports medicine. Applications of these materials can be found from head-to-toe throughout the human body as seen in Fig. 22 and further outlined in Table 2.

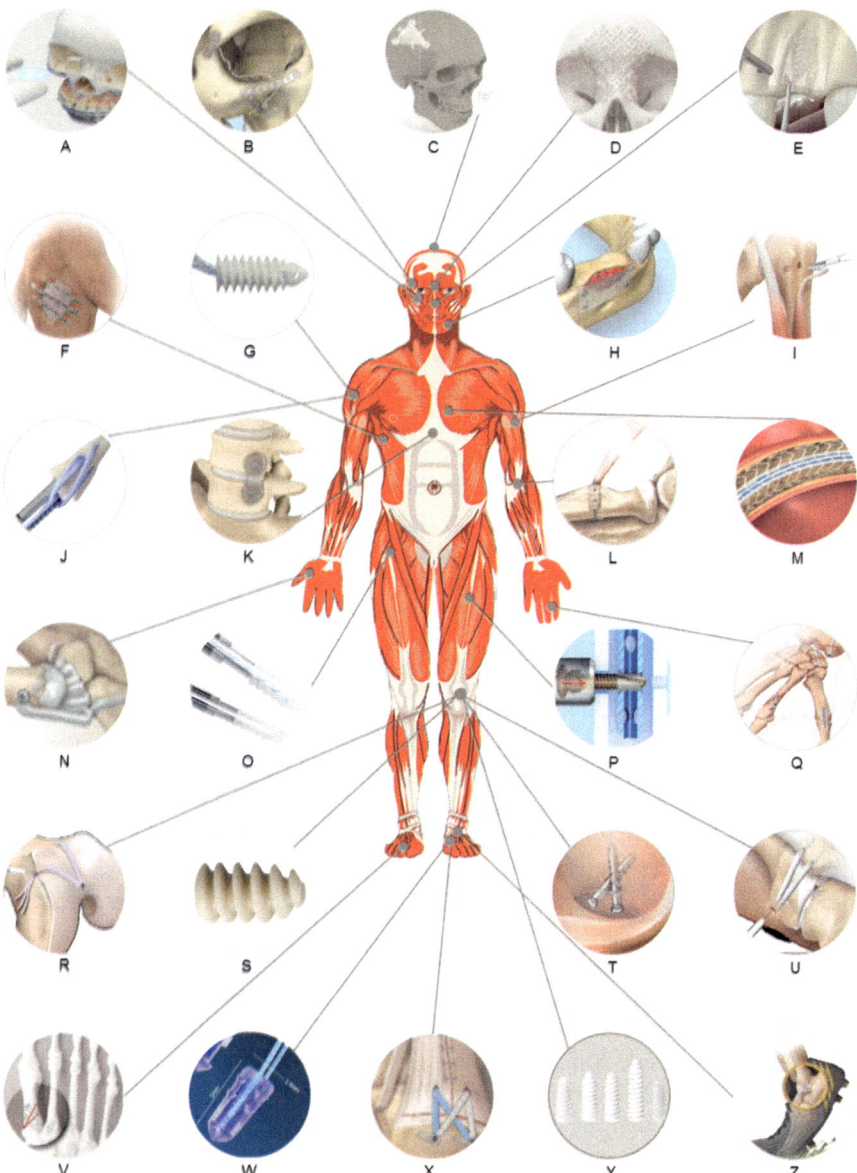

Fig. 22 Overview of typical bioresorbable implants used from head to toe. Images courtesy of respective companies outlined in the table below

Table 2 An overview of some commercially available bioresorbable implants

	Company	Product	Composition	Resorption	Ref.
A	Depuy Synthes	RAPIDSORB® Preshaped Orbital Floor Plate	85:15 PLGA	12 months	[143]
B	KLS Martin Group	Resorb x® and Resorb xG Plates	PDLLA, 85:15 PLLA/PGA	12–30 months	[145]
C	Stryker®	Delta System Plates	85:5:10 PLLA/PDLA/PGA	8–13 months	[144]
D	Depuy Synthes	RAPIDSORB® Contourable Mesh	85:15 PLGA	12 months	[143]
E	KLS Martin	Resorb x® Alveolar Protector	PDLLA	12–24 months	[145]
F	ACUTE Innovations®	BioBridge® Resorbable Chest Wall Stabilization Plate	70:30 PLDLA	18–24 months	[146]
G	Smith & Nephew, Inc.	TWINFIX® Ultra HA Suture Anchor	PLLA/HA	–	[147]
H	KLS Martin Group	Resorb x® Membrane	PDLLA	12–24 months	[145]
I	Arthrex	BioComposite™ SwiveLock® Tenodesis	85:15 PLLA/β-TCP	–	[148]
J	Depuy Synthes	Biocryl Rapide® Suture Anchors	70:30 PLGA/β-TCP	24 months	[37, 194]
K	Inion Inc.	S-2™ Biodegradable Anterior Thoraco-Lumbar Fusion System	PLLA and PDLA	24–48 months	[149]
L	Arthrex	BioComposite™ Distal Biceps Implant System	85:15 PLLA/β-TCP	–	[148]
M	Abbott	Absorb GT1™ Vascular Stent	PLLA/HA	36 months	[150]
N	Arthrex	BioComposite™ Tenodesis Screw Master Set	85:15 PLLA/β-TCP	–	[151]
O	Smith & Nephew, Inc.	OSTEORAPTOR® Suture Anchors	PLLA/HA	–	[152]
P	Depuy Synthes	Resorbable Sleeve for Screws Stabilizing Intramedullary Nails	70:30 PLDLA	18–24 months	[153]
Q	Arthrex	Micro-Compression FT Screws	PLLA	–	[148]
R	Arthrex	BioComposite™ Pushlock Suture Anchor for Patellofemoral Dysfunction	85:15 PLLA/β-TCP	>24 months	[67]
S	Depuy Synthes	MILAGRO® Advance Interference Screw	70:30 PLGA/β-TCP	24 months	[154]
T	ConMed	SmartNail®	96:4 PLDLA	–	[155]
U	Arthrex	BioComposite™ SwiveLock® for Medial Patellofemoral Ligament Procedures	85:15 PLLA/β-TCP	>24 months	[156]
V	Wright Medical	RFS™ Pins and Solid/Cannulated Screws	85:15 PLGA	24 months	[157]
W	Medtronic	Polysorb™ 2 mm Soft Tissue Anchor System	18:82 PGA/PLA	12–15 months	[158]

(continued)

Table 2 (continued)

	Company	Product	Composition	Resorption	Ref.
X	Arthrex	BioComposite™ SwiveLock® for Achilles SpeedBridge®	85:15 PLLA/β-TCP	>24 months	[159]
Y	Arthrex	BioComposite™ Interference Screw	70:30 PLDLA/ BCP	>24 months	[160]
Z	Arthrex	BioComposite™ SwiveLock® & BioComposite™ SutureTak Anchors	85:15 PLDLA/ BCP	>24 months	[161]

With advancements in composite bioresorbable materials due to the introduction of calcium-based inorganic additives, the osteoconductivity of these materials has been improved along with the added benefit of imparting a bone-like modulus to the material both of which promote faster healing times and better patient outcomes.

8.1 Craniomaxillofacial (CMF)

The term craniomaxillofacial (CMF) is an umbrella term that encompasses any tissue system located in the head. While the procedures performed for this specialty are varied, they can be grouped into those arising from two sources: (1) Anomalies resulting from genetic mutations or vitamin deficiencies (particularly folic acid) in the mother during pregnancy and (2) trauma induced defects. Among the treatable conditions occurring at birth, cleft lip, craniosynostosis, and deformational plagiocephaly are some of the more commonly occurring abnormalities [162]. These applications as well as those occurring from trauma are applications in which bioresorbable materials are highly advantageous. Since most of these applications do not require the implant to be placed under an elevated mechanical load, bioresorbable materials used for these treatments have focused on enhancing the biological response and ability to promote healthy bone regeneration without causing any adverse side effects upon degradation. The implants available for these types of applications include those such as screws, plates, and meshes with an example of commercially available PLLA based fixation plates shown in Fig. 23. The particular implants shown are not only bioresorbable also offer a unique advantage in that they are also capable of being shaped before implantation in the operating room by forming the plate to a desired geometry in hot water to ensure an ideal fit with the patient.

Among the most common applications for which a bioresorbable plate or mesh is used in CMF treatments is in the remediation of mandibular or orbital bone defects. For instance, the implantation of a resorbable mesh in the reconstruction of the orbital floor was shown to be a safe and effective method for reconstruction with full resorption occurring by the time of the two year follow-up period [163]. Another particular application in which a bioresorbable, custom-fitted mesh plate was used to treat enophthalmos (posterior displacement of the eye—the affected eye "sinks" posteriorly due to a loss of bone volume) which further shows the promise of these

Fig. 23 Plates and meshes such as the ones pictured above from DePuy Synthes—A Johnson & Johnson Company, are used extensively in craniomaxillofacial regeneration treatments. Image courtesy of DePuy Synthes

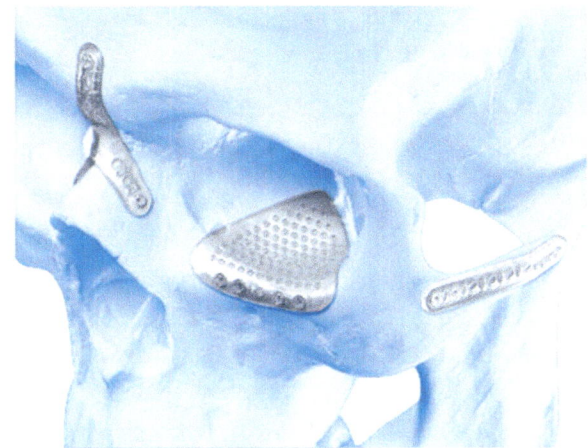

materials compared to typical metallic implants [164]. In this case, a PLGA based fixation system (RAPIDSORB®, DePuy Synthes) was compared to a titanium mesh on the effectiveness of repairing the defect in 78 patients with approximately 50% of the patients receiving the bioresorbable plate. The follow-up time for patients with a resorbable implant ranged from 3 months to 4 years with an average of 2 years while the follow-up time for those who received a titanium implant ranged from 7 months to 5.6 years. The study revealed no statistical difference between the patients who received the bioresorbable plate compared to those who received the titanium plate. A smaller study of ten patients requiring treatment and fixation of maxillofacial fractures was undertaken to analyze the efficiency of bioresorbable plates in this application showed similar results with all patients having fracture reduction and full degradation of the implant [165]. Building on the advances in processing technologies, one investigation of a 3D printed composite scaffold mesh of PCL/PLGA/β-TCP (20%:60%:20%) analyzed its efficiency in bone regeneration and osseointegration compared to a non-resorbable titanium mesh [166]. The 8-week histological sections of the control group which received no implant, the bioresorbable implant, and the titanium implant are shown in Fig. 24. After 8 weeks, the bioresorbable composite mesh was shown to perform comparably to the titanium mesh—even surpassing it in the formation of new bone area by approximately 50%.

8.2 Sutures and Suture Anchors

Recently, implant developments in the repair of large upper body joints, such as the treatment of torn rotator cuffs, have advanced to offer more ideal healing and increase patient satisfaction. One such device is an inflatable, bioresorbable balloon, shown in Fig. 25, which can be implanted above the torn rotator cuff during the

Control 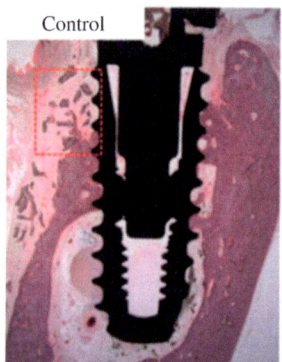 Bioresorbable 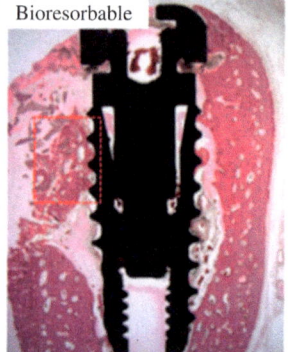 Titanium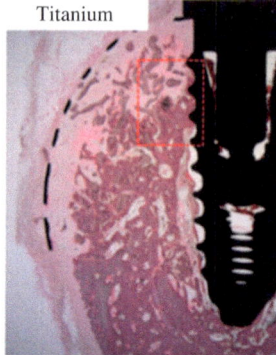

Fig. 24 A comparison of a 3D printed PLGA mesh to a typically used titanium mesh showed that the bioresorbable mesh was able to promote bone growth more effectively than the titanium mesh and the control sample which received no mesh. Image provided by Polymers, **7**(10) 1500 (2015), J.H. Shim et al.

Fig. 25 The OrthoSpace InSpace™ inflatable, bioresorbable balloon for use in torn rotator cuff repair. Image courtesy of OrthoSpace

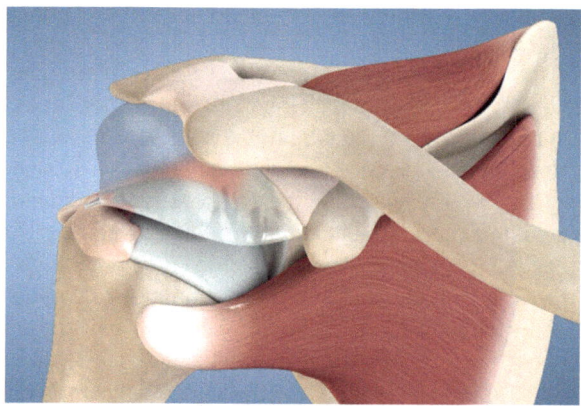

healing process [167]. This balloon is placed in order to allow for smooth, frictionless gliding of the humeral head against the acromion thereby lowering patient discomfort. This reduction in pain during the healing process allows the patient to undergo the physiotherapy required to restore shoulder balance as well as range of motion. While this balloon implant has its benefits, in place in order to heal, the torn tissues must be held together for an extended amount of time. Sutures are commonly used to achieve this by serving as a tether to hold to soft tissue in place. An ideal suture material must be able to hold to regenerating tissue system in place long enough to allow for full healing at which point the suture would dissolve away and be resorbed by the body. These types of sutures have been produced by Ethicon since the early 1970's with absorption times ranging from 21 to 238 days offering a wide-range of timeframes for various applications.

Holding these sutures in place, suture anchors are used when an affected soft tissue such as a tendon or muscle, needs to be kept stable and in contact to allow for

Fig. 26 An illustrative example of a Smith & Nephew, Inc. suture anchor used in the repair of rotator cuff trauma. Image courtesy of Smith & Nephew, Inc.

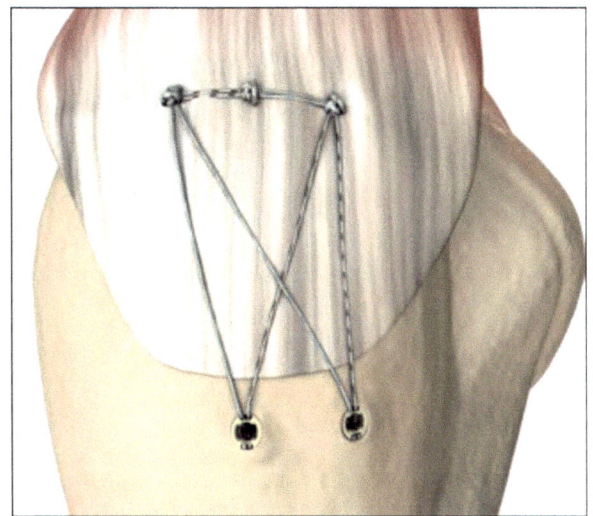

healing. In the instance of rotator cuff repair, these suture anchors are embedded into the local bone while the sutures are placed on each side of the torn ligament or muscle. Generally, a suture anchor will be placed in the bone near the torn tissue with suture material already within the anchor system. After placement, the sutures will be passed through each side of the torn tissue and then tightened so that each face of the torn tissue is in contact. The presence of the suture anchor not only makes the procedure more efficient and stable, it also aids in the healing of the torn tissue since the torn sections are continuously held in place near one another. Figure 26 shows a typical setup for the use of suture anchors in arthroscopic shoulder repair.

Although biocompatibility is an important factor to consider with all implant materials, in the case of suture anchor implants, an added factor of anchor stability must also be considered. This stability insures that the sutures remain in the local area of the torn tissue allowing for full regeneration. If the anchor does not have adequate stability at the implant site, the torn tissue could be subjected to tensile or shear forces before it is fully healed resulting in a failed procedure and requiring a secondary, revision procedure to reinitiate the healing process for the tissue. Since the stability of the anchor in the bone near the torn tissue is critical for achieving desired patient outcomes, companies such as Smith & Nephew, Inc. have fabricated PLLA/HA composite suture anchors such as the TWINFIX® Ultra HA Suture Anchor with improved bone response and a tailored degradation rate [168]. As stated previously in this chapter, by addition of calcium-phosphate based materials to polymer matrices, the bioactivity of the resulting composite is enhanced compared to the virgin polymer. Due to the addition of HA to the PLLA matrix, bone was able to infiltrate the suture anchor and ultimately overtake the screw space after full resorption of the screw. Further development of this controlled resorption with bone ingrowth has also been found in Smith & Nephew, Inc.'s REGENESORB™

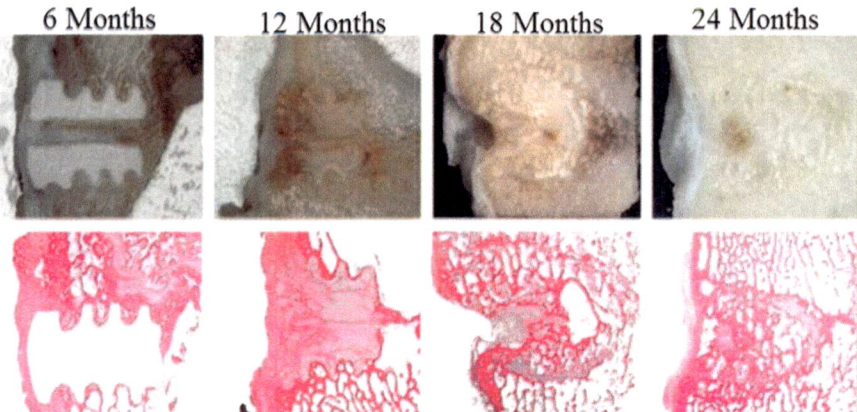

Fig. 27 The bioresorption and osteoconductivity of Smith & Nephew Inc.'s HEALICOIL™ REGENESORB suture anchor resulted in new bone growth at the implant site leading to enhanced fixation. Image courtesy of Smith & Nephew, Inc.

suture anchor. This particular suture anchor is able to promote long-term and short-term osteoconductivity due to the addition of calcium sulfate, the benefits which were outlined in above sections. Further, this suture anchor has an open-architecture allowing for the regenerating bone to grow into the inner section of the screw (known as the screw tunnel) as it is resorbed without sacrificing mechanical performance (e.g., pull-out or insertion strength). The resorption of the screw and accompanying bone ingrowth is shown in Fig. 27.

8.3 Interference Screw

One particular application that has seen wide clinical acceptance and thus accelerated market growth for bioresorbable materials is the use of bioresorbable interference screws in reconstructive surgery of the anterior cruciate ligament (ACL). Typically, this procedure is required when the ACL has been torn after being exposed to rapidly changing shear forces that can occur during physical activity. After taking an autograft from the patient (typically the semitendinosous tendon from the hamstring) the interference screw is used to hold the transplanted tendon in place to allow for full patient recovery. Previous materials that were used were either metallic or biostable materials, but the choice of bioresorbable materials as interference screws is becoming widely popular for more active patients. For this particular application, the advancements in both the ability to increase the mechanical properties of bioresorbable materials as well as the ability to prolong their degradation time is required positioning PLA a particularly advantageous choice [169–171]. With an ACL surgery being performed every five minutes in the United States alone, the need for implant success and post-procedure patient quality of life is evident. The advantage of bioresorbable screws however remains to be the lack of need for

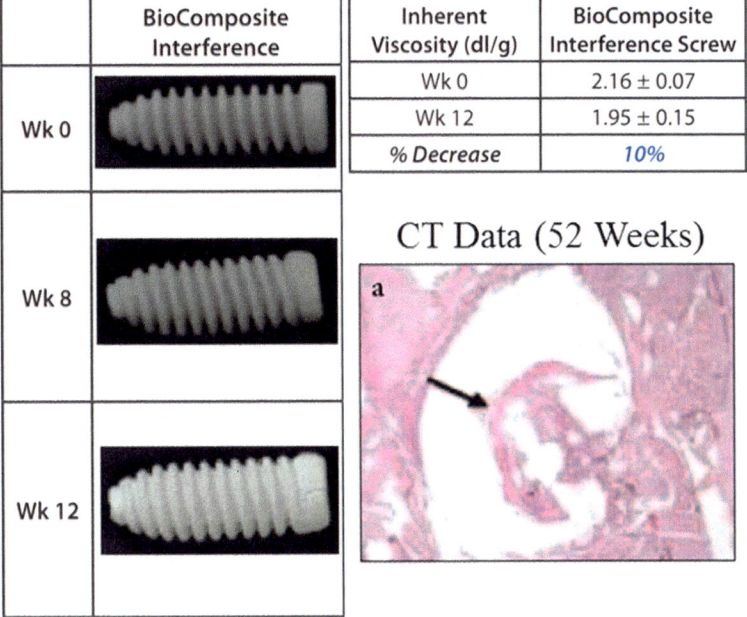

	BioComposite Interference	Inherent Viscosity (dl/g)	BioComposite Interference Screw
		Wk 0	2.16 ± 0.07
Wk 0		Wk 12	1.95 ± 0.15
		% Decrease	10%
Wk 8		CT Data (52 Weeks)	
Wk 12			

Fig. 28 The Arthrex BioComposite™ screw was able to maintain mechanical properties long enough for new bone growth to overtake the implant site due to the addition of calcium-phosphate based additives. Image courtesy of Arthrex, Inc.

a secondary surgery, the reduction of stress-shielding, and recent advancements to increase osteointegration.

Studies have shown that bioresorbable interference screws have potential to meet or surpass the characteristics of commonly used metallic implant in regards to range of motion, knee stability, and knee functional outcome scores [172]. The results from comprehensive studies on this topic have concluded that the patient outcomes from the use of a bioresorbable screw are not statistically different from those with a metallic screw. Further studies have agreed with this finding indicating that the advantages found in bioresorbable implants and their biological response are an added benefit to both patient and clinician in the application for ACL reconstruction [173].

To further understand the degradation of such implants, an accelerated, *in vitro* study was carried out on the BioComposite Interference Screw from Arthrex, a PLA-based screw with including 30%wt. BCP [174]. The BioComposite screw was able to retain a majority of its initial mass until week 12, at which point approximately 35% of its mass was lost confirming results from a follow up biological study where the BioComposite screw was able to last long enough to promote bone tissue formation with little to no inflammatory response at which point expected degradation was shown to occur [160]. The degradation characteristics, timeline, and the *in vivo* response at 52 weeks for the BioComposite™ Interference Screw is shown Fig. 28 showing new bone formation (black arrow) within the screw site.

This controlled degradation is highly beneficial in this application as the ingrowth of bone tissue into the interference screw region allows for the native fixation of the implanted tendon to occur resulting in effective patient outcomes once the bioresorbable screw is completely degraded. The selection of polymer as well as the inclusion of osteoconductive calcium-phosphate based additives was shown to increase the osteoconductivity of the implant as well as lead to a highly adjustable rate of degradation.

8.4   Distal Radius Plate

Distal radius fractures are among the most commonly occurring musculoskeletal injuries with nearly 600,000 occurring annually in the United States alone increasing the need for treatment methods and implants which offer quality patient outcomes with improved patient satisfaction [175]. These fractures are occasionally referred to as wrist fractures due to their close proximity to the wrist joint, but are in fact a break in the radial bone of the lower arm and are often treated by immobilization of the fracture. Although this immobilization can be treated without surgery, more complex cases require intraoperative placement of a plate to stabilize the bone and allow regeneration of the broken sections. Typical implants are metallic and have shown to have sufficient efficacy in the treatment of such fractures, but do require a second surgery for removal with patients occasionally reporting a sensation of foreign body or irritation at the implant site and wrist joint, respectively.

In order to show bioresorbable materials as being able to meet the same regenerative efficacy as metallic implants while adding the advantage of no secondary surgery, a direct comparison of distal radius fracture fixation and regeneration was made between a commonly used metallic plate and a PLLA-based resorbable implant to measure effectiveness and patient satisfaction from each implant [176]. Not only were no post-healing complications such as infection or carpal tunnel reported, the patients were overall more satisfied with the bioresorbable implants over the metallic implants. Representative images of the treated sites before and after implantation of both types of implants are shown in Fig. 29. The PLLA implant was resorbed by the body at the follow-up time of 2.5 months over which period it was able to steadily increase the transmission of mechanical loads required during healing thus allowing the healed bone to be restored to its original functionality and density. This result of natively healthy and dense bone regrowth is not always achievable when metallic implants are used for orthopedic implant devices, a result of stress-shielding which has been highlighted earlier in the chapter. Further investigations into the use of bioresorbable plates and fixation devices have shown similar results indicating promise for their use in future treatment of defects such as these [177, 178].

Fig. 29 The PLLA fixation plate was able to achieve comparable results to the metallic plate in the treatment of distal radius fractures. Image provided by the Journal of Dental Sciences, **8**, 44–52 (2013), I.L. Chang et al.

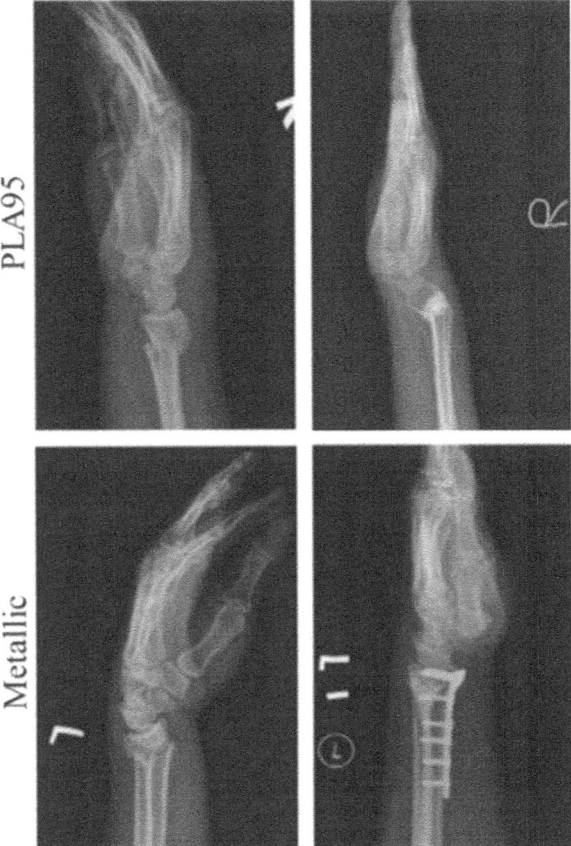

9 Regenerative Medicine

Regenerative engineering is a multi-disciplinary field that falls at the crossroads of stem cell research, materials science, molecular biology, and tissue engineering with the overall goal to regenerate damaged tissues as complex as entire organs and limbs [100, 179, 180]. Orthopedic regenerative engineering, specifically, focuses on the regeneration of bones, ligaments, and cartilage through the use of biomaterials that are naturally bioactive or through composite materials, such as the above high-lighted PLA/PGA/PLGA and inorganic additive materials that stimulate a positive response from native tissue and therefore promote tissue regeneration. These lactide and glycolide based materials are ideal implant materials in terms of biocompatibility and resorbability and are becoming more efficient in their biological response and mechanical properties as researchers develop novel approaches to modifying these polymers in order to promote tissue and bone regeneration once implanted [23, 74, 77–79, 81, 94, 129, 131, 134, 136, 137, 181–184].

Bone remodeling is a critical healing process within the human body, yet adult tissue often does not regenerate to full health and thus often require the use of biomaterial scaffolds to improve initial cell adhesion and proliferation resulting in enhanced bone regeneration. The degradation rates of these materials also provides a benefit in that adult bone cells tend to proliferate and grow at a much lower rate than young cells. As the degradation of the material can be controlled to match the rate of bone formation, the probability of success for full restoration of natural function with these materials is greatly improved when compared to other, non-bioresorbable implants.

Lactide and glycolide based materials not only offer benefits in the regeneration of damaged bone tissue, they also provide a potential platform for ligament regeneration, requiring scaffolds that differ from bone regeneration scaffolds and also vary based on the intended ligament in need of repair. Specific ligaments undergo different loads in different regions and at different points along the ligament. For example, the ACL has three different stress-strain regions, so scaffolds designed for ACL repair must have adequate biomechanical properties to withstand the stress in these three regions. The osteochondral ligament lacks these distinct phases and connects bone to cartilage, requiring simpler scaffolds than ACL [100, 180, 185]. Laurencin and coworkers created PLLA/fibronectin scaffolds that had higher cell adhesion, proliferation, and mechanical properties that made this material better suited for ACL regeneration than regular PLLA [186]. Sarukawa et al., [187] coated PLA scaffolds with chitosan and found that these scaffolds showed improved cell adhesion and proliferation as well as ECM matrix production, further indicating the benefit of PLA for ACL regeneration.

PLA, PGA, and PLGA-based polymer materials also show further promise in cartilage regeneration, especially for knee articular cartilage and knee meniscus regeneration [100, 188]. As long ago as 1993, studies on cartilaginous cell-polymer constructs with bovine articular chondrocytes on PGA meshes and PLLA sponges have shown that the PGA meshes induced higher cell growth rates and GAG deposition compared to PLLA [1, 189]. The mechanical properties of the scaffold not only provide adequate initial support of the tissue at the implant site, but these mechanical properties have also been shown to promote the regeneration of tissue systems with increased mechanical properties compared to those regenerated from weaker scaffold systems [190]. These types of studies reiterate and solidify the potential for PLA, PGA, and PLGA-based materials in the field of regenerative medicine ranging from applications in orthopedic regeneration to cartilage and meniscus repair.

While the advancement of materials has consistently shown that the future directions of regenerative medicine are promising, the development of the novel techniques highlighted in this chapter also offer advantages for future clinical applications. Specifically, the development of additive manufacturing as an optimized process for bioresorbable materials has shown the ability to rapidly produce the patient-to-patient specific requirements for ideal treatment. For instance, PCL is commonly used as a material for additive manufacturing due to its relative thermal

Fig. 30 Bioresorbable
interbody fusion cage,
lateral gutter implant, and
tracheal splint fabricated
through additive
manufacturing. Image
courtesy of DePuy Synthes

stability and ease of processing. The applications which can be impacted by the implementation of these types of processing methods are numerous with scaffolds being fabricated in any conceivable implant design as shown in the cervical inter-body fusion cage, the lateral gutter implant, and the tracheal implants from DePuy Synthes shown in Fig. 30 [191, 192]. The transition from laboratory to clinical settings is already taking place with the application as in the last few years, multiple lives have been saved using this same type of processing technology [193].

Although we have briefly covered current novel methods of additive manufacturing, a number of medical device manufacturers are pressing the envelope even more in an effort to create more geometrically controlled implants as well as reduce the time taken to produce the implant. The recent and continued advancements in additive manufacturing indicate that the future of regenerative medicine and the ability to create patient-specific treatment will require materials capable of being processed with these methods, such as those highlighted in this chapter.

10 Conclusion

There is an increasing need for advanced implant materials that are capable of improving both patient well-being and clinician efficiency in the field of orthopedic implants. The idea of a material which could be implanted to treat orthopedic defects and then have the capability to self-modify itself in accordance with the progressing needs of the surrounding biological environment is both novel and platform solution for diverse unmet needs both in orthopedic and also medical device industry overall. Lactide and glycolide based bioresorbable materials continue to gain acceptance as a replacement for permanent implants with active research and clinical studies focusing on new applications. With the development seen among bioresorbable materials and composites with increased bioactivity, the upcoming wave of innovation will be one of advanced processing such as additive manufacturing and patient specific implants. While these bioresorbable materials have rapidly evolved in their properties and biological abilities, the advanced processing techniques of these materials are at its infancy and the need for know-how development and application integration has significant potential for the medical community. Research is currently at the edge of taking the next big leap by allowing for the additive manufacturing of bioresorbable patient specific implants while the patient is being treated for life saving conditions. At the cross roads of robotic surgery, additive manufacturing, regenerative medicine and digitization, someday in the near future we can expect a patient specific smart bioresorbable implant material being able to continually assess the biological environment and healing process providing real time updates and monitoring to improve patient and clinician outcomes.

From Sutures to Screws to Stents to Scaffold to Skin someday soon….

Acknowledgement The authors wish to acknowledge the support provided by the medical device industry manufacturers, academic journals, Evonik Industries colleagues, and staff of the Evonik Project House Medical Devices for their help in the completion of this work.

References

1. Freed LE, et al. Biodegradable polymer scaffolds for tissue engineering. Bio/Technology. 1994;12:689–93.
2. Bischoff CA, Walden P. Ueber das Glycolid und seine Homologen. Chem Ber. 1893;26:262–5.
3. Frazza EJ, Schmitt EE. A new absorbable suture. J Biomed Mater Res Symp. 1971;1:43–58.
4. Buchanan FJ. Degradation rate of bioresorbable materials: prediction and evaluation. In: Buchanan FJ, editor. , vol. 1. Sawston: Woodhead Publishing; 2008.
5. Ahsan T, Sah RL. Biomechanics of integrative cartilage repair. Osteoarthr Cartil. 1999;7:29–40.
6. Yaszemski MJ, et al. In vitro degradation of poly(propylene fumarate)- based composite materials. Biomaterials. 1996;17:2127–30.
7. Kellomäki M, Törmälä P. Processing of resorbable poly-α-hydroxy acids for use as tissue-engineering scaffolds. In: Hollander AP, Hatton PV, editors. Biopolymer methods in tissue engineering. Totowa, NJ: Humana Press; 2004. p. 1–10.

8. Yu NYC, et al. Biodegradable poly(α-hydroxy acid) polymer scaffolds for bone tissue engineering. J Biomed Mater Res B Appl Biomater. 2010;93B(1):285–95.
9. Albertsson AC, Varma IK. Recent developments in ring opening polymerization of lactones for biomedical applications. Biomacromolecules. 2003;4(6):1466–86.
10. Gajria AM, et al. Miscibility and biodegradability of blends of poly(lactic acid) and poly(vinyl acetate). Polymer. 1996;37(3):437–44.
11. Seal BL, Otero TC, Panitch A. Polymeric biomaterials for tissue and organ regeneration. Mater Sci Eng R Rep. 2001;34(4):147–230.
12. Pitt CG. Poly-e-caprolactone and its copolymers. In: Chasin M, Langer R, editors. Biodegradable polymers as drug delivery systems. New York: Marcel Dekker; 1990. p. 71–120.
13. Woodruff MA, Hutmatcher DW. The return of a forgotten polymer polycaprolactone in the 21st century. Prog Polym Sci. 2010;35(10):1217–56.
14. Dziadek M, Stodolak-Zych E, Cholewa-Kowalska K. Biodegradable ceramic-polymer composites for biomedical applications: a review. Mater Sci Eng C. 2017;71:1175–91.
15. Miner MR, Berzins DW, Bahcall JK. A comparison of thermal properties between gutta-percha and a synthetic polymer based root canal filling material (resilon). J Endod. 2006;32(7):683–6.
16. Lowry KJ, et al. Polycaprolactone/glass bioabsorbable implant in a rabbit humerus fracture model. J Biomed Mater Res A. 1997;36:536–41.
17. Medlicott NJ, et al. Preliminary release studies of chlorhexidine (base and diacetate) from poly(ϵ-caprolactone) films prepared by solvent evaporation. Int J Pharm. 1992;84(1):85–9.
18. Middleton JC, Tipton AJ. Synthetic biodegradable polymers as orthopedic devices. Biomaterials. 2000;21(23):2335–46.
19. Hakkarainen M. Aliphatic polyesters: abiotic and biotic degradation and degradation products. In: Degradable aliphatic polyesters. Berlin: Springer; 2002. p. 113–38.
20. Sanchez JG, Tsuchii A, Tokiwa Y. Degradation of polycaprolactone at 50 °C by a thermotolerant Aspergillus sp. Biotechnol Lett. 2000;22(10):849–53.
21. Pitt CG, et al. Aliphatic polyesters. I. The degradation of poly(ϵ-caprolactone) in vivo. J Appl Polym Sci. 1981;26(11):3779–87.
22. Darney PD, et al. Clinical evaluation of the Capronor contraceptive implant: preliminary report. Am J Obstet Gynecol. 1989;160(5):1292–5.
23. Lee JW, et al. Bone regeneration using a microstereolithography-produced customized poly(propylene fumarate)/diethyl fumarate photopolymer 3D scaffold incorporating BMP-2 loaded PLGA microspheres. Biomaterials. 2011;32(3):744–52.
24. Ng KW, et al. In vivo evaluation of an ultra-thin polycaprolactone film as a wound dressing. J Biomater Sci Polym Ed. 2007;18(7):925–38.
25. Jones DS, et al. Poly(epsilon-caprolactone) and poly(epsilon-caprolactone)-polyvinylpyrrolidone-iodine blends as ureteral biomaterials: characterisation of mechanical and surface properties, degradation and resistance to encrustation in vitro. Biomaterials. 2002;23(23):4449–58.
26. Rai B, et al. Combination of platelet-rich plasma with polycaprolactone-tricalcium phosphate scaffolds for segmental bone defect repair. J Biomed Mater Res A. 2007;81:888–99.
27. Jin C, et al. Biodegradation behaviors of poly(p-dioxanone) in different environment media. J Polym Environ. 2013;21(4):1088–99.
28. Sabino MA, et al. Study of the hydrolytic degradation of polydioxanone PPDX. Polym Degrad Stab. 2000;69(2):209–16.
29. Goonoo N, et al. Polydioxanone-based bio-materials for tissue engineering and drug/gene delivery applications. Eur J Pharm Biopharm. 2015;97:371–91.
30. Boland ED, et al. Electrospinning polydioxanone for biomedical applications. Acta Biomater. 2005;1(1):115–23.
31. Barnes CP, et al. Nanofiber technology: designing the next generation of tissue engineering scaffolds. Adv Drug Deliv Rev. 2007;59(14):1413–33.

32. Jong SJD, et al. New insights into the hydrolytic degradation of poly(lactic acid): participation of the alchol terminus. Polymer. 2001;42(7):2795–802.
33. Zhang Y, et al. Effects of metal salts on poly(DL-lactide-co-glycolide) polymer hydrolysis. J Biomed Mater Res A. 1997;34(4):531–8.
34. Gajjar CR, King MW. Degradation process. In: Resorbable fiber-forming polymers for biotextile applications, springerbriefs in materials. Berlin: Springer International Publishing; 2014. p. 7–10.
35. Alexis F. Factors affecting the degradation and drug-release mechanism of poly(lactic acid) and poly[(lactic acid)-co-(glycolic acid)]. Polym Int. 2005;54(1):36–46.
36. Makadia HK, Siegel SJ. Poly lactic-co-glycolic acid (PLGA) as biodegradable controlled drug delivery carrier. Polymers. 2011;3(3):1377–97.
37. Nuo W, et al. Synthesis, characterization, biodegradation, and drug delivery application of biodegradable lactic/ glycolic acid oligomers: I. Synthesis and characterization. J Biomater Sci Polym Ed. 1997;8(12):905–17.
38. Tokiwa Y, Calabia BP. Biodegradability and biodegradation of poly(lactide). Appl Microbiol Biotechnol. 2006;72(2):244–51.
39. Reeve MS, et al. Polylactide stereochemistry: effect on enzymic degradability. Macromolecules. 1994;27(3):825–31.
40. Rezwan K, et al. Biodegradable and bioactive porous polymer/inorganic composite scaffolds for bone tissue engineering. Biomaterials. 2006;27(18):3413–31.
41. Li Y, et al. The effect of mechanical loads on the degradation of aliphatic biodegradable polyesters. Regen Biomater. 2017;4(3):179–90.
42. Russias J, et al. Fabrication and mechanical properties of PLA-HA composites: a study of in vitro degradation. Mater Sci Eng C Biomim Supramol Syst. 2006;26(8):1289–95.
43. Wang Z, et al. A comparative study on the in vivo degradation of poly(L-lactide) based composite implants for bone fracture fixation. Sci Rep. 2016;6:20770.
44. Huttunen M, Kellomaki M. Strength retention behavior of oriented PLLA, 96L/4D PLA, and 80L/20D,L PLA. Biomatter. 2013;3(4):e26395.
45. Suganuma J, Alexander H. Biological response of intramedullary bone to poly-L-lactic acid. J Appl Biomater. 1993;4(1):13–27.
46. Ciccone WJ, et al. Bioabsorbable implants in orthopaedics: new developments and clinical applications. J Am Acad Orthop Surg. 2001;9(5):280–8.
47. Bos RRM, et al. Bone-plates and screws of bioabsorbable poly (L-lactide) an animal pilot study. Br J Oral Maxillofac Surg. 1989;27(6):467–76.
48. Athanasiou KA, et al. Orthopaedic applications for PLA-PGA biodegradable polymers. Arthroscopy. 1998;14(7):726–7.
49. Roesler CRM, et al. Torsion test method for mechanical characterization of PLDLA 70/30 ACL interference screws. Polym Test. 2014;34:34–41.
50. Spenciner DB, Jr JM. Effect of thread profile and geometry on mechanical properties of interference screws. Raynham, MA: Johnson & Johnson; 2016.
51. Lipchitz J, Colleran D. Failure torque of bioabsorbable ACL interference screws. In: Endoscopy. Andover, MA: Smith & Nephew; 2007.
52. Kousa P, et al. Initial fixation strength of bioabsorbable and titanium interference screws in anterior cruciate ligament reconstruction. Am J Sports Med. 2001;29(4):420–5.
53. Martin K. SonicWeld Rx a new era in osteosynthesis. cited 2017.
54. Buijs GJ, et al. mechanical strength and stiffness of the biodegradable SonicWeld Rx osteofixation system. J Oral Maxillofac Surg. 2009;67(4):782–7.
55. Babiker H. Bone graft materials in fixation of orthopaedic implants in sheep. Dan Med J. 2013;60(7):B4680.
56. Navarro M, et al. Biomaterials in orthopaedics. J R Soc Interface. 2008;5(27):1137–58.
57. Currey JD. Bones: structure and mechanics. Princeton, NJ: Princeton University Press; 2002.
58. Webster TJ, Siegel RW, Bizios R. Osteoblast adhesion on nanophase ceramics. Biomaterials. 1999;20:1221–7.

59. Bouler JM, et al. Biphasic calcium phosphate ceramics for bone reconstruction: a review of biological response. Acta Biomater. 2017;53:1–12.
60. Wei G, Ma PX. Structure and properties of nano-hydroxyapatate/polymer composite scaffolds for bone tissue engineering. Biomaterials. 2004;25(19):4749–57.
61. Mondal S, et al. Studies on processing and characterization of hydroxyapatite biomaterials from different bio wastes. J Miner Mater Charact Eng. 2012;11(1):55–67.
62. LeGeros RZ. Calcium phosphates in oral biology and medicine. Monogr Oral Sci. 1991;15:1–201.
63. Daculsi G, et al. Formation of carbonate-apatite crystals after implantation of calcium phosphate ceramics. Calcif Tissue Int. 1990;46(1):20–7.
64. Heughebaert M, et al. Physicochemical characterization of deposits associated with HA ceramics implanted in nonosseous sites. J Biomed Mater Res. 1988;22(Suppl 3):257–68.
65. Daculsi G, et al. Transformation of biphasic calcium phosphate ceramics in vivo: ultrastructural and physicochemical characterization. J Biomed Mater Res. 1989;23(8):883–94.
66. Sheikh Z, et al. Biodegradable materials for bone repair and tissue engineering applications. Materials. 2015;8(9):5273.
67. Khojasteh A, et al. Development of PLGA-coated β-TCP scaffolds containing VEGF for bone tissue engineering. Mater Sci Eng C. 2016;69:780–8.
68. Roh H-S, et al. In vitro study of 3D PLGA/n-HAp/β-TCP composite scaffolds with etched oxygen plasma surface modification in bone tissue engineering. Appl Surf Sci. 2016;388:321–30.
69. Ogose A, et al. Comparison of HA and b-TCP as bone substitutes after excision of bone tumors. J Biomed Mater Res B Appl Biomater. 2005;72(1):94–101.
70. Ebrahimian-Hosseinabadi M, et al. Evaluating and modeling the mechanical properties of the prepared PLGA/nano-BCP composite scaffolds for bone tissue engineering. J Mater Sci Tech. 2011;27(12):1105–12.
71. Niu C-C, et al. Benefits of biphasic calcium phosphate hybrid scaffold-driven osteogenic differentiation of mesenchymal stem cells through upregulated leptin receptor expression. J Orthop Surg Res. 2015;10(1):111–20.
72. Peltier LF. The use of plaster of paris to fill large defects in bone. Am J Surg. 1959;97(3):311–5.
73. Barnes G. REGENESORB Absorbable biocomposite material a unique formulation of materials with long histories of clinical use 2013, London: Smith & Nephew.
74. Luo Y, et al. Enhanced proliferation and osteogenic differentiation of mesenchymal stem cells on graphene oxide-incorporated electrospun poly(lactic-co-glycolic acid) nanofibrous mats. ACS Appl Mater Interfaces. 2015;7(11):6331–9.
75. Jansen J, et al. Fumaric acid monoethyl ester-functionalized poly(D,L-lactide)/N-vinyl-2-pyrrolidone resins for the preparation of tissue engineering scaffolds by stereolithography. Biomacromolecules. 2009;10(2):214–20.
76. Mohan S, et al. Incorporation of 3D Coragraf® with poly(lactic-co-glycolic acid) microsphere loaded human platelet derived growth factor – BB enhance osteogenic differentiation of mesenchymal stromal cells in vitro. In: Orthopedic research society 2017 annual meeting. San Digeo, CA; 2017.
77. Yang X, et al. Poly(lactic-co-glycolic acid) scaffold coated with an antioxidative fullerene derivative for bone tissue engineering. In: Orthopedic research society 2017 annual meeting. San Digeo, CA; 2017.
78. Vergnol G, et al. In vitro and in vivo evaluation of a polylactic acid-bioactive glass composite for bone fixation devices. J Biomed Mater Res B Appl Biomater. 2016;104(1):180–91.
79. Simpson RL, et al. A comparative study of the effects of different bioactive fillers in PLGA matrix composites and their suitability as bone substitute materials: a thermo-mechanical and in vitro investigation. J Mech Behav Biomed Mater. 2015;50:277–89.
80. Rich J, et al. In vitro evaluation of poly(ε-caprolactone-co-DL-lactide)/bioactive glass composites. Biomaterials. 2002;23(10):2143–50.
81. Adegani FJ, et al. Coating of electrospun poly (lactic-co-glycolic acid) nanofibers with willemite bioceramic: improvement of bone reconstruction in rat model. Cell Biol Int. 2014;38(11):1271–9.

82. Wen Y, et al. 3D printed porous ceramic scaffolds for bone tissue engineering: a review. Biomat Sci. 2017;5(9):1690–8.
83. Zeimaran E, et al. Bioactive glass reinforced elastomer composites for skeletal regeneration: a review. Mater Sci Eng C. 2015;53:175–88.
84. Profeta AC, Prucher GM. Bioactive-glass in periodontal surgery and implant dentistry. Dent Mater J. 2015;34(5):559–71.
85. Lusquiños F, et al. Bioceramic 3D implants produced by laser assisted additive manufacturing. Phys Procedia. 2014;56:309–16.
86. Ignatius AA, Claes LE. In vitro biocompatibility of bioresorbable polymers: poly(L, DL-lactide) and poly(L-lactide-co-glycolide). Biomaterials. 1996;17(8):831–9.
87. Böstman O, Pihlajamäki H. Clinical biocompatibility of biodegradable orthopaedic implants for internal fixation: a review. Biomaterials. 2000;21(24):2615–21.
88. Majola A, et al. Absorption, biocompatibility, and fixation properties of polylactic acid in bone tissue: an experimental study in rats. Clin Orthop Relat Res. 1991;268:260–9.
89. Hollinger JO. Preliminary report on the osteogenic potential of a biodegradable copolymer of polyactide (PLA) and polyglycolide (PGA). J Biomed Mater Res. 1983;17(1):71–82.
90. Bos RRM, et al. Resorbable poly(L-lactide) plates and screws for the fixation of zygomatic fractures. J Oral Maxillofac Surg. 1987;45(9):751–3.
91. Bostman O, et al. Foreign-body reactions to fracture fixation implants of biodegradable synthetic polymers. J Bone Joint Surg Br. 1990;72-B(4):592–6.
92. Barber FA, et al. Biocomposite implants composed of poly(Lactide-co-Glycolide)/β-tricalcium phosphate: systematic review of imaging, complication, and performance outcomes. Arthroscopy. 2017;33(3):683–9.
93. Tracy MA, et al. Factors affecting the degradation rate of poly(lactide-co-glycolide) microspheres in vivo and in vitro. Biomaterials. 1999;20(11):1057–62.
94. Danoux CB, et al. In vitro and in vivo bioactivity assessment of a polylactic acid/hydroxyapatite composite for bone regeneration. Biomatter. 2014;4:e27664.
95. Mathieu LM, Bourban P-E, Manson J-AE. Processing of homogenous ceramic/polymer blends for bioresorbable composites. Compos Sci Technol. 2006;66(11–12):1606–14.
96. Nourbakhsh A, et al. Effects of particle size and coupling agent concentration on mechanical properties of particulate-filled polymer composites. J Thermoplast Compos Mater. 2009;23(2):169–74.
97. Widmer MS, et al. Manufacture of porous biodegradable polymer conduits by an extrusion process for guided tissue regeneration. Biomaterials. 1998;19(21):1945–55.
98. Niemela T, et al. Self-reinforced composites of bioabsorbable polymer and bioactive glass with different bioactive glass contents. Part I: initial mechanical properties and bioactivity. Acta Biomater. 2005;1(2):235–42.
99. Sadeghi-Avalshahr AR, et al. Physical and mechanical characterization of PLLA interference screws produced by two stage injection molding method. Prog Biomater. 2016;5(3–4):183–91.
100. Narayanan G, et al. Poly (lactic acid)-based biomaterials for orthopaedic regenerative engineering. Adv Drug Deliv Rev. 2016;107:247–76.
101. Butt MS, et al. Mechanical and degradation properties of biodegradable mg strengthened poly0lactid acid composite through plastic injection molding. Mater Sci Eng C Mater Biol Appl. 2017;1(70):141–7.
102. Leenslag JW, et al. Resorbable materials of poly(L-lactide). VI. Plates and screws for internal fracture fixation. Biomaterials. 1987;8(1):70–3.
103. Kau Y-C, et al. Compression molding of biodegradable drug-eluting implants for sustained release of metronidazole and doxycycline. J Appl Polym Sci. 2013;127(1):554–60.
104. Zhang J, et al. High-pressure compression-molded porous resorbable polymer/hydroxyapatite composite scaffold for cranial bone regeneration. ACS Biomater Sci Eng. 2016;2(9):1471–82.
105. Sherwood JK, et al. A three-dimensional osteochondral composite scaffold for articular cartilage repair. Biomaterials. 2002;23:4739–51.
106. Henkel J, et al. Bone regeneration based on tissue engineering conceptions - a 21st century perspective. Bone Res. 2013;1(3):216–48.

107. Yen H-J, et al. Evaluation of chondrocyte growth in the highly porous scaffolds made by fused deposition manufacturing (FDM) filled with type II collagen. Biomed Microdevices. 2009;11(3):615–24.
108. Hsu S-H, et al. Evaluation of the growth on chondrocytes and osteoblasts seeded into precision scaffolds fabricated by fused deposition manufacturing. Biomed Mater Res B. 2007;80B(2):519–27.
109. Yang J, et al. Cell-laden hydrogels for osteochondral and cartilage tissue engineering. Acta Biomater. 2017;57:1–25.
110. Bandyopadhyay A, et al. Three-dimensional printing of biomaterials and soft materials. MRS Bull. 2015;40(12):1162–9.
111. Youssef A, Hollister SJ, Dalton PD. Additive manufacturing of polymer melts for implantable medical devices and scaffolds. Biofabrication. 2017;9(1):012002.
112. Hao L, Yan C, Shi Y. Investigation into the differences in the selective laser sintering between amorphous and semi-crystalline polymers. Int Polym Process. 2011;26(4):416–23.
113. Mazzoli A. Selective laser sintering in biomedical engineering. Med Biol Eng Comput. 2013;51(3):245–56.
114. Shirazi SF, et al. A review on powder-based additive manufacturing for tissue engineering: selective laser sintering and inkjet 3D printing. Sci Technol Adv Mater. 2015;16(3):033502.
115. Kolan KC, et al. Effect of material, process parameters, and simulated body fluids on mechanical properties of 13-93 bioactive glass porous constructs made by selective laser sintering. J Mech Behav Biomed Mater. 2012;13:14–24.
116. Kundera C, Kozior T. Influence of printing parameters on the mechanical properties of polyamide in SLS technology. Tech Trans Mech. 2016;3:31–7.
117. Spierings AB, Herres N, Levy G. Influence of the particle size distribution on surface quality and mechanical properties in additive manufactured stainless steel parts. In: Proceedings of Solid Freeform Fabrication Symposium, Austin, TX; 2010.
118. Hapgood KP, et al. Drop penetration into porous powder beds. J Colloid Interface Sci. 2002;253(2):353–66.
119. Salmorioa GV, et al. Structure and mechanical properties of cellulose based scaffolds by selective laser sintering. Polym Test. 2009;28(6):648–52.
120. Mangano F, et al. Direct metal laser sintering titanium dental implants: a review of the current literature. Int J Biomater. 2014;2014:461534.
121. Savalani MM, et al. Fabrication of porous bioactive structures using the selective laser sintering technique. Proc Inst Mech Eng H. 2007;221(8):873–86.
122. Singh JP, Pandey PM. Fitment study of porous polyamide scaffolds fabricated from selective laser sintering. Procedia Eng. 2013;59:59–71.
123. Roskies M, et al. Improving PEEK bioactivity for craniofacial reconstruction using a 3D printed scaffold embedded with mesenchymal stem cells. Biomater Appl. 2016;31(1):132–9.
124. Dabbas F, et al. Selective laser sintering of polyamide/hydroxyapatite scaffolds. In: The minerals, metals, and materials series. Berlin: Springer; 2017. p. 95–130.
125. Kanczler JM, et al. Biocompatibility and osteogenic potential of human fetal femur-derived cells on surface selective laser sintered scaffolds. Acta Biomater. 2009;5(6):2063–71.
126. Bukharova TB, et al. Biocompatibility of tissue engineering constructions from porous polylactide carriers obtained by the method of selective laser sintering and bone marrow-derived multipotent stromal cells. Cell Tech Biol Med. 2010;1(1):148–53.
127. Leong KF, et al. Building porous biopolymeric microstructures for controlled drug delivery using selective laser sintering. Int J Adv Manuf Tech. 2006;31(5–6):483–9.
128. Tan KH, et al. Selective laser sintering of biocompatible polymers for applications in tissue engineering. Biomed Mater Eng. 2005;15(1–2):113–24.
129. Zhou WY, et al. Selective laser sintering of porous tissue engineering scaffolds from poly(L:-lactide)/carbonated hydroxyapatite nanocomposite microspheres. J Mater Sci Mater Med. 2008;19(7):2535–40.
130. Duan B, et al. Three-dimensional nanocomposite scaffolds fabricated via selective laser sintering for bone tissue engineering. Acta Biomater. 2010;6(12):4495–505.

131. Antonov EN, et al. Three-dimensional bioactive and biodegradable scaffolds fabricated by surface-selective laser sintering. Adv Mater. 2004;17(3):327–30.
132. Agrawal CM, Niederauer GG, Athanasiou KA. Fabrication and characterization of PLA-PGA orthopedic implants. Tissue Eng. 1995;1(3):241–52.
133. Gupta A, Kumar V. New emerging trends in synthetic biodegradable polymers–Polylactide: a critique. Eur Polym J. 2007;43(10):4053–74.
134. Zhou P, et al. Organic/inorganic composite membranes based on poly(L-lactic-co-glycolic acid) and mesoporous silica for effective bone tissue engineering. ACS Appl Mater Interfaces. 2014;6(23):20895–903.
135. Samavedi S, et al. Electrospun meshes possessing region-wise differences in fiber orientation, diameter, chemistry and mechanical properties for engineering bone-ligament-bone tissues. Biotechnol Bioeng. 2014;111(12):2549–59.
136. Whited BM, et al. Pre-osteoblast infiltration and differentiation in highly porous apatite-coated PLLA electrospun scaffolds. Biomaterials. 2011;32(9):2294–304.
137. Komaki H, et al. Effects of rhBMP-2 on bone formation and material resorption after implantation of the composite material consisting beta-TCP and PLLA. In: Orthopedic research society 2017 annual meeting. San Digeo, CA; 2017.
138. Chye Joachim Loo S, et al. Effect of isothermal annealing on the hydrolytic degradation rate of poly(lactide-co-glycolide) (PLGA). Biomaterials. 2005;26(16):2827–33.
139. Jiang L, et al. Study on the effect of annealing treatment on properties of nano-hydroxyapatite/poly-lactic-co-glycolic acid composites. Polym-Plast Technol Eng. 2014;53(10):1056–61.
140. Loo JSC, et al. Isothermal annealing of poly(lactide-co-glycolide) (PLGA) and its effect on radiation degradation. Polym Int. 2005;54(4):636–43.
141. Holy CE, et al. Optimizing the sterilization of PLGA scaffolds for use in tissue engineering. Biomaterials. 2000;22(1):25–31.
142. Loo JSC, Ooi CP, Boey FYC. Degradation of poly(lactide-co-glycolide) (PLGA) and poly(l-lactide) (PLLA) by electron beam radiation. Biomaterials. 2005;26(12):1359–67.
143. D.S. Companies. RAPIDSORB rapid resorbable fixation system. West Chester, PA: Synthes USA Products, LLC; 2015.
144. Delta system: resorbable implant technology. In: Stryker, editor. Caniomaxillofacial. Portage, MI: Stryker Craniomaxillofacial; 2015.
145. Martin K SonicWeld Rx Dental. 2018. cited 2018.
146. ACUTE Innovations. BioBridge Resorbable Chest Wall Stabilization Plate. Hillsboro, OR: ACUTE Innovations; 2017.
147. I. Smith and Nephew. HEALICOIL REGENESORB and HEALICOIL PK Suture Anchors. Andover, MA: Smith and Nephew; 2013.
148. The next generation in should & elbow repair and reconstruction technology, A. Inc.; 2017. https://www.arthrex.com/shoulder. Accessed on May 2017.
149. Inion S-2 graft containment system, I. Inc.; 2017. https://www.inion.com/Products/spine/Inion_S1_folder/en_GB/Inion_S-1_indications_USA/. Accessed on May 2017.
150. A. Vascular. Absorb GT1 bioresorbable vascular scaffold system. Santa Clara, CA: Abbott Vascular; 2016.
151. The next generation in hand and wrist repair and reconstruction technology, A. Inc.; 2017. https://www.arthrex.com/hand-wrist. Accessed on May 2017.
152. I. Smith and Nephew. OSTEORAPTOR 2.3mm & 2.9mm suture anchors. Andover, MA: Smith and Nephew, Inc.; 2016.
153. D.S. Companies. Angular stable locking system (ASLS). West Chester, OA: Synthes USA Products, LLC; 2009.
154. D.S. Companies. Anteromedial ACL reconstruction for bone-tendon-bone grafts. West Chester, OA: Synthes USA Products, LLC; 2015.
155. C. Corporation. SmartNail surgical technique. Largo, FL: ConMed Corporation; 2011.
156. The most advanced techniques in knee arthroscopy, A. Inc.; 2016. https://www.arthrex.com/knee. Accessed on May 2017.

157. W. Medical. RFS resorbable fixation system. Memphis, TN: Wright Medical; 2016.
158. Medtronic. Polysorb 2mm soft tissue anchor system. Waltham, MA: Covidien; 2009.
159. Achillies SpeedBridge, A. Inc.; 2016. https://www.arthrex.com/foot-ankle/achillesspeedbridge. Accessed on May 2017.
160. Development, A.R.a, Arthrex BioComposite interference screws for ACL and PCL reconstruction. In: A.R.a. Development, editor. Naples, Florida: Arthrex; 2010.
161. Arthrex, I. InternalBrace™ ligament augmentation repair; 2016. [cited 2018 January 25, 2018].
162. Cedars-Sinai. Cranio / Maxillofacial surgery; 2017. [cited 2017 August 6, 2017].
163. Tuncer S, et al. Reconstruction of traumatic orbital floor fractures with resorbable mesh plate. J Craniofac Surg. 2007;18(3):598–605.
164. Baek WI, et al. Comparison of absorbable mesh plate versus titanium-dynamic mesh plate in reconstruction of blow-out fracture: an analysis of long-term outcomes. Arch Plast Surg. 2014;41(4):355–61.
165. Bali RK, et al. To evaluate the efficacy of biodegradable plating system for fixation of maxillofacial fractures: a prospective study. Natl J Maxillofac Surg. 2013;4(2):167–72.
166. Shim J-H, et al. Comparative efficacies of a 3D-printed PCL/PLGA/β-TCP membrane and a titanium membrane for guided bone regeneration in beagle dogs. Polymers. 2015;7(10):1500.
167. ShoulderDoc. InSpace balloon for massive rotator cuff tears; 2017.
168. TWINFIX Ultra HA suture anchor enhancement of biocompatibility. Andover, MA: Smith & Nephew; 2010.
169. Rupp S, Krauss PW, Fritcsh EW. Fixation strength of a biodegradable interference screw an a press-fit technique in anterior cruciate ligament reconstruction with a BPTB graft. Arthroscopy. 1997;13(1):61–5.
170. Disegi JA, Wyss H. Implant materials for fracture fixation: a clinical perspective. Orthopedics. 1989;12(1):75–9.
171. Claes LE, et al. New bioresorbable pin for the reduction of small bony fragments - design, mechanical properties an in vitro degradation. Biomaterials. 1996;17:1621–6.
172. Papalia R, et al. Metallic or bioabsorbable interference screw for graft fixation in anterior cruciate ligament (ACL) reconstruction? Br Med Bull. 2014;109:19–29.
173. Chao S, et al. Bioabsorbable versus metallic interference screw fixation in anterior cruciate ligament reconstruction: a meta-analysis of randomized controlled trials. Arthroscopy. 2010;26(5):705–13.
174. Development, A.R.a, Arthrex 7mm x 23mm BioComposite Interference Screw vs. DePuy Mitek 7mm x 23mm Milagro Screw. In: A.R.a. Development, editor. Naples, Florida: Arthrex; 2009.
175. Nelson, D.L. Adult distal radius fractures (also known as a "Broken Wrist"); 2012. [cited 2017 August 8, 2017].
176. Chang IL, et al. Early clinical experience with resorbable poly-5D/95L-lactide (PLA95) plate system for treating distal radius fractures. J Dent Sci. 2013;8:44–52.
177. Manen CJ, et al. Bio-resorbable versus metal implants in wrist fractures: a randomised trial. Arch Orthop Trauma Surg. 2008;128(12):1413–7.
178. Waris E, et al. Use of bioabsorbable osteofixation devices in the hand. J Hand Surg Br Eur. 2004;29(6):590–8.
179. Gomes ME, et al. Tissue engineering and regenerative medicine: new trends and directions— a year in review. Tissue Eng Part B Rev. 2017;23(3):211–24.
180. Laurencin CT, Nair LS. Regenerative engineering: approaches to limb regeneration and other grand challenges. Regn Eng Tran Med. 2015;1(1):1–3.
181. Melchels FP, Feijen J, Grijpma DW. A poly(D,L-lactide) resin for the preparation of tissue engineering scaffolds by stereolithography. Biomaterials. 2009;30(23–24):3801–9.
182. Ronca A, Ambrosio L, Grijpma DW. Preparation of designed poly(D,L-lactide)/nanosized hydroxyapatite composite structures by stereolithography. Acta Biomater. 2013;9(4):5989–96.

183. Simpson RL, et al. Development of a 95/5 poly(L-lactide-co-glycolide)/hydroxylapatite and beta-tricalcium phosphate scaffold as bone replacement material via selective laser sintering. J Biomed Mater Res B Appl Biomater. 2008;84((1):17–25.
184. Holmes B, et al. A synergistic approach to the design, fabrication and evaluation of 3D printed micro and nano featured scaffolds for vascularized bone tissue repair. Nanotechnology. 2016;27(6):064001.
185. Hutmacher DW. Scaffolds in tissue engineering bone and cartilage. Biomaterials. 2000;21(24):2529–43.
186. Laurencin CT, Aronson MT, Nair LS. Mechanically competent scaffold for ligament and tendon regeneration. 2013, Google Patents.
187. Sarukawa J, Takahashi M, Abe M, Suzuki D, Tokura S, Furuike T & Tamura H. Effects of chitosan-coated fibers as a scaffold for three-dimensional cultures of rabbit fibroblasts for ligament tissue engineering. J Biomater Sci Polym Ed. 2011;22(4-6):717–732. https://doi.org/10.1163/092050610X491067.
188. Freed LE, Vunjak-Novakovic G. Culture of organized cell communities. Adv Drug Deliv Rev. 1998;33(1):15–30.
189. Freed LE, et al. Neocartilage formation in vitro and in vivo using cells cultured on synthetic biodegradable polymers. J Biomed Mater Res. 1993;27(1):11–23.
190. Zhang Y, et al. The impact of PLGA scaffold orientation on in vitro cartilage regeneration. Biomaterials. 2012;33(10):2926–35.
191. Gao M, et al. Tissue-engineered trachea from a 3D-printed scaffold enhances whole-segment tracheal repair. Sci Rep. 2017;7(1):5246.
192. Network, D.P.M. DePuy synthes acquires tissue regeneration systems' 3D printing technology. 3D bioprinting; 2017. [cited 2017 August 15].
193. System, U.o.M.H. Baby's life saved after 3D printed devices were implanted at U-M to restore his breathing; 2014. [cited 2017 August 15].
194. Implant: Depuy Synthes - Biocryl Rapide® Suture Anchors Publication: Biocryl Rapide has redefined our Suture Anchors as "Bio-Replaceable" Source: http://static1.1.sqspcdn.com/static/f/533579/8290811/1282822079543/lupine Publication Date: 2007, Access Date: May 2017

The Role of Polymer Additives in Enhancing the Response of Calcium Phosphate Cement

David K. Mills

Keywords Additives · Bone repair · Calcium phosphate · Functionalities · Nanoparticles · Natural and synthetic polymers · Regeneration · Tissue engineering

1 Introduction

Dental and orthopedic complaints (dysfunction, injury, pain) are the main reasons that over 12 million Americans seek clinical intervention each year. In many of these cases, revision procedures are needed due to complicating conditions arising due to bone infection and resorption, major bone tissue loss, bone disease, or failure tissue regeneration [1, 2]. Revision procedures and increased hospital stays can cost thousands of dollars for a single patient, significant lost time from work, altered and restricted lifestyles, and death. The number of high-risk individuals in this patient population has also led to an increase in the need for additional operations due to device failure, lack of healing, or infection.

Research into treatments for bone loss has increased significantly over the past ten years primarily due to the pressing need for repair or replacement of damaged or diseased tissue. As a tissue, bone has impressive self-healing capacities for repair, but above a certain critical size, it cannot completely heal without intervention [3–5]. Large-scale bone loss can result in aging populations, bone tumors and bone diseases, endoprosthetic surgery, failed arthroplasty, osteomyelitis, osteoporosis, and patient traumas, requiring major reconstructive surgery [6]. Autogenic bone graft from the iliac crest is the primary source of trabecular bone. It possesses an adequate osteoinductive potential and autografts remain the gold standard for treating many types of osseous defects [3, 4]. Autografts present some significant disadvantages that includes

D. K. Mills (✉)
School of Biological Sciences and the Center for Biomedical Engineering, Louisiana Tech University, Ruston, LA, USA
e-mail: dkmills@latech.edu

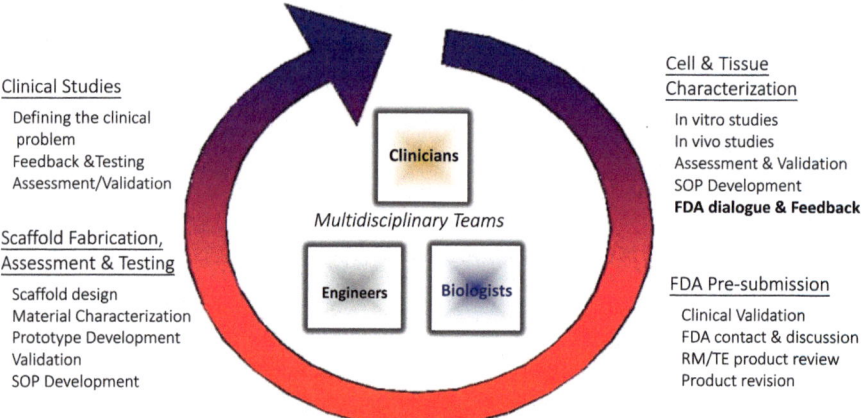

Fig. 1 Finding solutions to the critical need for viable treatment modalities for bone defect and disease repair requires a multidisciplinary approach. This graphics shows the respective strengths and skills that can be achieve through such an approach

donor site morbidity, disease transmission, and limited availability [4, 5]. Allogeneic bone is also widely employed due to its osteoconductive properties, but there remain risks for the transfer of disease. Inert implants often fail due to corrosion, mechanical failure, post-surgical infection, and incomplete healing.

Bone tissue repair or replacement designed to preserve an individual's health status, quality of life, and longevity, is a clinical necessity. Bone has a vast potential for regeneration from osteoblasts, mesenchymal stem cells, pre-osteoblasts (osteoprogenitor cells), and inactive bone-lining cells [7]. The bone healing process is initiated by mobilizing pre-osteoblasts and osteoblasts to migrate to the site of damaged or injured bone tissue. Osteoprogenitor cells differentiate into osteoblasts and with resident osteoblasts regenerate bone by producing an osteoid matrix, mineralizing this matrix, and then remodeling this tissue into compact or spongy bone [7, 8]. Bone formation stimulating regimes delivered via scaffolds of varying design and composition, which hold the promise of significant increases in bone density and mass, have not as yet become clinically available [9, 10]. Accordingly, the ability to generate new bone remains a pressing clinical need.

The search for new bone substitutes that combine porous biomaterial scaffolds, cells and bioactive factors is an area of extensive research and requires a multidisciplinary approach (Fig. 1) [11, 12]. Current strategies in bone tissue engineering involve the use of biodegradable polymers, either natural or of synthetic origin, designed for their physical strength, to serve as a supportive microenvironment for seeded bone or stem cells, and coupled with chemical or physical stimulus to facilitate bone tissue growth [8, 9]. Emphasis in these studies has been placed on the inclusion of additives that provide new functionalities that promote structural strength, bone tissue growth or impart antimicrobial or anti-cancer properties [7–9].

Fig. 2 Scanning electron micrograph showing bone cells attached to calcium phosphate bone cement. Source: www.nist.gov/ publicaffairs/techbeat/ tb20060413.htm#cell

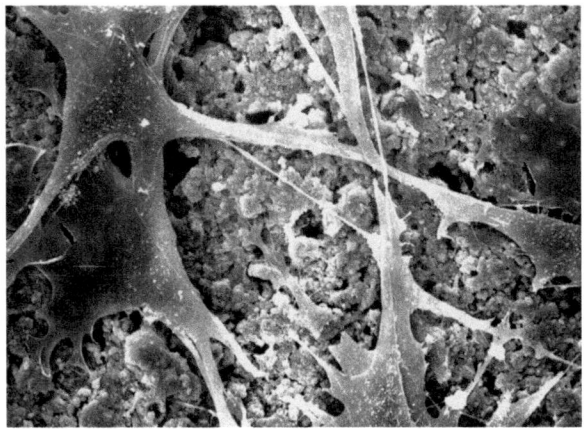

Bioactive glass, ceramics, metal, polymeric materials and hybrid composites have all been investigated to restore bone defects [10–12]. To date, however, many of these materials fail to integrate with the host tissue due to infection, limited bone regeneration capacity, or mechanical faults [6, 9, 13].

2   Advantages of Calcium Phosphate Cement

CPCs have been widely used for bone tissue augmentation, repair, and regeneration [14, 15]. First introduced in the 1980s, CPCs became available during the 1990s for treatment of craniofacial and maxillofacial defects [16, 17] and in fracture repair [18]. Following mixing, CPCs form a moldable paste and depending on the design, can be injected (or applied) during surgery using minimally invasive procedures [15, 16, 19, 20]. Clinically, this offers significant benefits and represents a clear advantage as compared with other bone substitutes.

CPCs possess intrinsic and vital properties for bone repair and regeneration. CPCs possess excellent osteoconductivity, mimic the bone's inorganic extracellular matrix (ECM), bind to neighboring bone, and are bioresorbable, in contrast to bio-inert implants where the bone-implant interface remains patent [21–23]. Furthermore, their degradation rate dependent on their composition and microstructural features [12, 14]. The CP minerals also provide an ideal substrate for cell attachment, proliferation and osteoblast differentiation (Fig. 2) [24–26]. They are osteoconductive, biocompatible, and harden *in vivo* via a low-temperature setting reaction that is not exothermic [16–18]. Compared to PMMA cement and other injectable biomaterials used as drug carriers, CPCs have many advantages [20, 27]. A low-temperature setting reaction permits the incorporation of different drugs, growth factors and other bioactive molecules [28–30]. Clinically, this offers significant benefits and represents a clear advantage with respect to conventional CPCs and PMMA [31]. CPCs are also resorbable, with a resorption rate that depends on their composition and microstructural features and forming a direct bond with surrounding bone [14, 31, 32]. Tricalcium phosphate (TCP), for example, is one of the most widely used materials due to its bioresorbable property with the ability to match degradation with bone tissue ingrowth [14]. CPCs further offer the advantage of being custom tailored for specific patient applications and directly applied to the affected site as a paste or in an injectable form.

3 Disadvantages of Calcium Phosphate Cement

CPCs have poor mechanical properties that limit their usage to non-load bearing or moderate-load-bearing sites [10, 14, 21]. An additional constraint is the fact that loading of CPCs with drugs at dosage required to combat infection, bone disorders, and disease, further reduces its mechanical properties [11, 14, 26, 33]. They also lack full injectability and their slow resorption rate limits bone repair [34, 35]. The inherent limitations of CPCs as led to a considerable body of work directed toward overcoming these limitations, enhancing CPCs' natural advantages [35], or imparting other functionalities through a variety of additives [36]. Selection of polymer additives has been limited as their degradation by-products may result in undesirable consequences such as inflammation, aseptic loosening, and implant failure [36]. Despite many efforts over the last three decades, the limitations of CPCs require further optimization, and the application to load-bearing sites has yet been achieved [9, 30, 34–36].

A strategy to overcome the intrinsic limitations of CPC or provide additional functionalities has been to incorporate natural and synthetic polymers. Current research efforts are focused on retaining CPCs bone regeneration potential of with an ability to incorporate drugs or other active molecules desirable for bone reinforcement [32, 37, 38], repair of cranial defects [39], and reinforcement of osteoporotic bones [40], appliances and implant fixation [41, 42], spinal fractures, and vertebroplasty [43, 44]. This chapter will examine the development of natural and synthetic additives polymers that enhance CPCs. The scope of this review is to provide a framework for understanding the role of polymers in the design of more enhanced CPC formulations. The use of natural and synthetic polymers is reviewed, and their effects on different CPC properties are evaluated. The different properties that can be enhanced by the addition of the specific polymers in the CPC are discussed in detail in the proceeding sections (Fig. 3).

4 Calcium Phosphate Applications

Over the last ten years, the need for biomedical applications of calcium phosphate has increased significantly [45, 46]. The inherent biological properties of CPCs, its biodegradability and biocompatibility, and osteoinductive materials and the total

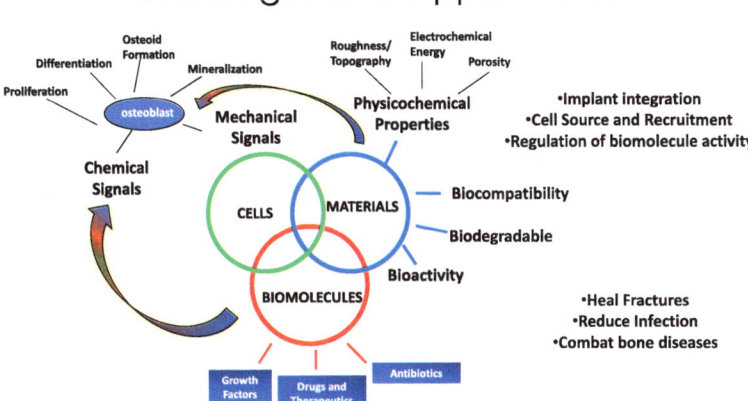

Fig. 3 The design of a bone repair biomaterial requires fabrication of the 'ideal package' consisting of biomolecules, cells, and materials. Attention to the many design constraints including the chemical, mechanical and physiochemical signals that are essential to cell functionality

absence of toxicity, have seen their increased use in recent years. All CPCs are formulated by mixing a solid and a liquid component [46]. The solid component typically consists of two or more calcium phosphate salts. The liquids can be water, alginates, chitosan (CS), sodium phosphates or other additives. To obtain maximum biological use, these components are mixed in predetermined proportions to form hydroxyapatite (HA). Modifications to the basic recipe have seen the development of different types of CPCs including resorbable and non-resorbable ceramics, biocoatings, injectable cement, CPC pastes, and osteoinductive powders [34, 46, 47]. Clinically CPCs are being used in the repair of bone defects, bone reconstruction, bone regeneration, as drug carriers (antibiotics, chemotherapeutics, growth factors), implant coatings, and in 3D printing bone tissue. The bio and material properties of CPCs ultimately depend on the end product (Fig. 4) [34].

5   Calcium Phosphate Additives and Setting Time

Resorbability of the CPCs completely depends on the end product and the physico-chemical reactions that occur during mixing are complex [48]. The basic recipe consists of a solid and a liquid phase. The solid phase typically is comprised of a basic and an acidic salt. When the salts react together in an aqueous medium (e.g., interstitial fluid, simulated body fluid) they precipitate HA as a final product [34, 36, 46]. During this process, water acts as a dissolving agent and is not a reactant in the setting process but allows dissolution of particles and precipitation of the products. A relatively low strength, high brittleness, inadequate adaptability in clinics, and a potential high rate of dissolution at the initial setting stages are common

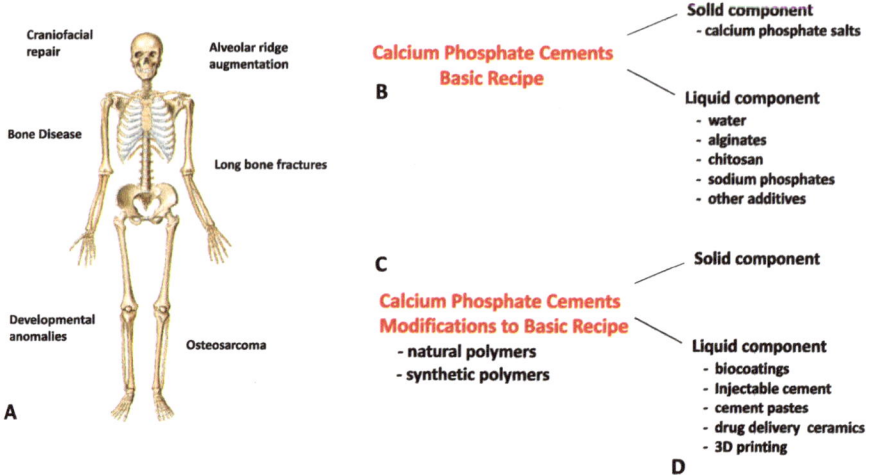

Fig. 4 (**a**) CPC application areas, (**b**) Basic calcium phosphate recipe, (**c**) Modifications to the basic calcium phosphate recipe, (**d**) Potential forms of CPS with polymer additions. The final bio- and material properties of CPCs ultimately depends on its end product

disadvantages of water used as the setting liquid [49].These properties of water limit its use in clinical applications and it cannot be used for treating defects that are bleeding or see a continuous flow of tissue liquids, like saliva. Furthermore, it cannot be used where in load-bearing bone sites.

However, the addition of water soluble biocompatible polymers in the CPC, either in a liquid or solid form, can increase cohesion, toughness, biological response and material resorption [50, 51]. If the cement does not set and disintegrates, it can elicit an inflammatory response leading to cell apoptosis at the site of application [51]. Cement setting time depends on factors such as solid-liquid composition, liquid-to-powder ratio, mixing liquid, and particle size of the powder. Setting conditions also influence the mechanical properties of the cement [48, 49].

5.1 Chitosan

Chitosan (CS) is natural biopolymer comprised of a linear copolymer of N-acetyl D-glucosamine and of D-glucosamine and has become a viable candidate for use in an osteogenic matri [48]. CS is cytocompatible and biocompatible, degrades easily, possesses antimicrobial properties, and can be produced in various forms such as membranes, films, fibrous mats and sponges [52]. CS's physicochemical and biological properties (adhesiveness, hydrophilicity, and solubility) have made it an excellent material for use as a component in drug delivery systems, in a diverse array of biomedical applications, and the bioengineering of skin, bone and cartilage [52, 53]. CS is also osteogenic and has been used to promote osteoblast growth and *in vivo* bone formation [52, 53]. In many studies, it has played an important role modifying CPC setting time [54, 55], in the attachment and differentiation of osteoblasts, and in bone tissue formation [56]. CS has been used in the surgical reduction of periodontal pockets [56] and as component materials for calcium phosphate compounds in various *in vitro* and *in vivo* studies [54–57]. Despite its tremendous promise in bone tissue engineering application, the poor mechanical properties of CS limits its clinical application to repairs of non-weight bearing bone. The addition of micro and nanofillers, such as silica, HA, and carbon nanotubes, halloysite, and montmorillonite have been used to reinforce CS in CS/CPC scaffolds [45–47].

5.2   Fibrin Glue

The use of natural polymers in CPCs such as collagen, fibrin and gelatin have been studied. Fibrin glue is a reaction product of fibrinogen with thrombin and *in vivo* its role is in maintaining vascular hemostasis [58, 59]. Fibrin glue has been shown to significantly increase CPC setting times and also increase its compressive strength significantly [60]. The effectiveness of fibrin glue alone in bone healing and tissue regeneration remains debatable [61]. Kneser *et al.* reported no bone formation after implantation of primary osteoblasts immobilized within fibrin gel-calcium phosphate bone cement [62]. In contrast, a 1:1 ratio of CPC/fibrin glue stimulated bone regeneration [63] in comparison to controls. Autologous fibrin glue seeded with stem cells increased bone formation two months post-operation and mature bone was present after three months [64]. Finally, fibrin-gelatin-HA composites also showed superior new bone formation, bone remodeling, and increased bone tissue volume [65, 66]. See Noori *et al.* for an excellent review of the potential of fibrin glue and bone tissue engineering [67].

5.3   Gelatin

Gelatin is a natural polymer derived from the partial hydrolysis of collagen and contains Arg-Gly-Asp (RGD)-like sequences. Gelatin facilitates bone cell adhesion and migration [67, 68]. Due to biocompatibility and biodegradability, it has gained significant interest in biomedical engineering [67, 68]. Gelatin's presence of in CPC accelerated the setting reaction and improved the mechanical properties of the cement [51, 69]. Gelatin added to HA via several methods had resulted in a porous-scaffold that has been shown to support bone tissue formation [70, 71]. CS and gelatin, in combination, have an enhanced effect on the biological activity of composite CPC scaffolds [71–73]. In a comparative study, Oryan *et al.* showed that gelatin alone or a CS-gelatin combination had positive effects on bone regeneration [74]. In contrast, CS alone was not able to promote measurable new bone formation [74].

5.4 Collagen

Collagen is a high molecular weight polymer (M_w = 300 kDa) with a long helical shaped structure (length = 300 nm) and it is a major component of all connective tissue matrices including bone [75]. Collagen comprises the bulk of the proteins that make up the organic matrix of bone [75]. Collagen's osteoinductive, osteoconductive and osteogenic properties are a major factor explaining its wide use in bone implant technology [76]. Collagen is often combined with other biomaterials for various tissue engineering applications [77]. In many of these investigations, collagen was combined with various calcium phosphates to obtain a biomimetic mimic for bone regeneration [78, 79]. Several studies have shown that collagen/CPC composites showed superior results in bone defect healing [80–82]. Incorporating collagen into CPC provides more cell recognition sites and speeds its biomaterial's degradation rate, thus allowing fast replacement by new bone [82, 83]. However, the application of collagen is limited by its cost and its inherent potential of antigenicity and pathogen transmission [84].

5.5 Polyethylene Glycol (PEG)

PEG is a polyether composed of glycerol monomers and it has been added to premixed CPCs to form more malleable CPCs (as pastes) that can be stored in an unreactive state for long time periods enabling prior preparation before surgery [85, 86]. A key property of premixed CPC/PEG pastes is after immersion in water, they set with setting times and material strengths comparable to the commercial-grade CPC [86, 87]. Numerous studies have combined PEG with other polymers that provide better cell adhesion, cell growth and functionality for bone tissue formation [88]. Other investigators have used PEG as a growth factor carrier in guided bone regeneration and in bone defect repair [89, 90].

6 Calcium Phosphate Additives: Material and Mechanical Properties

6.1 Natural Polymers

For osteointegration, a bone scaffold should simulate the mechanical properties, morphology, and composition of native bone tissue [88, 91]. Having the mechanical strength within the cancellous bone range and 100–800 µm pores with at least 50% porosity and an optimized degradation rate in balance with bone tissue regeneration rate are essential for the scaffolds applied in bone tissue engineering [92]. Due to the intrinsic porosity, CPC strength is reduced limiting their applicability to non-load-bearing applications [93–95]. The incorporation of inorganic nanofillers such as metal additives or polymers during the CPC fabrication imparts higher deformation before breaking. Moreover, polymer fiber reinforcement has been extensively explored as a strategy to increase composite toughness, CPC strength, and additional functionality [95, 96].

6.1.1 Alginate

Alginate, in the form of microbeads, has been added to CPC resulting as a means to introduce encapsulated cells and for enhanced biodegradation, mechanical properties, and setting time. Alginate is widely used as a mechanism for encapsulating tissue cells within gelled microcapsules or microbeads [97, 98]. A combined alginate and CPC microbead, capsules or hydrogels, seeded with pre-osteoblasts, osteoblasts or stem cells are now considered a viable means for tissue-engineering bone [99, 100]. Stem cell-encapsulated alginate beads added to CPCs remain viable and undergo normal cellular processes, such as cell mitosis and cell differentiation leading to histiogensis [101–103]. Alginate-CPC composites with doped (gentamicin, BMP-2) halloysite nanotubes added and formed as microbeads [104] and coatings [105] and were to enhance bone tissue formation and to provide an antibacterial material. The form in which alginate is added does, however, affect setting times, compressive strength and injectability. Wang *et al.* showed that setting times, compressive strength and injectability decreased when alginate powder particles were added [98]. In contrast, several alginate-composites including-polymer-based (PLGA, PEG, and CS), protein-based (collagen and gelatin), and ceramic-based alginate-showed impressive gains in mechanical strength, cell adhesion, proliferation and osteogenic differentiation [14, 100, 106].

6.1.2 Chitosan

CS has been added to many polymer composites due to its antimicrobial activity, biocompatibility, drug loading, and film forming capability, and ability to enhance mechanical properties [34, 107, 108]. When CS is introduced as a mixing liquid it can affect the kinetics of CPC setting and hardening [36, 108]. An increase in flexural strength of CPC composite was reported by Padois and Rodriguez after 15–20 wt% CS was incorporated within CPC and with no cytotoxicity reported [54]. A considerable increase in ALP activity was also observed when mesenchymal stem cells were cultured on a TTCP-DCPA composite [107].

CS and HA composites fabricated in other forms have also been promising as bone repair materials. HA- CS bilayers support bone marrow-derived stem cell (BMSC) growth and functionality [109]. CS nanofibers stimulated osteoblast proliferation and maturation via runt-related transcription factor 2-mediated regulation of the bone specific proteins, osteopontin, osteocalcin, and also increased ALP gene expression [110]. The CS nanofiber scaffolds further stimulated trabecular formation leading to enhanced bone healing [110]. Electrospun CS/HA fiber scaffolds can also facilitate the proliferation, differentiation and maturation of osteoblast-like and stem cells [111, 112]. CS/fibrin/collagen fiber scaffolds were shown to support osteoblast and stem cell growth and functionality [113, 114].

6.2 Synthetic Polymers

Due to their biodegradability, biocompatibility and processability, the most often used biodegradable synthetic polymers employed as CPC additives include poly-acrylic acid, polycaprolactone, polylactic acid, and poly(lactic-co-glycolic acid), among others [92–95, 115, 116]. These materials are favored because they are cur-rently in use in many commonly used medical implant devices and have been shown to be safe, non-toxic, and biocompatible [117]. Examples of medical devices that use these polymers include sutures, tissue screws and tacks, guided tissue regenera-tion membranes for dentistry, internal bone fixation devices, microspheres for implantable drug delivery systems, and systems for meniscus and cartilage repair [118]. Polymer characteristics can be manipulated and custom designed scaffolds can be produced for specific applications in bone tissue repair [91–93, 119]. However, the mechanical properties of these polyester materials are not ideal, espe-cially for bone tissue engineering scaffold materials. Also, the hydrophilicity and cytocompatibility of such materials are still needed to be optimized.

6.2.1 Polyacrylic Acid

An anionic polyelectrolyte, poly(acrylic acid) (PAA) is a non-biodegradable poly-mer and is easily polymerized and crosslinked into polymer composites and is a component widely used in biomedical applications due to its high swelling capabil-ity and polyanionic charge that permits the loading of various cationic molecules [117, 118]. PAA has excellent calcium binding property, which allows it to bond strongly with HA [120] Over the last ten years, many studies have been directed towards the study of PAA incorporation into the CPCs [10, 121–123]. PAA addition to CPC produces a composite with compressive strengths up to 90 MPa and tensile strengths up to 21 MPa, a tenfold increase when compared with commercial cement [10, 121]. Although the cements showed high mechanical properties they also had poor resorption properties [95, 121].

6.2.2 Polycaprolactone

Poly(o-caprolactone) (PCL) is a semicrystalline linear aliphatic polyester with a high degree of crystallinity and hydrophobicity and has been approved as a medical polymer by the Food and Drug Administration (FDA) [122]. Poly (ε-caprolactone)

(PCL) has been less frequently used because of its slow degradation kinetics [122]. However, for a long-term tissue recovery or drug delivery system, PCL is an ideal polymer and the degradation properties can be modified by copolymerizing with other polyesters [122, 123]. Cellular activities on the scaffolds are highly dependent on the stability of the structure of the scaffold as it was found that slower degradation rates (6 months to 2 years *in vivo*) provide better physiological and biological characteristics in bone tissue scaffolds as compared to polyglycolic acid(PGA), polycactic acid (PLA), and PGLA [123]. PCL is durable, possesses excellent biocompatibility and biodegradability, and has excellent mechanical properties, including anti-bending and tensile strength [122, 123].

Accordingly, with its cost-effectiveness, PCL-based biomaterials are widely used in medical and drug delivery devices and offer a promising strategy for designing temporary extracellular matrices for bone tissue engineering [124–126]. Composites of PCL and other polymers have evaluated the osteoinductive properties of the composite scaffold composite by culturing pre-osteoblasts and stem cells *in vitro* and assessing their differentiation into osteoblasts [125–127]. These studies have shown that PCL/PLGA [124], PCL/HA [125] or PCL/CPC scaffolds [126, 127] are suitable materials for bone repair and regeneration. There are three key factors that must be considered in using PCL to repair bone defects or regenerate bone tissue [10]. The selected material must have a degradation rate that matches new bone in-growth and must be able to absorb mechanical loads as it degrades. Recently, Bioink Solutions, Inc. has developed several PCL (and PLA) powder-based bioinks with controlled biodegradability, and excellent biomechanical and biological properties [128]. Finally, PCL-based biomaterials and composites often lack bioactivity which limits their application in bone tissue engineering and requires further customization or PCL functionalization.

6.2.3   Polylactic Acid (PLA)

Polylactic acid (polylactide, PLA) is a bioactive thermoplastic aliphatic polyester and has received significant interest as it is derived from renewable resources, such as corn starch polymer and is a "green" product [129]. PLA is biodegradable and quickly degrades into lactic acid. PLA is noted for its toughness and durability [122, 130]. PLA is used in the manufacture of sutures, stents, dialysis equipment, and in the form of anchors, screws, plates, pins, rods, and as a mesh [130]. PLA can be processed by extrusion in fused depositional modeling and through injection molding, film and, sheet casting. The electrospinning method enables fabrication of a wide range of materials [131]. PLA's biocompatible properties enables to be in contact with tissues and accordingly, PLA has also found many uses in tissue engineering where it is used as the artificial scaffold to support growth of cells and functionality [131]. In bone repair and the bioengineering of bone tissue, suitable biocompatible and biodegradable scaffolds are also required that are not only a durable material but also a compliant and porous material that does not require surgical removal after implantation [131, 132].

PLA scaffolds provide excellent mechanical support and CPC has been mixed with PLA in several forms to produce a composite bone scaffold [133–136]. PLA composite scaffolds enhanced with calcium sulfate hemihydrate [133], dicalcium phosphate [134, 137], HA [135] or TCP [136] can increase degradation time and improve the mechanical properties of the scaffolds. Several cell types (pre-osteoblasts and stem cells) were able to proliferate and differentiate into mature osteoblasts and produce mineralized tissue [133–137]. PLA composite bone scaffolds have been reinforced with many additives including ECM proteins and peptides [138] and growth factors [122, 131, 135]. The critical constraint in the design of PLA (or any polymer) combined scaffolds is to realize the advantageous properties from the combination while minimizing the respective disadvantages of each material [122, 130–132]. Lou *et al.* for example, showed that increases in CPC compressive strength were dependent on PLA fiber length [134, 137].

Modifications to its material properties can lead to changes in the rates of hydration (swelling) and hydrolysis (degradation), making the material suitable for specific applications. Like PGA, PLA is combined with other polymers to modify its degradation rates and improve their mechanical characteristics and osteoconductivity [122, 132].

6.2.4   Poly(lactic-co-glycolic) Acid

Poly(lactic-co-glycolic) (PLGA) has become one of the most popular polymers for use in the fabrication of drug delivery systems, medical devices and in tissue engineering applications. PLGA has many desirable properties. These include: biocompatibility; tunable biodegradation rate; ability to encapsulate bioactive factors and drugs; and it is FDA approved for clinical use in humans [138, 139]. PLGA, in particular, has been extensively studied for the development of controlled delivery of small molecule drugs, proteins and other macromolecules. PLGA has been widely used as a solid phase CPC additive as a means to deliver growth factors and antibiotics in a sustained fashion [138–142]. PLGA, various forms, are able to modulate cell behavior and create a microenvironment for improved bone tissue formation and bone tissue function [139–141]. CPCs that contain degradable PLGA microparticles have shown good biological properties [140, 141]; however, their incorporation was found to decrease the mechanical properties of the cement which may, in part, be due to poor bonding and insufficient dispersion and blending between the PLGA microparticles and CPC. Another issue with PLGA is that native PLGA has poor osteoconductivity properties and shows deficient mechanical properties restricting its use to non-load-bearing applications. Accordingly, PLGA is often complexed with other polymers, bioactive glass, or clay nanoparticles to enhance its bone regeneration potential [138].

6.3   Carbon Nanotubes, Clay Nanoparticles and Graphene

Nanoparticle composites derived from carbon nanotubes, clay nanoparticles and graphene are under intense investigation due to their intrinsic structural characteristics and potential functional applications, including commercial, environmental, industrial and biomedical uses [143]. Current research efforts are focused on using these materials to enhance mechanical properties, provide sustained drug-release, and support cellular behavior leading to tissue development, growth and repair [143, 144].

6.3.1   Carbon Nanotubes

The potential applications of carbon nanotubes (CNTs) in bone tissue engineering have been recently investigates in terms of toxicity and biological interactions but also how the different properties and features of CNTs (or CNT-CPC, CNT-HA and CNT-CPC-polymer) composites may affect bone tissue growth [145]. When CNTs, with BMP-2, were cultured in association with pre-osteoblasts or added as a bone defect filler, major improvements in toughness, flexural and impact strength were observed [142, 146]. Other investigators have also reported that CNTs alone or in polymer composites promoted bone regeneration, *in vitro* and *in vivo*, suggesting their strong potential as a scaffolding material in bone tissue regeneration and in other orthopedic applications [147, 148].

6.3.2   Clay Nanoparticles

The utility of clay for drug-delivery and scaffold design have seen many studies on halloysite nanotubes, Laponite and montmorillonite (MT) have been used in developed a variety of nanocomposites containing clay nanoparticles has attracted great research interest [149, 150]. Many of these research studies have focused on clay interactions with biomolecules, polymers, and cells and their potential for tissue engineering and regenerative medicine [148].

6.3.3   Halloysite Nanotubes

Halloysite nanotubes (HNTs) are a naturally occurring aluminosilicate clay with an external diameter of 50 nm, an inner lumen of 15 nm, and a length of 500–2000 nm [150]. HNTs are abundant and inexpensive (~$15/cubic ton) and have attracted significant interested since 2012 [151]. HNTs have been widely used in biomedical, cosmetic, environmental, and industrial applications [148]. These include drug delivery [152], wood and metal coatings [153], medical devices [154], tissue engineering [155], and wound healing [156]. This is due to their excellent cyto- and

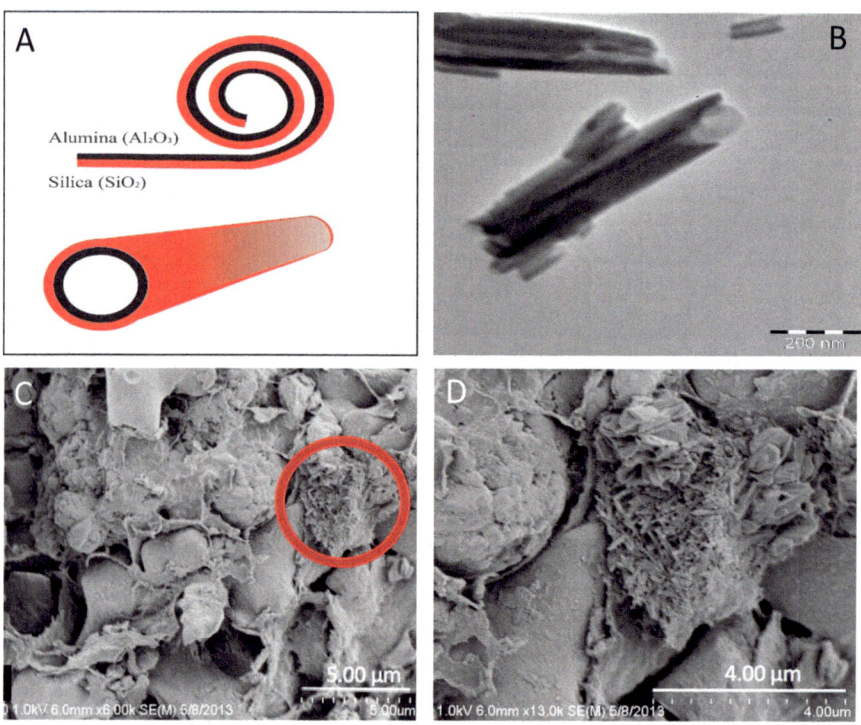

Fig. 5 Halloysite nanotubes. (**a**) Graphic of HNT structure; (**b**) TEM of HNTs; (**c**) SEM of CPC/ HNTs; (**d**) Higher power view of area contained in red circle in (**c**) showing HNTs. Micrographs courtesy of Dr. Uday Jammalamadaka

biocompatibility, ease at functionalization, and ability to load and release a variety of bioactive agents [148–150]. CPC and HNT composites for bone tissue engineering and orthopedic applications have shown that HNTs are osteoinductive [157], and facilitate bone tissue formation in alginate/HNT [105, 158], CPC/HNT [156, 159] and alginate/CS/HNT composites (Fig. 5) [104]. It is anticipated that use of HNTs in bone repair and regeneration will continue at a rapid pace.

6.3.4 Laponite

Laponite is an isomorphous substituted smectite clay which manufactured as a layered aluminosilicate disk-like clay material from naturally occurring mineral sources [160]. Laponite is currently used in household and industrial surface coatings, household and industrial cleansers, ceramic glazes, agrochemical, oilfield and horticultural products [160]. Given is conformation, it is currently being widely exploited as a drug carrier [161]. In CPCs, Laponite is used as a partial or complete replacement for commonly used organic polymers as it increases mechanical properties and enhances bone regeneration [162–164]. Laponite composites show much potential in *in vitro* studies to support osteoblast differentiation and matrix mineralization [163–165]. More importantly, *in vivo* cranial defect experiments support these findings as the addition of Laponite promoted bone defect healing [164, 165]. In summary, these findings indicate that a composite scaffold with laponite incorporation is a promising material for bone tissue engineering.

6.3.5 Montmorillonite (MT)

Montmorillonite (MT) is a three-layered clay nanoparticle derived from the smectite group of minerals [166]. It is structured as two silica tetrahedral layers sandwiched between a central octahedral sheet of aluminum hydroxide [166]. MT possesses many of advantages for use in bone repair or regeneration [166, 167]. MT offers a high surface area, a charge differential structure, is biocompatible, has a high adsorption and drug loading capacity. When MT was added to CPCs in the liquid component major increases in compressive and tensile strength of the bone cement were observed similar to HNTs and laponite [167, 168]. Many bone tissue engineering studies have added MT to CS, CPC, gelatin, HA, PLA, PCL, *etc.* [169–172] In these studies, the addition of MT improved mechanical and *in vitro* biological properties supporting osteoblast growth and functionality and suggesting that MT has a use in non-load bearing bone tissue engineering applications.

6.3.6 Graphene

Since its discovery in 2004, graphene has been intensively studies by researchers across disciplines primarily due to its extraordinary strength, high surface-to-mass ratio, superconducting properties and potential in biomedicine [173]. In the latter field, graphene has been applied to a diverse set of applications including biosensors, biomedical implants, tissue engineering, cancer therapy, and drug delivery [174]. These studies examined graphene, CPC or HA composites, suggesting that graphene has promise in enhancing bone tissue growth [175, 176]. Graphene did not provoke significant cytotoxicity and this is dependent on its mode of manufacture [175]. Graphene and its derivatives, in either 2D or 3D forms and as coatings, nanoparticles, or sheets, promote osteoblast and stem cell adhesion [177], stimulate osteogenic differentiation [178], and induce osteogenesis leading to bone formation [179].

6.4 Natural Fibrous Material

6.4.1 Cellulose

Bioabsorbable cellulose is derived from several sources including cotton, hemp, linen and wood [180]. Bacterial-derived cellulose obtained from the bacterium, *Acetobacter xylinum*, is a newer sources and in comparison to plant cellulose is identical in chemical structure, but it can be produced without intensive purification [92]. Cellulose has been intensively studies for use in many biomedical applications [92, 180]. It is not cytotoxic, possess high swelling capabilities, stable at various temperature and pH variations [92, 180]. Cellulose has been proven bioabsorbable carriers for delivering CPCs for bone defect repair [181–184]. The addition of cellulose to either HA or CPC increased porosity, enhanced mechanical properties and cell supportive properties. Cellulose fibers [181], cellulose hydrogels [182], sponges [183] and composite blends (*e.g.*, collagen, PLGA) [184], have all been shown to promote bone regeneration.

6.4.2 Collagen

Collagen added to CPC/HA adds both bioactive properties (cell adhesion, enhanced bioactivity) but also improvements in mechanical properties. HA/collagen cement improved setting time and mechanical properties and produced a form that was both moldable and injectable, and able to enhance bone regeneration in moderate stress-bearing applications [185]. Miyamoto *et al.* showed that collagen addition to CPS produced a composite with higher biocompatibility and improved mechanical properties [186]. Similarly, Kikuchi *et al.* [186] and Kikuchi [187]demonstrated that a HA/collagen composite produced a tissue with a bone-like nanostructure similar to autogenous bone [187]. In sum, collagen may be a desirable biomaterial when added to CPC or HA.

7 Calcium Phosphate: Injectability

For many preformed bioceramic scaffolds to completely fill a bone defect, the surgeon needs to machine the graft or shape the surgical site [188]. The results often leads to increases in bone loss, trauma, additional surgical time, and patient costs and discomfort [188, 189]. Preformed scaffolds have major disadvantages including seeding cells throughout the scaffold, post-surgical functionality, application in minimally-invasive surgeries, and lack of tunability [189]. Injectable scaffolds for cell or growth delivery have many advantages [189, 190]. Injectable CPCs can shorten the required surgical time, minimize damage during muscle retraction, reduce postoperative pain and patient costs, and advance rapid recovery for patients [191].

Extensive research is currently underway to develop injectable formulations to overcome these obstacles, have widespread applicability including in minimally invasive application, and provide for local strategies in delivering growth or other bioactive factors prolonged and targeted local delivery [190]. Hydrogels are often used in developing injectable bone scaffolds as they mimic the ECM facilitating the cell migration, adhesion, proliferation, and differentiation of osteoprogenitor cells to osteoblasts, can be embedded with growth factors, and create a microenvironment suitable for delivering nutrients and eliminating wastes.

Several injectable hydrogel and polymer CPC carriers have been developed for cell and growth factor delivery [24, 192–195]. Biomaterial choice and fabrication method are the key determinants in developing injectable hydrogels for bone repair. A review of the literature shows that numerous biomaterials (natural and synthetic) and fabrication methods including cross-linked (UV or chemical, click chemistry, Michael addition, pH and temperature-responsive hydrogels) have been studied as potential hydrogels including alginate, CS, collagen and gelatin, hyaluronic acid, PEG, and poly(vinyl alcohol). See Li *et al.* [192] for an excellent review of injectable hydrogels.

A critical consideration on bioengineering bone is that the bone substitute must be able to withstand stresses, maintain the shape of the regenerated tissues, and to fracture [196]. A major constraint in the use of injectable CPCs, particularly hydrogel-based scaffolds, is they do not possess the mechanical strength to be used in load-bearing applications [191, 196]. To date, a fully injectable, bioactive, and mechanically appropriate scaffold for osteoblast or stem cell encapsulation and bone engineering has yet to be developed. The current state of the art is directed toward developing novel and fully injectable and mechanically-strong gels or pastes that support stem cell migration to the defect site, cell growth and differentiation with production of viable bone tissue (Fig. 6). For example, Yasmeen *et al.* used CS gels, containing CS grafted CNTs and CS–HA composites, cross-linked with glycerol phosphate, to enhance the gels mechanical properties [197].

8   Calcium Phosphate: Biological Response

8.1   CPC/Growth Factor/Polymer Composites for Cell Growth and Functionality

There have been numerous attempts to develop bone substitute materials with osteoinductive stimulation for biomedical applications [14, 15, 20, 23, 29, 49, 92]. Low cell attachment and proliferation rates are usually observed when cells are cultured on CPCs *in vitro* and a common strategy is to incorporate polymers with either specific binding cell domain, a key component of their structure such as the RGD

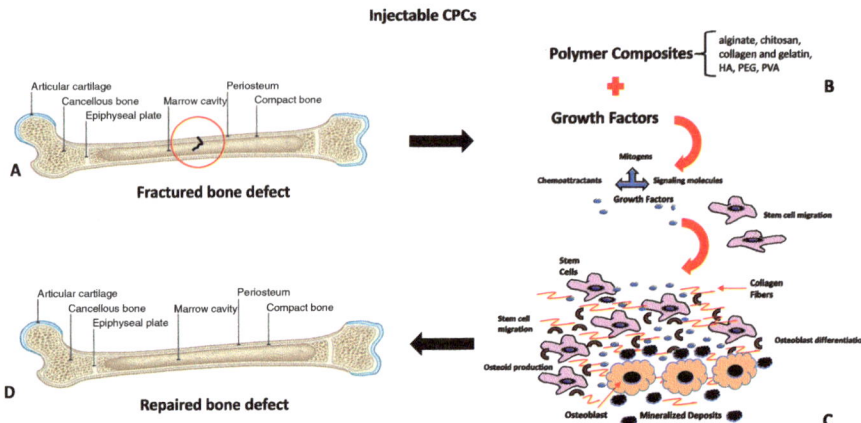

Fig. 6 Graphic depicting role of injectable CPCs in fracture repair. (**a**) Femur with a mid-diaphyseal fracture. (**b**) Reparative scheme using polymer composites with embedded growth factors, (**c**) Growth factor release recruits osteoprogenitor cells, induces osteoblast differentiation, and promotes bone tissue formation, (**d**) Femur with fracture completely healed

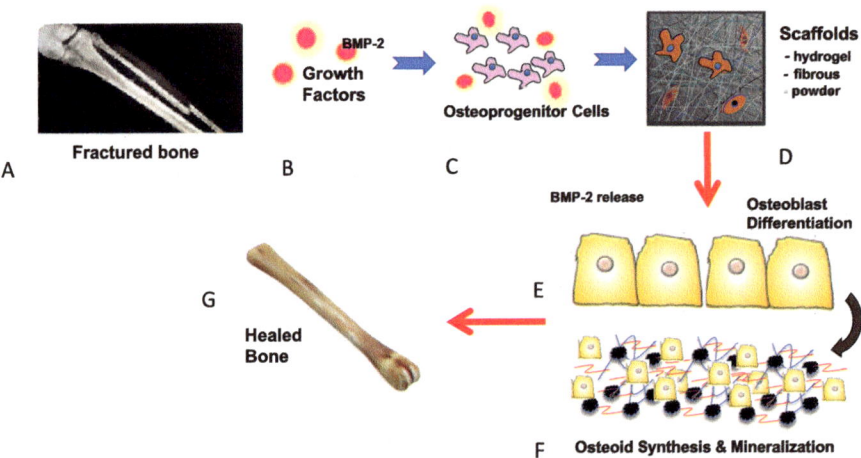

Fig. 7 Combinatorial design that complexes a growth factor and osteoprogenitor cells. (**a**) Fractured fibula, (**b**) BMP-2, (**c**) osteoprogenitor cells, (**d**) growth factors and cells are seeded into scaffolds, (**e**) BMP-2 release leads to osteoblast differentiation, (**f**) osteoblast synthesize an osteoid matrix and later mineralize this matrix, (**g**) fracture defect repaired

sequence, that is a component of their structure (i.e., collagen or gelatin) or has been added in some manner (ligand binding) [14, 29, 122]. Growth factors such as bone morphogenetic proteins (BMPs), connective tissue growth factor (CTGF), fibroblast growth factor (FGF), pleiotropin (PTN), platelet-derived growth factor, (PDGF), transforming growth factor-beta (TGF-B), have been used singly or in concert, and often embedded within a scaffold or cell-seeded scaffold (Fig. 7) [198].

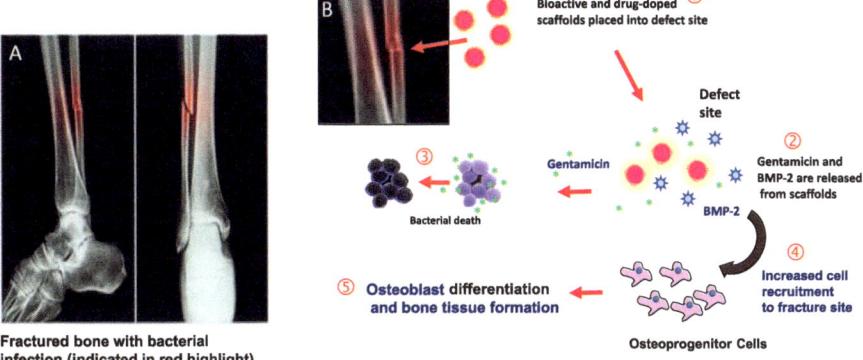

Fig. 8 Anti-infective and histogenic bone scaffold: mode of action. (**a** and **b**) Fractured and infected fibula (1). Bioactive and antibacterial doped scaffolds implanted into infected bone defect, (2) Release of gentamicin and BMP-2 from scaffold leads to: (3) Growth inhibition and reduction in infection, and (4) Osteoprogenitor recruitment leads to osteoblast differentiation and bone tissue formation

Growth factors have been packaged in bone scaffolds using various strategies for directing cellular behavior including chemotactic, differentiation, proliferation and synthetic agents during bone remodeling and fracture healing [198–200]. The BMPs have proved successful in bone regeneration in several clinical trials and are FDA approved [201]. A key design feature in such systems is to deliver growth factors to the proper target and in the right concentrations so as to avoid unwanted side effects [200–202]. Most design strategies that combine biomaterials and growth factors have focused on growth factor delivery (design of concentration, gradient, and release pattern and rate) and delivery vehicles [203]. The more effective designs have growth factors embedded with BMSCs for bone tissue formation combined with antibacterials (Fig. 8). In this manner, bone infections, a common occurrence resulting from traumatic inquires and contamination of the wound site as well as after surgical intervention or device implantation. A review of the field suggests that for effective bone formation, multiple growth factors must be delivered in a controlled and precise spatiotemporal manner for proper vascularization, chemotaxis, proliferation, and differentiation [203, 204].

8.2 CPC/polymer Composites for Cell Encapsulation

Cell encapsulation within biodegradable hydrogels offers numerous advantages in bioengineering a variety of tissues [203]. These include ease in fabrication, ability to manipulate inherent polymer properties, a highly hydrated microenvironment for cell growth and tissue formation, and the ability to transition development from an *in vitro* to a permanent *in vivo* situation [203, 204]. in addition, with appropriate

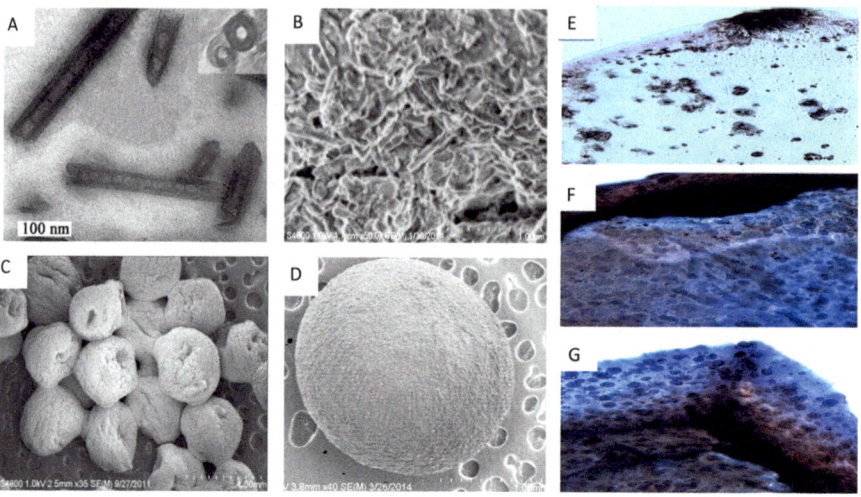

Fig. 9 (**a**) TEM micrograph of HNTs, (**b**) HNTs embedded in alginate, (**c**) Alginate doped with .5% HNT, (**d**) Alginate doped with 2.5% HNTs, (**e**) Pre-osteoblast-seeded alginate hydrogels without HNTs or BMP-2 after 7 days in culture, (**f**) Alginate hydrogels with osteoblasts and HNTs but without BMP-2 after 7 days in culture. (**g**) Experimental BMP-2 doped HNT/alginate hydrogel showing osteoblast clusters after 7 days in culture. Part labels (**e–g**) stained with Alcian blue. Micrographs courtesy of Dr. Sonali Karnik

polymer selection of polymers stealth cellular vehicles can be produced and thus, cell masking could make them immunologically invisible, a desirable property in the case of allogenic transplants [205]. A number of polymers have been used to encapsulate osteoblasts and BMSCs including alginate, CS, collagen, hyaluronic acid, and PEG-based hydrogels among others. The polymer protects the embedded cells and enables functionalization of the hydrogel and the inclusion of growth factors [206–210].

Alginate/CPC composites, alginate hydrogel beads and tubular hydrogels beads were used to pre-osteoblasts, osteoblast or BMSCs cells encapsulated in alginate beads. All cells were viable and pre-osteoblasts and stem cells differentiated and osteoblast functionality was enhanced [105, 207, 208]. CS composites have also been used as encapsulating agents and been shown to be osteosupportive [209]. Alginate/ CS CPC and alginate/ CS/TCP mixed with HNTs were also able to stimulate osteoblast differentiation and serve as an injectable vehicle or as a CPC coating (Fig. 9). Cell encapsulation systems (with or without growth factors) have been developed using collagen [209], hyaluronic acid [210], and PEG-based hydrogels [211].

8.3 Bioactive Glass and Silica Materials

8.3.1 Bioactive Glass

Several types of ceramic implants have been investigated such as bioactive glasses, glass-ceramics and silica materials [212, 213]. This field has become an important and emerging area of inquiry for bone tissue engineering applications [214]. Glass does not seem at first a viable material as it is brittle and friable, however, new fabrication methods have led to new formulations of bioactive glasses. Bioactive glass are glass/ceramic biomaterials which are bioactive and biocompatible and encompass the original formulation, "bioglass" and newer designs [215]. Bioglass, borate-based glasses, and bioactive glass doped with small additions of Cu, Zn and Sr are being intensively investigated as implantable materials for bone repair and replacement [214, 215].

The newer bioactive glasses, such as the borate-based bioglass, also have controllable degradation rates when compared to silicate glass, meaning their degradation rates can be matched with the rate of new bone formation [215]. Bioactive glasses is often doped with trace quantities of elements which can enhance new bone formation and vascularization, can be tailored to a broad range of range of mechanical properties making them suitable for both non-loaded and loaded skeletal sites [214, 215]. A recent study, has shown the potential for bioactive glass/silk fiber composite for tissue-engineered osteochondral constructs [216].

8.3.2 Silica Materials

Silicon (Si) is an important mineral in maintaining bone health and is essential for bone formation [217, 218]. It also has shown significant promise for bone formation when incorporated into calcium phosphate bioceramics [219]. As established, native CPCs are considered osteoconductive because they provide a template for bone tissue deposition and replacement for resorption and replacement through orchestrated interactions between osteoclasts (for resorption) and osteoblasts (new bone deposition). Inclusion of Si in CPCs adds additional functionalities and may aid in boosting bone formation [220]. The presence of Si in CPC and HA has been shown in several *in vitro* studies to stimulate osteogenic differentiation, increased osteoblast viability and bone mineral deposition [220–222]. Supporting this observation, several *in vivo* studies have also shown a more pronounced bone growth inside Si substituted HA leading to faster bone remodeling [223, 224]. When Si/CPC composites were deposited into bone defects in an animal model significantly greater amounts of new bone were observed as compared to CPC controls [225, 226].

It should be noted that the proposed roles of silicate in these materials has not been fully understood and remain in contention [225]. Several investigators have suggested that Si's more electronegative surface enhances Si bone formation, stimulates osteoblast differentiation and proliferation. Regardless of the mechanism of action, there is an increasing confirmation that silicon can plays major role in the treatment of bone defects [225].

8.4 Metal Nanoparticles

It is well known that metallic ions such as strontium, calcium, magnesium (Mg), and zinc (Zn) are necessary for bone growth and development [226]. These elements are major players during bone tissue development as they play a role in osteoblast activation and also in the inhibition of osteoblast-mediated bone degradation [226]. Numerous studies have demonstrated that the addition of metallic nanoparticles such as iron [227], magnesium [228], and titanium dioxide [229], into various scaffolds increase mechanical strength, promote cellular adhesion, and osteoblast functionality. These studies also showed potent increases in collagen synthesis, ALP and mineral deposition by osteoblasts, with gains in material properties, leading to enhanced tensile strength. One of the explanation for such a robust response is that nanotopographical cues that are critical for proper osteogenesis are facilitated by metal nanoparticle inclusion. Nanoscale surface features are known to critical for osteoblast adhesion and subsequent functionality [227]. A short review of the metal nanoparticles being explored as CPC additives is discussed below.

8.4.1 Copper and Zinc

Metallic nanoparticles of copper (Cu) and zinc (Zn) have been incorporated in scaffolds containing HA and showed increased antibacterial activity and no cytotoxicity to osteoprogenitor cells [230, 231]. When Cu was added to bioactive glass, CPC or HA, Cu-containing composites boosted new blood vessel formation and bone repair by providing the right conditions for new cell growth and tissue formation while suppressing bacterial infections [231, 232]. The constraints in using Cu directly or in Cu composite nanoparticles is that a layer of copper oxide forms that diminishes the Cu's antimicrobial activity [232].

8.4.2 Magnesium

Magnesium is one of the essential minerals for bone formation and studies support the concept that magnesium deficiencies lead to an increase in bone resorption coordinate with a decrease in bone formation [233]. Mg has been shown to influence adsorption of proteins known to promote the osteoblast function. Several studies have linked cellular cell proliferation, osteoid formation and mineralization and bone remodeling to the influence of Mg [234–236]. While current studies have described a relationship between Mg and Mg composites and bone formation, there remains no clear understanding regarding the effects of different concentrations of Mg ions on bone cells.

8.4.3   Zirconia

Finally, zirconia is being studied for inclusion in bone repair and replacement applications [237]. It has a number of properties that make it a beneficial material in bone composite materials. It is bioactive, biocompatible and mechanical strength. Its inherent properties can be substantially improved through surface modification or being combined with ceramics and bioactive glass [237, 238]. It is currently under active investigation as an implantable bioceramic, as a metallic implant coating, porous bone scaffold and, as a radiopacifier in bone cements [237]. Zirconia/TCP scaffolds were shown to significantly increase the compressive strength of composites without producing a cytotoxic effect [239]. Furthermore, CPC/HA/zirconia composites led to osteogenesis and induced bone tissue in the treatment of bony defects [240].

9   Future Studies

Bone diseases and bone injuries from trauma are a very common and increasing clinical problem. Whether occurring from bone cancer, developmental disorders, injury, osteomyelitis, or osteoporosis the most common occurrence is bone fracture [241]. Autogenic bone grafts are the most commonly used bone repair method. It does have a major disadvantage in that the quantity of extractable bone from this method is limited [242]. The alternative, allografts, also has a major disadvantage as they retain the potential as a further source of infection [242–244]. Because of the limitations of these methods and the increasing prevalence of bone injuries, there is a need to find better alternatives for bone repair [245].

Bone repair is a complex process, involving a cascade of cell-molecular interactions leading to osteoblast recruitment, stem cell differentiation and the production of mineralized matrix; subsequent remodeling events will produce the final bone tissue form [205]. CPC and HA have had a long history as biomaterial of choice for bone repair due to their biocompatibility, biodegradability and their high bioconductivity. CPCs are prone to fracture and this issue can be corrected through the addition of collagen, clay nanoparticles, fibers or polymer meshes [137]. Many approaches to developing CPC-based treatments are searching for new additives that can improve setting properties, injectability, material properties or in the case on many nanoparticles add new functionalities.

In this review chapter, the current therapy for bone regeneration was highlighted and promising approaches such as biodegradable and injectable hydrogels, growth factor-controlled delivery systems, BMSC cell-based therapy, nanomaterials, clay and metal nanoparticle polymer scaffolds that look promising for bone healing were explored. The advantages of these potential future clinical treatments foreshadow new treatment modalities that will see major improvement in patient care including reduced treatment times, decreased costs, acceleration in the bone-healing process and with the advent of 3D printing of medical devices and bioprinting patient-specific care. Further research is needed in several aspects in the field of bone regeneration and repair so that what is now considered novel becomes routine in the treatment of bone injuries. These include obtaining a detailed understanding of cell-material interactions, their influence in directing the course of bone tissue formation, finding the 'ideal' biomaterial or biomaterials, optimization of protein, and potentially, gene-based strategies, and further clinical studies that provide the testing and validation for these proposes treatment.

References

1. Liu Y, Lim, Teoh S-H. Review: development of clinically relevant scaffolds for vascularised bone tissue engineering. Biotechnol Adv. 2013;31:688–705.
2. Cortesini R. Stem cells, tissue engineering and organogenesis in transplantation. Transpl Immunol. 2005;15:81–9.
3. Amini AR, Laurencin CT, Nukavarapu SP. Bone tissue engineering: recent advances and challenges. Crit Rev Bioeng. 2012;40:363–408.
4. Cancedda R, Dozin B, Giannoni P, Quarto R. Tissue engineering and cell therapy of cartilage and bone. Matrix Biol. 2003;22:81–91.
5. Oryan A, Alidadi S, Moshiri A, Maffuli N. Bone regenerative medicine: classic options, novel strategies, and future directions. J. Orthop Surg Res. 2014;9(18):1–17.
6. Stevens MM. Biomaterials for bone tissue engineering. Mater Today. 2008;11(5):18–25.
7. Katagiri BT, Takahashi N. Regulatory mechanisms of osteoblast and osteoclast differentiation. Oral Dis. 2002;8:147–59.
8. Rose FR, Hou Q, Oreffo RO. Delivery systems for bone growth factors - the new players in skeletal regeneration. J Pharm Pharmacol. 2004;56:415–27.
9. Matassi F, et al. New biomaterials for bone regeneration. Clin Cases Min Bone Metab. 2011;8:21–4.
10. Khashaba RM, Moussa MM, Mettenburg DJ, Rueggeberg FA, Chutkan NB, Borke JL. Polymeric-calcium phosphate cement composites-material properties: in vitro and in vivo investigations. Int J Biomater. 2010; 2010: 691452, 14 pages.
11. Puppi D, Chiellini F, Piras AM, Chiellini E. Polymeric materials for bone and cartilage repair. Prog Poly Sci. 2010;35:403–40.
12. Griffin MF, Kalaskar DM, Seifalian A, Butler PE. An update on the application of nanotechnology in bone tissue engineering. Open Orthop J. 2016;10(Suppl-3, M4):836–48.
13. Rosa N, Simoes R, Magalhães FD, Marques AT. From mechanical stimulus to bone formation: a review. Med Eng Phys. 2015;37(8):719–28. https://doi.org/10.1016/j.medengphy.2015.05.015.
14. Ginebra MP. Cements as bone repair materials. In: Planell JA, editor. Bone repair biomaterials. Cambridge, England: Woodhead Publishing Limited; 2009.
15. Hollinger J, Einhorn TA, Doll F, Sfeir C. Bone tissue engineering. Boca Raton, FL: CRC Press; 2004.
16. Pilliar RM, Filiaggi M, Wells JD, Grynpas MD, Kandel RA. Porous calcium polyphosphate scaffolds for bone substitute applications in vitro characterization. Biomaterials. 2001;22:963–72.

17. Foppiano S, Marshall SJ, Marshall GW, Saiz E, Tomsia AP. The influence of novel bioactive glasses on in vitro osteoblast behavior. J Biomed Mater Res. 2004;71A:242–9.
18. Wang L, Singh M, Bonewald LF, Detamore MS. Signaling strategies for osteogenic differentiation of human umbilical cord mesenchymal stromal cells for 3D bone tissue engineering. J Tiss Eng Regen Med. 2009;3:398–404.
19. Dorozhkin SV, Epple M. Biological and medical significance of calcium phosphates. Angew Chem Int Ed. 2002;41:3130–46.
20. Barinov S, Komlev VS. Calcium phosphate bone cements. Inorg Mater. 2011;47(13):1470–85.
21. Bouler JM, Pilet P, Gauthier O, Verron E. Biphasic calcium phosphate ceramics for bone reconstruction: areview of biological response. Acta Biomater. 2017;53:1–12.
22. Dorozhkin SV. Calcium orthophosphates. J Mater Sci Mater Med. 2007;42(4):1061–95.
23. LeGeros RZ, Chohayeb A, Shulman A. Apatitic calcium phosphates: possible dental restorative materials. J Dent Res. 1982;61:343.
24. Link DP, van den Dolder J, van den Beucken JJ, Wolke JG, Mikos AG, Jansen JA. Bone response and mechanical strength of rabbit femoral defects filled with injectable CaP cements containing TGF-b1 loaded gelatin microspheres. Biomaterials. 2008;29:675–82.
25. Zhao L, Weir MD, Xu HHK. Human umbilical cord stem cell encapsulation in calcium phosphate scaffolds for bone engineering. Biomaterials. 2010;31:3848–57.
26. Ducheyne P, Qiu Q. Bioactive ceramics: the effect of surface reactivity on bone formation and bone cell function. Biomaterials. 1999;20:2287–303.
27. Bohner M. Reactivity of calcium phosphate cements. J Mater Chem. 2007;17(38):3980–92.
28. Friedman CD, Costantino PD, Takagi S, Chow LC. BoneSource hydroxyapatite cement: a novel biomaterial for craniofacial skeletal tissue engineering and reconstruction. J Biomed Mater Res. 1998;43:428–32.
29. Kühn KD. Properties of bone cement. In: Breusch S, editor. The well-cemented total hip arthroplasty. Heidelberg: Springer MedizinVerlag; 2005. p. 52–9.
30. Ginebra MP, Traykova T, Planell JA. Calcium phosphate cements: competitive drug carriers for the musculoskeletal system? Biomaterials. 2006;27:2171–7.
31. O'Dowd-Booth CJ, White J, Smitham P, Khan W, Marsh DR. Bone cement: perioperative issues, orthopaedic applications and future developments. J Perioper Pract. 2011;21(9):304–8.
32. Constantz BR, Ison IC, Fulmer MT, Poser RD, Smith ST, Van Wagoner M, et al. Skeletal repair by in situ formation of the mineral phase of bone. Science. 1995;267:1796–9.
33. DiMaio FR. The science of bone cement: a historical review. Orthopedics. 2002;25(12):1399–407.
34. Komlev VS, Fadeeva IV, Gurin N, Shvorneva LI, Bakunova NV, Barino SM. New calcium phosphate cements based on tricalcium phosphate. Dokl Chem. 2011;437(1):75–8.
35. Eliaz N, Metok N. Calcium phosphate bioceramics: a review of their history, structure, properties, coating technologies and biomedical applications. Materials. 2017;10(4):334. https://doi.org/10.3390/ma10040334.
36. Perez RA, Kim HW, Ginebra MP. Polymeric additives to enhance the functional properties of calcium phosphate cements. J Tiss Eng. 2012;3(1):2041731412439555. https://doi.org/10.1177/2041731412439555. published online 20 March.
37. Maestretti G, Cremer C, Otten P, Jakob RP. Prospective study of standalone balloon kyphoplasty with calcium phosphate cement augmentation in traumatic fractures. Eur Spine J. 2007;16:601–10.
38. Aral A, Yalçin S, Karabuda ZC, Anil A, Jansen JA, Mutlu Z. Injectable calcium phosphate cement as a graft material for maxillary sinus augmentation: an experimental pilot study. Clin Oral Implants Res. 2008;19:612–7.
39. Kroeses-Deitman HC, Wolke JG, Spauwen PH, Jansen JA. Closing capacity of cranial bone defects using porous calcium phosphate cement implants in a rabbit animal model. J Biomed Mater Res A. 2006;79:503–11.
40. Libicher M, Hillmeier J, Liegibel U, Sommer U, Pyerin W, Vetter M. Osseous integration of calcium phosphate in osteoporotic vertebral fractures after kyphoplasty: initial results from a clinical and experimental pilot study. Osteoporos Int. 2006;17:1208–15.

41. Mermelstein LE, Chow LC, Friedman CD, Crisco JJ. The reinforcement of cancellous bone screws with calcium phosphate cement. J Orthop Trauma. 1996;10:15–20.
42. Ooms E, Wolke J, Van der Waerden J, Jansen J. Use of injectable calcium-phosphate cement for the fixation of titanium implants: an experimental study in goats. J Biomed Mater Res B Appl Biomater. 2003;66:447–56.
43. Takemasa R, Kiyasu K, Tani T, Inoue S. Validity of calcium phosphate cement vertebroplasty for vertebral non-union after osteoporotic fracture with middle column involvement. Spine J. 2007;7:148S.
44. Lewis G. Injectable bone cements for use in vertebroplasty and kyphoplasty: state-of-the-art review. J Biomed Mater Res B Appl Biomat. 2006;76:456–68.
45. Calcium phosphate: structure, synthesis, properties, and applications. In: Robert B, editor. Heimann: Biochemistry Research Trends; 2012. 498pp. ISBN: 978-1-62257-299-1.
46. Habraken H, Habibovic P, Epple M, Bohner M. Calcium phosphates in biomedical applications: materials for the future? Mat Today. 2016;19(2):69–87.
47. Barinov SM. Trends in development of calcium phosphate-based ceramic and composite materials for medical applications: transition to nanoscale. Russian J Gen Chem. 2010;80:666–74.
48. Dutta PK. Chitin and chitosan for regenerative medicine. In: Springer series on polymer and composite materials. Berlin: Springer; 2015.
49. Zhang JT, Tancret F, Bouler JM. Fabrication and mechanical properties of calcium phosphate cements (CPC) for bone substitution. Mater Sci Eng. 2011;31(4):740–7.
50. Driessens FCM, Planell J, Boltong MG, Khairoun I, Ginebra MP. Osteotransductive bone cements. J Eng Med. 1998;212(6):427–35.
51. Bigi A, Bracci B, Panzavolta S. Effect of added gelatin on the properties of calcium phosphate cement. Biomaterials. 2004;25(14):2893–9.
52. Rinaudo M. Chitin and chitosan: properties and applications. Prog Poly Sci. 2006;31(7):603–32.
53. Shi C, Zhu Y, Ran X, Wang M, Su Y, Cheng T. Therapeutic potential of chitosan and its derivatives in regenerative medicine. J Surg Res. 2006;133(2):185–92.
54. Padois K, Rodriguez F. Effects of chitosan addition to self-setting bone cement. Biomed Mater Eng. 2007;17(5):309–20.
55. Sun L, Hockin H, Xu K, Takagi S, Chow LC. Fast setting calcium phosphate cement–chitosan composite: mechanical properties. J Biomat Appl. 2007;21(3):299–315. https://doi.org/10.1177/0885328206063687.
56. Chesnutt BM, Viano AM, Yuan Y, Yang Y, Guda T, Appleford MR, Ong JL, Haggard WO, Bumgardner JD. Design and characterization of a novel chitosan/nanocrystalline calcium phosphate composite scaffold for bone regeneration. J Biomed Mater Res A. 2009;88(2):491–502.
57. Al-Bayaty FH, Kamaruddin AA, Ismail MA, Abdulla MA. Formulation and evaluation of a new biodegradable periodontal chip containing thymoquinone in a chitosan base for the management of chronic periodontitis. J Nanomat. 2013;2013:397308., 5 pages. https://doi.org/10.1155/2013/397308.
58. Janmey PA, Winer JP, Weisel JW. Fibrin gels and their clinical and bioengineering applications. J R Soc Interface. 2009;6(30):1–10.
59. Balakrishnan B, Mohanty M, Umashanker PR, Jayakrishnan A. Evaluation of an in situ forming hydrogel wound dressing based on oxidized alginate and gelatin. Biomaterials. 2005;26:6335–42.
60. Cui G, Li J, Lei W, et al. The mechanical and biological properties of an injectable calcium phosphate cement-fibrin glue composite for bone regeneration. J Biomed Mater Res B. 2010;92(2):377–85.
61. Lopez-Heredia MA, Pattipeilohy J, Hsu S, van der Wieden B, Leewenburg SC, et al. Bulk physicochemical, interconnectivity, and mechanical properties of calcium phosphate cements-fibrin glue composites for bone substitute applications. J Biomed Mat Res A. 2013;101(2):478–90.

62. Kneser U, Voogd A, Ohnolz J, et al. Fibrin gel-immobilized primary osteoblasts in calcium phosphate bone cement: in vivo evaluation with regard to application as injectable biological bone substitute. Cells Tissues Organs. 2005;179(4):158–69.

63. Dong J, Cui G, Bi L, Lei W. The mechanical and biological studies of calcium phosphate cement-fibrin glue for bone reconstruction of rabbit femoral defects. Int J Med. 2013;8:1317–24. https://doi.org/10.2147/IJN.S42862.

64. Lee L-T, Kwan P-C, Chen Y-F, Wong Y-K. Comparison of the effectiveness of autologous fibrin glue and macroporous biphasic calcium phosphate as carriers in the osteogenesis process with or without mesenchymal stem cells. J Chin Med Assoc. 2008;1(2):66–73.

65. Gholipour H, Meimandi-Parizi A, Oryan A, Bigham SA. The effects of gelatin, fibrin-platelet glue and their combination on healing of the experimental critical bone defect in a rat model: radiological, histological, scanning ultrastructural and biomechanical evaluation. Cell Tiss Bank. 2017:1–16. Epub 2017 Dec 20.

66. Meimandi-Parizi A, Oryan A, Gholipour H. Healing potential of nanohydroxyapatite, gelatin, and fibrin-platelet glue combination as tissue engineered scaffolds in radial bone defects of rats. Conn Tiss Res. 2017;16:1–13.

67. Noori A, Ashrafi S, Vaez-Ghaemi R, Hatamian-Zaremi A, Webster TJ. A review of fibrin and fibrin composites for bone tissue engineering. Intern J Nanomed. 2017;12:4937–61. https://doi.org/10.2147/IJN.S12467183.

68. Gorgieva S, Kokol V. Collagen vs. gelatine-based biomaterials and their biocompatibility: review and perspectives, biomaterials applications for nanomedicine. In: Pignatello R, editor; 2011. ISBN: 978-953-307-661-4.

69. Unuma H, Matsuchima Y. Preparation of calcium phosphate cement with an improved setting behavior. J Asian Ceramic Soc. 2013;1(1):26–9.

70. Azami M, Mohamma RE, Fathollah M. Gelatin/hydroxyapatite nanocomposite scaffolds for bone repair. Plast Res. 2010. doi: https://doi.org/10.1002/spepro.003073.

71. Kim HW, Knowles JC, Kim HE. Hydroxyapatite and gelatin composite foams processed via novel freeze-drying and crosslinking for use as temporary hard tissue scaffolds. J Biomed Mater Res. 2005;A72:136–45.

72. Kim W, Kim HE, Salih V. Stimulation of osteoblast responses to biomimetic nanocomposites of gelatin-hydroxyapatite for tissue engineering scaffolds. Biomaterials. 2005;26:5221–30. https://doi.org/10.1016/j.biomaterials.2005.01.047.

73. Habraken WJ, Wolke JG, Mikos AG, et al. Porcine gelatin microsphere/calcium phosphate cement composites: an in vitro degradation study. J Biomed Mater Res B. 2009;91(2):555–61.

74. Oryan A, Alidadi S, Sadegh B, Mishiri A. Comparative study on the role of gelatin, chitosan and their combination as tissue engineered scaffolds on healing and regeneration of critical sized bone defects: an in vivo study. J Mater Sci Mater Med. 2016;27(10):155–61. https://doi.org/10.1007/s10856-016-5766-6.

75. Sionkowska A, Skrzyński S, Śmiechowski K, Kołodziejczak A. The review of versatile application of collagen. Polym Adv Technol. 2017;28:4–9. https://doi.org/10.1002/pat.3842.

76. Zhang J, Liu W, Schnitzler V, Tancret F, Bouler JM. Calcium phosphate cements for bone substitution: chemistry, handling and mechanical properties. Acta Biomater. 2013;10(3):1035–49.

77. Dong C, Lv Y. Application of collagen scaffold in tissue engineering: recent advances and new perspectives. Polymers. 2016;8(2):42. https://doi.org/10.3390/polym8020042.

78. Ferreira AM, Gentile P, Chono V, Ciardelli G. Collagen for bone tissue regeneration. Acta Biomat. 2012;8(9):3191–200.

79. Wang P, Zhao L, Liu J, Weir MD, Zhou X, Xu H. Bone tissue engineering via nanostructured calcium phosphate biomaterials and stem cells. Bone Res. 2014;2:14017. https://doi.org/10.1038/boneres.2014.17.

80. Perez RA, Altankov G, Jorge-Herrero E, Ginebra MP. Micro- and nanostructured hydroxyapatite–collagen microcarriers for bone tissue-engineering applications. J Tissue Eng Regen Med. 2013;7:353–61. https://doi.org/10.1002/term.530.

81. Walsh WR, Oliver RA, Christou C, Lovric V, Walsh ER, Prado GR, et al. Critical size bone defect healing using collagen–calcium phosphate bone graft materials. PLoS One. 2017;12(1):e0168883. https://doi.org/10.1371/journal.pone.0168883.

82. Palmer I, Nelson J, Schatton W, Dunne NJ, Buchanan F, Clarke SA. Biocompatibility of calcium phosphate bone cement with optimised mechanical properties. J Biomed Mater Res B Appl Biomater. 2015:1–8. https://doi.org/10.1002/jbm.b.33370.

83. Maas M, Guo P, Keeney M, et al. Preparation of mineralized nanofibers: collagen fibrils containing calcium phosphate. Nano Lett. 2011;11:1383–8.

84. Kikuchi M, Itoh S, Ichinose S, Shinomiya K, Tanaka J. Self-organization mechanism in a bone-like hydroxyapatite/collagen nanocomposite synthesized *in vitro* and its biological reaction *in vivo*. Biomaterials. 2001;22:1705–11.

85. Aberg J, Brisby H, Henriksson HB, et al. Premixed acidic calcium phosphate cement: characterization of strength and microstructure. J Biomed Mater Res B. 2010;93(2):436–41.

86. Carey LE, Xu HH, Simon CG, et al. Premixed rapid-setting calcium phosphate composites for bone repair. Biomaterials. 2005;26(24):5002–14.

87. Takagi S, Chow LC, Hirayama S, et al. Premixed calcium–phosphate cement pastes. J Biomed Mater Res B. 2003;67(2):689–96.

88. Tozzi G, Mori A, Oliveira A, Roldo M. Composite hydrogels for bone regeneration. Materials. 2016;9:267. https://doi.org/10.3390/ma9040267.

89. Short AR, et al. Hydrogels that allow and facilitate bone repair, remodeling, and regeneration. J Mater Chem B Mater Biol Med. 2015;3(40):7818–30.

90. Planell JA, Best SM, Lacroix D, Merolli A. Bone repair biomaterials. Amsterdam: CRC Press, Elsevier; 2009.

91. Nedde AT, Julich-Gruner KK, Leindlein A. Combinations of biopolymers and synthetic polymers for bone regeneration. Chapter 4. In: Dubruel P, Vlierberghe SV, editors. Biomaterials for bone regeneration: novel techniques and applications. Amsterdam: Elsevier; 2014. p. 87–110. https://doi.org/10.1533/9780857098104.1.87.

92. Susana Cortizo M, Soledad Belluzo M. Biodegradable polymers for bone tissue engineering. In: Goyanes N, D'Accorso NB, editors. Industrial applications of renewable biomass products. Berlin: Springer International Publishing AG; 2017. p. 47–74S. https://doi.org/10.1007/978-3-319-61288-1_2.

93. Kroeze RJ, Helder MN, Govaert LE, Smit TH. Biodegradable polymers in bone tissue engineering. Mater. 2009;2:833–56. https://doi.org/10.3390/ma2030833.

94. Sheikh Z, Najeeb S, Khurshid Z, Verma V, Rashid H, Glogauer M. Biodegradable materials for bone repair and tissue engineering applications. Mater. 2015;8(9):5744–94. https://doi.org/10.3390/ma8095273.

95. Engstrand J, Persson C, Engqvist H. Influence of polymer addition on the mechanical properties of premixed calcium phosphate cement. Biomatter. 2013;3(4):e27249. https://doi.org/10.4161/biom.27249.

96. Geffers M, Groll J, Gbureck U. Reinforcement strategies for load-bearing calcium phosphate biocements. Materials. 2015;8:2700–17. https://doi.org/10.3390/ma8052700.

97. Sun J, Tan H. Alginate-based biomaterials for regenerative medicine applications. Materials. 2013;6:1285–308.

98. Wang L, Wang P, Weir MD, et al. Hydrogel fibers encapsulating human stem cells in an injectable calcium phosphate scaffold for bone tissue engineering. Biomed Mater. 2016;11:065008.

99. Venkstesan J, Nithya R, Kim SK. Role of alginate in bone tissue engineering. Adv Food Nutr Res. 2014;73:45–57.

100. Venkatesan J, Bhatnagarb I, Manivasagana P, Kanga K-H, Kima SK. Alginate composites for bone tissue engineering: a review. Int J Biol Macromol. 2015;72:269–81.

101. Zhao L, Weir MD, Xu HHK. An injectable calcium phosphate-alginate hydrogel-umbilical cord mesenchymal stem cell paste for bone tissue engineering. Biomaterials. 2010;31:6502–10.

102. Thein-Han WW, WahWah MD, Weir CG, Wu HH. Novel non-rigid calcium phosphate scaffold seeded with umbilical cord stem cells for bone tissue engineering. J Tiss Eng Regen. 2013;7(10):777–87.

103. Wang X, Chen L, Xiang H, et al. Influence of anti-washout agents on the rheological properties and injectability of a calcium phosphate. J Biomed Mater Res B. 2007;81(2):410–8.

104. Karnik S, Jammalamadaka U, Tappa K, Mills DK. Performance evaluation of nanoclay enriched anti-microbial hydrogels for biomedical applications. Heliyon. 2016;2(2):e00072. https://doi.org/10.1016/j.heliyon.2016.e00072.
105. Karnik S, Mills DK. Nanoenhanced hydrogel system with sustained release capabilities. J Biomed Mater Res A. 2015;103(7):2416–26.
106. Wang P, Song Y, Weir MD, Sun J, Zhao L, Simon CG, Xu HH, et al. A self-setting iPSMSC-alginate-calcium phosphate paste for bone tissue engineering. Dental mater. 2018;32(2):252–63.
107. Costa-Pinto AR, Reis RL, Neves NM. Scaffolds based bone tissue engineering: the role of chitosan. Tiss Eng B Rev. 2011;17:331–47.
108. Muzzarelli RAA. Chitins and chitosans for the repair of wounded skin, nerve, cartilage and bone. Carbohydr Polym. 2009;76:167182.
109. Oliveira JM, Rodrigues MT, Silva SS, Malafaya PB, Gomes ME, Viegas CA, et al. Novel hydroxyapatite/chitosan bilayered scaffold for osteochondral tissue engineering applications: scaffold design and its performance when seeded with goat bone marrow stromal cells. Biomaterials. 2006;27:6123–37.
110. Zeng S, Liu L, Shi Y, Qiu J, Fang W, Rong M, Guo Z, Gao W. Characterization of silk fibroin/chitosan 3D porous scaffold and in vitro cytology. PLoS One. 2015;10(6):e0128658. Epub 2015 Jun 17.
111. Li J, Wang Q, Gu Y, Zhu Y, Chen L, Chen Y. Production of composite scaffold containing silk fibroin, chitosan, and gelatin for 3D cell culture and bone tissue regeneration. Med Sci Monit. 2017;23:5311–20.
112. Frihberg ME, Katsman A, Mondrinos MJ, Stabler CT, Hankenson KD, Oristaglio JT, Lelkes PI. Osseointegrative properties of electrospun hydroxyapatite-containing nanofibrous chitosan scaffolds. Tissue Eng Part A. 2015;21(5–6):979–81.
113. Zhang Y, Reddy VJ, Wong SY, Li X, Su B, Ramakrishna S, et al. Enhanced biomineralization in osteoblasts on a novel electrospun biocomposite nanofibrous substrate of hydroxyapatite/collagen/chitosan. Tissue Eng Part A. 2010;16:1949–60.
114. Venugopal J, Low S, Choon AT, Sampath Kumar TS, Ramakrishna S. Mineralization of osteoblasts with electrospun collagen/hydroxyapatite nanofibers. J Mater Sci Mater Med. 2008;19:2039–46.
115. Wnek GE, Bowlin GL. Encyclopedia of biomaterials and biomedical engineering. New York: Marcel Dekker; 2004.
116. Bleek K, Taubert A. New developments in polymer-controlled, bioinspired calcium phosphate mineralization from aqueous solution. Acta Biomater. 2013;9(5):6283–321.
117. Ratner BD, Hoffman AS, Schoen FJ, Lemons JE. Biomaterials science: aintroduction to materials in medicine. 2nd ed. London: Elsevier Academic Press; 2004.
118. Rebelo R, Fernandesa M, Fangueiroa R. Biopolymers in medical implants: a brief review. Process Eng. 2017;200:236–43.
119. Tereshchenko VP, Kirlova A, Sadavoy MA, Larionov PM. The materials used in bone tissue engineering. In: AIP Conference Proceedings. vol. 1688, 030022; 2015. doi: https://doi.org/10.1063/1.4936017
120. Melinda Molnar R, Bodnar M, Hartmann JF, Borbely J. Preparation and characterization of poly(acrylic acid)-based nanoparticles. Coll Poly Sci. 2009;287(6):739–44.
121. Verma D, Katti K, Mohanty B. Mechanical properties of biomimetic composites for bone tissue engineering. MRS Proc. 2004;844:Y6.2. https://doi.org/10.1557/PROC-844-Y6.2.
122. Stevens B, Yang Y, Mohandas A, Stucker B, Nguyen KT. A review of materials, fabrication method and strategies used to enhance bone regeneration. J Biomed Mater Res. 2008;85B:573–82.
123. He H, Qiao Z, Liu C. Accelerating biodegradation of calcium phosphate cement. In: Liu C, He H, editors. Developments and applications of calcium phosphate bone cements, Chapter. 5. Singapore: Springer; 2018. p. 227–56.

124. Shim J-H, Moon T-S, Yun M-J, Jeon Y-C, Jeong C-M, Cho D-W, Huh J-B. Stimulation of healing within a rabbit calvarial defect by a PCL/PLGA scaffold blended with TCP using solid freeform fabrication technology. J Mater Sci Mater Med. 2012;23:2993–3002.

125. Park SA, Lee SH, Kim WD. Fabrication of porous polycaprolactone/hydroxyapatite (PCL/HA) blend scaffolds using a 3D plotting system. Bioprocess Biosyst Eng. 2011;34:505. https://doi.org/10.1007/s00449-010-0499-2.

126. Liao HT, Lee MY, Tsai WW, Wang HV, Lu WC. Osteogenesis of adipose-derived stem cells on polycaprolactone-β-tricalcium phosphate scaffold fabricated via selective laser sintering and surface coating with collagen type I. J Tissue Eng Regen Med. 2016;10(10):E337–53. https://doi.org/10.1002/term.1811. Epub 2013 Aug 16.

127. Nyberg E, Rindone A, Dorafshar A, Grayson WL. Comparison of 3D-printed poly-ε-caprolactone scaffolds functionalized with tricalcium phosphate, hydroxyapatite, Bio-Oss, or Decellularized Bone Matrix. Tissue Eng Part A. 2017;23(11–12):503–14.

128. http://www.bioinksolutions.com/

129. Ghosh SB, Bandyopadhyay-Ghosh S, Sain M. Composites. Chapter 18. In: Auras R, Lim L-T, Selke SEM, Tsuji H, editors. Poly(lactic acid): synthesis, structures, properties, processing, and applications. Hoboken, NJ: Wiley; 2010. https://doi.org/10.1002/9780470649848.

130. Adeosun SO, Lawal GI, Gbenebor P. Characteristics of biodegradable implants. J Min Mater Charact Eng. 2014;2:88–106.

131. Rajendran T, Venugopalan S. Role of polylactic acid in bone regeneration–a systematic review. J Pharm Sci Res. 2015;7(11):960–6.

132. Rezwan K, Chen QZ, Blaker JJ, Boccaccini AR. Biodegradable and bioactive porous polymer/inorganic composite scaffolds for bone tissue engineering. Biomaterials. 2006;27(18):3413–31.

133. Tanataweethum N, Liu W, Goebel W, Li D, Chu T. Fabrication of poly-l-lactic acid/dicalcium phosphate dihydrate composite scaffolds with high mechanical strength—implications for bone tissue engineering. J Funct Mater. 2015;6(4):1036–53. https://doi.org/10.3390/jfb6041036.

134. Liu X, Liu H-Y, Lian X, Shi X-L, Wang W, Cui F-Z, Zhang Y. Osteogenesis of mineralized collagen bone graft modified by PLA and calcium sulfate hemihydrate: in vivo study. J Biomater Appl. 2012;28(1):12–9.

135. Montjovent M-O, Silke S, Mathieu L, Scaletta C, Scherberich S, et al. Human fetal bone cells associated with ceramic reinforced PLA scaffolds for tissue engineering. Bone. 2008;42:554–64.

136. Lou C-W, Chen W-C, Luo C-T, Huang C-C, Lin JH. Compressive strength of porous bone cement/polylactic acid composite bone scaffolds. Appl Mech Mater. 2013;365-366:1062–5.

137. Danoux CB, Barberi D, Yuan H, de Brulin JD, van Blitterswilk CA, et al. In vitro and in vivo bioactivity assessment of a polylactic acid/hydroxyapatite composite for bone regeneration. Biomaterials. 2014;4:e27664. PMC Web 22 Feb. 2018.

138. Gentile P, Chiono V, Carmagnola I, Hatton PV. An overview of poly(lactic-co-glycolic) acid (PLGA)-based biomaterials for bone tissue engineering. Intern J Mol Sci. 2014;15(3):3640–59. https://doi.org/10.3390/ijms15033640.

139. Sun X, Chu X, Ye Q, Wang C. Poly(lactic-co-glycolic acid): applications and future prospects for periodontal tissue regeneration. Polymers. 2017;9:189. https://doi.org/10.3390/polym9060189.

140. Ortega-Oiler I, Padial-Molina M, Galindo-Moreno P, O'Valle F, et al. Bone regeneration from PLGA micro-nanoparticles. Biomed Res Int. 2015;2015:415289. https://doi.org/10.1155/2015/415289.

141. Kane RJ, Weiss-Bilka HE, Meagher MJ, et al. Hydroxyapatite reinforced collagen scaffolds with improved architecture and mechanical properties. Acta Biomater. 2015;17:16–25. https://doi.org/10.1016/j.actbio.2015.01.031.

142. Demirci DS, Bayir Y, Halici Z, Karakus E, et al. Boron containing poly-(lactide-co-glycolide) (PLGA) scaffolds for bone tissue engineering. Mater Sci Eng C. 2014;44:246–53. https://doi.org/10.1016/j.msec.2014.08.035.

143. Mousa M, Evans ND, Oreffo ROC, Lawson JI. Clay nanoparticles for regenerative medicine and biomaterial design: a review of clay bioactivity. Biomaterials. 2018;159:204–14. https://doi.org/10.1016/j.biomaterials.2017.12.024.

144. Ruiz-Hitzky E, Aranda P, Dardera M, Rytwobc G. Hybrid materials based on clays for environmental and biomedical applications. J Mater Chem. 2010;20:9306–21.

145. Newman P, Minett HA, Ellis-Behnke R, Zreiqat H. Carbon nanotubes: their potential and pitfalls for bone tissue regeneration and engineering. Nanomedicine. 2013;9(8):1139–58.

146. Tanaka M, Sato Y, Haniu H, Sato H, et al. A three-dimensional block structure consisting exclusively of carbon nanotubes serving as bone regeneration scaffold and as bone defect filler. PLoS One. 2017;12(2):e0172601. https://doi.org/10.1371/journal.pone.0172601.

147. Mukharjee S, Kumar S, Kundu B, Chanda A, Sen S, Das PK. Enhanced bone regeneration with carbon nanotube reinforced hydroxyapatite in animal model. J Mech Behav Biomed Mater. 2016;60:243–55.

148. Venkatesan J, Pallela R, Kim SK. Applications of carbon nanomaterials in bone tissue engineering. J Biomed Nanotechnol. 2014;10:3105–23.

149. Liu M, Jia Z, Jia D, Zhou C. Recent advance in research on halloysite nanotubes-polymer nanocomposite. Prog Polym Sci. 2014;39:1498–525.

150. Leporatti S. Halloysite clay nanotubes as nano-bazookas for drug delivery. Polym Int. 2017;66:1111–8. https://doi.org/10.1002/pi.5347.

151. Fan L, Zhang J, Wang A. In situ generation of sodium alginate/hydroxyapatite/halloysite nanotubes nanocomposite hydrogel beads as drug-controlled release matrices. J Mater Chem B. 2013;1:6261–70.

152. Lvov Y, Wang W, Zhang L, Fakhrullin R. Halloysite clay nanotubes for loading and sustained release of functional compounds. Adv Mater. 2016;28:1227–50.

153. Mills DK, Jammalamadaka U, Tappa UK, Weisman JA. Studies on the cytocompatibility, mechanical and antimicrobial properties of 3D printed poly(methyl methacrylate) beads. Bioactive Mater. 2018;3(1):157–66.

154. Naumenko EA, Guryanov ID, Yendluri R, Lvov YM, Fakhrullin RF. Clay nanotube-biopolymer composite scaffolds for tissue engineering. Nanoscale. 2016;8:7257–71.

155. Liu M, Wu C, Jiao Y, Xiong S, Zhou C. Chitosan-halloysite nanotubes nanocomposite scaffolds for tissue engineering. J Mater Chem B. 2013;1:2078–89.

156. Massaro M, Lazzara G, Milioto S, Noto R, Riela S. Covalently modified halloysite clay nanotubes: synthesis, properties, biological and medical applications. J Mater Chem B. 2017;5:2867–82.

157. Jammalamadaka U, Tappa K, Mills DK. Calcium phosphate/clay nanotube bone cement with enhanced mechanical properties and sustained drug release. In: Zoveidavianpoor M, editor. Clay science and engineering. London: InTech Publishers. (in press) Publication date: May 2018.

158. Karnik S, Mills DK. Clay nanotubes as growth factor delivery vehicle for bone tissue engineering. J Nanomed Nanotechnnol. 2013;4(6):102.

159. Tappa K, Jammalamadaka U, Mills DK. Formulation and evaluation of nanoenhanced antibacterial calcium phosphate bone cements. In: Webster T, Li B, editors. Orthopedic biomaterials. New York, NY: Springer. (in press) May 2018.

160. Tomas H, Alves CS, Rodrigues J. Laponite®: a key nanoplatform for biomedical applications?. Nanomed Nanotech Biol Med. 2017, in press.

161. Jung H, Kim HM, Choy YB, Hwang SJ, Choy JH. Itraconazole-laponite: kinetics and mechanism of drug release. Appl Clay Sci. 2008;40(1–4):99–107.

162. Wang C, Wang S, Li K, Lu Y, Li J, Zhang Y, Li J, Liu X, Shi X, Zhao Q. Preparation of laponite bioceramics for potential bone tissue engineering applications. PLoS One. 2014;23:e99585. https://doi.org/10.1371/journal.pone.0099585.

163. Xavier JR, Thakur T, Desai P, Jaiswal MK, Sears N, Cosgriff-Hernandez E, Kaunas R, Gaharwar AK. Bioactive nanoengineered hydrogels for bone tissue engineering: a growth-factor-free approach. ACS Nano. 2015;9(3):3109–18. https://doi.org/10.1021/nn507488s.

164. Thorpe A, Freeman C, Farthing P,Hatton P, Brook I, Sammon C, Le Maitre CL. Osteogenic differentiation of human mesenchymal stem cells in hydroxyapatite loaded thermally triggered, injectable hydrogel scaffolds to promote repair and regeneration of bone defects. In: Frontiers in bioengineering and biotechnology. Conference abstract: 10th world biomaterials congress; 2016. doi: 10.3389/conf.FBIOE.2016.01.00636

165. Tao L, Zhonglong L, Ming X, Zezheng Y, Zhiyuan L, Xiaojun Z. In vitro and in vivo studies of a gelatin/carboxymethyl chitosan/LAPONITE® composite scaffold for bone tissue engineering. RSC Adv. 2017;7:54100.

166. Jayrajsinh S, Shankar G, Agrawal YK, Bakre L. Montmorillonite nanoclay as a multifaceted drug-delivery carrier: a review. J Drug Delivery Sci Technol. 2017;39:200–9.

167. Aguzzi C, Cerezo P, Viseras C, Caramella C. Use of clays as drug delivery systems: possibilities and limitations. Appl Clay Sci. 2007;36:22–36. https://doi.org/10.1016/j.clay.2006.06.015.

168. Baker KC, Maerz T, Saad H, Shaheen P, Kannan RM. In vivo bone formation by and inflammatory response to resorbable polymer-nanoclay constructs. Nanomedicine. 2015;11(8):1871–81. https://doi.org/10.1016/j.nano.2015.06.012. Epub 2015 Jul 26.

169. Olad A, Azhar FF. The synergetic effect of bioactive ceramic and nanoclay on the properties of chitosan–gelatin/nanohydroxyapatite–montmorillonite scaffold for bone tissue engineering. Ceram Int. 2014;40(7):10061–72.

170. Kar S, Kaur T, Thirugnanam A. Microwave-assisted synthesis of porous chitosan–modified montmorillonite–hydroxyapatite composite scaffolds. Inter J Biol Macromol. 2016;82:628–36.

171. Kwon SY, Cho EH, Kim SS. Preparation and characterization of bone cements incorporated with montmorillonite. J Biomed Mater Res. 2007;83B:276–84. https://doi.org/10.1002/jbm.b.30793.

172. Sharma C, Dinda AK, Potdar PD, Chu C-F, Mishra NC. Fabrication and characterization of novel nano-biocomposite scaffold of chitosan–gelatin–alginate–hydroxyapatite for bone tissue engineering. Mater Sci Eng C. 2016;64:416–27.

173. Hamzah AA, Selvarajan RS, Majlis BY. Graphene for biomedical applications: a review. Sains Malaysiana. 2017;46(7):1125–39. https://doi.org/10.17576/jsm-2017-4607-16.

174. Pattnaik S, Swain K, Linc Z. Graphene and graphene-based nanocomposites: biomedical applications and biosafety. J Mater Chem B. 2016;4:7813–31.

175. Nasrin S, Hasanzadeh M. Graphene and its nanostructure derivatives for use in bone tissue engineering: recent advances. J Biomed Mater Res Part A. 2016;104A:1250–75.

176. Reina G, Criado A, Prato M, Gonzalez-Domınguez JM, Vazques E, Bianco A. Promises, facts and challenges for graphene in biomedical applications. Chem Soc Rev. 2017;46:4400–16.

177. Kalbacova M, Bronz A, Kong J, Kalbac M. Graphene substrates promote adherence of human osteoblasts and mesenchymal stromal cells. Carbon. 2010;48:4323–9.

178. Nayak TR, Andersen H, Makam VS, Khaw C, Bae S, Xu X, P-LR E, Ahn JH, Hong BH, Pastorin G. Graphene for controlled and accelerated osteogenic differentiation of human mesenchymal stem cells. ACS Nano. 2011;5(6):4670–8.

179. Dubey N, Bentini R, Islam I, Cao T, Neto AHC, Rosa V. Graphene: a versatile 531 carbon-based material for bone tissue engineering. Stem Cells Int. 2015;2015:804213.

180. Tommila M, Jokilammi A, Penttinen R, Ekholm E. Cellulose—a biomaterial with cell-guiding property. In: van de Ven T, Godbout L. Cellulose-medical, pharmaceutical and electronic applications, chapter 5. Croatia: InTech. ISBN: 978-953-51-1191-7. 314 pages.

181. Beladia F, Saber-Samandarib S, Saber-Samandaric S. Cellular compatibility of nanocomposite scaffolds based on hydroxyapatite entrapped in cellulose network for bone repair. Mater Sci Eng C. 2017;75:385–92.

182. Teti G, Orsini G, Mazzotti A, Belmonte M, Ruggeri A. 3D polysaccharide based hydrogel for bone tissue engineering. Ital J Anat Embryol. 2015;120(1):129. https://doi.org/10.13128/IJAE-17000.

183. Novotna K, Havelka P, Sopuch T, Kolarova K, et al. Cellulose-based materials as scaffolds for tissue engineering. Cellulose. 2013;20(5):2263–78.

184. Aravamudhan A, Ramos DM, Nip J, Kalajzic I, Kumbar SG. Micro-nanostructures of cellulose-collagen for critical sized bone defect healing. Macromol Biosci. 2018;18(2). https://doi.org/10.1002/mabi.201700263. Epub 2017 Nov 27.
185. Moreau JL, Weir MD, Xu HH. Self-setting collagen-calcium phosphate bone cement: mechanical and cellular properties. J Biomed Mater Res. 2009;91A:605–13.
186. Kikuchi M, Ikoma T, Itoh S, Matsumoto HN, Koyama Y, Takakuda K, Shinomiya K, Tanaka J. Biomimetic synthesis of bone-like nanocomposites using the self-organization mechanism of hydroxyapatite and collagen. Compos Sci Technol. 2004;64(6):819–25.
187. Kikuchi M. Hydroxyapatite/collagen bone-like nanocomposite. Biol Pharm Bull. 2013;36(11):1666–9.
188. Sarkar SK, Lee BT. Hard tissue regeneration using bone substitutes: an update on innovations in materials. Korean J Intern Med. 2015;30:279–93. https://doi.org/10.3904/kjim.2015.30.3.279.
189. Bohner M, Baroud G. Injectability of calcium phosphate pastes. Biomaterials. 2005;26:1553–63.
190. Blokhuis TJ. Formulations and delivery vehicles for bone morphogenetic proteins: latest advances and future directions. Injury. 2009;40(Suppl 3):S8–11.
191. Kretlow JD, Young S, Klouda L, Wong M, Mikos AG. Injectable biomaterials for regenerating complex craniofacial tissues. Adv Mater. 2009;21:3368–93.
192. Liu M, Zeng X, Ma C, Yi H, Zeeshan A, et al. Injectable hydrogels for cartilage and bone tissue engineering. Bone Research. 2017;5:17014–32. PMC. Web 27 Feb. 2018.
193. Chen L, Shen R, Komasa S, Xue Y, et al. Drug-loadable calcium alginate hydrogel system for use in oral bone tissue repair. In: Hardy JG, editor. Inter J Mol Sci. 2017; 8(5): 989. PMC. Web. 27 Feb 2018.
194. Bi L, Cheng W, Fan H, Pei G. Reconstruction of goat tibial defects using an injectable tricalcium phosphate/chitosan in combination with autologous platelet-rich plasma. Biomaterials. 2010;31(12):3201–11.
195. Martínez-Sanz E, Ossipov DA, Hilborn J, Larsson S, Jonsson KB, Varghese OP. Bone reservoir: injectable hyaluronic hydrogels for minimal invasive bone augmentation. J Cont Rel. 2011;152(2):232–40.
196. Hanninka G, Chris Arts JJ. Bioresorbability, porosity and mechanical strength of bone substitutes: what is optimal for bone regeneration? Injury. 2011;42(Suppl. 2):S22–5.
197. Yasmeen S, lo MK, Bajarcharya S, Roldo M. Injectable scaffolds for bone regeneration. Langmuir. 2014;30(43):12977–85. https://doi.org/10.1021/la503057w.
198. Polo-Corrales L, Latorre-Esteves M, Ramirez JE. Scaffold design for bone regeneration. J Nanosci Nanotechnol. 2014;14(1):15–56.
199. Devescovi V, Leonardi E, Ciapetti G, Cenni E. Growth factors in bone repair. Musculoskel Surg. 2008;92:161–8.
200. Vo TN, Kasper FK, Mikos AG. Strategies for controlled delivery of growth factors and cells for bone regeneration. Adv Drug Deliv Rev. 2012;64(12):1292–309. https://doi.org/10.1016/j.addr.2012.01.016.
201. Rahman CV, Ben-David D, Dhillon A, Kuhn G, Gould TW, et al. Controlled release of BMP-2 from a sintered polymer scaffold enhances bone repair in a mouse calvarial defect model. J Tissue Eng Regen Med. 2014;8(1):59–66.
202. Santo VE, Gomes ME, Mano JF, Reis RL. Controlled release strategies for bone, cartilage, and osteochondral engineering—Part II: challenges on the evolution from single to multiple bioactive factor delivery. Tissue Eng B Rev. 2013;19(4):327–52. https://doi.org/10.1089/ten.teb.2012.0727.
203. Majewski RL, Zhang W, Ma X, Ciu A, Ren W, Markel DC. Bioencapsulation technologies in tissue engineering. J Appl Biomater Funct Mater. 2016;14(4):e395–403. https://doi.org/10.5301/jabfm.5000299.
204. Nicodemus GD, Bryant SJ. Cell encapsulation in biodegradable hydrogels for tissue engineering applications. Tissue Eng B Rev. 2008;14(2):149–65. https://doi.org/10.1089/ten.teb.2007.0332.

205. Jimi E, Hirata S, Osawa K, Terashita M, Kitamura C, Fukushima H. The current and future therapies of bone regeneration to repair bone defects. Int J Dent. 2012;2012: 148261, 7 pages. doi:https://doi.org/10.1155/2012/148261.

206. Kolambkara YM, Dupont KM, Boerckle JD, Huebsch N, Mooney DJ, et al. An alginate-based hybrid system for growth factor delivery in the functional repair of large bone defects. Biomaterials. 2011;32(1):65–74.

207. Bendtsen ST. Alginate hydrogels for bone tissue regeneration. 2017. Doctoral Dissertations. 1409.http://digitalcommons.uconn.edu/dissertations/1409

208. Kim J, Kim IS, Cho TH, Lee KB, Hwang SJ, et al. Bone regeneration using hyaluronic acid-based hydrogel with bone morphogenic protein-2 and human mesenchymal stem cells. Biomaterials. 2007;28(10):1830–7.

209. Włodarczyk-Biegun MK, Farbod K, Werten MWT, Slingerland CJ, de Wolf FA, van den Beucken JP, et al. Fibrous hydrogels for cell encapsulation: a modular and supramolecular approach. PLoS One. 2016;11(5):e0155625. https://doi.org/10.1371/journal.pone.0155625.

210. Hamlet SM, Vaquette C, Shah A, Hutmacher DW, Ivanovski S. 3-Dimensional functionalized polycaprolactone-hyaluronic acid hydrogel constructs for bone tissue engineering. J Clin Periodontol. 2017;44(4):428–37. https://doi.org/10.1111/jcpe.12686.

211. Burdick JA, Anseth K. Photoencapsulation of osteoblasts in injectable RGD-modified PEG hydrogels for bone tissue engineering. Biomaterials. 2002;23(22):4513–23.

212. Yamamuro Y, Hench LL, Wilson J. Bioactive glasses and glass ceramics. In: Handbook of bioactive ceramics, vol. 1. Boca Raton: CRC Press; 1990.

213. Vallet-Regí M, Ruiz-González L, Izquierdo-Barba I, et al. Revisiting silica based ordered mesoporous materials: medical applications. J Mater Chem. 2006;16:26–31.

214. Gerhardt L-C, Boccaccini AR. Bioactive glass and glass-ceramic scaffolds for bone tissue engineering. Materials. 2010;3:3867–910. https://doi.org/10.3390/ma3073867.

215. Kim HW, Kim HE, Knowles JC. Production and potential of bioactive glass nanofibers as a next-generation biomaterial. Adv Funct Mater. 2006;16(12):1529–35. https://doi.org/10.1002/adfm.200500750.

216. Christkiran, Reardon PJ, Konwarh R, Knowles JC, Mandal BB. Mimicking hierarchical complexity of the osteochondral interface using electrospun silk–bioactive glass composites. ACS Appl Mater Interfac. 2017;9(9):8000–13.

217. Price CT, Koval KJ, Langford JR. Silicon: a review of Its potential role in the prevention and treatment of postmenopausal osteoporosis. Int J Endocrinol. 2013;2013:316783., 6 pages. https://doi.org/10.1155/2013/316783.

218. Price CT, Langford JR, Liporace FA. Essential nutrients for bone health and a review of their availability in the average North American diet. Open Orthopaed J. 2012;6:143–9.

219. Rodrigues AI, Reis RL, van Blitterswijk CA, Leonor IB, Habibović P. Calcium phosphates and silicon: exploring methods of incorporation. Biomater Res. 2017;21(6):1–11.

220. Izquierdo-Barba I, Colilla M, Vallet-Regí M. Nanostructured mesoporous silicas for bone tissue regeneration. J Nanomat. 2008, . 2008: 106970, 14 pages. doi: 10.1155/2008/106970.

221. Yan X, Yu C, Zhou X, Tang J, Zhao D. Highly ordered mesoporous bioactive glasses with superior in vitro bone-forming bioactivities. Angew Chem Int. 2004;43(44):5980–4.

222. Parra J, García Páez IH, De Aza AH, Baudin C, Rocío Martín MM, Pena P. In vitro study of the proliferation and growth of human fetal osteoblasts on Mg and Si co-substituted trical-cium phosphate ceramics. J Biomed Mater Res Part A. 2017;105A:2266–75.

223. Aparicio JL, Rueda C, Manchon A, et al. Effect of physicochemical properties of a cement based on silicocarnotite/calcium silicate on in vitro cell adhesion and in vivo cement degradation. Biomed Mater. 2016;11:045005.

224. Yu L, Li Y, Zhao K, Tang Y, Cheng Z, Chen J, Wu Z. A novel injectable calcium phosphate cement-bioactive glass composite for bone regeneration. PLoS One. 2013;8(4):e62570. https://doi.org/10.1371/journal.pone.0062570.

225. Zhou X, Zhang N, Mankoci S, Sahai N. Silicates in orthopedics and bone tissue engineering materials. J Biomed Mater Res Part A. 2017;105A:2090–102.

226. Bose S, Fielding G, Tarafder S, Bandyopadhyay A. Understanding of dopant-induced osteogenesis and angiogenesis in calcium phosphate ceramics. Trends in Biotechnol. 2013;31:594–605. https://doi.org/10.1016/j.tibtech.2013.06.005.
227. Tran N, Webster TJ. Increased osteoblast functions in the presence of hydroxyapatite-coated iron oxide nanoparticles. Acta Biomater. 2011;7(3):1298–306.
228. Midde S. Osteoblast functionality on bioactive TiO_2 nanosubstrates. MS Thesis, Louisiana Tech University, Ruston LA. 71272.
229. Goto K, et al. Bioactive bone cements containing nano-sized titania particles for use as bone substitutes. Biomaterials. 2005;26(33):6496–505.
230. Shiad M, Chen Z, Farnaghib S, Friis T, Mao X, et al. Copper-doped mesoporous silica nanospheres, a promising immunomodulatory agent for inducing osteogenesis. Acta Biomater. 2015;30:334–44.
231. Swetha M, Sahithi K, Moorthi A, Saranya N, Saravanan S, et al. Synthesis, characterization, and antimicrobial activity of nano-hydroxyapatite-zinc for bone tissue engineering applications. J Nanosci Nanotechnol. 2012;12:167–72.
232. Baria A, Bloisebec N, Firilla S, Novajraa G, Vaellet-Regid M, et al. Copper-containing mesoporous bioactive glass nanoparticles as multifunctional agent for bone regeneration. Acta Biomater. 2017;55:493–504.
233. Ishimi Y. Nutrition and bone health. Magnesium and bone. Clin Calc. 2010, 20(5): 762–7. CliCa1005762767.
234. Weng L, Webster TJ. Nanostructured magnesium has fewer detrimental effects on osteoblast function. Int J Nanomedicine. 2013;8:1773–81. https://doi.org/10.2147/IJN.S39031.
235. Staiger MP, Pietak AM, Huadmai J, Dias G. Magnesium and its alloys as orthopedic biomaterials: a review. Biomaterials. 2006;27:1728–34. https://doi.org/10.1016/j.biomaterials.2005.10.003.
236. Malladi L, Mahapatro A, Gomes AS. Fabrication of magnesium-based metallic scaffolds for bone tissue engineering. Mater Technol. 2017;33(2):173–82. https://doi.org/10.1080/10667857.2017.1404278.
237. Denry I, Kelly JR. State of the art of zirconia for dental applications. Dent Mater. 2008;24:299–308.
238. Al-Amleh, Lyons K, Swain M. Clinical trials in zirconia: a systematic review. J Oral Rehabil. 2010;37:641–52.
239. Hulbert SF. The use of alumina and zirconia in surgical implants. In: Hench LL, Wilson J, editors. An Introduction to bioceramics. Singapore: World Scientific; 1993. p. 25–40.
240. Padovan LEM, Ribero Junior MA, Sartori EM, Caludio M. Bone healing in titanium and zirconia implants surface: a pilot study on the rabbit tibia. RSBO. 2013;10(2):110–5.
241. Ham AW, Harris WR. Repair and transplantation of bone. Biochem PhysiolBone. 2012;3:337.
242. Somaiya R, Kaur G. Future of bone repair. Bone Tissue Regen Insight. 2015;6:107. https://doi.org/10.4137/BTRi.s12333.
243. Bohner B. Resorbable biomaterials as bone graft substitutes. Mat Today. 2009;13(1):24–30. https://doi.org/10.1016/S1369-7021(10)70014-6.
244. Lee KY, Park M, Kim HM, Lim YJ, Chun HJ, Kim H, et al. Ceramic bioactivity: progresses, challenges and perspectives. Biomed Mater. 2006;1:R31–7.
245. Fernandez-Yaguea MA, Abba SA, McNamarab L, Zeugolisa D, Manus AP, Biggs MJ. Biomimetic approaches in bone tissue engineering: Integrating biological and physicomechanical strategies. Adv Drug Del Rev. 2015;84:1–19.

Biological Fixation: The Role of Screw Surface Design

Robert S. Liddell and John E. Davies

Keywords Osseointegration · Implant topography · Nanotopography · Microtopography · Macrotopography · Implant geometry · Simple machines · Screw implants · Screw design · Screw components · Force vectors

1 Introduction

A screw is a type of simple machine, which is a device that has the ability to transform the magnitude and/or direction of an applied force, resulting in a changed output force vector. There are generally considered to be six simple machine types: inclined plane, lever, pulley, screw, wedge, and wheel and axel. They all have the advantage of amplifying an applied force while reducing the overall distance that the output force (or load) can act upon an object. The product of these two values (applied force and distance) is equivalent to the amount of work that is done (with units being energy). Simple machines can also be used in "reverse" resulting in an increase in distance over which an output force can act, accompanied by a proportional reduction in the magnitude of the output force. The efficiency of a simple machine is calculated by the ratio of the output work to the input work. Maximal efficiency would have a value of 1, though in reality this never occurs as there are always some resistive forces, such as friction, that need to be overcome resulting in loses in energy between the input and output, and thus a drop in efficiency.

R. S. Liddell
Faculty of Dentistry, University of Toronto, Toronto, ON, Canada
e-mail: rob.liddell@mail.utoronto.ca

J. E. Davies (✉)
Faculty of Dentistry, University of Toronto, Toronto, ON, Canada

Institute of Biomaterials and Biomedical Engineering, University of Toronto, Toronto, ON, Canada
e-mail: jed.davies@utoronto.ca; davies@ecf.utoronto.ca

© Springer International Publishing AG, part of Springer Nature 2018
B. Li, T. Webster (eds.), *Orthopedic Biomaterials*,
https://doi.org/10.1007/978-3-319-89542-0_15

Screws are typically characterized by a helical (or spiral) ridge which is wrapped around a central rod that is commonly cylindrical, although sometimes conical, in shape. This ridge is referred to as the thread of the screw and can be described by its cross-sectional shape, the axial distance between successive threads (the pitch), and the number of individual helices that wrap around the screw (the number of starts). The thread engages with a corresponding thread that is formed in the substrate material by a process called tapping. Screws can either be self-tapping, creating the required matching threads during the placement of the screw, or the threads can be tapped by the use of a separate tool with a matching thread pattern. As the screw is rotated around its long axis, the thread of the screw follows the groove pattern that has been formed in the material, like a track, and results in the screw being driven in the direction of the long axis of the screw. In effect, the thread transforms the applied rotational force, or torque, to a linear force that acts to further imbed or extract the screw. The distance the load moves with respect to the screw and to what degree the applied torque is amplified can be determined from the pitch and number of starts. Specifically, a reduction of the pitch or a reduction in the number of threads has the effect of increasing how much the applied torque is amplified. This is calculated by the following relationship: $\left(\dfrac{2 * \pi}{pitch * Number\ of\ Starts} \right)$.

As discussed above, there will always be a loss of energy between the input and output of the screw, as a result of needing to overcome the resistive forces which are generated due to interactions between the screw and the substrate. Usually this resistance to motion is the result of friction, which is caused by the interaction of the features present on the surfaces of the two objects that are in contact [1, 2]. If friction is the only force resisting the motion between the two objects, then a displacement of one surface with respect to the other, which leaves the two in contact, would not result in a change in the frictional force [3]. Numerically the interaction between the surfaces is described by the coefficient of friction, which relates the frictional force to the compressive force at the interface of the sliding objects. If the force of friction is high enough to reduce the efficiency of a screw to <0.5 then it is considered self-locking and the screw will only rotate if a torque, which is not the result of axial loading, is applied. Geometrically this is the case if the coefficient of friction divided by the tangent of the screw thread angle is larger than 1, as it indicates that the resistive frictional force will always be greater than the rotational torque which is generated from an applied axial load. Fig. 1 depicts how this is the case given the same axial load is applied to two screws with differing pitches. The self-locking property ultimately shifts the mode of failure from being one of slippage, as is the case with nails, to one of material failure, as either the screw or the substrate would have to break or deform for the two entities to come apart as a result of an axial load. This difference is most clearly demonstrated when using the claw of a hammer to remove a nail which, if not bent, will be removed from the wood without causing further deformation or tear out damage in the substrate. A screw on the other hand could not be removed in such a fashion.

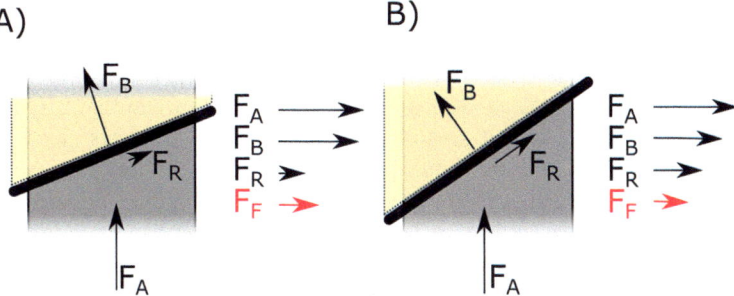

A)   B)

F_A: Axially applied force
F_B: Force supported by the bone ingrowth volume
F_R: Rotational Torque that needs to be resisted by friction
F_F: Force of friction required to overcome for rotation

Fig. 1 Two screws with threads of differing pitch, one smaller and self-locking (**a**) and one larger and not self-locking (**b**) have the same load applied axially (F_A). In both cases, a portion of F_A, which is dependent on the angle of the screw, is applied directly to the material which is in contact with the thread (F_B). As the angle of the thread increases less of F_A is transferred to the substrate material and a greater proportion goes towards rotating the screw (F_R). The only force that resists this rotation, in the case of a non-adherent surface is friction (F_F), which is proportional to F_B. The screw with the smaller pitch (A) has a larger F_B which results in an F_F that is greater than F_R preventing rotation. With a larger pitch, a smaller proportion of F_A goes towards F_B and a larger proportion goes towards F_R. The F_F is then smaller than F_R meaning that rotation can happen. Since F_B, F_R, and F_F are all a function of F_A and the pitch of the screw, increasing F_A will never make a self-locking screw rotate

2 History

The first known use of a screw was not as a fastener, but was as a method of raising water from lower elevations. This type of screw is commonly known as an Archimedes' screw after the Greek mathematician, who is generally considered to be the inventor, in the third century BCE. However, there is evidence that such screws were used by the Assyrians some 350 years earlier to water the Hanging Gardens of Babylon. Interestingly, the gardens may not have actually been in Babylon but in Nineveh (Old Babylon), which is located in the outskirts of Mosul in modern day Iraq [4]. Later on in the first century BCE screws would be used to press olives and grapes in the production of olive oil and wine. Around this time screws were also being used in the design of medical equipment, such as the opening mechanism for gynaecological specula, as found in archeological sites in Pompeii [5, 6]. The screws used in devices like these, and as fasteners for jewelry would all have had to be handmade, making them less common than other simple machines like wedges and pins which could also be used for similar tasks and were simpler to produce. The sixteenth century saw efforts to simplify screw production. Screw cutting lathes were designed by Leonardo Da Vinci (c. 1500) and Jacques Besson (in 1586) who were both court engineers for the King of France. However, there is no evidence that either machine was actually built [7]. It wasn't until the mid eighteenth century that innovations would lead to the mass production of screws. The first fully automated screw cutting lathes were created by Job and William Wyatt in England (1760) whose machines allowed for the production of 1 screw in <10 s as opposed to minutes as was the case previously, when they all had to be made by hand. This mechanization of the manufacturing of screws meant that the prices dropped to less than two pennies for a dozen in the early 1800s [8], or approximately 1 USD in today's currency according to the composite price index published by the UK Office for National Statistics.

At the time, the most commonly available structural metals were wrought iron and cast iron. Cast iron has a high carbon content (approx. 3–4%) and as a result is very hard and brittle, preventing it from being forged. Wrought iron on the other hand has a lower carbon content (<1%) and, as a result, is softer and much less brittle allowing blacksmiths to shape and forge it by hand. Thus, from the renaissance onwards the majority of the screws were made from wrought iron. It is likely that screws similar to this would have been the first screws to be placed in bone, in the 1850s, by Cucel and Riguad to fix an olecranon fracture [9]. Later, in 1886, Carl Hansmann introduced plate fixation as a method of fixing bone fractures, which itself was inspired by the work of Bernard Langenbeck. For this, Hansmann used nickel-plated steel screws and plates, presumably to prevent corrosion of the metal while *in vivo* [10], although corrosion still occurred [11]. In 1912, W.O. Sherman described the benefits of using screws that were more appropriate to metal work in comparison to those being used in wood. Wood screws are designed to have a sharp thread which cuts into the fibrous structure of the wood so that the screw can be anchored. When used in bone, such sharp metal threads would often strip the bone threads resulting in loss of anchorage. It was also suggested that implants were "blued" (growing the oxide layer of a metal such that it appears blue due to thin-film interference) in an attempt to make the surface interfacing with the biology less reactive [12]. The biocompatibility of these early metals was quite poor, as a result of continued susceptibility to corrosion, metallosis, and the potential for adverse reactions [13, 14]. With time came the introduction of new metals such as stainless steel in 1926 [15] and vitallium, a cobalt-chromium alloy, in 1938 [16]. Titanium was first identified as a potential implant material in 1940 [17], although it would not be until the 1950's that titanium screws were first tested *in vivo* [18].

With the creation of novel biomaterials as well as novel methods of modifying the surface of implants, it is important to evaluate the effectiveness of screw implants, and measure how well they anchor in bone. The evaluation of screws in engineering applications is typically done via some form of pullout test, where the embedded screw is loaded to failure. However, in typical structural engineering applications, the substrate in which a screw embedded is stable and the mechanical properties of which do not change drastically as long as the construct is properly maintained. As such, the mechanical linkage which is formed between the screw and substrate should also be stable and not change significantly. In contrast, screws placed in bone are being put into a living tissue which can remodel. By altering the biological environment it is possible to modify the steady state, or homeostasis, of the bony tissue with either positive or negative consequences. This variability means

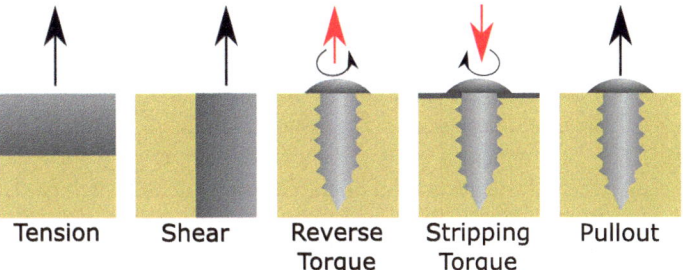

Fig. 2 The diagram outlines the different testing methods, with the direction of the measured force for each testing method being shown with a black arrow. Of these five testing methods, only the tensile testing method, also known as a pulloff test, results in a test of the tensile strength of the bone-implant construct, and will typically have the smallest measured disruption forces. The four other testing methods all test the shear strength of anchorage. Shear tests result in greater measured anchorage due to a greater amount of mechanical interlock that can occur which can resist the shear forces applied. Both the reverse torque test and stripping torque test require the measurement of the force required to rotate the implant with the direction of the screw movement being shown with a red arrow. The reverse torque measures the strength of anchorage between the implant and bone and results in minimal disruption of the surrounding bone. The stripping torque measures the strength of the bone which resides between the threads of the screw, and requires a plate between the head of the screw and the bone to prevent sinking of the screw into the bone. This test is similar to the pullout test, which would also test the strength of the bone between the threads of a screw type implant. Both the pullout test of a screw and the stripping torque test of a screw would cause significant damage to the bone resulting in a shearing of the bone between the screw threads

that a screw which appeared stable upon initial placement can become loose over time. As such it is important to evaluate both the overall strength, and the interfacial strength, of the bone-screw construct at both short and long time points. If used as a method to fasten a plate to bone for fixation of a bone fracture, screws are often loaded either radially, or by an axial load pulling the screw out of the bone. In contrast, the screws that are used as dental implants to support crowns, bridges, and dentures must be able to withstand occlusal loading and bending forces. Focusing on tests that target the axial and radial anchorage of the implant allows for the design of the screw, such as thread shape, to be specialized for an intended purpose, insuring proper primary (initial) stability. In addition, the response of the bone to the implant should also be investigated as it may not be immediately noticeable in a pullout test. In this case, it would be useful to also record the interfacial strength, as measured by a test which avoids the effects of the mechanical interlocking between the screw thread and bone. By specifically targeting the interfacial bone strength and how it changes with time, one can attain an indication of the biocompatibility of the implant and whether it is likely to have a positive or negative effect on the bone supporting the implant. Of the available testing methods for screws, pullout tests and stripping torque tests measure the shear strength of the bulk bony material; whereas reverse torque tests measure the bone implant interfacial strength. Diagrams of these tests can be seen in Fig. 2, with explanations of how they compare to simpler shear and tensile tests.

3   A Brief Review of Common Orthopedic Materials

The majority of modern metallic implants, including screws, that are used surgically today are made from stainless steel, cobalt chromium alloys, titanium, and titanium alloys [19]. The biocompatibility of these materials is largely attributed to their stability, due to the passive formation of stable oxide layers, which prevent further corrosion [11, 20, 21]. The spontaneous oxide layer formed in the case of cobalt chromium alloys and stainless steel is chromium oxide, whereas titanium and its alloys form titanium oxide. Of these two oxide layers, the stability of titanium oxide far exceeds that of chromium oxide. Indeed, titanium is commonly used as a "getter", which is a material used to scavenge residual gases from a vacuum chamber to allow for the maintenance of extremely low pressures. Chromium does have some niche uses as a getter in Tokamak reactors as it is less reactive with hydrogen [22]. The protective oxide surface is important as the implant will be immersed in a warm, moist environment which is surrounded by corrosive chloride ions, as well as being exposed to the acidic pH produced by osteoclasts and activated macrophages.

As discussed, stainless steel is one of the older metal biomaterials that are still commonly used. Stainless steel represents a range of alloys which require that the chromium content is at least 10.5%, while iron is the majority component. Typically, other elements such as nickel and molybdenum are added to further enhance corrosion resistance and to aid with the mechanical properties of the steel. The most common stainless steel compositions for use as biological implants are 316 and 316L stainless steels which differ in their carbon content. Though they are called stainless, stainless steels are not in fact stainless, and will slowly corrode, especially when in a wet, chloride rich environment. To keep corrosion to a minimum the surface finish of stainless steels are typically left smooth, reducing the surface area. The limiting of corrosion is important because it reduces the release of metallic ions, particularly nickel and chromium ions which are associated with toxic and carcinogenic effects [23, 24]. Although the smooth surface limits the amount of anchorage that is achievable by stainless steel implants, the reduction of topography makes them simpler to remove and reduces the likelihood of complications that may occur when removing the implant [25]. These materials are also among the cheapest of the current metallic biomaterials.

Improved corrosion resistance can be found with the cobalt-chromium alloys. Vitallium was one of the first cobalt-chromium alloys used for implant purposes and is the trademark name for an alloy of 65% cobalt, 30% chromium, and 5% molybdenum, and is the most common cobalt-chromium alloy used. The cobalt chromium alloys demonstrate a great resistance to wear which is why they are commonly used in sliding surfaces, such as implants used in joint replacement surgery. Though the implants that are placed in bone are not subject to the same sliding forces that joints are subjected to, there is still the potential for micromotion, particularly at the junctions of screws and fracture healing plates [26]. Similar to stainless steel, the corrosion by-products of cobalt-chromium alloys are chromium and nickel.

Titanium and titanium alloys show significantly greater resistance to corrosion than either the cobalt-chromium alloys or the stainless steels. In addition, they have a much lower stiffness, and are lighter than either of the other groups of metals. This lower stiffness, known as Young's modulus, is closer to that of bone, which aids with reducing the phenomenon of stress shielding which results in the resorption of bone due to disuse. The increased corrosion resistance and lower Young's modulus make titanium and titanium alloys some of the most commonly used metals for implants. The major drawback of titanium, and particularly commercially pure titanium, is its susceptibility to wear as a result of components sliding against each other. For this reason titanium and its alloys are rarely used as the contacting components of joint replacement implants. Commercially pure titanium also has a low fatigue strength, being prone to cracking under cyclical loading. This is improved by alloying, the most popular of alloys being one of titanium, aluminum, and vanadium. As with the other alloys discussed, concern over the release of metallic ions has been raised as aluminum is associated with Alzheimer's, and both vanadium and aluminum may be cytotoxic given high enough concentrations. The release of these ions can be controlled by increasing the thickness of the oxide layer. Care should be taken to verify that the method used to create the oxide layer is appropriate for the material being used. A nitric acid treatment ("passivation") was developed to increase the oxide layer thickness of stainless steel and cobalt chromium alloys and has been applied to titanium. However, nitric acid passivation treatment of titanium actually decreased the oxide layer thickness and increased the release of titanium, aluminum, and vanadium ions from Ti6Al4V alloy implants [27]. Later, an optimized treatment appears to have favorably affected these results and increased corrosion resistance [28].

4 A Brief Overview of Peri-implant Bone Healing

Bone has the ability to heal a fracture completely without leaving a scar, as long as the two ends of bone are stable and are in close enough juxtaposition. Although micromotions can be tolerated, typically the motion should be restricted to less than 150 microns or else a fibrous connection, or pseudo-arthrosis, will be formed [29]. Similarly, a mechanically and chemically stable implant, which does not poison surrounding tissue, would also eventually be fully incorporated into bone. The time required for this to happen and the eventual strength of the connection is dependent on the surface properties of the implant, since it is the implant surface that contributes to the interface with the biological environment. Interactions between implant surfaces and the biological environment occur immediately upon implant placement in the prepared osteotomy. As the implant comes into contact with blood, ion exchange and protein adsorption occurs. Platelets quickly come into contact with the implant becoming activated and releasing their stores of cytokines and growth factors. The major factors released with respect to bone growth being platelet derived growth factor (PDGF) and transforming growth factor beta (TGF-β) [30].

Both of these factors act as chemotactic attractants, promoting the migration of osteoprogenitor cells from the marrow and blood supply to the surface of the implant [31–33]. In addition, the platelets initiate the cleaving of fibrinogen into fibrin which forms an aggregate with the platelets and becomes adsorbed on the implant surface [34]. As osteoprogenitor cells follow the chemical gradient to the implant surface they use the fibrin network of the blood clot to migrate. The migrating cells, and the fibrin crosslinking that occurs as a result of factor XIII, result in contractile forces being transmitted through the fibrin network which can pull the clot away from the implant if not suitably anchored to the surface, preventing the cells from reaching the surface of the implant [35–37].

If the osteoprogenitor cells reach the implant surface, they then differentiate into osteoblasts, which initiate new bone apposition directly on the implant surface, known as contact osteogenesis. The first tissue to be deposited during contact osteogenesis is the cement line, a non-collagenous layer which intervenes between the implant surface and the collagen component of new bone. Contact osteogenesis is the preferred form of bone apposition with the alternative being distance osteogenesis, where bone is first deposited on the peripheries of the osteotomy and bone deposition moves inwards towards the implant [38, 39].

Structurally, the major difference between these two methods of bone formation is the location of the cement line, which is on the implant surface for contact osteogenesis and on the bony surface of the osteotomy for distance osteogenesis. The cement line is formed from non-collagenous SIBLING (small integrin-binding ligand, N-linked glycoprotein) proteins, primarily osteopontin and bone sialoprotein, in addition to proteoglycans [40]. These SIBLING proteins are responsible for the nucleation and mediation of the hydroxyapatite crystal growth which makes up the mineral component of bone. Mineralization of the cement line occurs after the SIBLING proteins have had time to diffuse and spread into the interstices of the implant surface and thus the mineralized cement line conforms intimately to the surface contours of the implant. [For more on SIBLING proteins, see section on Nanotopography below.] As the cement line is mineralizing, collagen is assembled extracellularly following secretion by osteoblasts marking the next stage of bone formation: osteoid deposition. The mineralization of collagen contributes to the strength of the bone providing support to both compressive and shear loads. Rapid deposition of new bone matrix has little organization and is referred to as woven bone.

The final step of bone repair is remodeling, a normal process that is happening throughout life. Typically, remodeling occurs to replace bone tissue that has developed micro-cracks as a result of fatigue. In addition, remodeling can also occur to optimize the architecture of the bone to more efficiently support the loads to which the bone is subjected [41]. Following bone fracture, and implant placement, remodeling also occurs to replace the woven bone which was initially deposited. To do so, osteoclasts first form a tunnel along the long axis of the bone by resorbing old bone. The osteoclasts aid in the initiation of angiogenesis as a by-product of the release of TGF-β from the old mineralized tissue and PDGF from pre-osteoclasts [42, 43]. The novel vasculature allows for the transport of nutrients as well as perivascular osteoprogenitor cells to the site of bone remodeling. The cells migrate to the surface of the bone and starting with the deposition of the non-collagenous cement line matrix, follow the same steps of bone deposition as outlined above. The major difference between bone regeneration and bone remodeling is time; the latter being a slower process and the formed bone is organized into lamellae which are parallel arrangements of collagen which form nested cylinders along the length of the tunnel. As a group, this nested cylindrical structure which is outlined by the cement line is called an osteon, and the organization of the bone into such structures gives bone its improved fracture toughness and strength compared to woven bone [44]. It also replaces any bone which may have been affected by necrosis, as bone death can occur up to a millimeter away from the implant [45] as a result of applied pressure and being cut off from the blood supply. At this point, barring the effects of aging, disease, or changes in implant loading, the bone will have reached homeostasis, and the strength of the bone implant interface would have reached a plateau in terms of strength.

5 How Topography Affects Anchorage of an Implant in Bone

Implant anchorage in bone is dependent on the implant surface design and how the surface interacts with the surrounding biology. Of considerable importance is screw surface topography, of which there are three major scale ranges: nano, micro and macro [36, 46]. Nano-topography describes surface features less than 1 μm (typically around 100 nm), microtopography describes features between 1 and 30 μm, and macrotopography describes features that are larger than 30 μm. Each of these scale ranges of topography contributes to obtaining long term stable anchorage of an implant in bone, and each of these scale ranges of topography can be modified to meet the goals intended for a specific implant.

In evaluating the role of implant surface design, studies often look at how osteo-progenitor cells interact with implant surfaces *in vitro*. However, ultimately an implant needs to be functional and provide anchorage when placed in bone.

Functional anchorage of an implant in bone is called osseointegration and is best evaluated by an *in vivo* test of the strength of the bone implant interface, which is formed as a result of healing following implant placement. Often this type of assessment is done over a few experimental time points, and provides little information concerning the biological relevance of any differences observed. Recently, we have suggested that a curve fitting method, which has been applied to a wide range of natural phenomena, such as the recovery of bone mineral density in astronauts, the recovery of fish stocks in protected waters, and heart rate recovery [47–49], could be used to provide a more meaningful evaluation of implant osseointegration [50]. Such an evaluation requires disruption force data being measured at many time points, which can then be analyzed using an exponential recovery curve $(1-e^{-x/\tau})$ which is scaled by some value (C), indicative of the peak strength of the bone implant interface when fully healed, and some function (B(x)) which describes the

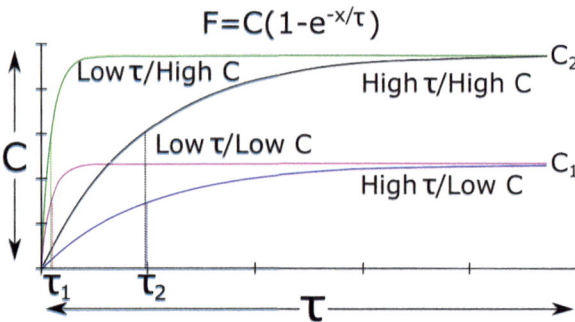

Fig. 3 Curves which reach the plateau quickly have small values of *tau* (τ_1), and curves that take longer to reach the plateau have larger values of *tau* (τ_2). In this way the value of *tau* is an indication of the rate of healing and the rate of osseointegration. The height of the plateau changes the value of C, which then indicates the anchorage that will be achievable once the bone-implant interface has reached homeostasis

relative thickness of bone in contact with the implant, resulting in the equation: $F=C*B(x)*(1-e^{-x/\tau})$. In its most basic form, when the peri-implant bony architecture remains relatively constant, the function B(x) would be equal to 1 resulting in the equation: $F=C*(1-e^{-x/\tau})$. Fitting this curve to the recorded experimental data provides two parameters, C and τ (*tau*), which describe the overall peak anchorage reached when the disruption force values plateau with time, and a time constant— the time required to reach a percentage (approximately 63%) of that plateau— respectively. This allows for comparison of the osseointegration of different implants, where changes in C indicate a difference in the strength of anchorage and changes in *tau* indicate a difference in the rate of osseointegration (Fig. 3). In addition, by choosing an appropriate formulation for B(x), one can account for the effects of the changing bony architecture observed during the lifespan of a placed implant. This is particularly relevant for implants placed in the appendicular skeleton of smaller animals, as one can account for the significant bone growth and remodeling that can occur during the experimental period. We have previously illustrated how such "relative drift" of an implant, from metaphyseal to diaphyseal bone, can occur due to the differential growth of the proximal and distal epiphyseal growth plates [50].

5.1 Implant Surface Nanotopography

Nano-scale surface features can have a significant effect on the cells and tissues with which they interact, influencing the adsorption of proteins as well as the proliferation and differentiation of cells. Mendonça et al. found that human mesenchymal stem cells that were cultured on various surfaces which had apparent complex nanoscale surfaces had increased mRNA expression of osterix (OSX) and bone sialoprotein (BSP) [51]. OSX is a transcription factor which leads to osteoblastic differentiation and BSP is an initiator for hydroxyapatite crystal growth [40]. As stated previously, BSP is classed a SIBLING protein, which in combination with other SIBLING proteins, aid in the regulation of hydroxyapatite crystal growth. These proteins are secreted once the osteoprogenitor cells have differentiated into committed osteoblasts. They then locally diffuse into the aqueous tissue fluid peri-implant environment and the interstices of the nano-scale surface features. To give an idea of scale, BSP has a form similar to a ball and chain where the overall length is about 20 nm and the diameter of the ball is 5.5 nm and can easily occupy any available sites on an implant surface at the nanometer scale [52]. For relatively smooth nano features this may have little to no effect. However for more complex surfaces with features that create undercuts there are significant advantages to be gained. Once the SIBLING proteins mineralize in these undercut features, a mechanical interlock is formed between the cement line matrix and the implant surface, which restricts relative motion [53]. An analogy would be the pouring of concrete into the interstices of a three-dimensional network of steel reinforcement bars—initially the concrete flows, but once set, the concrete is locked into the steel network. As a result, the true interface, which is the surface created by the boundary between the implant and the bone, is no longer the weakest link in the bone implant interface and the plane of failure is pushed outward, into the bony structure itself [54, 55]. Tests which disrupt the interface formed between bone and a nanotopographically complex implant result in residual bone being observed on the implant surface, and in these cases the implant is said to be bone bonding [56]. Using the described mathematical method to analyze the implant anchorage over time it was found that some nanotopographies significantly increase bone anchorage at early time points. However, the effect of such nanotopography may diminish with time due to the increasing predominance of the larger topographical features of the implant surface [50].

Another nanoscale feature which has been created on implant surfaces and has been shown to increase implant anchorage are nanotubes [57]. However, unlike the nanofeatures described above where the beneficial effect on bone anchorage comes from an accelerated osseointegration (decreased *tau*) with no appreciable effect on the maximum disruptive force (C), our preliminary data suggest that nanotubes can increase the strength of anchorage (increase C) but may not reduce *tau* in comparison to other nano-surfaces (Fig. 4). This was determined using a similar curve fitting method as described previously, though it is clearly evident from the shape of the curve that the precise formulation has been modified. This was because instead of the simplified version of the curve being used, where B(x) is equal to "1", a more complex formulation had to be used $(B(x) = (1 - e^{-\max (x,d) * \beta})$, Fig. 4b). This additional term was included to account for the initial implant only having mono-cortical engagement in metaphyseal bone at the time of implant placement, which changed due to "relative drift", of an already osseointegrated implant, during the experimental time frame, to bi-cortical engagement in the tibial diaphysis. The values of the parameters in this expression, $B(x)$, were kept constant for all implant surfaces tested since they are dependent on the mechanical properties of cortical

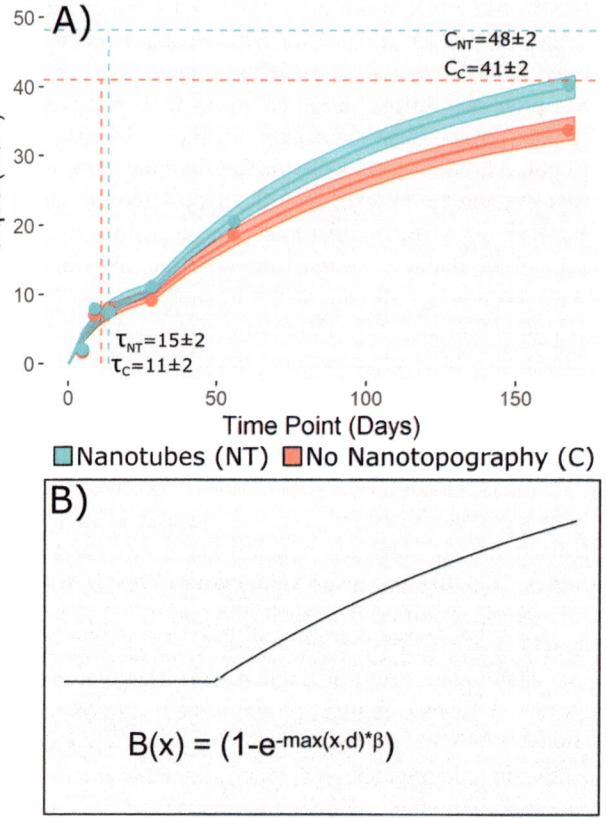

Fig. 4 Reverse torque data recorded from mini-screw implants implanted in rat tibias. The surfaces had nanotubes (or not) superimposed on microtopographically complex implants and were analyzed with the equation $F = C * B(x) * (1 - e^{-\frac{x}{\tau}})$. The second part of the curve ($B(x)$) represents an increase in bone anchorage due to distal drift of the osseointegrated implant from metaphyseal to diaphyseal bone, as explained in the text. The value and shape of $B(x)$ is shown in (**b**). (**a**) On a microtopographically complex surface no difference was seen in the value of *tau*, with the addition of nanotubes to the surface. However, the implant with nanotubes exhibited a significantly higher C value compared to the surface without nanotubes. Error values included next to the parameter values are standard error. (**b**) The function $B(x)$ comprises two parameters. The first, "d" indicates the location where the horizontal line should start to increase and become the curved line. In this case "d" was equal to 28 days, which was determined to be the best fit with our data. The second parameter "β" is used to describe the shape of the curved line. Both of these parameters were the same for all implants as the function of $B(x)$ is dependent on the mechanical properties of the cortical bony architecture rather than the osseointegration of the implant

bony architecture rather than the osseointegration of the implant *per se*. These data, from a reverse torque study, illustrate the important relationship between the two key parameters, *tau* and C, since at specific early time points the same force may be required to disrupt the anchorage of two implants, yet they will ultimately have statistically different maximum disruption values when homeostasis is achieved, with the higher C value implant also having a higher *tau* value. These rate differences, at the earlier time points, are potentially attributable, among other reasons, to 100 nm nanotube surfaces having been reported to decrease platelet adhesion and activation compared to machined implant surfaces [58].

5.2 Implant Surface Microtopography

The initial effects of microtopography on the biology of bone healing result in increased platelet activation [59]. Additional work has shown that this effect was independent from the presence of calcium phosphates indicating that mimicking the surface chemistry of bone is not a requirement for bone anchorage [60].

In addition to affecting platelets, microtopography also affects macrophages which are in the vicinity of the wound to clear debris created during implant placement. An *in vitro* study by Refai et al. found that the macrophages cultured on microrough titanium disks had increased expression of genes associated with initiating angiogenesis [61] which is crucial for healing and the recruitment of osteoprogenitor cells, as the latter are perivascular cells [62]. Once the osteoprogenitor cells reach the implant surface, Brett et al. has shown that the cells have increased viability which extended up to three days *in vitro*, on rougher implant surfaces [63]. The rougher surfaces also appear to upregulate genes associated with osteogenesis, particularly in the case of surfaces which also have nanotopography [64].

Surface treatments such as grit blasting and/or acid etching have been found to be effective at increasing implant microtopography and increasing implant anchorage [65]. Laser etching has also been found effective at creating finely controlled localized topographical changes which result in greater resistance to reverse torque [66]. The effect of the surface topography on bone anchorage has been found to increase with the size of the features. A study by Gotfredson et al. found that by increasing the size of the micron scale features, the reverse torque forces required to disrupt the bone implant interface significantly increased [67]. By replotting and applying the curve fitting method described above to this data (Fig. 5) it was found that all of the microtopographically complex implants had similar tau's, indicating that the size of the microtopographies employed in their study did not affect the rate of healing. However, the value of C did increase with the size of the topography. In contrast, the implant without microtopography had a much slower rate of healing. We have also discussed the importance of the size of topography [53, 56]. The cement line which forms on the implant in the case of contact osteogenesis almost completely engulfs the micro-features which are 1–10 μm. The larger features, which are on the scale of 20 μm, extend through this layer and integrate with the collagenous bony matrix. The collagen which is deposited during osteoid formation becomes wrapped around these larger features, which increases the strength of the anchorage achievable by such implants.

In many of these models, the implants are not subjected to the loads which they would experience in a clinical environment, being placed in the long bones

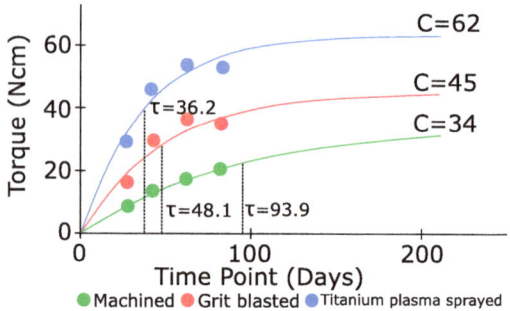

Fig. 5 Data replotted from data by Gotfredson et al. [67]. Comparison of three implants, two (Grit blasted and Titanium Plasma Spayed) with significant microtopographical complexity have *tau* values (36.2 and 48.1) that appear much smaller than the implant which was smooth (machined, *tau*: 93.9). The value of C increased with the size of the topography. The best curve fitting, requires multiple early time points as well as time points which extend into the plateau region of healing, which represents when homeostasis is gained and occurs when the healing process is complete and is no longer a major contributor to increasing bone-implant interface strength

of rabbits or rats. The loading of an implant can have a profound effect on osseointegration as some work shows that the effect of microtopography is attenuated by loading [68].

5.3   *Implant Macrotopography and Geometry*

Features on the macro scale play a large role in the transmission of forces from the implant to the surrounding bone. The most immediately evident macro feature of screws are their thread, which has a significant effect on the transmission of loads from the implant to the bone, and results in greater long term implant success compared to non-thread implants [69]. Efficacy of the transfer of load from the implant to the bone can significantly alter the long term outcome and health of the bone. Though some compressive forces are beneficial in terms of bone remodeling, excessive loads can cause significant damage to the bone which then leads to extensive bone resorption and a loss of implant anchorage [70]. To best ensure that stresses on the bone do not exceed the strength of the bone, it is best to distribute any forces applied to the implant evenly along the entire length. These stresses can also be further reduced by increasing the overall area of contact between the implant and the bone, and ensuring that the forces that are transferred to the bone are compressive loads, which bone is most suited to resist. This does not mean that an ideal implant will be completely covered in bone with a bone-to-implant contact (BIC) ratio of 1 because some bone-bonding implant surfaces can more efficiently transfer the load across the interface. In this case, less bone will be deposited at the interface. As such, one cannot effectively judge the quality of osseointegration achieved by an

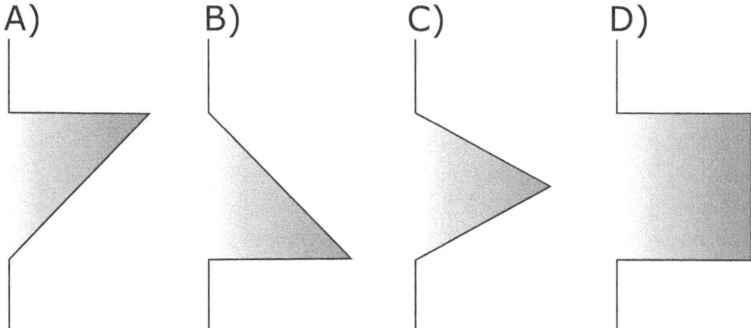

Fig. 6 The common thread shapes, assuming the bottom of the figure is the apical end of the screw: (**a**) buttress thread, (**b**) reverse buttress thread, (**c**) v-thread, and (**d**) square thread. The sharp points of the buttress thread, reverse buttress thread, and v-thread, make screw placement easier due to lower rotational torque requirements, however the surfaces that are not horizontal provide less resistance to axial forces. The square thread provides good resistance to axial loads in either direction along the long axis. The buttress and reverse buttress threads provide good resistance to pull out and push in loads respectively while also having the sharp edge to aid in implant insertion

implant from measuring the BIC, as it gives no relevant information on the functionality of the connection between the implant and bone.

The simplest ways of increasing the area of the implant which is in contact with the surrounding bone is to increase either the length and/or thread diameter of the screw. Walter et al. found that screws with larger thread diameters placed in artificial bone, analogous to highly cancellous bone, could resist greater disruption forces, than similar implants with smaller diameters [71]. Finite element analysis (FEA) showed that by increasing the diameter of the thread the stresses exerted on the bone were decreased [72]. The same study found similar results for implant length [72]. If the implant cannot be made wider, shrinking the diameter of the shaft that the thread wraps around will also increase the implant surface area, and results in greater pull out loads as demonstrated in artificial bone [73] as well as cadaveric bovine bone [74]. The reduction of the shaft diameter however has the disadvantage of reducing the cross sectional area of the screw making it more likely to fracture [71].

Another possibility of increasing the surface area of a screw is by reducing the pitch of the screw thread. By reducing the thread pitch, the number of threads is increased along the length of the screw and more of the screw becomes engaged with the bone. The reduction of thread pitch was observed to result in increased pull out forces [74], and analysis by FEA found that decreased thread pitch also resulted in a more even distribution of stresses in the bone [72]. In addition to increasing the number of threads, decreasing the pitch also decreases the angle of the thread. With decreasing thread angle, less of the axial load is converted into shear loading and has been found to result in a more stable implant [75]. The last factor to be discussed which affects the distribution of stresses around a screw is the shape of the thread.

Of the four main thread designs (Fig. 6), the v-thread and the square thread have been found to result in the most even distribution of stresses by FEA [76], with preference given to the square thread [77]. However, the dimensions of the thread play a significant role in the reduction of stress as Geng et al. showed that a thin square thread results in the greatest peak stresses in bone [76].

In addition to stress distribution, the shape of a screw thread can also influence the distribution and anchorage of bone. Hall et al. found that the inclusion of a 110 μm wide and 70 μm deep groove resulted in greater bone deposition and greater reverse torque values in rabbits [78], and Scarano et al. found that bone was preferentially deposited into the concavities of the thread [79]. Work by Moreo et al. on the modeling of bone growth around implants attribute this increase bone growth in implant concavities to an increase in platelet concentration, as the surfaces surrounding the volume increase the surface area with respect to the volume [80, 81]. It is uncertain whether these grooves have a significant effect on the rate of osseointegration.

In the evaluation of the shape of the implant, both cylindrical and conical implants have advantages. Conical implants have been found to be more stable in terms of their initial stability [82]. FEA found that the conical shape also provided greater resistance to forces which pressed the implant into the bone, such as those commonly exerted by dental implants [83]. In contrast, when extractive forces were applied to the implant, it was found that cylindrical implants required a greater force to be applied to pull the implant out of an artificial bone substitute [71].

6 Conclusion

The value of a screw is its ability to resist significant axial forces, while at the same time being simple to place, and remove when required. A screw accomplishes this by having a small pitch and few thread starts which results in the torsional forces applied to the screw being amplified and the direction of the force vector moved to an axial direction. The low pitch of the screw also means that the screw is self-locking, meaning that any axial loads applied to the screw are transmitted through the bony substrate instead of just simply unscrewing the screw. However, as bone is a living tissue which can remodel, it is of particular interest to make sure that the screw placed in the bone does not result in any adverse reactions. In doing so, it can be assured that instead of weakening over time; the anchorage of the screw in the bone will increase with time. This is possible if the implant surface promotes the growth of new bone. Implant geometry which promotes even distribution of the stresses around the screw, and surfaces that do not corrode will result in an implant being held fast in the bone. Importantly, changes in interfacial strength are an early indicator for eventual function of the implant.

References

1. Sung IH, Lee HS, Kim DE. Effect of surface topography on the frictional behavior at the micro/nano-scale. Wear. 2003;254(10):1019–31.
2. Gachot C, Rosenkranz A, Reinert L, Ramos-Moore E, Souza N, Müser MH, Mücklich F. Dry friction between laser-patterned surfaces: role of alignment, structural wavelength and surface chemistry. Tribol Lett. 2013;49(1):193–202.
3. Kydd WL, Daly CH. Bone-titanium implant response to mechanical stress. J Prosthet Dent. 1976;35(5):567–71.
4. Dalley S, Oleson JP. Sennacherib, archimedes, and the water screw: the context of invention in the ancient the context of invention in the ancient world. Source Technol Cult. 2003;44(1):1–26.
5. Kravetz RE. Vaginal speculum. Am J Gastroenterol. 2006;101(11):2456.
6. Medical and Surgical Reporter. Speculum from pompeii. Am Period. 1886;55(16):509.
7. Rolt LTC. A short history of machine tools. Cambridge: The MIT Press; 1965.
8. Rybczynski W. One good turn: a natural history of the screwdriver and the screw. New York: Scribner; 2000.
9. Cucel L, Rigaud R. Des vis métalliques enfoncées dans les tissues des os, pour le traitement de certaines fractures. Rev Med Chir Paris. 1850;8:113–4.
10. Hansmann C. Eine neue methode der fixirung der fragmente bei complicirten fracturen. Verhandlungen der Dtsch Gesellschaft fur Chir. 1886;15:134–7.
11. Gotman I. Characteristics of metals used in implants. J Endourol. 1997;11(6):383–9.
12. Sherman WO. Vanadium steel bone plates and screws. Surg Gynocol Obstet. 1912;12:629–34.
13. Key JA. Stainless steel and vitallium in internal fixation of bone: a comparison. Arch Surg. 1941;43(4):615–26.
14. Fisher AA, Shapiro A. Allergic eczematous contact dermatitis due to metallic nickel. J Am Med Assciation. 1956;161(8):717–21.
15. Roberts TT, Prummer CM, Papaliodis DN, Uhl RL, Wagner TA. History of the orthopedic screw. Orthopedics. 2013;36(1):12–4.
16. Venable CH, Struck WG. Electrolysis controlling factor in the use of metals in treating fractures. J Am Med Assoc. 1938;15:1349–52.
17. Bothe R, Beaton L, Davenport H. Reaction of bone to multiple metallic implants. Surg Gynocol Obstet. 1940;71:598–602.
18. Beder OE, Stevenson JK, Jones TW. A further investigation of the surgical application of titanium metal in dogs. Surgery. 1957;41(6):1012–5.
19. Manam NS, Harun WSW, Shri DNA, Ghani SAC, Kurniawan T, Ismail MH, Ibrahim I. Study of corrosion in biocompatible metals for implants: a review. J Alloys Compd. 2017;701:698–715.
20. Mavrogenis AF, Papagelopoulos PJ, Babis GC. Osseointegration of Cobalt-chrome alloy implants. J Long-Term Eff Med Implants. 2011;21(4):349–58.
21. Geetha M, Singh AK, Asokamani R, Gogia AK. Ti based biomaterials, the ultimate choice for orthopaedic implants - a review. Prog Mater Sci. 2009;54(3):397–425.
22. Simpkins JE, Mioduszewski P, Stratton LW. Studies of chromium gettering. J Nucl Mater. 1982;111:827–30.
23. Salnikow K, Zhitkovich A. Genetic and epigenetic mechanisms in metal carcinogenesis and cocarcinogenesis: nickel, arsenic, and chromium. Chem Res Toxicol. 2008;21(1):28–44.
24. Costa M, Zhuang Z, Huang X, Cosentino S, Klein CB, Salnikow K. Molecular mechanisms of nickel carcinogenesis. Sci Total Environ. 1994;148:191–9.
25. Pearce AI, Pearce SG, Schwieger K, Milz S, Schneider E, Archer CW, Richards RG. Effect of surface topography on removal of cortical bone screws in a novel sheep model. J Orthop Res. 2008;26(10):1377–83.
26. Hallab NJ, Jacobs JJ. Orthopedic implant fretting corrosion. Corros Rev. 2003;21(2–3):183–213.
27. Callen BW, Lowenberg BF, Lugowski S, Sodhi RNS, Davies JE. Nitric acid passivation of Ti6Al4V reduces thickness of surface oxide layer and increases trace element release. J Biomed Mater Res. 1995;29(3):279–90.

28. Masmoudi M, Capek D, Abdelhedi R, El Halouani F, Wery M. Application of surface response analysis to the optimisation of nitric passivation of cp titanium and Ti6Al4V. Surf Coatings Technol. 2006;200(24):6651–8.

29. Szmukler-Moncler S, Salama H, Reingewirtz Y, Dubruille JH. Timing of loading and effect of micromotion on bone – dental implant interface : review of experimental literature. J Biomed Res. 1998;43(2):192–203.

30. Oryan A, Alidadi S, Moshiri A. Platelet-rich plasma for bone healing and regeneration. Expert Opin. Biol Ther. 2016;16(2):213-32.

31. Fiedler J, Etzel N, Brenner RE. To go or not to go: migration of human mesenchymal progenitor cells stimulated by isoforms of PDGF. J Cell Biochem. 2004;93(5):990–8.

32. Fiedler J, Röderer G, Günther KP, Brenner RE. BMP-2, BMP-4, and PDGF-bb stimulate chemotactic migration of primary human mesenchymal progenitor cells. J Cell Biochem. 2002;87(3):305–12.

33. Lind M. Growth factors: possible new clinical tools. A review. Acta Orthop Scand. 1996;67(4):407–17.

34. Green D. Coagulation cascade. Hemodial Int. 2006;10:S2–4.

35. Oprea WE, Karp JM, Hosseini MM, Davies JE. Effect of platelet releasate on bone cell migration and recruitment in vitro. J Craniofac Surg. 2003;14(3):292–300.

36. Anil S, Anand PS, Alghamdi H, Jansen JA. Dental implant surface enhancement and osseointegration. In: Turkyilmaz I. Implant dentistry - a rapidly evolving practice; 2011. p. 83–108.

37. Davies JE. Understanding peri-implant endosseous healing. J Dent Educ. 2003;67(8):932–49.

38. Davies JE. Bone bonding at natural and biomaterial surfaces. Biomaterials. 2007;28(34):5058–67.

39. Davies JE. Mechanisms of endosseous integration. Int J Prosthodont. 1998;11(5):391–401.

40. Baht GS, Hunter GK, Goldberg HA. Bone sialoprotein-collagen interaction promotes hydroxyapatite nucleation. Matrix Biol. 2008;27(7):600–8.

41. Martin TJ. Bone remodelling : its local regulation and the emergence of bone fragility. Best Pract Res Clin Endocrinol Metab. 2008;22(5):701–22.

42. Xie H, Cui Z, Wang L, Xia Z, Hu Y, Xian L, Li C, Xie L, Crane J, Wan M, Zhen G, Bian Q, Yu B, Chang W, Qiu T, Pickarski M, Duong LT, Windle JJ, Luo X, Liao E, Cao X. PDGF-BB secreted by preosteoclasts induces angiogenesis during coupling with osteogenesis. Nat Med. 2014;20(11):1270–8.

43. Tang Y, Wu X, Lei W, Pang L, Wan C, Shi Z, Zhao L, Nagy TR, Peng X, Hu J, Feng X, Van Hul W, Wan M, Cao X. TGF-β1–induced migration of bone mesenchymal stem cells couples bone resorption with formation. Nat Med. 2009;15(7):757–65.

44. Burr DB, Schaffler MB, Frederickson RG. Composition of the cement line and its possible mechanical role as a local interface in human compact bone. J Biomech. 1988;21(11):939–45.

45. Roberts WE. Bone tissue interface. J Dent Educ. 1988;52(12):804–9.

46. Saghiri MA, Asatourian A, Garcia-Godoy F, Sheibani N. The role of angiogenesis in implant dentistry part I: review of titanium alloys, surface characteristics and treatments. Med Oral Patol Oral Cir Bucal. 2016;21(4):e514–25.

47. Sibonga JD, Evans HJ, Sung HG, Spector ER, Lang TF, Oganov VS, Bakulin AV, Shackelford LC, LeBlanc AD. Recovery of spaceflight-induced bone loss: bone mineral density after long-duration missions as fitted with an exponential function. Bone. 2007;41(6):973–8.

48. McClanahan TR, Graham NAJ, Calnan JM, MacNeil MA. Toward pristine biomass: reef fish recovery in coral reef marine protected areas in Kenya. Ecol Appl. 2007;17(4):1055–67.

49. Savin WM, Davidson DM, Haskell WL. Autonomic contribution to heart rate recovery from exercise in humans. J Appl Physiol. 1982;53(6):1572–5.

50. Liddell R, Ajami E, Davies JE. Tau (τ): a new parameter to assess the osseointegration potential of an implant surface. Int J Oral Maxillofac Implants. 2017;32(1):102–12.

51. Mendonça G, Mendonça DBS, Simões LGP, Araújo AL, Leite ER, Duarte WR, Aragão FJL, Cooper LF. The effects of implant surface nanoscale features on osteoblast-specific gene expression. Biomaterials. 2009;30(25):4053–62.

52. Vincent K, Durrant MC. A structural and functional model for human bone sialoprotein. J Mol Graph Model. 2013;39:108–17.
53. Davies JE, Mendes VC, Ko JCH, Ajami E. Topographic scale-range synergy at the functional bone/implant interface. Biomaterials. 2014;35(1):25–35.
54. Coelho PG, Zavanelli RA, Salles MB, Yeniyol S, Tovar N, Jimbo R. Enhanced bone bonding to nanotextured implant surfaces at a short healing period : a biomechanical tensile testing. Implant Dent. 2016;25(3):322–7.
55. Yamada M, Ueno T, Minamikawa H, Ikeda T, Nakagawa K, Ogawa T. Early-stage osseointegration capability of a submicrofeatured titanium surface created by microroughening and anodic oxidation. Clin Oral Implants Res. 2013;24(9):991–1001.
56. Davies JE, Ajami E, Moineddin R, Mendes VC. The roles of different scale ranges of surface implant topography on the stability of the bone/implant interface. Biomaterials. 2013;34(14):3535–46.
57. Bjursten LM, Rasmusson L, Oh S, Smith GC, Brammer KS, Jin S. Titanium dioxide nanotubes enhance bone bonding in vivo. J Biomed Mater Res A. 2010;92(3):1218–24.
58. Huang Q, Yang Y, Zheng D, Song R, Zhang Y, Jiang P, Vogler EA, Lin C. Effect of construction of TiO2 nanotubes on platelet behaviors: structure-property relationships. Acta Biomater. 2017;51:505–12.
59. Park JY, Gemmell CH, Davies JE. Platelet interactions with titanium: modulation of platelet activity by surface topography. Biomaterials. 2001;22(19):2671–82.
60. Kikuchi L, Park JY, Victor C, Davies JE. Platelet interactions with calcium-phosphate-coated surfaces. Biomaterials. 2005;26(26):5285–95.
61. Refai AK, Textor M, Brunette DM, Waterfield JD. Effect of titanium surface topography on macrophage activation and secretion of proinflammatory cytokines and chemokines. J Biomed Mater Res A. 2004;70:194–205.
62. Kalajzic Z, Li H, Wang LP, Jiang X, Lamothe K, Adams DJ, Aguila HL, Rowe DW, Kalajzic I. Use of an alpha-smooth muscle actin GFP reporter to identify an osteoprogenitor population. Bone. 2008;43(3):501–10.
63. Brett PM, Harle J, Salih V, Mihoc R, Olsen I, Jones FH, Tonetti M. Roughness response genes in osteoblasts. Bone. 2004;35(1):124–33.
64. Zhao L, Mei S, Chu PK, Zhang Y, Wu Z. The influence of hierarchical hybrid micro/nano-textured titanium surface with titania nanotubes on osteoblast functions. Biomaterials. 2010;31(19):5072–82.
65. Halldin A, Jimbo R, Johansson CB, Gretzer C, Jacobsson M. Improved osseointegration and interlocking capacity with dual acid-treated implants: a rabbit study. Clin Oral Implants Res. 2016;27(1):22–30.
66. Shah FA, Johansson ML, Omar O, Simonsson H, Palmquist A, Thomsen P. Laser-modified surface enhances osseointegration and biomechanical anchorage of commercially pure titanium implants for bone-anchored hearing systems. PLoS One. 2016;11(6):e0157504.
67. Gotfredsen K, Berglundh T, Lindhe J. Anchorage of titanium implants with different surface characteristics: an experimental study in rabbits. Clin Implant Dent Relat Res. 2000;2(3):120–8.
68. Vandamme K, Naert I, Vander Sloten J, Puers R, Duyck J. Effect of implant surface roughness and loading on peri-implant bone formation. J Periodontol. 2008;79(1):150–7.
69. Kan JYK, Rungcharassaeng K, Kim J, Lozada JL, Goodacre CJ. Factors affecting the survival of implants placed in grafted maxillary sinuses: a clinical report. J Prosthet Dent. 2002;87(5):485–9.
70. Chamay A, Tschantz P. Mechanical influences in bone remodeling. Experimental research on Wolff's law. J Biomech. 1972;5(2):173–80.
71. Walter A, Winsauer H, Marcé-Nogué J, Mojal S, Puigdollers A. Design characteristics, primary stability and risk of fracture of orthodontic mini-implants: pilot scan electron microscope and mechanical studies. Med Oral Patol Oral Cir Bucal. 2013;18(5):e804–10.
72. Bianco R-J, Arnoux P-J, Wagnac E, Mac-Thiong J-M, Aubin C-E. Minimizing pedicle screw pullout risks: a detailed biomechanical analysis of screw design and placement. Clin Spine Surg. 2017;30(3):226–32.

73. Migliorati M, Benedicenti S, Signori A, Drago S, Cirillo P, Barberis F, Silvestrini Biavati A. Thread shape factor: evaluation of three different orthodontic miniscrews stability. Eur J Orthod. 2013;35(3):401–5.
74. Alkaly RN, Bader DL. The effect of transpedicular screw design on its performance in vertebral bone under tensile loads: a parametric study. Clin Spine Surg. 2016;29(10):433–40.
75. Ma P, Liu H, Li D, Lin S, Shi Z, Peng Q. Influence of helix angle and denisty on primary stability of immediatelt loaded dental implants: three-dimensional finite element analysis. Chinese J Stomatol. 2007;42(10):618–21.
76. Geng JP, Ma QS, Xu W, Tan KBC, Liu GR. Finite element analysis of four thread-form configurations in a stepped screw implant. J Oral Rehabil. 2004;31(3):233–9.
77. Chun H-J, Cheong S-Y, Han J-H, Heo S-J, Chung J-P, Rhyu I-C, Choi Y-C, Baik H-K, Ku Y, Kim M-H. Evaluation of design parameters of osseointegrated dental implants using finite element analysis. J Oral Rehabil. 2002;29(1998):565–74.
78. Hall J, Miranda-Burgos P, Sennerby L. Stimulation of directed bone growth at oxidized titanium implants by macroscopic grooves: an in vivo study. Clin Implant Dent Relat Res. 2005;7(Suppl 1):S76–82.
79. Scarano A, Degidi M, Perrotti V, Degidi D, Piattelli A, Iezzi G. Experimental evaluation in rabbits of the effects of thread concavities in bone formation with different titanium implant surfaces. Clin Implant Dent Relat Res. 2014;16(4):572–81.
80. Moreo P, García-Aznar JM, Doblaré M. Bone ingrowth on the surface of endosseous implants. Part 2: theoretical and numerical analysis. J Theor Biol. 2009;260(1):13–26.
81. Moreo P, García-Aznar JM, Doblaré M. Bone ingrowth on the surface of endosseous implants. Part 1: mathematical model. J Theor Biol. 2009;260(1):1–12.
82. Kim JW, Baek SH, Kim TW, Il Chang Y. Comparison of stability between cylindrical and conical type mini-implants. Angle Orthod. 2008;78(4):692–8.
83. Geng JP, Xu W, Tan KBC, Liu GR. Finite element analysis of an osseointegrated stepped screw dental implant. J Oral Implantol. 2004;30(4):223–33.

Fracture Fixation Biomechanics and Biomaterials

Scott M. Tucker, J. Spence Reid, and Gregory S. Lewis

Keywords Fracture fixation · Orthopedic biomechanics · Implant mechanics · Bone healing · Fatigue · Failure analysis · Implant materials · Screw fixation · Biocompatibility · Computational modeling · Finite element method · Intramedullary nail · Internal plating · External fixation · Translational research

1 Clinical Aspects

1.1 Introduction

Fracture management constitutes a very large portion of modern orthopedic care. Many fractures can be managed by non–surgical methods such as splinting, casting, and functional bracing. This type of fracture management is particularly applicable in nondisplaced or minimally displaced fractures – often in the pediatric population whose healing times are quite short and the remodeling potential after healing can correct residual deformity. Nonetheless, operative fracture care is being used increasingly as it has been shown to reduce disability, improve outcome, and improve the quality of life following a significant fracture. There are three areas in which surgical facture care has been shown to make a major impact on patient recovery from injury, and is pivotal in the restoration of function and quality of life.

Displaced fractures of the femur and tibia in the adult are now usually treated surgically as it has been well-demonstrated that this method is safe, promotes the rapid return to function including ambulation, preserves joint function, and avoids the condition associated with non-operative treatment previously coined "fracture disease" which is a complex of problems such as joint stiffness, pressure ulcers, disuse osteopenia, formation of deep vein thrombosis, and muscle atrophy.

The second area in which operative fracture care has resulted in dramatic improvement in clinical outcomes is the fracture of the articular surfaces of joints.

S. M. Tucker · J. S. Reid · G. S. Lewis (✉)
Department of Orthopedics & Rehabilitation, Penn State College of Medicine,
Hershey, PA, USA
e-mail: greglewis@psu.edu

© Springer International Publishing AG, part of Springer Nature 2018
B. Li, T. Webster (eds.), *Orthopedic Biomaterials*,
https://doi.org/10.1007/978-3-319-89542-0_16

401

The surfaces of most joints (e.g. hip, knee, ankle, and elbow) are highly congruent with each other such that distortion of the shape of either side of a joint from a fracture can dramatically alter the load transmission across the surface and raise cartilage contact pressures to the point that the cartilage will breakdown and lead to post-traumatic degenerative arthritis. Often, this will result in joint destruction within several years of the injury. Modern orthopedic care of the injured joint rests on the concept of restoration of the shape of bones on both sides of the joint as well as restoration of axial alignment and ligamentous stability of the joint. This approach restores the normal pressures across the joint during functional use of the limb. A joint that has been restored surgically in this way can undergo very early motion which prevents stiffness and optimally promotes recovery of the injured articular cartilage.

The final arena in which surgical treatment of fractures has made a dramatic improvement in outcome is the multiple injury trauma patient. In this group, rapid stabilization of long bone and pelvis fractures is viewed as part of the overall resuscitation of the patient since it helps control bleeding, pain, and facilitates overall management of the multisystem injuries. Often, this is accomplished initially using external fixation in a "damage control" mode. Then, after the patient is stable and through the initial period of bleeding, these external fixation frames are replaced with more definitive internal fixation to allow progressive mobilization and ambulation.

1.2   Types of Implants

The common element in the entire spectrum of surgical fracture care is the use of orthopedic implants to restore skeletal stability that has been lost by the fracture or joint dislocation. It is the thoughtful application of these devices that allows the modern fracture surgeon to create conditions at the fracture site that will promote healing as well as allow functional use of the extremity during the period of healing (Fig. 1).

Four basic types of implants are used in the surgical treatment of fractures: screws, plates, nails, and external fixators. The most basic type of orthopedic implant is the bone screw (Fig. 2). Modern orthopedic bone screws are available in a vast variety of sizes (1–8 mm diameter) and thread designs which allow use in a wide variety of clinical situations. Screws can be used as stand-alone fracture fixation devices which are placed across fracture fragments to generate compression and create stability and promote fracture union. These "compression screws" employ two different design techniques to achieve compression. The partially threaded bone screw gains purchase in the bone fragment away from where the screw is placed with no screw threads in the near fragment. Compression is generated when the head of the screw contacts the near bone fragment and the threads engage the far bone fragment in essence "pulling" them together. A fully threaded bone screw can also generate compression by the method of application (compression

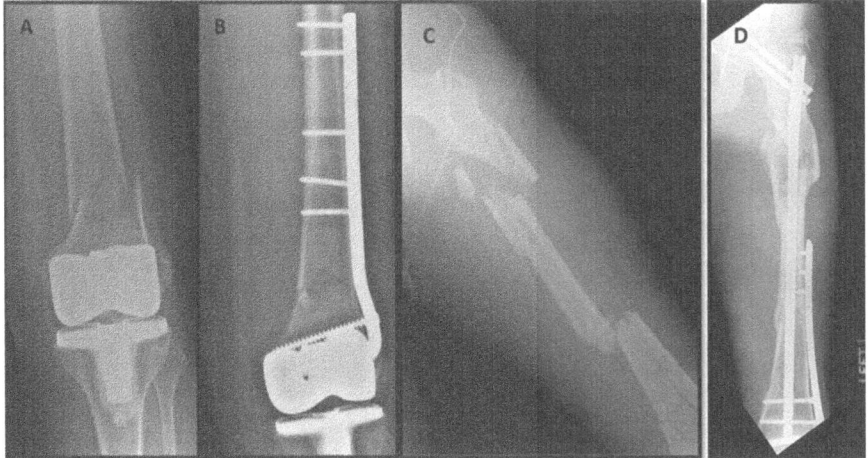

Fig. 1 Applications of internal plating (**b**) and combined internal plating and intramedullary nailing (**d**) to fix complex femur fractures (**a**, **c**). (Images from cases in our medical center)

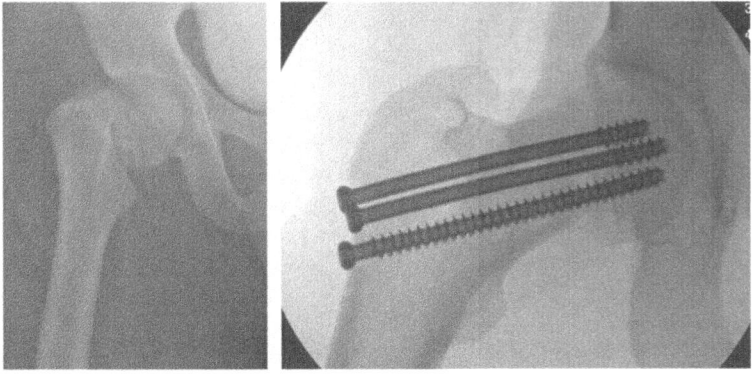

Fig. 2 Bone screws used to stabilize a femoral neck fracture. Two of the screws employ partial threading to achieve compression of the bone fragments. (Images from cases in our medical center)

by technique). A hole is drilled in the near bone fragment that is slightly larger than the outer diameter of the screw (gliding hole) and the far bone fragment is drilled with a size equal to the core diameter of the screw (threaded hole). When the screw is placed, the threads gain purchase only in the far fragment and as the screw head contacts the bone surface, compression is generated. The same fully threaded screw can be placed across bone fragments without the use of a gliding hole. In this situation, the screw engages threads on both sides of the fracture. Compression cannot be generated, and this construct is termed a "static" or "position" screw.

Plates are placed on the outer aspect of the bone surface and are mechanically linked to the bone via the use of screws. These bone screws can be mechanically

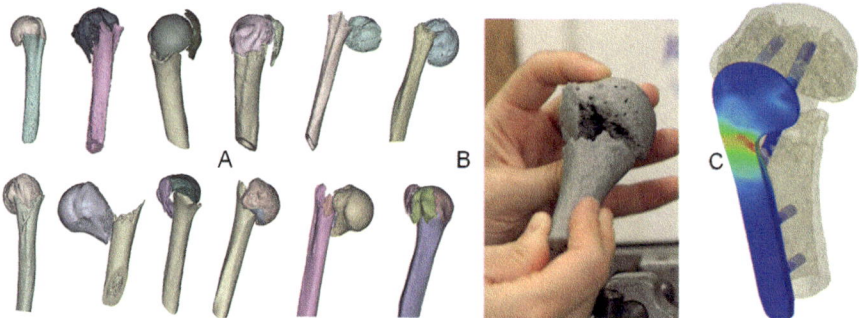

Fig. 3. (**a**) 3D CT-based models created in our lab of proximal humerus fracture patients. (**b**) 3D-printed models were anatomically reduced to simulate intraoperative reduction, prior to (**c**) computer modeling of stresses and biomechanical stability with plate and screw fixation (Images from our laboratory)

Fig. 4 (**a**) Periprosthetic internal plate fixation on a synthetic bone model. (**b-d**) Intramedullary nail implant with helical blade demonstrating nail curvature and screw holes (Images from our laboratory)

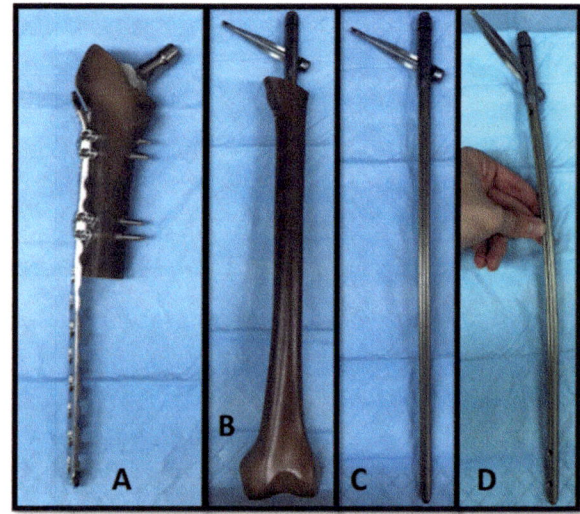

joined to the plate either by a frictional method (non-locked), or designed in such a way to allow the screw head to engage the holes in the plate via a series of threads in the head of the screw (locked) such that movement of the screw-plate interface is highly constrained. A plate designed to stabilize a fracture must fit within the local anatomy such that muscles and tendons and the adjacent joints can function normally while the plate is in place (Figs. 3 and 4a).

The third major type of fixation device is the intramedullary (IM) nail (Fig. 4b–d). These devices are placed into the medullary cavities of diaphyseal bones (e.g. femur, tibia, and humerus), and act as internal splints across the fractures. IM nails can span the length of the bone and be large devices, up to 50 cm. in length. Modern IM nails allow screws to be placed perpendicular to the nail through the ends of the

A

B

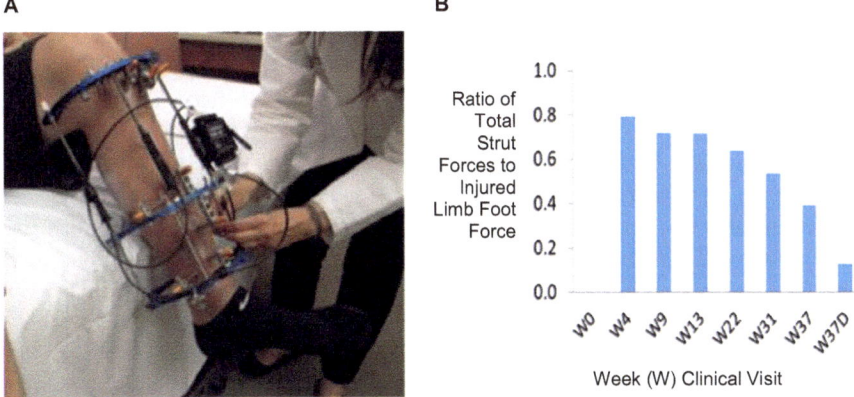

Fig. 5 We are currently conducting a clinical study in which the forces in vertical struts of ring-type external fixators are measured. Force data from the struts, as well as foot forces, are transmitted wirelessly (**a**) during the patient's gait. A decrease over time in normalized strut forces was observed in this patient (**b**), and in other patients, consistent with offloading due to bone healing. (In (**b**), 'W37D' indicates dynamization of the frame.) (Image from our medical center)

bone and the ends of the nail on either side of the fracture to control motion and displacement. This type of device is termed an "interlocked" nail and is the work-horse device in the treatment of adult femur and tibia fracture.

The fourth category of fixation device is the external fixator (Fig. 5). In this type of device, threaded bone pins (3–6 mm in diameter) are percutaneously placed into the bone on either side of the fracture and then connected outside the body using clamps and bars. External fixation can be constructed across an entire spectrum of stability (under the surgeon's control), to solve many clinical problems in orthopedic fracture care up to the creation of stability to allow definitive fracture healing. By its very design, external fixation is temporary and will have to be removed. Screws, plates and intramedullary nails do not have to be removed per se, but may be at a secondary procedure for a large variety of clinical reasons.

Variations in fixation implant design and surgical application may enhance overall mechanics and promote healing. Features such as improved stability, reduced anatomical interference, and more robust fixation to bone may be achieved through implant design modification, and careful planning by the surgeon.

1.3   Anatomical Constraints

The biomechanical and anatomic design constraints on orthopedic fracture implants are quite rigorous. They must be biologically compatible, and meet the biomechanical requirements of the situation in which they are used. Since they are used to support the bone and not replace it during the period of healing, the implants must fit the local anatomy which varies among patients. IM nails must fit within the intramedullary canal. Plates must fit on the external surface of bone and not impede important muscles, tendons, and ligaments surrounding them. Implants are also designed with consideration of the surgical exposure needed to apply them. Although many plates and screws are applied with an open approach that exposes the bone fragments, IM nails, external fixators and certain plates can be applied with a minimally invasive approach that avoids disturbing the soft tissues surrounding the fracture that are important for blood supply and the healing process. Plates designed to reduce the area of contact with bone via geometric modification of the contact surface are termed minimum contact plates. Reducing contact area at the bone-plate interface may enhance fracture healing by mitigating disruption to the periosteum and bone blood supply [1].

As a result of the aforementioned stringent geometric design requirements, fracture fixation implants typically do not have an infinite fatigue life. Since the fixation device may initially (at the time of the acute fracture and before bone healing) be load bearing to allow the patient to begin functional use of the extremity or ambulation, there will be accumulated damage in the microstructure of the implant. It is well understood by both orthopedic clinicians and implant designers that there is a "race" between bone healing and implant fatigue fracture. The clinical presentation of a fatigue fracture of an orthopedic implant may imply some pathology of bone healing.

2 Fracture Healing Biology

2.1 Fracture Healing

Fracture healing is a complex biomechanical process with multiple possible endpoints. The most desirable endpoint, termed 'union' in the medical vernacular, is complete fusion of the fractured components into one continuous osseous structure. A union may be achieved through one of two well-described mechanisms [2]. Primary or intramembranous bone healing occurs through Haversian remodeling which requires absolute stability between bony fragments. On a cellular level, primary healing involves direct healing of bone without the presence of intermediary tissues. A very stable interface between two bone fragments is necessary because there is creation of bridging capillaries as part of the Haversian remodeling which is intolerant of shear forces due to its mostly linear morphology [3]. Primary healing also requires contact between the fracture fragments, or only a very small gap (on the order of 1 mm). The degree of stability required for primary bone healing can usually only be achieved via compression between fragments. Thus, primary bone healing usually only occurs for simpler fractures in which the fragments can readily be fixed back in their anatomical position, usually with plates and/or lag screws.

Secondary healing or enchondral ossification is characterized by formation and evolution of progressively stiffening intermediary tissues in a relative stability mechanical environment. Secondary fracture healing progresses through three biological stages. The inflammation phase begins the moment fracture occurs and can last up to 5 days. This first phase is characterized by an immune response in which damaged, necrotic tissues are scavenged by immune cells and the fibrinous hematoma formed during injury is organized into granulation tissue. Fibroblasts produce type III collagen. Inflammatory mediators are released resulting in pain, swelling, and an increase in local blood flow. A chondroid stage then begins enhancing the vascularity and cellularity of the granulation tissue via mesenchymal stem cells from the nearby periosteum. The chondroid matrix is then replaced by osteoblasts creating type I collagen. The replacement and removal of the cartilage intermediate is the hallmark of endochondral ossification. At this point in the healing process, mechanical loads can be transmitted from one fragment to the other, but there is no internal organization of the callus. As normal loading returns, the woven bone characteristic of early callus is remodeled into lamellar bone in response to the loading pattern. This remodeling process can take months to years and is accompanied by a decrease in the size of the callus mass often closely re-approximating the diameter of the normal surrounding bone. Coincident with the decrease in size during remodeling, is an optimization of the mechanical properties of the bone.

The biological stages of secondary bone healing are intimately linked to the local mechanical environment. The interfragmentary strain theory, first described by Stephen Perren in 1991 [4], states that a tissue type cannot exist in a region in which the mechanical forces exceed its strain tolerance. The elegant interplay of biology and mechanics described in the strain theory provides the fracture surgeon and implant designer a framework to understand these seemingly unrelated concepts. For example, recent research on bone healing has found immature myeloid cell-mediated angiogenic cascade to enhance bone healing in a mouse model [5].

By definition, a fracture involves damage to the local blood supply of the bone often extending to the local tissue. As a result, bleeding occurs around the bone ends creating a fracture hematoma. As might be expected, conditions within this clotted hematoma are acidotic, hypercarbic, and have a low oxygen tension. The first tissue to appear histologically in this region is type III collagen made by fibroblasts. The syncytia of loosely organized collagen fibers and fibroblasts within the clotted fracture hematoma is called granulation tissue. Mechanically, granulation tissue has a strain tolerance of about 100%. Metabolically, granulation tissue survives in this region because it has a low oxygen requirement which can be met via diffusion. The granulation tissue organizes into an early fracture callus and conditions for the second stage of fracture healing are created. The second tissue type to appear is cartilage, which is composed of type II collagen, is made by chondroblasts. The presence of chondrocytes implies that the region has become mechanically stiffer as the strain tolerance of cartilage is about 10%. At this time, the local blood supply is improving in response to the release of angiogenic pyrogens. This larger more organized callus composed of granulation tissue and cartilage further stiffens the region to the point where enchondral ossification can occur. In regions of exceptionally low strain, osteoblasts will differentiate from mesenchymal stem cells and begin production of an osteoid. The histological appearance of osteoblasts implies significant mechanical stability has been achieved, as this tissue type has a strain tolerance of only about 2%. An important concept is that osteoblasts are quite metabolically active and have a high oxygen requirement. This requires the co-development of a rich capillary network. Bone cannot form in regions of high strain because the supplying capillaries cannot survive. The strain theory informs that each tissue type that appears in a healing fracture prepares the region both mechanically and biologically for the next tissue. Bone formation implies that stability to maintain a capillary network has been achieved. A histological hallmark of fracture healing is capillaries crossing the fracture gap. In the case of primary bone healing, a surgical procedure has created stability between bone fragments such that a capillary can cross the gap without the appearance of precursor tissues.

The different stability requirements between primary and secondary fracture healing are functions of the strain tolerance of the local biological tissues. Although the strain tolerances for relative stability are less defined, secondary healing has been shown to be induced by interfragmentary motion on the order of millimeters [6].

Impaired fracture healing can have biological and mechanical origins. Focusing on mechanical causes, nonunions or delayed unions are generally a result of insufficient stability at the healing site. Excessive strain during healing can rupture nascent capillaries and prevent nutrient transport to the metabolically active healing site. Poor stability can have multiple origins including inadequate initial surgery, implant hardware failure, or an unforeseen traumatic mechanical event. Biologically, a patient may have decreased bone density that will affect the stability that can be achieved at surgery. Tissue damage, infection [7], medications, smoking [8], nutrition, and genetic factors [9] influence the healing response following a fracture. Surgically, a risk of nonunion is created if the local tissues are damaged by the procedure especially if secondary bone healing is relied upon. In this situation, callus formation is impaired as a result of the tissue damage, and the healing may not be able to proceed to union. In nonunion cases, stresses on the implant are never relieved by bone unions, eventually resulting in a fatigue failure of the implant.

Residual bone deformity can result from a failed fracture fixation or improper bone healing. Residual deformity can occur when the fracture is not reduced to its original anatomical position or orientation, bone migrates through partial implant loosening, or implants partially fail through yield. Most common in comminuted and bone loss fractures, residual deformity can leave a fracture patient with pain, joint stiffness, limb-length discrepancies, and posttraumatic arthritis (in addition to the deformity) associated with insufficient reduction of fractures affecting a joint. Residual deformity outcomes may require medical or even surgical attention and reduce a patient's quality of life. The close relationship of local mechanics to successful bone healing necessitates that proper care and consideration be given to all fractures.

2.2 Infection

Infection is a devastating complication following surgery, and implanted materials lead to greater risk of infection. Open fractures where the skin is broken increase the risk for infection. This risk can reach 30% in certain high grade open fractures with severe contamination and damage to muscle and bone. Surgery further increases the risk for infection, although most infection rates during surgery on non-immunocompromised patients are below 1%. Diabetes mellitus, HIV, and rheumatoid arthritis are common examples of chronic conditions which can also increase the risk for infection during fracture fixation. Additionally, lifestyle characteristics such as smoking, obesity, and poor nutrition can also predispose patients to higher infection risks. Novel approaches for reducing bacterial colonization and biofilm formation on fracture fixation biomaterials are an active area of research.

3 Biomechanics

The biomechanics of the fracture fixation construct are integral to understanding how to develop an optimal implant and surgical approach to fracture repair. The term 'implant' herein will refer to the large fixation component that stabilizes the fracture often via attachment with screws. The term 'construct' refers to the entire system encompassing the implant, screws, fracture site, bone, and healing tissues. 'Configuration' will refer to surgical variables, i.e. number of screws used, implant position, type of screw fixation, etc.

3.1 Implant Loading

The initial stress born by a fracture implant following surgery is quite variable across patients. At one end is the scenario in which bone fragments have been anatomically reduced and compressed (absolute stability). Load transmission will occur from one bone fragment to the other directly and result in low implant stresses. At the other end of the stability spectrum is the situation in which there is no initial contact between bone fragments and the implant carries the entire initial load of the construct. A bridge plating across a large zone of comminution would be an example of this.

Fracture fixation constructs are subjected to a variety of loading conditions ranging from singular high force loading, such as due to an accidental patient fall, to repeated functional activities. These load patterns are classified as either static or cyclic, respectively, and may lead to either material yield or fatigue failure mechanisms. Orthopedic implants can be subject to surprisingly large mechanical loads within the body. For example, instrumented arthroplasty implant studies show that implant reaction forces in the knee, hip, and spine can reach nearly three times the bodyweight during simple gait [10]. Furthermore, instrumented fracture fixation implant studies can show the strain in an implant and use strain data as a surrogate for fracture healing over time [11]. A study using an instrumented femoral nail demonstrated that the load in that implant was almost exclusively parallel to the implant's primary axis in four different postures [12].

Before a fractured bone has begun healing, the fixation construct may be responsible for up to all of the load transferred across the fracture gap. This is particularly true in situations in which there is no contact between the major bone fragments of the fracture. As time progresses, intermediate bone tissues form across the fracture site which gradually reduce the load carried by the implant. The construct must maintain mechanical integrity during the course of bone healing as the mechanical environment at the fracture site plays a large role in achieving union. As discussed previously, excessive motion at the fracture gap can result in local tissue strain that exceeds the tolerance of nascent capillary networks that are attempting to bridge these gaps [3]. Thus, we must understand the mechanical responses of fixation constructs to physiological loads within the scope of the stability tolerance required for a biological fracture healing process.

Bone, as a living material, adapts its composition and shape over time in response to its mechanical stimulus history. Stress shielding refers to decreased bone density in regions that are subjected to lower stress levels due to loads borne by the implant. The fixation implant effectively offloads some of the forces that normally pass through the bone. Such a decrease in stress is believed to engage biomechanical signals and either stimulate bone resorption or inhibit bone formation.

Although limited or non-weightbearing is often appropriate in the early stages of fracture healing, sometimes to protect the implant itself, prolonged limited weightbearing in a patient can result in disuse osteopenia. Disuse osteopenia is a decrease in bone mineral density as a result of prolonged periods of unloading in the bone. Bone remodeling processes, perhaps with sensing by osteocyte cells, detect the relatively low loading profile of bone and favor resorption over deposition, resulting in a bone density that decreases with time. Osteopenia decreases bone mineral density and increases the risk of implant-bone loosening or bone fracture during activity [13, 14].

3.2 Implant Stress and Failure

Stress is a measure of internal forces in a localized region of material and has units of force divided by area. Most often in orthopedic biomaterials stresses are generated in an implant in response to applied external loads. Stress causes a material to deform, creating strain which is a measure of internal displacement in a localized region of material. Strain is a unitless ratio of final length to initial length. Basic yield and fatigue failure theories for engineering materials such as orthopedic implants primarily utilize stress to predict failure. Most current fracture fixation implants are usually considered as linear elastic materials, meaning that while they undergo deformation below failure thresholds, a linear relationship between stress and strain is maintained. Young's Modulus or elastic modulus is equivalent to the ratio of stress to strain. Examples of linear elastic materials include steel, carbon fiber, and glass. Once the yield stress is reached, the material begins to fail plastically and the Young's Modulus no longer accurately describes the stress-strain relationship.

Stress in a fracture fixation construct, like in other mechanical applications, is subject to concentration effects. Stress concentration describes the localized magnification of stress due to geometric features such as holes and sharp corners. Changes in cross sectional area are common along fracture fixation implants and are frequently caused by variable implant design, holes in the implant intended for screw placement, tapers, fillets, and edge characteristics. Most fracture fixation plates have unique underside geometries that limit areas of contact with the periosteum in order to reduce damage to blood supply. Variable cross sections are introduced in some cases to avoid interference with anatomical structures and reduce the weight of the implant. In most modern plates, the removal of material under the plate to limit bone contact area is done in such a way as to keep the cross-sectional moment area of inertia of the implant the same throughout the plate length. This allows the surgeon to easily bend the plate if needed during the surgical procedure.

Fracture fixation construct failure can occur within the implant. If the failure occurs under a static load, yield can occur at the location of maximum shear stress in ductile materials. The yield strengths for common fracture fixation implant materials such as cold-worked stainless steel, hot-forged CoCr alloy, and forged, heat treated titanium are 792, 1600, and 1034 MPa, respectively. Bone screws are typically made of annealed stainless steel which has a yield strength of 331 MPa. Typically, shear stress is maximized in the implant region near an unsupported fracture gap where there are fewer points of fixation and where the largest bending moments are likely to occur. When the applied load is not sufficient to cause yield failure, the implant may still accumulate damage in the form of microcracks. Cyclic loading at sub-yield stress levels causes fatigue failure via progressive crack propagation and eventual brittle fracture within the implant. At stresses above a material's endurance limit (a cyclic stress amplitude below which failure will never occur), one can calculate the number of fully reversed loading cycles required to reach fatigue failure. The resulting data are plotted on an S-N curve to characterize the material's response to cyclic loading. A Goodman Diagram may be used to consider the fact that implant mechanical loading is often not fully reversed, and is instead repeated with a non-zero mean stress.

However, the amplitude of cyclic loading of fracture fixation implants is often variable, dependent on a variety of functional activities (e.g. walking, climbing stairs, jumping, etc.). Under Miner's rule, the total damage is defined as the sum of incremental damages calculated for each load amplitude [15]. Each incremental damage can be calculated as the ratio of the number of cycles at a given stress or strain range to the total number of cycles required for failure at that range. Some commercial finite element software include more advanced algorithms for predicting

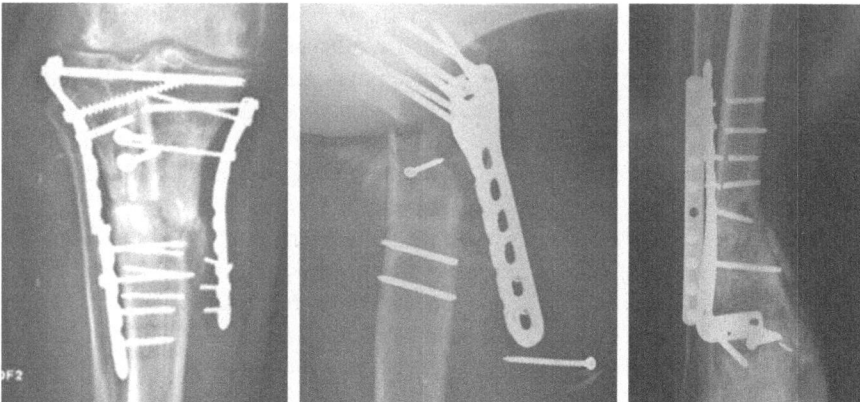

Fig. 6 Careful considerations of biomechanics is needed when treating difficult fractures. Surgeons attempt to create conditions at the fracture site that will promote healing while allowing functional use of the extremity. Radiographs from our center show a loss of fixation, plate, and/or screw failure in the (left) proximal tibia, (middle) proximal humerus, and (right) distal dibia. (Images from our medical center)

fatigue, e.g. 'fe-safe', a durability finite element software suite from Dassault Systemes. Fixation and contact points between the implant, screws, and bone are of special interest for biomechanical characterization as these locations are common sites of construct failure. Screw head shearing has been documented to occur. In this case, screw fractures typically occur near the screw-implant-bone junction (Fig. 6). Shear stresses in this region can be influenced by decreasing the distance between the fracture gap and the closest screws (working length), increasing the screw count, and biological stiffening across the fracture site via healing [16]. Local bone micro-fracture at the screw-bone interface can also occur, typically leading to screw pull-out from the bone [17].

3.3 Fracture Gap Strain

Bone is remarkable in its ability to regenerate following a fracture. Unlike most other tissues which produce scar tissue, a fractured bone, adequately stabilized and with sufficient vascularization and other biological factors, is able to ultimately regain its near-original form [18]. Nonoperative treatment relies on some degree of stability from soft tissues including muscles, ligaments, fat, and skin, whereas operative treatment increases stability through implant fixation.

Various theories have been proposed for how mechanical stimulus affects differentiation and adaptation of healing tissues. These theories are based on mechanical stimulus in the form of hydrostatic pressure, deviatoric stress, and even fluid flow velocity, the latter viewing the tissue as a poroelastic material.

Interfragmentary displacement or strain (displacement divided by fracture gap width) measures provide a useful and intuitive means of characterizing mechanics at the fracture gap. Based on available evidence, moderate strains of approximately 10–50% have been proposed as being beneficial to healing [19, 20]. This strain is often split into two components: axial or longitudinal, acting along the long bone axis, and shear, acting perpendicular to the long bone axis. There is some consensus that shear strains can be detrimental to fracture healing, whereas moderate axial strains are beneficial [21, 22]. Dynamized intramedullary nails and certain new plate technologies promote additional axial strain [23]. Excessively high strains are inhibitory to fracture healing and may result in nonunions [24–26]. Compressive interfragmentary displacements are superior to distractive displacements in creating callus [27]. Bone formation is linked with vascularization, and interfragmentary movements early in fracture healing to promote revascularization, but in the later stages of healing can inhibit blood flow [24, 28]. Some investigators have demonstrated clinically the influence of biomechanical and biomaterial choices on fracture healing and callus formation [29]. However much of the evidence supporting these theories are from studies in animals, especially sheep. In these studies, fractures, often in the tibia, were simulated with osteotomy cuts leaving a gap. The fractures were stabilized with external fixators. In one such study [21], five sheep were fixed with a device that allowed only axial interfragmentary displacements, and five sheep were fixed with a device that allowed only shear displacements. Displacement magnitudes were 1.5 mm with a 3 mm fracture gap (50% strain) in both groups. The group with shear displacements experienced significantly delayed healing of the osteotomies, with one third stiffness of the healing site after 8 weeks.

Although there is some consensus that axial and shear strain influence fracture healing differently, unfortunately control of these two strain components by changes in fracture fixation construct are not always intuitive. For example, increasing the working length (or bridge span) between the inner-most screws in a plate construct seems that it would mostly affect axial interfragmentary strain; however recent predictions from finite element models of 66 supracondylar femoral fracture fixations in patients (with supporting data from synthetic bone experiments) revealed that increasing working length primarily affected interfragmentary shear motions, not axial motions [22]. Changing the plate from stainless steel to titanium increased both types of motion. Callus formation in the patients was not associated with comorbidities including smoking or diabetes, but was promoted by longitudinal motions and inhibited by shear motions at the fracture site [22]. Additionally, improvements in one aspect of biomechanics may lead to concerns in another aspect; the scenario of large longitudinal strains combined with small shear strains may be achieved with smaller plate bridge spans, but unfortunately these smaller spans lead to larger plate stresses and a risk of plate failure [6, 22].

3.4   Biomechanical Variables

Bone strain and displacement at the fracture gap are influenced by the stiffness of the construct, interfragmentary contact, and load transfer at the fracture gap. The stiffness of a fixation construct is its resistance to deformation in response to applied loads. Stiffness is a function of several construct variables: design, orientation, material composition, working length, screw count, and biological material properties in the fracture gap. Construct stiffness can be measured in multiple modes: bending, torsion, and compression. The bending stiffness for plate constructs in long bones often dictates most of the displacement at the fracture gap, whereas circular-type external constructs rely more on the compression stiffness mode. Torsional stiffness can become important clinically in all types of fracture repair constructs (plate, intramedullary nails, and external fixation).

Various fracture fixation devices provide stiffness to the fracture. For example, a bridge plate is a construct in which a plate implant acts as an extramedullary splint spanning a complex fracture and bends and compresses in response to axial loads on the bone. The bending stiffness of a bridge plate can be calculated as an applied bending moment divided by lateral deflection. Similarly, the compressive stiffness of the bridge plate can be determined by dividing the applied axial load by axial displacement. All loading modes (compression, bending, and torsion) work in unison to dictate the bony displacement and strain at the fracture site, critical for fracture healing, under a given load scenario.

Graded-stiffness and composite material compositions can modify construct stiffness although currently they do not see much clinical use. The working length of a fixation construct is the distance between the two fixation points closest to the fracture gap. The stiffness of the construct can be approximately inversely proportional to the working length [30]. Surgical decisions such as screw placement can cause construct stiffness to be too extreme in either direction (too stiff or too compliant) to promote healing.

Use of either locking or nonlocking screws will influence the fracture fixation biomechanics. Locking screws maintain a fixed angle with the plate and do not enable motion at the screw-plate interface. Nonlocking screws compress the plate and the bone together creating a frictional interface between them. Over time, especially in unicortical screw fixation, normal stresses in the cancellous bone surrounding screws may cause loosening at the bone-screw interface, allowing the screws to toggle at the implant-screw junction. Thus, the biomechanics of nonlocking screws allow for some motion at the screw-plate interface, potentially resulting in a decreasing construct stiffness over time [31].

4   Biomaterials

4.1   Stainless Steel Vs. Titanium alloys & Other Materials

Although a vast array of metallic alloys are available for titanium and stainless steel, the mechanical, chemical, and biocompatibility properties dictate that a small subset of these alloys are appropriate for use in fracture fixation applications. Metallurgical alloying can influence multiple material properties of the base metal. Specifically relevant to fracture fixation are the toughness, manufacturability, oxidative potential, corrosion resistance, and biocompatibility. Most modern implants are manufactured through forging [32].

The current most common stainless steel alloy used in orthopedic fracture fixation applications is designated as 316L. This low carbon content alloy is manufactured to follow ASTM F138 and F139 standards which specify: $\leq 0.03\%$ carbon, $\leq 2\%$ manganese, $\leq 0.03\%$ phosphorous, $\leq 0.75\%$ silicon, 17–20% chromium, 12–14% nickel, 2–4% molybdenum, with the remainder being iron. The carbon and chromium content contribute to corrosion resistance while nickel and molybdenum improve the mechanical toughness of the alloy. 316L is amenable to cold-working strengthening which can potentially increase the yield strength from 330 MPa to a maximum of 1200 MPa [32]. The elastic modulus for stainless steel is 190 GPa.

Titanium and titanium alloys have an elastic modulus of 110 GPa, considerably lower than that of stainless steel and closer to that of bone, which explains why titanium alloy implants mitigate the risk of bone loss associated with stress shielding. Titanium is commonly alloyed with aluminum and vanadium, the most common being Ti-6AL-4V (5.5–6.5% aluminum and 3.5–4.5% vanadium).

4.2   Biocompatibility

All components of internal fracture fixation implants and percutaneous components of external implants must be biocompatible. Because fracture healing typically occurs on the order of weeks to months, biocompatibility must be maintained long after the materials have been implanted. Although a complex topic, the biocompatibility of an implant refers to the local and systemic physiological changes undertaken by host tissues in response to the implant's presence. Potential biocompatibility hazards of orthopedic implants include toxicity, immunogenicity, mutagenicity, and infection propensity. Most assessments of biocompatibility of fracture fixation constructs focus on immune responses to the implant. Immune responses can be triggered by material composition, surface topography, as well as size, depending on the local implant environment, and can activate traditional innate, complement, and humoral immunological pathways.

The implant surface-tissue interface is the single most important driver of biocompatibility in metallic implants [33]. Morphological features such as porosity, surface modification, and smoothness as well as chemical characteristics such as hydrophobicity, wettability, surface charge, and polarity influence an implant's immunogenicity [34]. Furthermore, corrosion (see below) may modify an implant's morphology, produce small particles from the implant, and ultimately reduce long term biocompatibility. Soluble corrosion products can be immunogenic and can also be associated with local cytotoxic and tissue toxicity reactions.

Biomaterials may be classified as inert, interactive, or viable [35]. Inert biomaterials such as cobalt-chromium alloys generate little or no biological response in the absence of wear and corrosion. Titanium and stainless steel are classified as inert biomaterials. Titanium and titanium alloys typically exceed the biocompatibility of stainless steel [36] as well as have lower infection rates over steel [37]. Both titanium and stainless steel alloys commonly used in fracture fixation have very good biocompatibility modulated by an inert, insoluble oxide layer that forms on the surface and is chemically impermeable.

Interactive biomaterials differ in that they are designed to trigger biological responses such as osteoinduction or osteointegration. Viable biomaterials contain a biological component and may be resorbed or biodegraded. Fracture fixation implants are currently most commonly comprised of inert biomaterials, but a construct that can adapt its mechanical properties as a fracture heals may be desired. Early investigation into resorbable plates and screws shows potential in a rising field of research [38, 39].

Biodegradable polymeric biomaterials, such as synthetic polyesters, are emerging as candidates for orthopedic implant applications as biodegradation may eliminate the need for surgical removal of implants. Degradation of synthetic polymeric biomaterials causes foreign-body reactions that are measured using histopathology and currently their immunogenic hazards have little clinical significance [40]. Current research finds biomaterials such as polyetheretherketone (PEEK) to have good strength and radiolucent properties that enable improved visualization of fracture healing by computed tomography and standard radiographs. However, PEEK suffers from inferior fixation strength and stability compared to titanium [41]. Advances in polymer manufacturing and fracture fixation technology may yet have clinical applications in fracture fixation.

4.3 Corrosion

Metallic materials used for fracture fixation in the body are susceptible to chemical attack, termed corrosion. Corrosion is a collection of degradative processes resulting from metals reacting with charged particles in solution. An oxidation reaction in which the metal loses electrons to the surrounding solution starts a corrosive process. Metallic surface particles then either dissolve or form an oxide on the material surface. Corrosion initiation is driven by the thermodynamics of redox reactions and inhibited by kinetic barriers. For a given implant material, the thermodynamics of reduction/oxidation reactions are fixed within a biological milieu. Therefore, a metallic implant's defense against corrosion is through development of kinetic barriers through passivation. Passivation typically characterizes formation of a passive metal-oxide layer on the outer surface of the metal. A continuous passive film presents a physical obstacle for the ion transfer necessary to begin a corrosive process.

Broadly, corrosion can occur through several main mechanisms described as uniform attack, galvanic corrosion, intergranular corrosion, crevice corrosion, and fretting corrosion [32]. Fracture fixation implants, however, are typically susceptible to galvanic, crevice, and fretting corrosion pathways.

Galvanic corrosion is driven by an electrochemical potential gradient of metals in contact with each other. While a larger problem for modular implants, galvanic corrosion can occur in fracture fixation implants, for example, if the screws are made of a metal with a different electrochemical potential than that of the implant metal. For this reason it is uncommon for a fracture fixation construct to vary in composition between screws and implant.

Stress corrosion cracking is a subset of crevice corrosion in which pits grow on the surface of the material. The ionic microenvironment in pits serves as a barrier between the local implant surface and the surrounding tissues. The net effect of the pit microenvironment is to facilitate corrosion through ionic exchange between the implant metal and nearby tissue. Additionally pit geometry concentrates stress and promotes crack formation. When the stress intensity in a pit reaches the critical value for crack propagation under corrosive conditions, a crack can propagate and cause component failure. Unfortunately, the critical value for crack propagation under corrosive conditions is typically lower than the fracture toughness of the material, making stress corrosion cracking difficult to predict.

Fretting corrosion is corrosive damage resulting from cyclic motion between two opposing surfaces. In this mechanism, grooves and oxide debris develop on the surface as a result of toggling between contacting components. Fretting corrosion exposes the material beneath the passive oxide layer. For example, nonlocking fracture fixation designs are at risk for fretting corrosion whereas locking screws do not experience motion at the screw-plate interface and are less susceptible.

5 Experimental and Computational Modeling of Fracture Fixation Mechanics

The growth of orthopedic fracture knowledge facilitates a shift from hypothesis-based medicine to evidence based medicine. This change in approach accompanies technological developments in computer science, manufacturing, materials research, and mathematical methods which combine to create more accurate laboratory models of interactions between biology and medical devices than past iterations. Medical decisions should be made based on the best known available models which, sometimes, are still clinically observation-based. As research efforts expand to meet the evidence needs of modern fracture fixation, two distinct modalities emerge.

Experimental research in this context will refer to physical experiments, often involving mechanical loading, kinematics, kinetics, or chemical phenomena. Computational experiments will encompass those experiments conducted predominantly in silica, although experimental research accompanies them as validation frequently.

Mechanical and biological environments involved in clinical fracture fixation are very complex. Despite substantial advances in lab-based experimental and computational models, clinical studies remain a vital part of modern investigation.

5.1   Experimental

Most lab-based experimental research in fracture fixation biomechanics and biomaterials focuses on mechanical testing and/or bone healing. Mechanical testing involves a direct application of load or displacement and is valuable because it enables direct measurement of a construct's response. The mechanical testing setup includes a loading apparatus, implant hardware, a fixation mechanism, and a bone model. The most commonly used loading machines are either uniaxial or biaxial standard mechanical testing frames retrofitted with fixtures to accommodate a fracture construct. Implant hardware used during testing is commercially available hardware or, in the case of testing new devices, custom made. Regardless of hardware, most fixation mechanisms selected for testing are those that would be also used in clinical cases analogous to the experimental model. Depending on the research question, the bone model can range in complexity from cylindrical plastic pipe to compound synthetics to human cadaveric specimens. Such studies are typically focused on the 'time-zero' mechanical response before any fracture healing occurs, unless the fracture callus or bone healing is artificially simulated. Advanced composite fiber glass, epoxy resin, and polyurethane synthetic bones can adequately represent certain bone mechanical properties, but adequacy in representation of screw or implant fixation and loosening is still controversial. A question regarding specific implant stresses and strains may be answered without the need for expensive cadaveric bone specimens but a question focused on screw-bone fixation or involving a complex fracture pattern may be best answered by recreating the fracture in a true bone in the laboratory setting. These cadaver bones are typically stored frozen, not formalin-fixed because of potential changes in mechanical properties associated with chemical preservation.

Experimental research can be used to test fracture construct performance during functional activities or to simulate traumatic construct yield events. Generally of interest are failure scenarios in the implant, the fixation mechanism, or in the bone. Static loading patterns are effective ways to test yield failure modes, whereas cyclic loads can be applied to simulate fatigue failure. Both failure modes are usually possible in a patient; yield if a patient has an accidental fall or other high load event, and fatigue associated with long term physiological loading, especially if the fracture does not heal quickly and the implants bear larger loads for longer time (Fig. 7).

Fig. 7 Plot to quantify the race between bone healing (callus stiffness: blue) and implant fatigue (stresses: orange), with data based on simulations with callus. The red line represents the fatigue strength of the material (Data from our laboratory)

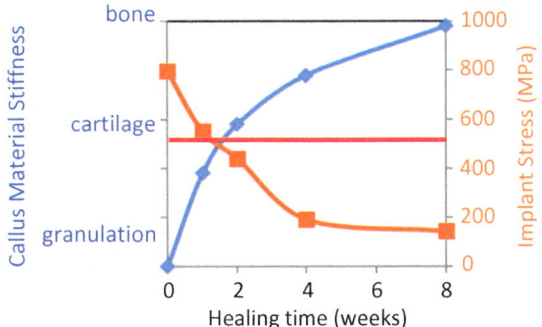

Outcomes such as construct stiffness and strength are commonly measured under loading conditions including axial compression, torsion, bending, or a physiological combination thereof [42]. Motion of bone fragments or implant components during mechanical testing can be measured using infrared cameras, DVRTs, or simple optical measurement methods using a digital camera.

Lab-based experimental research aimed at fracture healing is often conducted in animal models. Animal models can be used to assess the impact of novel fixation constructs or biological therapies on the healing response. For example, a recent study demonstrates a faster and stronger healing response with active plating over conventional compression plating in sheep [43]. Although larger animal models are costly and time consuming they have the ability to simulate a biological environment with similarities to that of a fractured human bone. Animal models can be exposed to loading protocols in custom loading devices to simulate the mechanics of activities with ties to human function [44, 45]. Furthermore, studies in smaller species enable the use of genetic modification to understand mechanotransduction of fracture healing on a cellular and molecular level [46, 47], although the metabolism and anatomy of these animals is often very different than humans, a hurdle to clinical translation.

5.2   Computational

Computational fracture fixation research aims to simulate the mechanical environment of a fixation construct. Due to limitations in computational power and modern characterizations of biological materials it is not yet possible to accurately model all biological, chemical, and physical properties of implant hardware and biological tissue, however these features may have limited impact on specific aspects of construct behavior. Therefore, major assumptions are made to simplify development and run time of computational models of fracture fixation. The most commonly used computational approach is the finite element method.

Computational models from previous studies focus on the effects of varying implant design [48], implant material composition [49–51], implant alignment and positioning [52, 53], implant fixation [54–56], and bone and fracture characteristics [57–59]. Similar to experimental models, computational models often focus on predicting implant failure, fracture gap strain, and implant load transmission. Computational models have an important advantage of being able to predict displacement, stress, and strain throughout all 3D points of a construct, whereas experimental measurements are often limited to selected external points of a construct. Computational models may also simulate the time-dependent effects of fracture healing on local mechanics, a challenging scenario to recreate in an experimental model. Computational models enable comparison of results associated with variation in a single model parameter, which may not be easily facilitated by experimental models, especially if expensive biological specimens are involved. Computational models allow for multiple levels of data interrogation while the physical nature of experimental models limit analysis to specifically measured outcomes. However, computational models depend on user-defined inputs to run simulations whereas experimental models allow for direct measurement of a physical system. A computational model developer must understand and justify all inputs to the model. Experimental models generally can provide better realism, especially when using cadaver tissues, and are thus often used to validate computer models. In some cases though, experimental setups are less accurate such as when measurements are sensitive to boundary conditions that are difficult to precisely control experimentally but not computationally. Both computational and experimental modeling modes can be very time consuming depending on setup and research question.

6 Internal Plating

Internal plating is a common method of treatment for many types of fractures. Plate fixation can provide stability by transferring all of the load across a fracture or by sharing some of the load burden with bone. Depending on the technique of application and design, plates can be used for multiple functions: compression, bridging, buttress, protection, and tension band [60].

Screw fixation between the plate and the bone must have sufficient strength to resist failure. Screws can be locked or compression, uni- or bi- cortical, fixed angle or variable angle, and can be inserted into as many or as few screw holes as desired. Locking screws are threaded into the implant as well as the bone to maintain a gap between the two, whereas conventional screws rely on frictional compression to hold the plate to the bone. Unicortical screws anchor in the cortex closest to the implant and into any trabecular bone present beyond the cortex. Unicortical fixation is necessary around articular joints and pre-existing prostheses, but is more susceptible to loosening over time and may lead to construct failure via screw pullout from the bone [61]. Bicortical fixation is more rigid because the screws pass through the other dense cortex after crossing trabecular bone. Variable angle locking screw holes allow for user-directed angulation of the screw relative to the plate up to ~17°. Varying screw angle can allow the surgeon to accommodate for local anatomy and

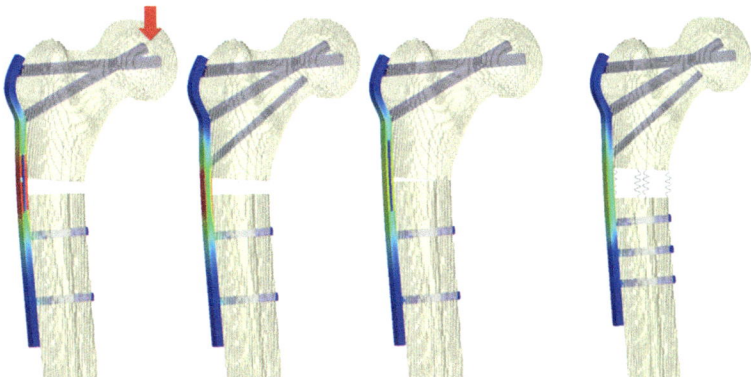

Fig. 8 Examples of finite element models of subtrochanteric fracture fixation with a lateral plate and screws subject to axial loading, showing effects of screw configuration and fracture gap size on resulting implant stresses (colors) and fracture gap strains. (Right) Callus was represented with springs having various stiffness. Blue represents low stress, red represents high stress. (Data from our laboratory)

other implants, or to avoid regions of poor bone quality for screw fixation. Unfortunately variable angle locking plate-screw interfaces are not as strong as standard threaded locking mechanisms [62], presenting an opportunity for additional improvements in plate systems. Additionally, screw holes within a plate can be elastically suspended to enable dynamization of the fracture in what are called active plates. Furthermore, the screw holes in the implant may be designed to be slots instead of circular holes. Slot-shaped screw holes allow the plate to slide around the screw which, like active plating, enables motion at the fracture.

Internal fixation plates are most commonly made of either stainless steel or titanium alloys. It has been shown that stainless plates create a stiffer construct with a longer fatigue life than titanium plates for applications at the distal femur [63]. Other plate materials such as carbon fiber PEEK composites have been shown to have comparable torsional stiffness to stainless steel but inferior failure characteristics for distal fibula fractures [64]. New hybrid materials such as glass/flax/epoxy composites demonstrated higher ultimate strengths than conventional metals in tension, compression, and bending while also having a lower axial stiffness [65], however, they are not yet commonly used in the clinic.

Considering some of the subjective variables involved in internal plating construct implantation (plate length, plate material, number of screws, screw fixation, fracture shape, plate working length, etc.), it is evident that an optimization scheme could facilitate mechanically-informed surgical decisions [66]. Parametric models that iterate construct design variables have been employed to observe the 3D mechanical effects of design changes under axial as well as combined axial, torsional, and bending loads (Fig. 8) [16, 67]. Although the number of possible design variable combinations is exponentially large, consideration of the surgical environment and local anatomy and biology can focus the solution space and eliminate analysis of unlikely or impossible construct designs such as one in which no screws are implanted. Model outputs such as implant and screw stresses and fracture gap strain can also be used to predict construct failure and to observe the mechanical influences of simulated healing. Thus, the utility of such models extends beyond surgical decision support and preoperative planning into training applications for clinicians and implant designers.

7 Intramedullary Nailing

Intramedullary (IM) nailing is a common method of treatment for adult shaft fractures in diaphyseal bones. Nails are commonly used to provide relative stability for secondary bone healing for fractures in the femur, tibia, and sometimes even the humerus. IM nailing entails multiple decision points which may influence the healing course and even success of a fixation construct. For example, the nail implant diameter can be selected to achieve a tight or loose fit within the medullary canal. Additionally, the canal can be reamed to accommodate a larger nail diameter and increase the contact area between the nail and the bone. IM nails may be solid or cannulated. Solid nails were shown to have a lower risk of infection in rabbits [68], whereas the more common cannulated nails may be surgically guided by wire via their cannula during installation. IM nail length can dictate whether or not the nail engages bone in the distal metaphysis for stability. IM nails are often constrained relative to the bone both proximal and distal to the fracture with screws (static locking). Screws confer rotational as well as longitudinal stiffness to the construct.

Some femur nail designs accommodate two screws both proximal and distal to the fracture. A common design to stabilize a proximal femur fracture incorporates a helical blade which embeds within cancellous bone in the femoral head. Another design feature common to femur nails is a dynamic interlocking screw slot. A dynamic screw slot allows for guided translation of the fracture gap along the direction of the slot (inducing desirable axial interfragmentary strain) while still conferring torsional shear stability with bicortical screw fixation. Sometimes screws may not be installed on one side of the fracture to allow for dynamic compression at the fracture site, although care must be taken in these cases to ensure the nail does not perforate the bone on the unfixed side.

The first IM nail implants were made of stainless steel for their strength and biocompatibility. However, stainless steel nails were too stiff to accommodate shape mismatch between the nail and bone. These features were enhanced by a change to a Ti-6AL-7NB titanium alloy which is now standard as a nail material. Titanium nails demonstrate less slipping, more even stress distributions, and increased contact area within the medullary canal over stainless steel nails in a computational model [49]. Furthermore, titanium nails reduce interfragmentary shear motion to better promote fracture healing over stainless steel nails [69]. Another study found that titanium nails are more stable than stainless steel in torsion and axial compression, although both nail materials resisted failure at non-weight-bearing loads [70].

The mechanics of IM nails spanning a fracture are complex. The type and comminution of the fracture dictates how much of the load applied through the bone must pass through the nail. A fully comminuted fracture will require the nail to transfer all of the load across the gap via the locking screws and represents the worst-case scenario in terms of implant load bearing. Other factors that influence nail construct mechanics are the material properties of the nail and screws, the cross sectional shape and anterior bow of the nail, nail diameter, nail length, medullary canal reaming, and screw configuration. For example, it has been shown that longer nails can generate higher contact stresses with the bone medullary canal surface than shorter nails, and that the contact stresses can be mitigated by increasing the flexibility of the distal end of the nail [71]. An experimental model in cadavers indicated that distal screws significantly increase maximum rotational load to failure in unstable intertrochanteric fractures and recommended their use due to the improved torsional strength [72].

Contact within the medullary cavity is a significant mechanism for nail load transfer [73]. The location and area of contact are influenced by nail size and shape, canal reaming, and implant position and orientation. If the contact area is small, the load transferred to the bone is concentrated and can generate high stress. This is thought to be a primary cause of bone pain which some patients report after IM nail implantation. Increasing the area of contact between the implant and the bone may reduce pain through design of the nail cross section and radius of curvature, selecting an appropriate diameter nail for the bone, and canal reaming. It has been shown that an 11 mm-diameter nail with static interlocking reduces motion at the fracture site up to 59% compared with a 9 mm-diameter nail [74]. Nail implant modifications such as diameter, material, and cannulation as well as screw material and area have been shown to reduce interfragmentary shear motion up to 54% [69]. Prior work combining computational and experimental modeling has validated finite element models of an IM nailed femur at four stages of gait and suggested future improvements in implant design and surgical implantation [75].

Research in our laboratory using the finite element method has demonstrated some effects of surgical and implant alterations on IM nail biomechanics (Figure 9). These models utilize idealized bone and implant geometries. Additionally, more realistic bone and IM nail geometries have been modeled for the femur. The femur model was built using Mimics software suite from a patient CT scan. Separate materials are defined to represent cortical bone and adjacent cancellous bone. The nail geometry is generated in Solidworks software using the manufacturer's design specifications. The nail implant is aligned with the bone to represent surgical positioning and then the bone canal is reamed. The implant and bone models are then imported into Abaqus software for finite element analysis. The models are meshed and boundary conditions, contact constraints, and applied loads are defined. Screws elements are positioned to align with screw holes in the nail and are embedded within the surrounding bone mesh. A fracture can then be modeled by defining a cutting plane and removing elements that intersect the plane. A complex 3D model with physiological boundary conditions and applied loads can provide insight into the location and magnitude of stresses in the implant, screws, and bone (Figure 10).

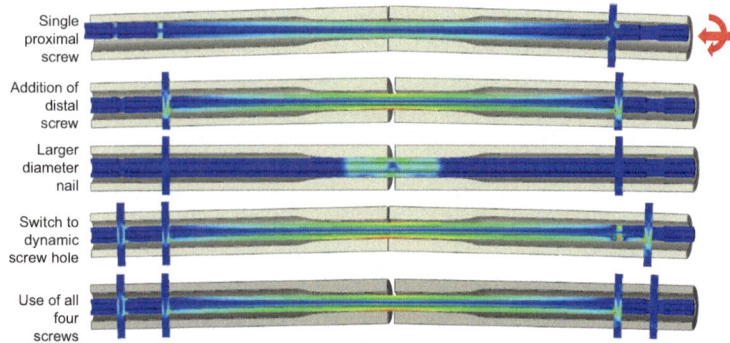

Fig. 9 Finite element models of intramedullary nailed long bones subject to axial force and bending. Effects on fracture gap motions and stresses due to construct configuration are evident. Blue represents low stress, red represents high stress (Data from our laboratory)

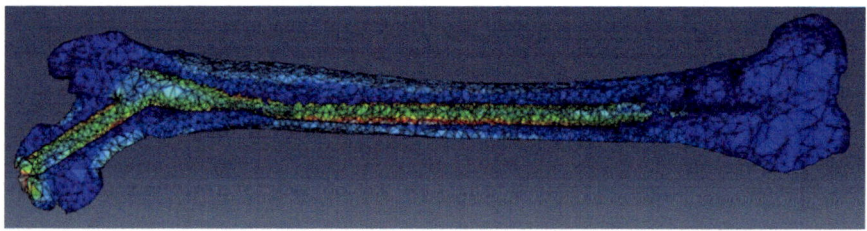

Fig. 10 Stress color map of an implanted IM nail within an unfractured femur under loading simulating the stance phase of gait. Blue represents low stress, red represents high stress (Data from our laboratory)

Surgical variables such as nail positioning, bone quality, type and number of fixation screws, applied loads, and fracture type may all be varied to view the influence of these features on resultant stresses.

8 Perspective

Advances in fracture fixation rely heavily on the intersecting disciplines of biomaterials, mechanics, biology, and clinical sciences. The twenty-first century holds promise for advancement in personalized medicine for the field. As computational technologies become more accessible and advanced for medical applications, researchers identify novel challenges that can drive improvements in patient care. Developed computational methods can be used to iterate implant design concepts and suggest geometric, positional, and material variations to achieve better optimization of construct mechanics. Furthermore, the detailed, accurate approximations of mechanics now available from simulated fracture fixation experiments can be used in conjunction with fabrication techniques such as additive manufacturing to create implant concepts that were previously impractical to develop. As the one-implant-fits-all approach loses traction in lieu of case-specific construct designs, it is expected that patient outcomes will continue to improve, resulting in reduced time required for fixation, enhanced healed bone mechanics, and fewer negative outcomes such as residual deformity and implant or fixation failure.

Acknowledgments The authors gratefully acknowledge support from the AO Foundation, Switzerland (Project S-15-196 L), and the National Science Foundation/Penn State Center for Health Organization Transformation. Hwabok Wee, PhD performed many of the finite element simulations shown in figures. We also acknowledge contribution from April D. Armstrong, MD.

References

1. Perren S, Buchanan J. Basic concepts relevant to the design and development of the point contact fixator (PC-fix). Injury. 1995;26:1–4.
2. Runyan C, Gabrick K. Biology of bone formation, fracture healing, and distraction osteogenesis. J Craniofac Surg. 2017;28:1380–9.
3. Maggiano I, et al. Three-dimensional reconstruction of haversian systems in human cortical bone using synchotron radiation-based micro-CT: morphology and quantification of branching and transverse connections across age. J Anat. 2016;228:719–32.
4. Cheal EJ, Mansmann KA, DiGioia AM, Hayes WC, Perren SM. Role of interfragmentary strain in fracture healing: ovine model of a healing osteotomy. J Orthop Res Off Publ Orthop Res Soc. 1991;9:131–42.
5. Levy S, et al. Immature myeloid cells are critical for enhancing bone fracture healing through angiogenic cascade. Bone. 2016;93:113–24.
6. Duda GN, et al. Interfragmentary motion in tibial osteotomies stabilized with ring fixators. Clin Orthop. 2002;396:163–72.
7. Garnavos C. Treatment of aseptic non-union after intramedullary nailing without removal of the nail. Injury. 2017;48:S76–81.
8. Santiago H, Zamarioli A, Neto M, Volpon J. Exposure to secondhand smoke impairs fracture healing in rats. Clin Orthop. 2017;475:894–902.
9. Sathyendra V, et al. Single nucleotide polymorphisms in osteogenic genes in atrophic delayed fracture-healing: a preliminary investigation. J Bone Joint Surg Am. 2014;96:1242–8.
10. Damm P, Kutzner I, Bergmann G, Rohlmann A, Schmidt H. Comparison of in vivo measured loads in knee, hip and spinal implants during level walking. J Biomech. 2017;51:128–32.
11. Kienast B, et al. An electronically instrumented internal fixator for the assessment of bone healing. Bone Jt Res. 2016;5:191–7.
12. Schneider E, et al. Loads acting in an intramedullary nail during fracture healing in the human femur. J Biomech. 2001;34:849–57.
13. Schlecht SH, Pinto DC, Agnew AM, Stout SD. Brief communication: the effects of disuse on the mechanical properties of bone: what unloading tells us about the adaptive nature of skeletal tissue. Am J Phys Anthropol. 2012;149:599–605.
14. Akhter MP, Alvarez GK, Cullen DM, Recker RR. Disuse-related decline in trabecular bone structure. Biomech Model Mechanobiol. 2011;10:423–9.
15. Miner MA. Cumulative damage in fatigue. J Appl Mech. 1945;12:A159–64.
16. Wee H, Reid JS, Chinchilli VM, Lewis GS. Finite element-derived surrogate models of locked plate fracture fixation biomechanics. Ann Biomed Eng. 2017;45:668–80.

17. Lenz M, Perren SM, Gueorguiev B, Höntzsch D, Windolf M. Mechanical behavior of fixation components for periprosthetic fracture surgery. Clin Biomech Bristol Avon. 2013;28:988–93.

18. Fratzl P, Weinkamer R. Nature's hierarchical materials. Prog Mater Sci. 2007;52:1263–334.

19. Cowin S. Pathology of functional adaptation of bone remodeling and repair in vivo. In Bone mechanics handbook. Taylor and Francis Group; 2001.

20. Comiskey DP, MacDonald BJ, McCartney WT, Synnott K, O' Byrne J. The role of interfragmentary strain on the rate of bone healing: a new interpretation and mathematical model. J Biomech. 2010;43:2830–4.

21. Augat P, et al. Shear movement at the fracture site delays healing in a diaphyseal fracture model. J Orthop Res. 2003;21:1011–7.

22. Elkins J, et al. Motion predicts clinical callus formation: construct-specific finite element analysis of supracondylar femoral fractures. J Bone Jt Surg. 2016;98:276–84.

23. Bottlang M, et al. Far cortical locking can improve healing of fractures stabilized with locking plates. J Bone Joint Surg Am. 2010;92:1652–60.

24. Claes L, Recknagel S, Ignatius A. Fracture healing under healthy and inflammatory conditions. Nat Rev Rheumatol. 2012;8:133–43.

25. Augat P, et al. Early, full weightbearing with flexible fixation delays fracture healing. Clin Orthop. 1996;328:194–202.

26. Schell H, et al. Mechanical induction of critically delayed bone healing in sheep: radiological and biomechanical results. J Biomech. 2008;41:3066–72.

27. Hente R, Füchtmeier B, Schlegel U, Ernstberger A, Perren SM. The influence of cyclic compression and distraction on the healing of experimental tibial fractures. J Orthop Res. 2004;22:709–15.

28. Claes L, Eckert-Hübner K, Augat P. The effect of mechanical stability on local vascularization and tissue differentiation in callus healing. J Orthop Res. 2002;20:1099–105.

29. Lujan TJ, et al. Locked plating of distal femur fractures leads to inconsistent and asymmetric callus formation. J Orthop Trauma. 2010;24:156–62.

30. Nassiri M, Macdonald B, O'Byrne JM. Computational modelling of long bone fractures fixed with locking plates - how can the risk of implant failure be reduced? J Orthop. 2013;10:29–37.

31. Beltran MJ, Collinge CA, Gardner MJ. Stress modulation of fracture fixation implants: J. Am Acad Orthop Surg. 2016;24:711–9.

32. Pruitt LA, Chakravartula AM. Mechanics of biomaterials: fundamental principles for implant design. Cambridge University Press, 2011.

33. Morehead J, Holt G. Soft-tissue response to synthetic biomaterials. Otolaryngol Clin N Am. 1994;27:195–201.

34. Nuss KM, von Rechenberg B. Biocompatibility issues with modern implants in bone—a review for clinical orthopedics. Open Orthop J. 2008;2:66–78.

35. Wright T, Maher S. Biomaterials. In: Orthopaedic basic science 65–85 American Academy of Orthopaedic Surgeons, 2007.

36. Ungersboeck A, Geret V, Pohler O, Schuetz M, Wuest W. Tissue reaction to bone plates made of pure titanium: a prospective, quantitative clinical study. J Mater Sci Mater Med. 1995;6:223–9.

37. Arens S, et al. Influence of materials for fixation implants on local infection. J Bone Jt Surg Br. 1996;78:647–51.

38. Uckan S, Veziroglu F, Soydan SS, Uckan E. Comparison of stability of resorbable and titanium fixation systems by finite element analysis after maxillary advancement surgery. J Craniofac Surg. 2009;20:775–9.

39. Marasco SF, Liovic P, Šutalo ID. Structural integrity of intramedullary rib fixation using a single bioresorbable screw. J Trauma Acute Care Surg. 2012;73:668–73.

40. Böstman O, Pihlajamäki H. Clinical biocompatibility of biodegradable orthopaedic implants for internal fixation: a review. Biomaterials. 2000;21:2615–21.

41. Schliemann B, et al. PEEK versus titanium locking plates for proximal humerus fracture fixation: a comparative biomechanical study in two- and three-part fractures. Arch Orthop Trauma Surg. 2017;137:63–71.

42. Bottlang M, Doornink J, Fitzpatrick DC, Madey SM. Far cortical locking can reduce stiffness of locked plating constructs while retaining construct strength. J Bone Jt Surg Am. 2009;91:1985–94.
43. Bottlang M, et al. Dynamic stabilization of simple fractures with active plates delivers stronger healing than conventional compression plating. J Orthop Trauma. 2017;31:71–7.
44. Piccinini M, Cugnoni J, Botsis J, Ammann P, Wiskott A. Influence of gait loads on implant integration in rat tibiae: experimental and numerical analysis. J Biomech. 2014;47:3255–63.
45. Gardner MJ, Ricciardi BF, Wright TM, Bostrom MP, van der Meulen MCH. Pause insertions during cyclic in vivo loading affect bone healing. Clin Orthop. 2008;466:1232–8.
46. Tsuji K, et al. BMP2 activity, although dispensable for bone formation, is required for the initiation of fracture healing. Nat Genet. 2006;38:1424–9.
47. McBride-Gagyi SH, McKenzie JA, Buettmann EG, Gardner MJ, Silva MJ. BMP2 conditional knockout in osteoblasts and endothelial cells does not impair bone formation after injury or mechanical loading in adult mice. Bone. 2015;81:533–43.
48. Helwig P, et al. Finite element analysis of four different implants inserted in different positions to stabilize an idealized trochanteric femoral fracture. Injury. 2009;40:288–95.
49. Perez A, Mahar A, Negus C, Newton P, Impelluso T. A computational evaluation of the effect of intramedullary nail material properties on the stabilization of simulated femoral shaft fractures. Med Eng Phys. 2008;30:755–60.
50. Fouad H. Assessment of function-graded materials as fracture fixation bone-plates under combined loading conditions using finite element modelling. Med Eng Phys. 2011;33:456–63.
51. Feerick EM, Kennedy J, Mullett H, FitzPatrick D, McGarry P. Investigation of metallic and carbon fibre PEEK fracture fixation devices for three-part proximal humeral fractures. Med Eng Phys. 2013;35:712–22.
52. Favre P, Kloen P, Helfet DL, Werner CML. Superior versus anteroinferior plating of the clavicle: a finite element study. J Orthop Trauma. 2011;25:661–5.
53. Tupis TM, Altman GT, Altman DT, Cook HA, Miller MC. Femoral bone strains during antegrade nailing: a comparison of two entry points with identical nails using finite element analysis. Clin Biomech Bristol Avon. 2012;27:354–9.
54. Shih K-S, Hsu C-C, Hsu T-P. A biomechanical investigation of the effects of static fixation and dynamization after interlocking femoral nailing: a finite element study. J Trauma Acute Care Surg. 2012;72:E46–53.
55. Brown CJ, Sinclair RA, Day A, Hess B, Procter P. An approximate model for cancellous bone screw fixation. Comput Methods Biomech Biomed Engin. 2013;16:443–50.
56. MacLeod AR, Pankaj P, Simpson AHRW. Does screw-bone interface modelling matter in finite element analyses? J Biomech. 2012;45:1712–6.
57. Wang H, et al. Accuracy of individual trabecula segmentation based plate and rod finite element models in idealized trabecular bone microstructure. J Biomech Eng. 2013;135:044502.
58. Zdero R, Olsen M, Bougherara H, Schemitsch EH. Cancellous bone screw purchase: a comparison of synthetic femurs, human femurs, and finite element analysis. Proc Inst Mech Eng H. 2008;222(1175–1183):1175–83.
59. Leonidou A, et al. The biomechanical effect of bone quality and fracture topography on locking plate fixation in periprosthetic femoral fractures. Injury. 2015;46:213–7.
60. Ruedi T, Buckley R, Moran C. AO principles of fracture management. 1. In: AO publishing; 2007.
61. Lewis G, et al. Tangential biocortical locked fixation improves stability in Vancouver B1 periprosthetic femur fractures: a biomechanical study. J Orthop Trauma. 2015;29:e364–e370.
62. Tidwell JE, et al. The biomechanical cost of variable angle locking screws. Injury. 2016;47:1624–30.
63. Kandemir U, et al. Implant material, type of fixation at the shaft and position of plate modify biomechanics of distal femur plate Osteosynthesis. J Orthop Trauma. 2017;1:e241–6. https://doi.org/10.1097/BOT.0000000000000860.

64. Wilson WK, Morris RP, Ward AJ, Carayannopoulos NL, Panchbhavi VK. Torsional failure of carbon Fiber composite plates versus stainless steel plates for comminuted distal fibula fractures. Foot Ankle Int. 2016;37:548–53.
65. Manteghi S, Mahboob Z, Fawaz Z, Bougherara H. Investigation of the mechanical properties and failure modes of hybrid natural fiber composites for potential bone fracture fixation plates. J Mech Behav Biomed Mater. 2017;65:306–16.
66. Smith WR, Ziran BH, Anglen JO, Stahel PF. Locking plates: tips and tricks. JBJS. 2007;89:2298–307.
67. Wee H, Reid J, Lewis G. Parametric and surrogate modeling of internal fixation of femur fractures demonstrate influence of surgical and patient variables. Ann Biomed Eng. 2016;44:3719–49.
68. Horn J, Schlegel U, Krettek C, Ito K. Infection resistance of unreamed solid, hollow slotted and cannulated intramedullary nails: an in-vivo experimental comparison. J Orthop Res. 2005;23:810–5.
69. Wehner T, Penzkofer R, Augat P, Claes L, Simon U. Improvement of the shear fixation stability of intramedullary nailing. Clin Biomech Bristol Avon. 2011;26:147–51.
70. Mahar AT, Lee SS, Lalonde FD, Impelluso T, Newton PO. Biomechanical comparison of stainless steel and titanium nails for fixation of simulated femoral fractures. J Pediatr Orthop. 2004;24:638–41.
71. Wang CJ, Brown CJ, Yettram AL, Procter P. Intramedullary nails: some design features of the distal end. Med Eng Phys. 2003;25:789–94.
72. Gallagher D, et al. Is distal locking necessary? A biomechanical investigation of intramedullary nailing constructs for intertrochanteric fractures. J Orthop Trauma. 2013;27:373–8.
73. Simpson DJ, Brown CJ, Yettram AL, Procter P, Andrew GJ. Finite element analysis of intramedullary devices: the effect of the gap between the implant and the bone. Proc Inst Mech Eng II. 2008;222(333 345):333–45.
74. Penzkofer R, et al. Influence of intramedullary nail diameter and locking mode on the stability of tibial shaft fracture fixation. Arch Orthop Trauma Surg. 2009;129:525–31.
75. Cheung G, Zalzal P, Bhandari M, Spelt J, Papini M. Finite element analysis of a femoral retrograde intramedullary nail subject to gait loading. Med Eng Phys. 2004;26:93–108.

Biomaterials for Bone Tissue Engineering: Recent Advances and Challenges

Kanchan Maji

Keywords Biomaterial · Tissue engineering · Stem cell · Scaffold · Composite · Mechanical strength · Embryonic · Adult stem cell · Particulate-leaching · Phase separation

1 Introduction

Millions of people are suffering today due to bone disease caused by bone injury and trauma. Bone tissue engineering is the interdisciplinary research field of biomaterial and tissue engineering to address these problems for the improvement of better quality of life. Due to several complexity in conventional approach like limited supply of autograft and donor side morbidity in case of allograft, researcher opted for tissue engineering scaffold to counter these problems. Tissue engineering scaffold is design to grow and proliferate bone cell in a three dimensional platform which mimics the extra cellular matrix of bone. Bone is composed of 30% organic (collagen fiber) and 70% inorganic (carbonated apatite) material designed in an artistic manner to withstand the load provided by our body. Scaffold seeded with mesechymal stem cells (MSCs) is considered to be a very usefull technique in the field of biomedical engineering. MSC's along with osteoblast and chondrocytes derived from various locations of patient's hard and soft tissue body can be expanded by *in vitro* culture, and then seeded on the scaffold, which will gradually degrade and resorb as the tissue structure developed *in vitro* and *in vivo*. The scaffold provides the necessary support for proliferation and differentiation of MSCs into osteoblastic lineage. The architecture of the scaffold determines the final shape of bone and cartilage tissue. Different types of scaffold materials have been developed and investigated for tissue engineering bone and cartilage, including hydroxyapatite, tri-calcium phosphate (β-TCP), bioactive glass and natural biopolymer such as collagen, chitosan, and alginate [1–5]. Many reviews are available regarding general properties and design feature of biodegradable and bioresorbable polymer

K. Maji (✉)
Department of Ceramic Engineering, NIT, Rourkela, India

© Springer International Publishing AG, part of Springer Nature 2018
B. Li, T. Webster (eds.), *Orthopedic Biomaterials*,
https://doi.org/10.1007/978-3-319-89542-0_17

429

and scaffold. The main objective of this chapter is to provide an overall work done so far in this field, with a special focus on the evolution of biomaterials and their characteristics that are specific for biopolymer scaffold based strategies of bone and cartilage tissue engineering.

2 Tissue Engineering

Tissue engineering is a trans-disciplinary approach that applies the knowledge of engineering and life science in order to develop biological substitute to repair or replace damaged and diseased tissues. There are three important issues need to be considered for successful tissue regeneration: Isolation of cells from a particular source and *in vitro* culture of the cells, then seed the cultured cells on a suitable 3D platform which supports 3D tissue regeneration as shown in Fig. 1 [7]. In conventional tissue engineering strategy, cultured cells are implanted on the scaffold and implanted into defect site for generation of new tissue [8]. In this strategy, cells proliferate and differentiate in a three dimensional scaffold to develop new tissue, whereas, scaffold materials degraded away leaving the cells embedded by extracellular matrix as the tissues are regenerated [9]. Though the promising earlier research outcomes [10, 11], the donor cell limitation and low quality of regenerated tissues remain two hurdles still prevails in tissue engineering application [12].

Scaffolds are needed to provide various others function besides maintaining basic structural support with proper degradation kinetics on the site of application. In other approaches, cellular scaffold implanted at defect site can deliver appropriate biomolecules in a controlled manner, which in turn manage to accelerate the proliferation and differentiation to successfully regenerate tissue [13]. Nowadays,

Fig. 1 The basic principle of tissue engineering [6]

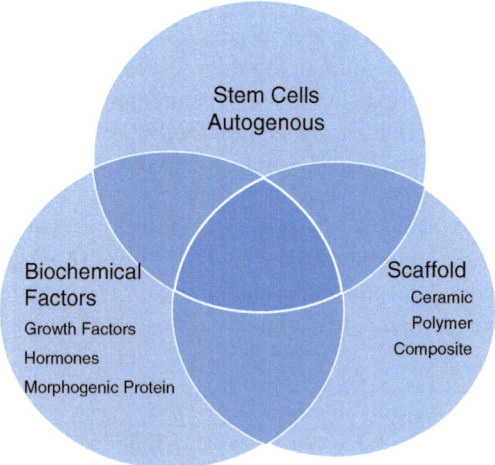

Table 1 Composition of
bone

Component	Amount (wt.%)
Hydroxyapatite	69
Organic matrix	22
Water	9

growth factor releasing cell supportive scaffolds are much more superior and have also been used recently with incorporated cells [14, 15]. These types of scaffolds are called bioactive scaffolds [16, 17].

3   Bone

Bone is a highly vascularised mineralized tissue which basically provides a skeletal support by protecting different internal organ system inside our body [18, 19]. Therefore, entire body functionality and thus lifestyle are affected by major defects in its structure [20]. Bone tissue regeneration provides intricate factors, one of which is to mimic the natural microenvironment of bone. The intrinsic ability of bone to regenerate is extent up to a certain limit, but when the defect is large in bony site it is very difficult to heal properly and as a result in long run it may fail [21].

3.1   Structure and Composition of Bone

Bones are consisted of three types of bone cells having specific functions with surrounding extracellular matrix. Among other components, mainly Collagen I occupies almost 95% of bone's organic part of Extra Cellular Matrix (ECM). Another 5% contributed by proteoglycans and non-collagenous proteins [22]. Glycosaminoglycan [GAG] like proteoglycans can be found in bone and cartilage tissue as well. Calcium phosphate minerals are responsible for providing rigidity ad strength in bone matrix [23]. The composition of bone is presented in Table 1.

3.2   Types of Bone

Cortical (hard) and cancellous (trabecular) bones are the two types of bone exist in our body. The cortical bone typically occupies the outermost part of most of the bones with a 80–90% of mineralization [24]. This is mainly responsible for providing proper mechanical strength to the whole skeleton system. Cancellous bone generally occupies the interior parts of bone with 15–20% of mineralization [25]. All the metabolic action inside bone are taken care by cancellous bone due to its

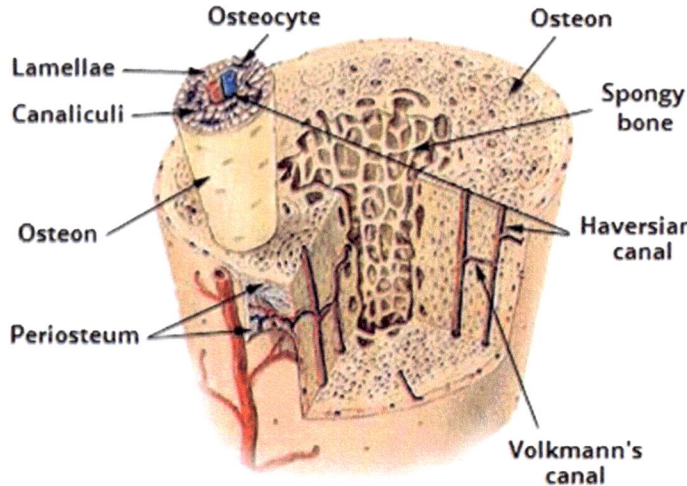

Fig. 2 Structure of mature bone. Note the differing structure of compact and spongy bone types [27]

highly vascularised internal architecture. The calcium phosphate containing collagen fibers form lamellar sheets which are arranged in concentric rings to form osteon [26] as shown in Fig. 2.

4 Stem Cells for Tissue Engineering

Stem cells are playing a major role in tissue regeneration and orchestrate tissue remodelling. MSCs have multilineage differential ability which makes them important source of cell to be used in tissue regeneration practice [28]. MSCs can be isolated from several sources like placenta, umbilical cord blood (UCB), adipose tissue and bone marrow. Among them, UCB is easily available as a hospital waste product [29]. As if now, the treatment of damaged or diseased bone using stem cell tissue engineering technique is counted as the foremost research area for a successful tissue engineering application.

4.1 Embryonic Stem Cells

Embryonic stem cells (ESCs) are found in the inner cell mass of blastocyst, which later develop into embryo tissue. ESCs have a unique ability to differentiate to adipocytes, chondrocytes, neuron and osteocytes. The self renewal property of ESCs combined with capability to generate any cells, tissue or organ for tissue regeneration, promises its potential application in treatment of human disease [30].

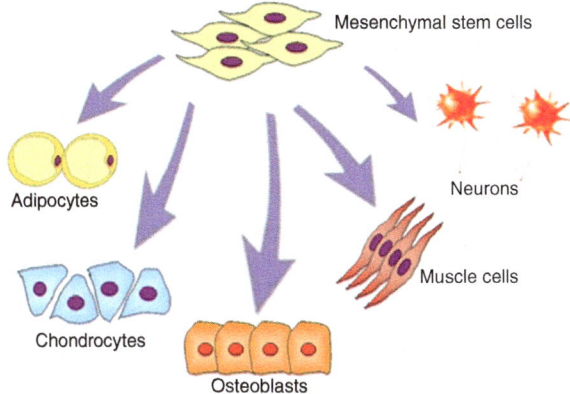

Fig. 3 Mesenchymal stem cells differentiation in osteogenic, chondrogenic, adipogenic and neuronal lineages [33]

4.2   Adult Stem Cells

Apart from ESC, adult stem cells (ASCs) have been identified as a potential cell source for the treatment of bone diseases. These ASCs can be found in bone marrow, periosteum, muscle, fat and skin. The main function of ASCs is to maintain and regenerate damage tissues in which it is found [31].

4.3   Mesenchymal Stem Cells (MSCs)

MSCs are categorized as undifferentiated adult stem cells found in any tissue and organ inside body. These cells are derived from mesoderm and differentiate into several other types of tissues like bold, muscle, bone, cartilage, and ligament [32]. MSCs have the ability of self renewal including differentiating into other specialized cells of different tissues [Fig. 3]. These cells can be isolated easily and cultured under *in vitro* conditions for use in the cell based therapies. Morphology of MSCs under *in vitro* culture conditions has been shown as fibroblast like spindle shape having good affinity towards tissue culture plate. So, These are the most suitable cells for the use in tissue engineering application.

5   Scaffold

Due to similarities with the artificial ECM of various tissue (bone, cartilage, skin), scaffold plays a crucial role in developing particular tissue of want shape, size and functionality. In this way, the fabrication method and design of the scaffold are the main consideration of biomaterial research area and therefore identified as a relevant subjects with tissue engineering and regenerative medicine [34].

The behavior of the scaffold is intensely subjects of few outline factors, for example microstructure, pore size and porosity, mechanical properties and surface science. For effective tissue regeneration, scaffold microstructure should impersonate local tissue structure containing ECM. It ought to have adequate porosity and the pore should be interconnected all around the scaffold matrix for transport of supplement and waste [35].While tissue regeneration is in advance stage, scaffold should have enough porosity to accommodate cells as well as support the transportation of nutrient and removal of toxic product from the system through 200–300 μm capillary transport channel. Additionally, scaffold ought to have pore size depending upon type of tissue engineering. For bone, muscle and skin the required pore sizes are 100–300 μm, 100–200 μm and 20–120 μm, respectively [36]. Another essential part of designing of scaffold is its mechanical properties. The scaffold must have satisfactory mechanical properties to withstand stress and physiological load and to mediate cues for new tissue regeneration. For the most part, scaffold mechanical properties should be matched with concerned tissue for which the scaffold is expected to.

Surface topography of scaffold is likewise critical and actuate cell attachment, proliferation and differentiation. In addition, scaffold surface should posses particular functional group for the attachment of various biomolecules which in turn govern the signalling cues for tissue regeneration. The topology of the scaffold should have the capacity to control the orientation of cells inside pore for effective cell based therapy. Therefore, scaffold material should be investigated with some of the above properties and in this way, material choice is imperative for intended tissue engineering using scaffold.

6   Scaffold Fabrication Techniques

To develop desired functional tissues and organs successfully, scaffold should be cultured with specialized cells such that it facilitates cell distribution and guide regeneration of tissue in three dimension platform. The potential for scaffolding techniques in the spectrum of tissue engineering is still in its initial stage which required to grow further to achieve the excellence in this field. Though, in terms of their working principle, each method is unique, but they can be different according to their prerequisite application and fabrication techniques. Currently, several fabrication methods are used which include solvent casting, Freeze drying, electrospinning, gas foaming and rapid prototyping [37]. These are describing below.

6.1 Particulate-Leaching Technique

This technique has been extensively used over a large area of tissue engineering for designing and fabrication of three dimensional scaffolds [38]. Among various porogens, salt is more commonly used in the particulate leaching technique. Briefly, first salt (such as NaCl) is ground into small particle of desired size and put into a mould, a polymer solution then is poured over the salt inside that mold and freezed. The freezed samples are then washed in a non-solvent (such as water or ethanol) but solvent for porogen. To create pores in scaffold, the scaffold is dipped in water which washes away the salt crystal to form a three dimensional porous scaffold. The process is very simple where porosity and pore size can be controlled according to the size of salt crystals and salt/polymer ratio. Other critical parameters like pore shape and pore interconnectivity are difficult to control in this technique.

6.2 Gas Foaming

This is a very unique technology to develop microcellular foams [39] of thermoplastic polymers like polymethyle methacrylate and polystyrene, however only in 1994. Mooney *et al.* had use this technique for the fabrication of Poly (L-lactic acid) (PLLA) and polyglycolic acid (PGA) PLLA50PGA50 scaffold to apply in tissue engineering [40]. Since then it has became a popular method for development of microporous scaffolds.

6.3 Lyophilization

Lyophilization or freeze drying is one of the most promising techniques to fabricate microporous scaffold where scaffolding materials are dissolved and cast into a mould followed by freezing and drying at lower temperature and vacuumed. Finally the whole slurry mixture is divided in two separates phases, a polymer rich phase and a polymer lean phase. The polymer rich phase can be developed into a three dimensional porous scaffold, when the solvent remove from the system via freeze drying process. The following two methods [41] have been used to fabricate porous membrane for filtration and separation via phase separation techniques.

6.3.1 Solid-Liquid Phase Separation

These techniques can be achieved at low temperatures, which induce solvent crystallization within polymer solution. And the process is defined as solid liquid phase separation (formation of solid phase in a liquid phase). Pores are generated within the system after removal of the solvent crystal via sublimation or solvent exchange process. This technique is very useful for the fabrication of scaffold using polymers and composite materials [42].

6.3.2 Liquid-Liquid Phase Separation

Liquid-liquid phase separation occurs when the temperature of the polymer solution is reduced below its upper critical solution temperature. In this condition, the process develops a bi-continuous structure of polymer rich and polymer lean phases. An open-pore structure 3D scaffold can be obtained after evaporation of solvent from the system. For example, Poly (D,L-lactic acid) (PLA) and poly (D,L-lactic-co-glycolic acid) (PLGA) scaffold fabricated using liquid-liquid phase separation techniques by using a mixture of water and di-oxane as a solvent [43].

6.4 Electro-Spinning

Electro-spinning method was employed in the early 1930, when it was used to fabricate non-woven fabric textile products. Afterwards, this technique has been upgraded further over a few decades to process biocompatible polymer in order to fabricate polymer fibers having a diameter at micro and nanometer level for the development of tissue engineering scaffolds. In this process, the polymer solution is dragging through a small capillary to form a drop of polymer solution at the end of the capillary channel (Fig. 4a). The solvent evaporates as the polymer jet travels through the air under high applied voltage, simultaneously a non-woven fibre is deposited on the target (Fig. 4b). Though, this technique faces a lot of challenges in its application but its sub micron range fiber has the potential application in the field of wound healing, drug delivery and tissue engineering [46].

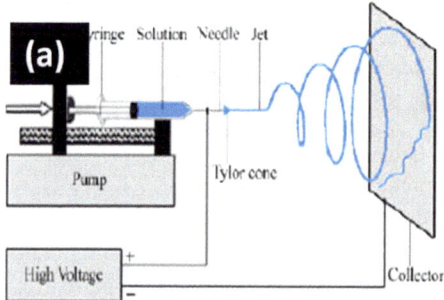
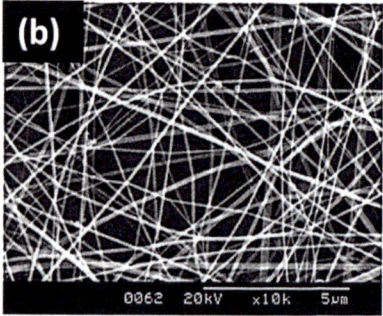

Fig. 4 (**a**) Schematic diagram of polymer nanofiber formation using electro spinning method [44] Adapted from Ref. [44]. Copyright IOP Publishing Ltd., 2007. (**b**) SEM image of electro-spun nanofiber [45]

6.5 Solid Freeform Fabrication Technique (SFFT)

Solid freeform technology or rapid prototyping has been successfully developed to generate highly precise and reproducible artificial ECM scaffolds with fully interconnected porous network (Fig. 5b) for tissue engineering application [48]. Digital data produced from computer tomography or magnetic resonance has been required to design the accurate and controllable scaffold structure. As this technology relies on the photo-polymerized biomaterial, the material property is very essential measure for the successful application of this technology. Thus, researchers have further investigated to modify or improve the current processing condition for the fabrication of optimum scaffold suitable for desired application in bone tissue reconstruction [49]. Some other types of scaffold fabrication methods are indicated in Table 2.

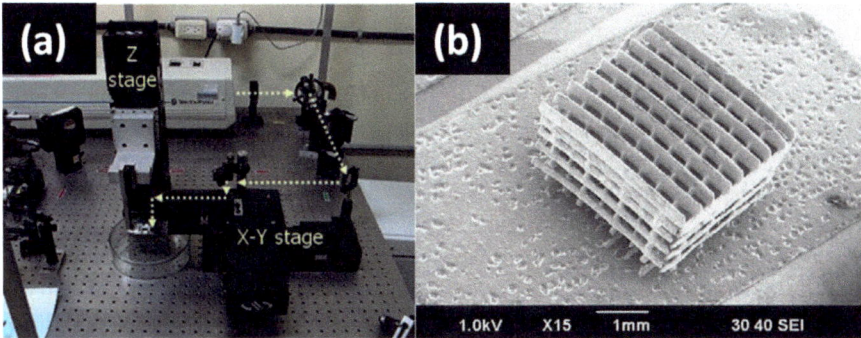

Fig. 5 (**a**) Photograph of the newly developed scaffold fabrication system using the axiomatic approach. (**b**) SEM images of fabricated microstructure using the newly designed fabrication system. Adapted from Ref. [47]. Copyright IOP Publishing Ltd., 2007

Table 2 Fabrication method for 3D composite scaffolds and their advantages and disadvantages

Fabrication route	Advantages	Disadvantages
Thermally induced phase separation (TIPS) [50]	Highly interconnected pore structures (porosity >95%). Anisotropic and channel pores. Pore size controlled by varying preparation conditions.	Long time to sublime solvent (48 h) Shrinkage issues. Small scale production. Use of organic solvents.
Solvent casting/ particle leaching [51]	Controlled porosity and Controlled interconnectivity	Structures generally isotropic. Use of organic solvents
Solid free-form [52]	Porous structure can be tailored to host tissue. Protein and cell encapsulation possible. Good interface with medical imaging.	Resolution needs to be improved to the micro-scale. Some methods use organic solvents.
Microsphere sintering [53]	Graded porosity structures possible Controlled porosity Can be fabricated into complex shapes.	Interconnectivity is an issue Use of organic solvents.
Scaffold coating [54]	Quick and easy.	Clogging of pores, sometimes organic solvents used,coating adhesion to substrate can be too weak.

7   Structural Design

It is very much supported that scaffold with a interconnected porous network favours tissue ingrowth, nutrient transport and osteointigration with the host tissue, and likewise maximize the long-term stability of the implant [55]. To encourage recovery of damaged bone tissue, the parameters for structural configuration of the scaffold like porosity, size and shape of the pore, orientation of interconnected pore and a progressive control over structural design parameter are often considered.

7.1   Porosity

Porosity is counted as a morphological property of the scaffold which is independent of used material property. Pores is basic for outlining bone tissue designing framework since it give important spaces for the proliferation and differentiation of bone cells to facilitate the transport phenomenon (transport of nutrients and waste material) inside scaffold and support to improved vascularization. In addition osteogenesis or interlocking in between implant and surrounding host bone, which is more prominent in case of porous surface. Moreover, these phenomena are also responsible for the mechanical stability of the scaffold *in vivo*. Several techniques have been modified to created porosity in biomaterials like salt leaching, gas foaming, freeze drying and rapid prototyping [56]. The selection of the fabrication techniques depends upon the property of the biomaterial. Higher porosity in the scaffold does not have any potential impact over cellular event but as the porosity increases the effective pore space which directly enhances the cell proliferation because unhindered supply of nutrient and oxygen transport to the cellular colony [57].

7.2   Pore Size

Impact of porosity and pore structure on the relative degree of osteogenesis has been applied *in vitro* in osteoblast and undifferentiated MSCs. For example, composite scaffold containing hydroxyapatite (HAp) and collagen having pore size ranging between 50 and 300 μm, found to be diminished the porosity with increasing HAp content, yet not significant contrast were observed in MC3T3-E osteoblast proliferation [58]. An extremely fascinating part of the impact of pore dimension on bone regrowth is the effect on progressive osteoinduction. Regular honeycomb shaping pores were observed in HAp scaffold used for subcutaneous delivery of Bone morphogenic protein (BMP2) into rats [59]. In general, macropores and micropores played the key role in the permeability of drug to the application site. Smaller pores were more favorable to chondrogenesis than osteogenesis, whereas in large pores bone was formed directly.

8 Mechanical Properties

Though higher porosity and pore size encourage bone regeneration, the outcome is however a decrease in mechanical strength. As this encompasses the constructural coherence of three dimensional matrixes. In tissue engineering applications, permeable scaffolds with higher porosity must have adequate mechanical quality to hold their underlying structures after implantation, especially in the reconstruction of hard, load-bearing tissues, for examples, bones and ligaments. The structure must not pack together under physiological pressure, which causes damage cells inside the scaffold. Chitosan based sponges with a pore size of 100 μm were developed inside hydroxyapatite/β-tricalcium phosphate 3D scaffolds with macropores between 300 and 600 μm having both compressive modulus and yield stress increased about four times [60]. Porous foams like structure were fabricated after sintering poly(lactide-co-glycolide) PLGAmicrospheres and increase in larger median pore size from 72 to 164 μm had no significant effect on total porosity (>30%) [61]. Similarly, higher porosity leads to the lower mechanical properties of porous poly (L-lactide-co-D,L-lactide) scaffolds, with compressive strength decreased from 11.0 to 2.7 MPa and modulus from 168.3 to 43.5 MPa [62].

The stability of various biomaterial based scaffolds after implantation depends on factors such as strength, elasticity, absorption at the material interface and chemical degradation. In this way, the examination of mechanical properties, for example, compressive strength is of essential significance in deciding the appropriateness of the designed scaffold. Firstly, it should be considered that at the end the scaffolds would be utilized in the physiological environment where the primary loading is compressive in case of bone or cartilage. Secondly, the presence of numerous tiny voids in the porous solid poses significant structural flaws to magnify the effect of crack propagation in stretching or bending. In this manner, the larger part of research discoveries on tissue engineering scaffolds are centered around their compressive properties when scaffold's mechanical properties.

The deformation behaviors of porous solids under compressive loads for honeycombs are shown in Figs. 6 and 7. The mechanical properties of honeycombs are arranged in two groups in-plane properties and out-of-plane properties. The in-plane properties are those identifying with loads applied in the X1 and X2 plane. Responses to loads applied to the faces normal to X3 are referred to as out-of-plane properties.

At the point when a honeycomb is compacted in-plane, the pore walls at first bend (Fig. 6a), giving linear elastic deformation (shown on stress-strain curves in Fig. 7a).

Past a critical strain the pores collapse by elastic buckling, plastic yielding (Fig. 6b), creep or brittle fracture, depending on the nature of the pore wall material. Pore collapse ends once the opposing pore walls begin to touch each other; as the pores close up, the structure is densified and its stiffness increases rapidly. On loading the honeycomb out-of-plane, the pore walls experience compression under both axial and bending stresses. The moduli and collapse stresses are much larger. Fig. 7b shows the family of curves of honeycombs with different relative density, compressed out-of-plane.

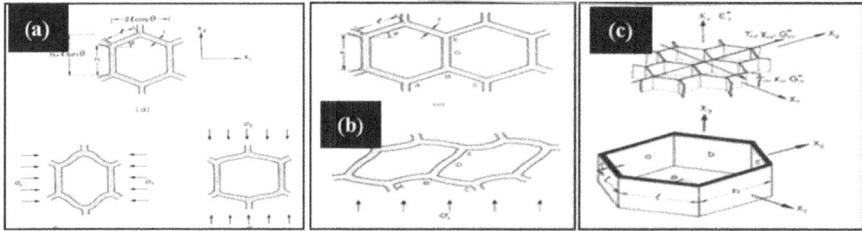

Fig. 6 In-plane compression of honeycomb pores; (**a**) Initial elastic bending of pore walls, (**b**) buckling of pore edges at higher stress levels and (**c**) out-of plane compression of honeycomb pores

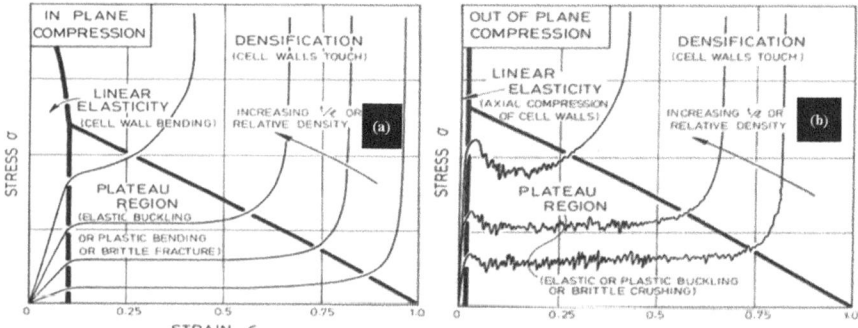

Fig. 7. A schematic diagram for a honeycomb loaded in compression, showing linear elastic, collapse and densification regimes, and the way the stress-strain curves changes with t/l; (**a**) compressed in X1–X2 plane, (**b**) compressed in axial (X3) direction [63]

9 Composite Scaffold Material

The term composite is typically considered for materials that contain at least two distinctive constituent materials at a huge scale. The physical and mechanical properties of these materials are altered in a general sense, and hereafter, these might be more valuable in comparison with homogeneous material. Biocomposites can be manufactured utilizing nontoxic material and has alluring scaffold properties ready to reproduce new bone recovery. Hence biocompatibility assumes a key part in determination of biocomposite material than some other compatibility [64]. In this manner, bioceramics based biocomposites and hybrid functional biomaterials has all the earmarks of being new for proficient tissue recovery capacity. The most normal characteristic of the bio-natural and inorganic materials are consolidated in biocomposites and have been summerized in Table 3.

Different human tissues, for example, human hard tissues like bone, comprise of inorganic-natural part [66]. In particularly, the property of a composite material depends upon the size and morphology of the heterogeneities, relative percentage of the components and nature of interfaces among the components. Among composite

Table 3 General respective properties from the bioorganic and inorganic domains, to be combined in various composites and hybrid materials [65]

Inorganic	Bioorganic
Hardness, brittleness	Elasticity, plasticity
High density	Low density
Thermal stability	Permeability
Hydrophilicity	Hydrophobicity
High refractive index	Selective complexion
Mixed valance state	Chemical reactivity
Strength	Bioactivity

Table 4 Important properties and applications of the nanocomposite

Name of the composite	Properties	Application	References
NanoHAp/ collagen	HAp nanocrystals regularly aligned along collagen fibrils	Synthetic bone materials in medical applications	Kikuchi *et al.* [67]
HAp/sodium alginate (SA) nanocomposite beads	The well-swelling, drug loading and controlled release behavior	Drug- controlled release	Zhang *et al.* [68]
NanoHAp/ chitosan composite scaffold	Good biocompatibility and sufficient porosity	Bone tissue engineering	Kong *et al.* [69]
Chitosan/ nanocrystalline calcium phosphate	Increased proliferation of osteoblast cells and improved mechanical strength	Bone regeneration	Chesnutt *et al.* [70]
HAp/chitosan– gelatin (CG) composite	Enhanced protein and calcium ion adsorption and improved initial cell adhesion and long term growth	Development of human stem cell	Zhao et al. [71]

materials, bioceramics and biopolymer based composites have been found to have significant biological characteristic and observed to be reasonable for tissue designing applications as a biomaterial. The essential properties and utilizations of some of the composites are summerized in Table 4.

9.1 Synthetic Biopolymer/CaP Composite Scaffold

Synthetic polymer such as poly(glycolic acid) (PGA), poly(lactic acid) (PLA), and their copolymers poly(lactic acid-co-glycolic acid) (PLGA) are commonly considered in tissue engineering to fabricate artificial ECM [72]. These polymers degrade through hydrolysis of the ester bonds that act as backbone of polymer structure [73].

Furthermore, various linear aliphatic polyesters like polycaprolactone (PCL) and polyhydroxybutyreate (PHB) are commonly considered in bone tissue engineering research. However, the degradation rate of PCL is significantly slower than PLA, PGA, and PLGA [74]. Thus slow degradation significantly affects utility of PCL for tissue engineering applications, however suitable for long-term implants and controlled release of functional molecules. Therefore, PCL-based copolymers are also synthesized to improve degradation properties.

Porous 3D scaffold of calcium phosphate ceramics with interconnected macropores (> 200 μm), micropores (~5 μm) and high porosities (~80%) have been produced by firing polyurethane (PU) foams coated with calcium phosphate bioceramic at 1200 °C [75]. The micropores of the developed scaffold were infiltrated with poly (lactic-co-glycolic acid) (PLGA) to achieve an interpenetrating bioactive ceramic/ biodegradable polymer composite structure. Miao et al. [76] reported development of highly porous HA/TCP composite scaffolds having 87% porosity, infiltrated with PLGA to form bioceramic-polymer interpenetrating microstructures. However, in these composites PLGA provides significant improvement of the compressive strength. Furthermore, PCL based scaffold also reported to be coated with HA and possess appropriate cell supportive property.

[77]. Chen et al. [78] demonstrated fabrication of Bioglass®-based scaffolds coated with poly-DL-lactic acid PDLLA that provided appropriate osteogenic potential for healing of bone damage. Polyhydroxyalkanoate (P(3HB) has also been investigated for preparation of scaffolds for tissue engineering application [79]. Bretcanu et al. [80] reported the use of natural bacteria derived P(3HB) to infiltrate into 45S5 Bioglass® scaffolds. The mechanical properties of these novel scaffolds were investigated. It was shown qualitatively that the work of fracture increased dramatically with the P(3HB) coating [81].

9.2   Natural Biopolymer/Bioactive Ceramic Based Composite

As a result of excellent biocompatibility and biological characteristics that are similar to human bone, natural biopolymer and CaP based composites were researched broadly. Agarose, alginate, silk, hyaluronic acid, collagen, chitosan, gelatin are some of the well known natural biopolymers that have been used for tissue engineering application. Agarose, a polysaccharide, was used for making injectable scaffold that provided a three dimensional environment to maintain the round shape of the chondrocyte. Furthermore, BCP (HA + β-TCP)/agarose macroporous scaffolds with high degree of interconnected open porosity and tailored pore size were prepared for application in bone tissue engineering [82]. Biocompatibility of this BCP/agarose system was studied using mouse L929 fibroblasts and human SAOS-2 osteoblasts during different colonization times [83]. In another study, a composite hydrogel based on Hap/alginate biomaterials were biommetically prepared using freeze-drying techniques.

The prepared HA-alginate scaffold exhibited high degree of *in vitro* biocompatibility and controlled biodegradability. Osteoinductive properties of porous hybrid scaffolds prepared from β-TCP, alginate-gelatin was reported elsewhere [84].

A wide range of properties of silk fibroin such as their size, shape, crystallinity and mechanical properties are previously reported [85]. A functionally graded

HA/silk fibroin biocomposite was prepared by pulse electric current sintering [86, 87] that showed excellent mechanical strength and osteoconductivity. The 3D mesh of fibrin sealant was found to interpenetrate the macro- and microporous structure of calcium orthophosphate ceramics [88]. Le et al. proposed that the physical, chemical and biological properties of calcium orthophosphate bioceramics and the fibrin glue might cumulate in biocomposites suitable for preparation of advanced bone grafts [89]. Hyaluronic acid and derivatives have been used as therapeutic aids in the treatment of osteoarthritis as a means of improving lubrication of articulating surfaces and thus reducing joint pain [90]. Chitosan-gelatin-hyaluronic acid based scaffolds were found to be suitable for preparing a bilayer skin substitute [91].

Gea et al. [92] prepared HAp/chitin composite materials with HAp content shifting between 25 and 75% wt% and observed *in vivo* bone recovery capacity in rodent. Also, Lee et al. [93] proposed in their stdudy, the utilization of chitosan/TCP used as tissue building scaffolds for bone recovery. Moreover, bioglass-chitosan composite displayed the possibility to help the development of osteoprecursor cells *in vitro* and to support separation of osteoblast in the process of encouraging the synthesis of phenotypic markers, for example, alkaline phosphatase, Type I collagen, and osteocalcin [94]. Table 5 demonstrates the properties of different chitosan based composite with different basic arrangement like three dimensional platform, fibrous network, thin film and so forth were set up for tissue engineering application.

HAp/collagen composite materials are promising candidates for bone replacement purposes because of their resemblance to bone in point of view of their bioactivity and biodegradability. Chang et al. [105] reported a simple method to produce CaP/collagen composites by adjusting the reaction conditions and the ratio of components. The mechanical properties of these types of composites are still very poor and require further improvement for its effective use in bone tissue engineering [106]. Collagen based implants are crosslinked to improve its mechanical properties and regulate cellular behavior when they are used in as processed form to prepare scaffolds [107]. Various crosslinking treatments including physical, [108], chemical [109], enzymatic [110] and combination treatments have been developed to tailor their mechanical and degradation properties. A composite scaffold formed by combining gelatin with bioceramics can yield tailored degradation rates, while also having improved biological and mechanical and physical properties. Yaylaoglu et al. [111] developed a CaP/gelatine composite implant that released drugs and growth hormone into the implant site to assist in bone healing.

Porous gelatin/collagen scaffolds especially gelatin sponges containing bioceramic particles were synthesized by the lyophilisation of a gelatin/calcium phosphate mixture [112]. Usually, gelatin/collagen scaffolds are highly cross-linked for better mechanical stability, but this can reduce the biocompatibility since crosslinking agents like glutaraldehyde leads to a cytotoxic reaction [113]. To prevent this limitation, a colloidal β-TCP/collagen composite was prepared for which further treatment was not required [114]. Further, collagen type I was also used as a matrix for CaP mineralization to obtain a collagen/CaP composite [115]. In another study,

Table 5 Chitosan-bioactive ceramic based composite scaffolds with enhanced properties

Composite	Outcome
CS/ALRGDN-peptide	Promoted biocompatibility [95]
CS/Gelatin/β-TCP	Increased osteoblast attachment and proliferation [96]
CS/Alginate/carboxy methylcellulose	Increased compressive modulus [97]
CS/RGD/UV cross-linking	Increased proliferation [GAG andDNA] [98]
CS/nHAp	Increased mechanical strength [99]
CS/PLGA	Increased stiffness of scaffold, biocompatibility [100]
CS/gelatin/TCP/ Glutraldehyde	Increased cell spreading [101]
CS/PLAGA/Fibrin	Increased mechanical strength, Osteoblast cell proliferation [102]
CS/Fibroin	Enhanced stiffness andbiocompatibility [103]
CS/Ca₃PO₄	Increase in cell attachment,proliferation and phenotypic expression of osteoblastic markers [104]

a 10% gelatin composite scaffold was formed by soaking a macroporousHAp to allow gelatin penetration to the bulk of the ceramic [116]. To maintain biocompatibility of the scaffold, cross-linker was used at a low concentration to prevent diffusion of gelatin out of the scaffold.

With the expectations of improved cytocompatibility and desirable cellular response, a porous chitosan/gelatin network scaffold was developed for ligament tissue engineering and artificial skin [117]. Further, biomimetic 3D HA/chitosan–gelatin network composite scaffolds were developed for bone tissue engineering [118]. The point of this work was centered around enhancing the mechanical and biological properties of chitosan based scaffolds through incorporation of bioceramics for example, HA, β-tricalcium phosphate, calcium phosphate [119] and biopolymers like gelatin [120], alginate [121] or inorganic material such as wollastonite [122]. Incorporation of calcium phosphate into the chitosan matrix improved biocompatibility and hard tissue integration and assisted in tailoring degradation and resorption kinetics.

A biodegradable composite framework was created utilizing β-tricalcium phosphate (β-TCP) with chitosan (CS) and gelatin (Gel) as a hybrid polymer arrangement (HPN) by means of co-cross linking with glutaraldehyde (Fig. 8). The macroporous composite scaffold showed diverse pore structures with enhanced compressive modulus from 3.9 to 10.9 MPa

Designing scaffold using straightforward and economic way is a major challenge that should be investigated for the betterment of tissue engineering scaffold. Freeze drying strategy observed to be best and basic way of manufacturing scaffold as far as sparing both time and vitality. Residual solvent and advancement of surface skin onto scaffold are the two key points which should be altered for the improvement of the current procedure.

Futhermore, freeze dried scaffold generally had interconnected porosity and better mechanical properties. Previous report recommend that freeze drying procedure

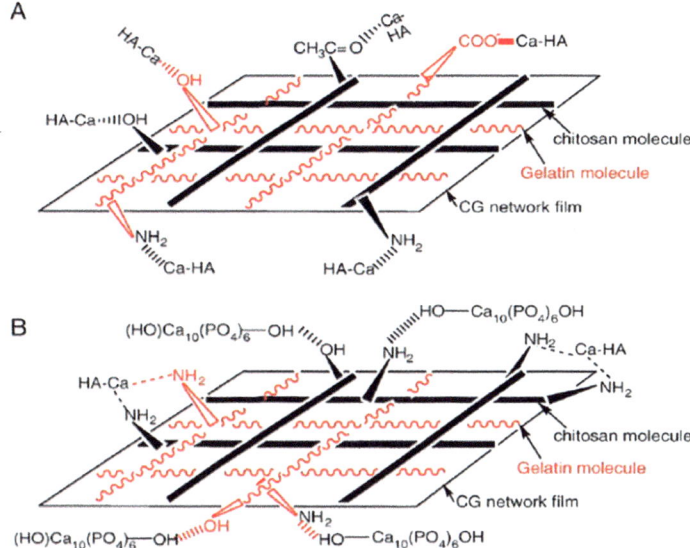

Fig. 8 Possible interactions between a chitosan-gelatin (CG) network and HA crystals in HA/CG biocomposites: (**a**) In the case of a nano-dimensional HA (nHA); (**b**) In the case of a micro-dimensional HA (mHA) [119]. Adapted from Ref. [119]. Copyright Elsevier, 2008

is valuable in creating different biopolymeric 3D scaffold utilizing collagen [123], carboxy methyl cellulose [124], poly (d,l-lactic-co-glycolic corrosive) [125], chitosan [126], pectin [127] and silk fibroin [85]. Chitosan can be easily incorporated in combination with other bio-materials to accomplish required mechanical and surface properties of scaffold [128].

10   Challenges and Opportunities

10.1   *Mechanical Integrity of Porous Scaffolds*

Till date, manmade porous scaffolds suffer from insufficient mechanical integrity and various other properties relevant for bone tissue engineering. For comparison, data of elastic modulus and the compressive strength of dense bioactive ceramic, biodegradable polymers, cancellous and cortical bone and their porous monophasic scaffolds and composites were taken from Ref [129]. From various data source, it was found out that some dense biopolymers and bioactive ceramics can match the properties of cancellous bones. Porous 3D scaffold appears to be one order weaker than cancellous bone and far behind from the mechanical properties of cortical bone.

10.2   In vitro *Degradation*

Since tissue designing goes for recovery of new tissues, biomaterials are believe to be degradable and absorbable with an appropriate rate to coordinate the speed of new tissue development. The degradation phenomena crucially affects the long term progression of a tissue engineered cell/polymer constract. The degradation kinetics may influence organization of process, for example, cell development, tissue recovery, and host response. Gelatin containing chitosan scaffold has quicker degradation rate because of the hydrophilicity of gelatin [130]. Bioglass and calcium phosphates have been incorporated to hinder the degradation [131].

10.3   In vitro *and* In vivo *Characterization*

Although several *in vitro* and *in vivo* studies have already been carried out with bioresorbable polymers and bioactive ceramics separately, *in vivo* characterization of biopolymer/bioceramic composite have been started recently. Till date very limited number of composites have been investigated *in vivo*. More research required to be directed towards determining the suitability of bioactive ceramic composite scaffolds to be used in hard tissue engineering. Further study on the effect of degradation products of bioactive polymer/ceramic composite on vascularization and *in vivo* bone formation need to be carried out.

11   Discussion and Future Aspects

From past few decades, bioactive ceramic based composites will be promising biomaterials for bone tissue designing. The mix of bioactive ceramic, like, Hap, bioactive glass and β-TCP with common biopolymer like gelatin and chitosan is excellent choice compared to other biomaterial to deal with progressive development of artificial bone. Blend of biopolymer with bioactive ceramic nano-particle can bring about the normal three dimensional frameworks as bone graft substitute, associated with adequate bone tissue develop properties, for example, mechanical quality, pore size and osteoconductivity. Moreover, cell material collaboration, mechanical integrity and response of bioactive ceramic based nano-composites should be examined in detail and further be explored for their biomedical importance. However, clinical examinations should have been performed on the bioactive ceramic based composite platform to encourage biological and physichochemical problems.

References

1. Rehm BHA. Bacterial polymers: biosynthesis, modifications and applications. Nat Rev Microbiol. 2010;8(8):578–92.
2. Haugh MG, et al. Crosslinking and mechanical properties significantly influence cell attachment, proliferation, and migration within collagen glycosaminoglycan scaffolds. Tissue Eng A. 2011;17(9–10):1201–8.
3. Zhang W, et al. Nucleation sites of calcium phosphate crystals during collagen mineralization. J Am Ceram Soc. 2003;86(6):1052–4.
4. Meinel L, et al. Silk based biomaterials to heal critical sized femur defects. Bone. 2006; 39(4):922–31.
5. Zimmermann KA, et al. Biomimetic design of a bacterial cellulose/hydroxyapatite nanocomposite for bone healing applications. Mater Sci Eng C. 2011;31(1):43–9.
6. Stock UA, Vacanti JP. Tissue engineering: current state and prospects. Annu Rev Med. 2001;52(1):443–51.
7. Sundelacruz S, Kaplan DL. Stem cell-and scaffold-based tissue engineering approaches to osteochondral regenerative medicine. In: Seminars in cell & developmental biology. Berlin: Elsevier; 2009.
8. Cancedda R, Giannoni P, Mastrogiacomo M. A tissue engineering approach to bone repair in large animal models and in clinical practice. Biomaterials. 2007;28(29):4240–50.
9. Griffith LG, Naughton G. Tissue engineeringDOUBLEHYPHENcurrent challenges and expanding opportunities. Science. 2002;295(5557):1009–14.
10. Cancedda R, et al. Tissue engineering and cell therapy of cartilage and bone. Matrix Biol. 2003;22(1):81–91.
11. Discher DE, Janmey P, Wang Y-l. Tissue cells feel and respond to the stiffness of their substrate. Science. 2005;310(5751):1139–43.
12. Bruder SP, Fox BS. Tissue engineering of bone: cell based strategies. Clin Orthop Relat Res. 1999;367:S68–83.
13. Schantz J-T, et al. Repair of calvarial defects with customised tissue-engineered bone grafts II. Evaluation of cellular efficiency and efficacy in vivo. Tissue Eng. 2003;9(4, Suppl. 1): 127–39.
14. De Boer R, et al. Rat sciatic nerve repair with a poly-lactic-co-glycolic acid scaffold and nerve growth factor releasing microspheres. Microsurgery. 2011;31(4):293–302.
15. Whitaker M, et al. Growth factor release from tissue engineering scaffolds. J Pharm Pharmacol. 2001;53(11):1427–37.
16. Dankers PY, et al. A modular and supramolecular approach to bioactive scaffolds for tissue engineering. Nat Mater. 2005;4(7):568–74.
17. Weng J, Wang M, Chen J. Plasma-sprayed calcium phosphate particles with high bioactivity and their use in bioactive scaffolds. Biomaterials. 2002;23(13):2623–9.
18. Shi S, Gronthos S. Perivascular niche of postnatal mesenchymal stem cells in human bone marrow and dental pulp. J Bone Miner Res. 2003;18(4):696–704.
19. Nguyen LH, et al. Vascularized bone tissue engineering: approaches for potential improvement. Tissue Eng Part B Rev. 2012;18(5):363–82.
20. Engh GA, Herzwurm PJ, Parks NL. Treatment of major defects of bone with bulk allografts and stemmed components during total knee arthroplasty. J Bone Joint Surg Am. 1997;79(7):1030–9.
21. Younger EM, Chapman MW. Morbidity at bone graft donor sites. J Orthop Trauma. 1989;3(3):192–5.
22. Buckwalter J, Mankin H. Instructional course lectures, the american academy of orthopaedic surgeons-articular cartilage. Part I: tissue design and chondrocyte-matrix interactions*†. J Bone Joint Surg Am. 1997;79(4):600–11.

23. Clarke B. Normal bone anatomy and physiology. Clin J Am Soc Nephrol. 2008;3(Suppl. 3): S131–9.
24. Carter DR, Spengler DM. Mechanical properties and composition of cortical bone. Clin Orthop Relat Res. 1978;135:192–217.
25. Taicher GZ, et al. Quantitative magnetic resonance (QMR) method for bone and whole-body-composition analysis. Anal Bioanal Chem. 2003;377(6):990–1002.
26. Buckwalter J, et al. Bone biology. J Bone Joint Surg Am. 1995;77(8):1256–75.
27. Nicolson R, Johal J. Ultrastructure of bone. TechMe Anatomy info.
28. Baksh D, Song L, Tuan R. Adult mesenchymal stem cells: characterization, differentiation, and application in cell and gene therapy. J Cell Mol Med. 2004;8(3):301–16.
29. Kern S, et al. Comparative analysis of mesenchymal stem cells from bone marrow, umbilical cord blood, or adipose tissue. Stem Cells. 2006;24(5):1294–301.
30. Dani C, et al. Differentiation of embryonic stem cells into adipocytes in vitro. J Cell Sci. 1997;110(11):1279–85.
31. Alison MR, et al. Cell differentiation: hepatocytes from non-hepatic adult stem cells. Nature. 2000;406:257.
32. Pittenger MF, et al. Multilineage potential of adult human mesenchymal stem cells. Science. 1999;284(5411):143–7.
33. Kidney disease syndrome, Stem cell therapy, http://www.kidney-symptom.com/stem-cell-therapy.html.
34. Hollister SJ. Porous scaffold design for tissue engineering. Nat Mater. 2005;4(7):518–24.
35. Hing KA. Bioceramic bone graft substitutes: influence of porosity and chemistry. Int J Appl Ceram Technol. 2005;2(3):184–99.
36. Harley BA, Gibson LJ. In vivo and in vitro applications of collagen-GAG scaffolds. Chem Eng J. 2008;137(1):102–21.
37. Yeong W Y, et al. Rapid prototyping in tissue engineering: challenges and potential. Trends Biotechnol. 2004;22(12):643–52.
38. Liao CJ, et al. Fabrication of porous biodegradable polymer scaffolds using a solvent merging/particulate leaching method. J Biomed Mater Res. 2002;59(4):676–81.
39. Harris LD, Kim B-S, Mooney DJ. Open pore biodegradable matrices formed with gas foaming. J Biomed Mater Res. 1998;42(3):396–402.
40. Mooney S. Bioinformatics approaches and resources for single nucleotide polymorphism functional analysis. Brief Bioinform. 2005;6(1):44–56.
41. Wang W. Lyophilization and development of solid protein pharmaceuticals. Int J Pharm. 2000;203(1):1–60.
42. Lloyd DR, Kinzer KE, Tseng H. Microporous membrane formation via thermally induced phase separation. I. Solid-liquid phase separation. J Membr Sci. 1990;52(3):239–61.
43. Hua FJ, Park TG, Lee DS. A facile preparation of highly interconnected macroporous poly (D, L-lactic acid-co-glycolic acid)(PLGA) scaffolds by liquid–liquid phase separation of a PLGA–dioxane–water ternary system. Polymer. 2003;44(6):1911–20.
44. Haghi AK, Akbari M. Trends in electrospinning of natural nanofibers. Phys Status Solidi A. 2007;204(6):1830–4.
45. Jarcho M. Calcium phosphate ceramics as hard tissue prosthetics. Clin Orthop Relat Res. 1981;157:259–78.
46. Zeugolis DI, et al. Electro-spinning of pure collagen nano-fibres–just an expensive way to make gelatin? Biomaterials. 2008;29(15):2293–305.
47. Lee S-J, et al. Development of a scaffold fabrication system using an axiomatic approach. J Micromech Microeng. 2006;17(1):147.
48. Billiet T, et al. A review of trends and limitations in hydrogel-rapid prototyping for tissue engineering. Biomaterials. 2012;33(26):6020–41.
49. Leong K, Cheah C, Chua C. Solid freeform fabrication of three-dimensional scaffolds for engineering replacement tissues and organs. Biomaterials. 2003;24(13):2363–78.
50. Nam YS, Park TG. Biodegradable polymeric microcellular foams by modified thermally induced phase separation method. Biomaterials. 1999;20(19):1783–90.

51. Sin D, et al. Polyurethane (PU) scaffolds prepared by solvent casting/particulate leaching (SCPL) combined with centrifugation. Mater Sci Eng C. 2010;30(1):78–85.

52. Hutmacher DW, Sittinger M, Risbud MV. Scaffold-based tissue engineering: rationale for computer-aided design and solid free-form fabrication systems. Trends Biotechnol. 2004;22(7):354–62.

53. Jiang T, Abdel-Fattah WI, Laurencin CT. In vitro evaluation of chitosan/poly (lactic acid-glycolic acid) sintered microsphere scaffolds for bone tissue engineering. Biomaterials. 2006;27(28):4894–903.

54. Dupont KM, et al. Synthetic scaffold coating with adeno-associated virus encoding BMP2 to promote endogenous bone repair. Cell Tissue Res. 2012;347(3):575–88.

55. Venugopal J, et al. Interaction of cells and nanofiber scaffolds in tissue engineering. J Biomed Mater Res B Appl Biomater. 2008;84(1):34–48.

56. Bundela H, Bajpai A. Designing of hydroxyapatite-gelatin based porous matrix as bone substitute: correlation with biocompatibility aspects. Express Polym Lett. 2008;2:201–13.

57. Murphy CM, Haugh MG, O'Brien FJ. The effect of mean pore size on cell attachment, proliferation and migration in collagen–glycosaminoglycan scaffolds for bone tissue engineering. Biomaterials. 2010;31(3):461–6.

58. Loh QL, Choong C. Three-dimensional scaffolds for tissue engineering applications: role of porosity and pore size. Tissue Eng Part B Rev. 2013;19(6):485–502.

59. Kuboki Y, Jin Q, Takita H. Geometry of carriers controlling phenotypic expression in BMP-induced osteogenesis and chondrogenesis. J Bone Joint Surg Am. 2001;83(1 suppl 2): S105–15.

60. Di Martino A, Sittinger M, Risbud MV. Chitosan: a versatile biopolymer for orthopaedic tissue-engineering. Biomaterials. 2005;26(30):5983–90.

61. Borden M, et al. Tissue engineered microsphere-based matrices for bone repair: design and evaluation. Biomaterials. 2002;23(2):551–9.

62. Yucel D, Kose GT, Hasirci V. Polyester based nerve guidance conduit design. Biomaterials. 2010;31(7):1596–603.

63. Gibson LJ. The mechanical behaviour of cancellous bone. J Biomech. 1985;18(5):317–28.

64. Bledzki A, Gassan J. Composites reinforced with cellulose based fibres. Prog Polym Sci. 1999;24(2):221–74.

65. Sanchez C, et al. Applications of hybrid organic–inorganic nanocomposites. J Mater Chem. 2005;15(35–36):3559–92.

66. Green AA, et al. A transformation for ordering multispectral data in terms of image quality with implications for noise removal. IEEE Trans Geosci Remote Sens. 1988;26(1):65–74.

67. Yoneyama M, et al. The RNA helicase RIG-I has an essential function in double-stranded RNA-induced innate antiviral responses. Nat Immunol. 2004;5(7):730–7.

68. Zhang J, Yedlapalli P, Lee JW. Thermodynamic analysis of hydrate-based pre-combustion capture of CO2. Chem Eng Sci. 2009;64(22):4732–6.

69. Wang Y, et al. The Holocene Asian monsoon: links to solar changes and North Atlantic climate. Science. 2005;308(5723):854–7.

70. Chesnutt BM, et al. Design and characterization of a novel chitosan/nanocrystalline calcium phosphate composite scaffold for bone regeneration. J Biomed Mater Res A. 2009;88((2):491–502.

71. Chen D-M, Zhao H. Strong lensing probability for testing TeVeS theory. Astrophys J Lett. 2006;650(1):L9.

72. Athanasiou KA, Niederauer GG, Agrawal CM. Sterilization, toxicity, biocompatibility and clinical applications of polylactic acid/polyglycolic acid copolymers. Biomaterials. 1996; 17(2):93–102.

73. Middleton JC, Tipton AJ. Synthetic biodegradable polymers as orthopedic devices. Biomaterials. 2000;21(23):2335–46.

74. Ma PX. Scaffolds for tissue fabrication. Mater Today. 2004;7(5):30–40.

75. Miao X, et al. Porous calcium phosphate ceramics modified with PLGA–bioactive glass. Mater Sci Eng C. 2007;27(2):274–9.

76. Balandin AA, et al. Superior thermal conductivity of single-layer graphene. Nano Lett. 2008;8(3):902–7.
77. Agarwal S, Wendorff JH, Greiner A. Use of electrospinning technique for biomedical applications. Polymer. 2008;49(26):5603–21.
78. Xu H, et al. Chronic inflammation in fat plays a crucial role in the development of obesity-related insulin resistance. J Clin Invest. 2003;112(12):1821–30.
79. Messerli D, et al. Multipiece allograft implant; 2009. Google Patents.
80. Chen Q, Boccaccini A. Poly (D, L-lactic acid) coated 45S5 Bioglass®-based scaffolds: Processing and characterization. J Biomed Mater Res A. 2006;77(3):445–57.
81. Misra SK, et al. Polyhydroxyalkanoate (PHA)/inorganic phase composites for tissue engineering applications. Biomacromolecules. 2006;7(8):2249–58.
82. Sánchez-Salcedo S, Nieto A, Vallet-Regí M. Hydroxyapatite/β-tricalcium phosphate/agarose macroporous scaffolds for bone tissue engineering. Chem Eng J. 2008;137(1):62–71.
83. Puértolas J, et al. Compression behaviour of biphasic calcium phosphate and biphasic calcium phosphate–agarose scaffolds for bone regeneration. Acta Biomater. 2011;7(2):841–7.
84. Samavedi S, Whittington AR, Goldstein AS. Calcium phosphate ceramics in bone tissue engineering: a review of properties and their influence on cell behavior. Acta Biomater. 2013;9(9):8037–45.
85. Min B-M, et al. Electrospinning of silk fibroin nanofibers and its effect on the adhesion and spreading of normal human keratinocytes and fibroblasts in vitro. Biomaterials. 2004;25(7):1289–97.
86. Miroiu F, et al. Composite biocompatible hydroxyapatite–silk fibroin coatings for medical implants obtained by matrix assisted pulsed laser evaporation. Mater Sci Eng B. 2010;169(1):151–8.
87. Bhumiratana S, et al. Nucleation and growth of mineralized bone matrix on silk-hydroxyapatite composite scaffolds. Biomaterials. 2011;32(11):2812–20.
88. Spotnitz WD. Fibrin sealant: past, present, and future: a brief review. World J Surg. 2010;34(4):632–4.
89. Le Guéhennec L, Layrolle P, Daculsi G. A review of bioceramics and fibrin sealant. Eur Cell Mater. 2004;8(13):1–11.
90. Walker-Bone K, et al. Regular review: medical management of osteoarthritis. Br Med J. 2000;321(7266):936.
91. Liu H, et al. A study on a chitosan-gelatin-hyaluronic acid scaffold as artificial skin in vitro and its tissue engineering applications. J Biomater Sci Polym Ed. 2004;15(1):25–40.
92. Yodsuwan N, et al. Effect of carbon and nitrogen sources on bacterial cellulose production for bionanocomposite materials. In: 1st Fah Luang University international conference, Thailand; 2012.
93. Lee J-Y, et al. Transforming growth factor (TGF)-β1 releasing tricalcium phosphate/chitosan microgranules as bone substitutes. Pharm Res. 2004;21(10):1790–6.
94. Samira J, et al. Cytocompatibility, gene-expression profiling, apoptotic, mechanical and 29Si, 31P solid-state nuclear magnetic resonance studies following treatment with a bioglass-chitosan composite. Biotechnol Lett. 2014;36(12):2571–9.
95. Huang D, et al. Optical coherence tomography. Science (New York, NY). 1991;254(5035):1178.
96. Mohammadi Y, et al. Nanofibrous poly (epsilon-caprolactone)/poly (vinyl alcohol)/chitosan hybrid scaffolds for bone tissue engineering using mesenchymal stem cells. Int J Artif Organs. 2007;30(3):204.
97. Liu X, et al. Characterization of structure and diffusion behaviour of Ca-alginate beads prepared with external or internal calcium sources. J Microencapsul. 2002;19(6):775–82.
98. Kim I-Y, et al. Chitosan and its derivatives for tissue engineering applications. Biotechnol Adv. 2008;26(1):1–21.
99. Saravanan S, et al. Preparation, characterization and antimicrobial activity of a bio-composite scaffold containing chitosan/nano-hydroxyapatite/nano-silver for bone tissue engineering. Int J Biol Macromol. 2011;49(2):188–93.
100. Chronopoulou L, et al. Chitosan-coated PLGA nanoparticles: a sustained drug release strategy for cell cultures. Colloids Surf B: Biointerfaces. 2013;103:310–7.

101. Mohammadi Y, et al. Osteogenic differentiation of mesenchymal stem cells on novel three-dimensional poly (L-Lactic Acid)/Chitosan/Gelatin/Beta-Tricalcium phosphate hybrid scaffolds. Iran Polym J. 2007;16(1):57.
102. Wang Z, Qin T-W. Review: vitreous cryopreservation of tissue-engineered compositions for tissue repair. J Med Biol Eng. 2013;33(2):125–32.
103. Yeo JH, et al. The effects of Pva/chitosan/fibroin (PCF)-blended spongy sheets on wound healing in rats. Biol Pharm Bull. 2000;23(10):1220–3.
104. Wang M. Bioactive calcium phosphates and nanocomposite scaffolds for bone tissue engineering. Ceram Trans. 2010;218:175–83.
105. Chang MC, Tanaka J. FT-IR study for hydroxyapatite/collagen nanocomposite cross-linked by glutaraldehyde. Biomaterials. 2002;23(24):4811–8.
106. Li X, et al. Collagen-based implants reinforced by chitin fibres in a goat shank bone defect model. Biomaterials. 2006;27(9):1917–23.
107. Spilker M, et al. The effects of collagen-based implants on early healing of the adult rat spinal cord. Tissue Eng. 1997;3(3):309–17.
108. Weadock KS, et al. Physical crosslinking of collagen fibers: comparison of ultraviolet irradiation and dehydrothermal treatment. J Biomed Mater Res. 1995;29(11):1373–9.
109. Lahav J, Schwartz MA, Hynes RO. Analysis of platelet adhesion with a radioactive chemical crosslinking reagent: interaction of thrombospondin with fibronectin and collagen. Cell. 1982;31(1):253–62.
110. Reiser K, McCormick R, Rucker R. Enzymatic and nonenzymatic cross-linking of collagen and elastin. FASEB J. 1992;6(7):2439–49.
111. Chang CH, et al. Cartilage tissue engineering on the surface of a novel gelatin–calcium-phosphate biphasic scaffold in a double-chamber bioreactor. J Biomed Mater Res B Appl Biomater. 2004;71(2):313–21.
112. Schonauer C, et al. The use of local agents: bone wax, gelatin, collagen, oxidized cellulose. Eur Spine J. 2004;13(1):S89–96.
113. Li M, et al. Electrospun protein fibers as matrices for tissue engineering. Biomaterials. 2005;26(30):5999–6008.
114. Sawyer A, et al. The stimulation of healing within a rat calvarial defect by mPCL–TCP/collagen scaffolds loaded with rhBMP-2. Biomaterials. 2009;30(13):2479–88.
115. Habraken W, Wolke J, Jansen J. Ceramic composites as matrices and scaffolds for drug delivery in tissue engineering. Adv Drug Deliv Rev. 2007;59(4):234–48.
116. Ko C-C, et al. Mechanical properties and cytocompatibility of biomimetic hydroxyapatite-gelatin nanocomposites. J Mater Res. 2006;21(12):3090–8.
117. Vacanti JP, Langer R. Tissue engineering: the design and fabrication of living replacement devices for surgical reconstruction and transplantation. Lancet. 1999;354:S32–4.
118. Mohamed KR, Beherei HH, El-Rashidy ZM. In vitro study of nano-hydroxyapatite/chitosan–gelatin composites for bio-applications. J Adv Res. 2014;5(2):201–8.
119. Li J, et al. Surface characterization and biocompatibility of micro-and nano-hydroxyapatite/chitosan-gelatin network films. Mater Sci Eng C. 2009;29(4):1207–15.
120. Choi YS, et al. Study on gelatin-containing artificial skin: I. Preparation and characteristics of novel gelatin-alginate sponge. Biomaterials. 1999;20(5):409–17.
121. Li J, et al. Effect of nano-and micro-hydroxyapatite/chitosan-gelatin network film on human gastric cancer cells. Mater Lett. 2008;62(17):3220–3.
122. Sharma S, et al. Bone healing performance of electrophoretically deposited apatite–wollastonite/chitosan coating on titanium implants in rabbit tibiae. J Tissue Eng Regen Med. 2009;3(7):501–11.
123. Liotta L, et al. Metastatic potential correlates with enzymatic degradation of basement membrane collagen. Nature. 1980;284(5751):67–8.
124. Pals D, Hermans J. Sodium salts of pectin and of carboxy methyl cellulose in aqueous sodium chloride. I. Viscosities. Recueil des Travaux Chimiques des Pays-Bas. 1952;71(5):433–57.
125. Shikinami Y, Okuno M. Bioresorbable devices made of forged composites of hydroxyapatite (HA) particles and poly-L-lactide (PLLA): Part I. Basic characteristics. Biomaterials. 1999;20(9):859–77.

126. Kumar MNR. A review of chitin and chitosan applications. React Funct Polym. 2000;46(1): 1–27.
127. Willats WG, et al. Pectin: cell biology and prospects for functional analysis. In: Plant cell walls. Berlin: Springer; 2001. p. 9–27.
128. Uragami T, et al. Structure of chemically modified chitosan membranes and their characteristics of permeation and separation of aqueous ethanol solutions. J Membr Sci. 1994;88(2):243–51.
129. Rezwan K, et al. Biodegradable and bioactive porous polymer/inorganic composite scaffolds for bone tissue engineering. Biomaterials. 2006;27(18):3413–31.
130. Ghasemi-Mobarakeh L, et al. Electrospun poly (ε-caprolactone)/gelatin nanofibrous scaffolds for nerve tissue engineering. Biomaterials. 2008;29(34):4532–9.
131. Boccaccini AR, Maquet V. Bioresorbable and bioactive polymer/Bioglass® composites with tailored pore structure for tissue engineering applications. Compos Sci Technol. 2003;63(16):2417–29.

Progress of Bioceramic and Bioglass Bone Scaffolds for Load-Bearing Applications

Jingzhou Yang

Keywords Load-bearing bone scaffold · Bioceramic · Bioglass · Bone tissue engineering · 3D printing · Freeze casting · Biomaterial · Regenerative medicine · Bone repair · Bone regeneration · Microstructure design · Biofabrication · Orthopedic material · Calcium phosphate · Hydroxyapatite

1 Introduction

Massive load-bearing bone defects, which often result from trauma (e.g., battles and accidents) or diseases (e.g., tumor, osteonecrosis, vertebral fractures, and spine degeneration), remain a grand challenge to regenerate and represent a major financial burden to our healthcare system. Although autologous/allogenic live bone transplantation is still the primary treatment strategy for reconstruction of load-bearing bones, considerable drawbacks for these exist, including painful and costly additional surgeries, limitation in implant quantity, increased risk of infection and disease transfer, and technical difficulties for fragile children and the elderly. Tissue engineering aims to promote regeneration by using artificial scaffolding (macroporous) materials as the substitutes combing appropriate cells and growth factors to restore the biofunctions of the defective or focal tissues including bones [1–7]. In particular, an ideal engineered tissue scaffold for the regeneration of load-bearing bones should possess appropriate mechanical functions to provide structural support, share biomechanical load, and distribute stress that stimulates bone growth and remodeling. It should also present excellent biological functions to deliver bioactive factors, preserve tissue volume, degrade gradually, and induce new bone formation.

J. Yang (✉)
Affiliated Hospital of Hebei University, Health Science Centre, Hebei University, Baoding, P. R. China

Research Centre of Biomedical Materials 3D Printing, National Engineering Laboratory for Polymer Complex Structure Additive Manufacturing, Baoding, P. R. China

Hebei Dazhou Smart Manufacturing Technology Co., Ltd., Baoding, P. R. China

© Springer International Publishing AG, part of Springer Nature 2018
B. Li, T. Webster (eds.), *Orthopedic Biomaterials*,
https://doi.org/10.1007/978-3-319-89542-0_18

Due to excellent bioactivity, bioresorbability, apatite formation ability, and osteoconductivity, bioceramic/bioglass (e.g., calcium phosphate, calcium carbonate and calcium silicate) scaffolds have been successfully used for non-load-bearing bone restoration in the recent decades [7–11]. However Current bioceramic/bioglass scaffolds cannot reestablish massive load-bearing bones, because they do not have sufficient mechanical properties [12, 13] for immediate load sustaining and cyclic load bearing to stimulate bone growth and remodeling. It is well known that the mechanical properties of materials decrease with increasing the porosity and pore size. Bioceramic/bioglass scaffolds normally require an interconnected macroporous structure with a high porosity of over 90% and a pore size ranging from 100 to 1000 μm [2, 10] to allow transport of nutrients and migration of cells, as well as formation of new bones and blood vessels. Such porous constructs typically have low mechanical properties, such as compressive/bending strength of a few MPa, and fracture toughness of nearly zero, far below those of natural bones.

This review will highlight the recent progress in the development of load-bearing bioceramic/bioglass bone tissue scaffolds. It focuses on the improvement of mechanical properties other than biological properties. High porosity leads to low mechanical properties. However through tailoring the microstructure and macrostructure (e.g., pore size/shape, porosity, interconnectivity, grain size, and grain morphology) researchers are able to enhance the mechanical properties of bioceramic/bioglass scaffolds. This review will discuss the fundamental design concepts and manufacturing methods that are required for the fabrication of load-bearing bioceramic/bioglass bone scaffolds. *In vitro* mechanical characterization and *in vivo* assessment methods/models will also be discussed. There are strong and tough dense/porous materials in the nature, such as coral, nacre, bamboo, and animal shells. Such strong natural materials have well-aligned multiple scale micro/macrostructures. At last, this review will also give our perspective on design concepts of strong and tough bioceramic/bioglass bone scaffolds by bioinspiration.

2   Design Concepts

Tissue scaffolds normally require an interconnected macroporous structure with a high porosity of over 90%, and a pore size ranging from 100 to 1000 μm [2, 10] to allow transport of nutrients and migration of cells, as well as formation of new tissues and blood vessels. Such porous constructs typically do not have sufficient mechanical properties for load-bearing applications. Recent *in vivo* studies have proven that excellent bone formation and regeneration could be achieved with implantation of bone tissue scaffolds with micro/nano pores (<100 μm) and relatively lower porosity (<70%) [14, 15]. So it is feasible to design and manufacture stronger bioceramic/bioglass bone scaffolds with load-bearing capacity through tailoring the microstructure and macrostructure (pore size, porosity, pore shape, etc.). In addition, the scaffolds can also be strengthened by optimization of grain size and morphology, and incorporation of stronger second phases.

2.1 Microstructure Design: Micropore Size, Microporosity, Grain Size/Morphology and Second Phase

2.1.1 Pore Size

To achieve better mechanical properties, the pore size of bioceramic/bioglass bone scaffolds can be decreased to some extent, e.g., below 100 μm. Although pores over 100 μm are required for bone tissue scaffolds to provide excellent biological properties for osteoconductivity and bone unit ingrowth from a tissue engineering view, recent *in vivo* results demonstrated that scaffolds with smaller pores are capable to form new bone tissues [14, 16, 17]. Therefore pioneers have designed strong microporous and nanoporous bioceramic/bioglass scaffolds for the repair of load-bearing bone defects. Zhang, et al. [18] designed and manufactured a micro-porous hydroxyapatite (HA)/barium titanate composite bioceramic scaffold that had a compressive strength of 14.5 MPa and a porosity of 57.4%. The lamellar pore diameter was around 32 μm and the lamellar thickness was around 9 μm. This microporous bioceramic scaffold is promising to reconstruct light-load-bearing bone defects. Sprio, et al. [19] designed another load-bearing microporous bioc-eramic bone scaffold of HA/calcium silicate (CS). It had a pore diameter of below 10 μm, a bending strength of over 35 MPa, and a porosity of around 30%. Only the microstructure and general material properties were characterized in this study. It would be suggested to evaluate the load-bearing capacity more comprehensively, to assess the *in vitro* osteogenesis, and to investigate the *in vivo* bone formation in a load-bearing animal model. Xu, et al. [20] studied a microporous 45S5-calcium borosilicate bioglass bone scaffold that showed a compressive strength of 8 MPa and a porosity of 70%.

In general, micro/nanoporous bioceramic/bioglass scaffolds are stronger than the macroporous ones. So, some scientists care more about the *in vivo* bone regenera-tion behavior of strong microporous tri-calcium phosphate (TCP)/HA scaffolds in load-bearing models [14, 16, 17]. Yuan, et al. and H.O. Mayr, et al. studied the *in vivo* bone formation in a dog spinal fusion model and sheep osteochondral model, respectively. The scaffolds had a compressive strength of over 75 MPa and a poros-ity of below 45%. The micropores had a diameter of below 10 μm, as shown in Figs. 1 and 2. Histological results indicated that the microporous bioceramic scaf-folds regenerated 25% new bone in 12 weeks and almost 100% new bone in 52 weeks in load-bearing defect models, as displayed in Figs. 3 and 4. These studies demonstrated a design concept of strong microporous bioceramic/bioglass bone scaffolds for load-bearing applications. Yang, et al. [21] developed a new family of nanoporous calcium aluminate/TCP bioceramic scaffolds. The pore size was 200–500 nm. The compressive strength ranged from 40 to 195 MPa. The nanoporous structure was interconnected. Such strong nanoporous bioceramic scaffolds are promising for the repair of load-bearing bone defects.

Fig. 1 SEM images showing various microporous structures of HA, TCP, and BCP (composite of HA and TCP) bioceramic scaffolds for implantation in a dog spine defect. Reproduced with permission [14] *Copyright © 2010 National Academy of Science*

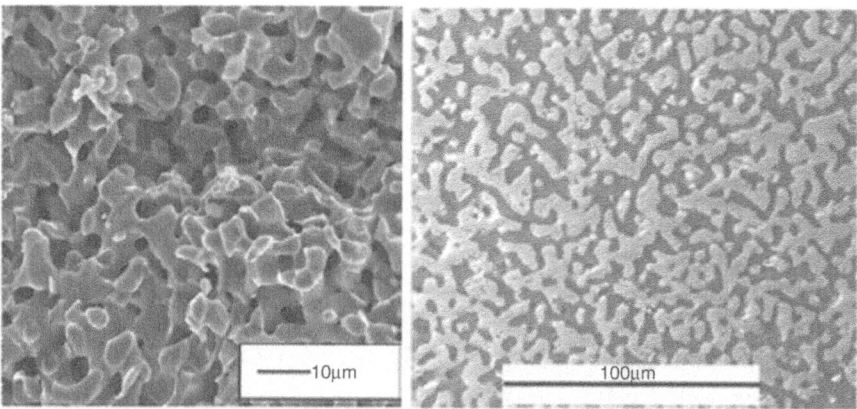

Fig. 2 SEM images showing the microporosity and homogeneous interconnected pores of TCP bioceramic scaffolds for implantation in a sheep osteochondral defect (left: fracture surface; right: polished surface). Reproduced with permission [16] *Copyright © 2012 Elsevier B.V*

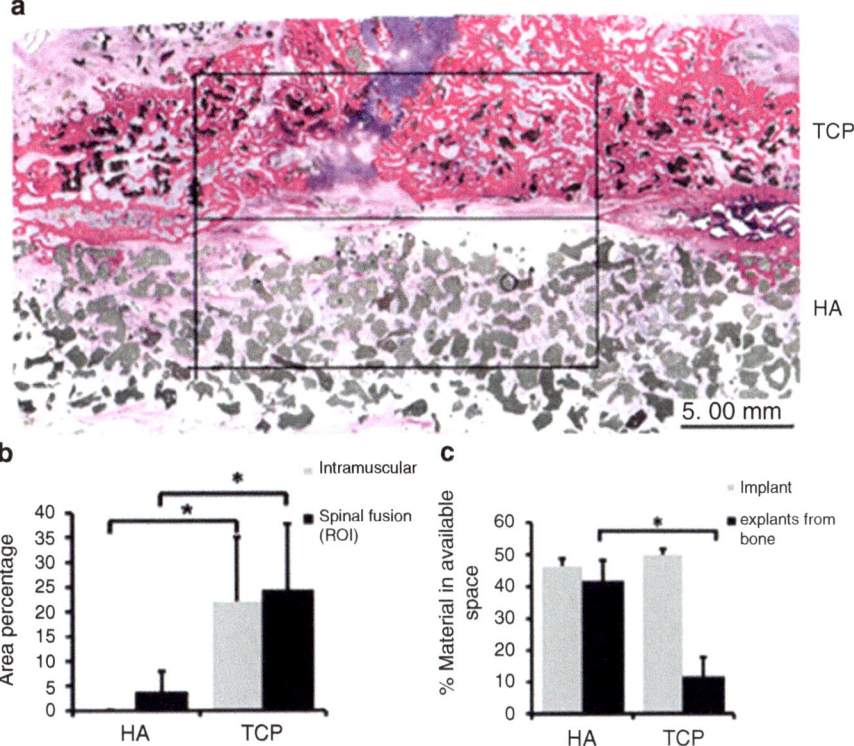

Fig. 3 Posterolateral spine fusion with HA and TCP microporous bioceramic scaffolds in dogs after 12-week post-implantation. (**a**) Histological overviews showing the newly formed bone; (**b**) The new bone percentage for HA and TCP scaffolds; (**c**) The material available before implantation. An asterisk (*) denotes statistical difference (Student's paired t test. $P < 0.05$). Reproduced with permission [14] *Copyright © 2010 National Academy of Science*

2.1.2 Porosity

Bone tissue scaffolds have a porous structure that is required for nutrient transport, cell migration, bone tissue ingrowth, vasculature formation, bioactive factor delivery, etc. In addition, porosity is also helpful to increase biodegradability. Scaffolds with a higher porosity normally show better osteoconductivity which is crucial for the restoration of critical sized bone defects. However, mechanical properties of materials decrease with increasing the porosity. The bioceramic/bioglass scaffolds had a compressive strength from over 150 MPa down to several MPa with the porosity increasing from 40 to 80% [15, 22–27], as shown in Fig. 5. Recent studies have also shown that bone scaffolds with a low porosity also performed excellent during *in vivo* bone formation studies [14, 17]. So it is feasible to sacrifice some mechanical properties to design strong bone scaffolds with a low porosity for load-bearing applications.

Bi, et al. [15] designed a family of low-porosity but strong bioglass scaffolds and studied their ability to repair critical sized segmental bone defects in a rat femur

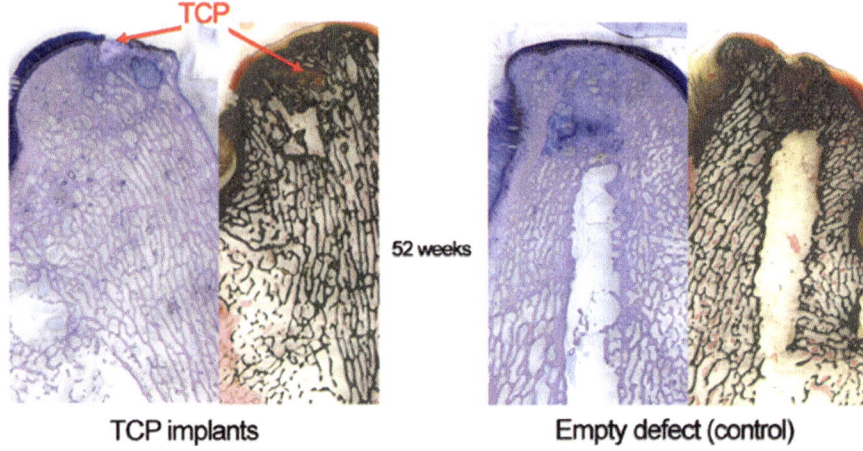

Fig. 4 Bone regeneration of TCP microporous bioceramic scaffolds in osteochondral defect after 52-week post-implantation (left: Giemsa staining; right: Safranin O staining). Reproduced with permission [17] *Copyright © 2013 Elsevier B.V*

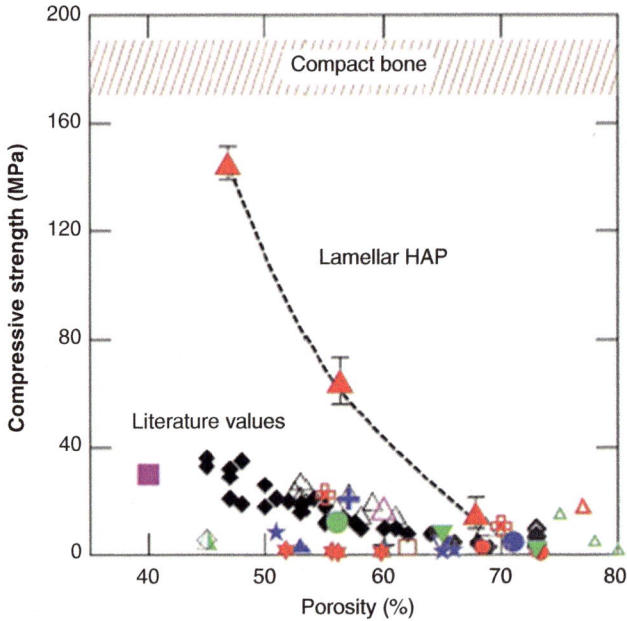

Fig. 5 Relationship between compressive strength and porosity of bioceramic scaffolds. Reproduced with permission [22] *Copyright © 2006 Elsevier B.V*

load-bearing model. Autografts were used as the control materials. The bone scaffolds were made of 13-93 silicate bioglass (mol%: 54.6 SiO_2, 6.0 Na_2O, 7.9 K_2O, 7.7 MgO, 22.1 CaO, 1.7 P_2O_5) and 13-93B3 borate bioglass (mol%: 18.0 SiO_2, 36.0 B_2O_3, 6.0 Na_2O, 8.0 K_2O, 2.1 MgO, 6.0 SrO, 22.0 CaO, 2.0 P_2O_5). The porosity and pore diameter were 47–50% and 300–500 µm. The compressive strength was 86 MPa and 40 MPa, respectively. At 12-week post-implantation, histological results showed that the percentage of new bone area (von Kossa-positive area) in the defects implanted with the bioglass scaffolds was 32–38%, not significantly different from that for the autografts (40%). New blood vessel area generated by the bioglass scaffolds was 4–8%, also not significantly different from that for the autografts (5%). Jia, et al. [28] investigated strong bioglass scaffolds of 13-93 and 2B6Sr for the repair of load-bearing bone defects in a critical sized rabbit femoral model. The compressive strength was 80.4 MPa and 35.6 Ma, respectively. The porosity was 50% and the pore diameter was 200 µm. At nine-month post-implantation, the 10 mm × 6 mm segmental defects were reconstructed with new mature bone tissues in the two glass scaffolds and granule autografts, as shown in Fig. 6. In addition, it

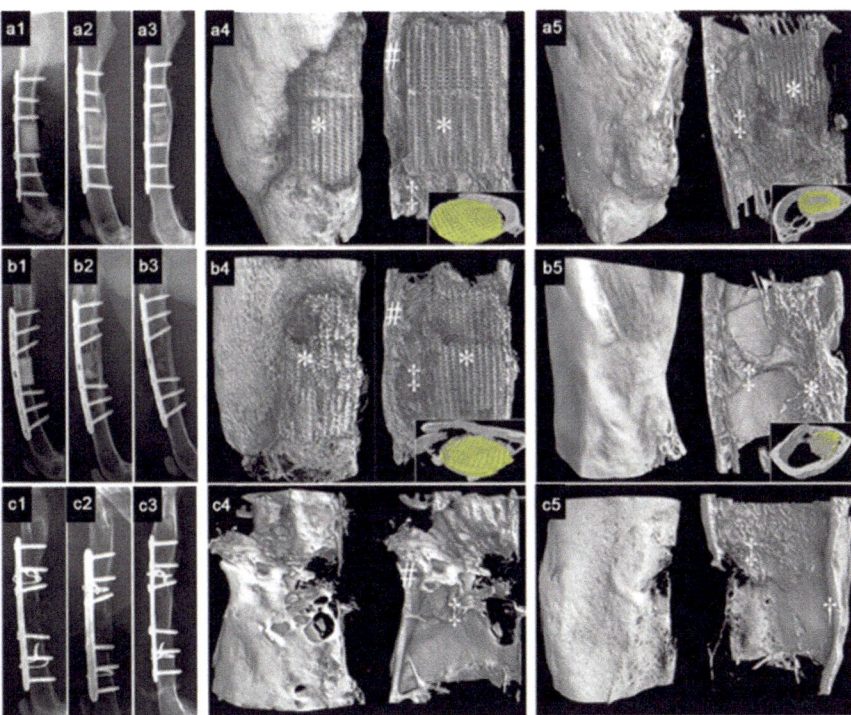

Fig. 6 Lateral radiographs and micro-CT images showing the bone regeneration of low-porosity bioglass scaffolds in a load-bearing rabbit femur defect: (**a**) 13-93 scaffold; (**b**) 2B63 scaffold; and (**c**) ABG scaffold. Reproduced with permission [28] *Copyright © 2015 Wiley-VCH Verlag GmbH & Co. KGaA, Weinheim*

can be seen that the tubular bone structure remodeling was satisfied. These studies demonstrated that low-porosity bioglass bone scaffolds can not only provide load-bearing capacity but also promote new bone formation/remodeling and vasculature formation. Huang, et al. [24] fabricated another strong 13-93 bioglass scaffold with the compressive strength of 140 ± 70 MPa, a Young's modulus of 5 ± 0.5 GPa, porosity of 50%, and pore diameters of 300 μm. Fu, et al. [29] optimized the porous structure of 6P53B bioglass scaffold and achieved a high compressive strength of 136 ± 22 MPa comparable with that of human cortical bone (100–150 MPa). The porosity was 60%, in the range of that of cancellous bone (35–95%). After immersion in simulated body fluid (SBF) for 3 weeks, the compressive strength was still as high as 77 MPa. The strong bioglass bone scaffolds open a new pathway for repair of heavy-load-bearing bone defects through tissue engineering. Deville, et al. [22] developed a low-porosity and high-strength HA bioceramic bone scaffold. The compressive strength was up to 145 MPa for 47% porosity and 65 MPa for 56% porosity. The pore diameter was 20–100 μm. Such strong bone scaffolds could be considered for load-bearing applications. Zhang, et al. [30] designed another type of HA bioceramic scaffold with a low porosity of 52.5% and a high bending strength of 73 MPa. The pore diameter ranged from 100 μm to 300 μm. It is an effective strategy to improve the mechanical properties by decreasing the porosity of bioceramic/bioglass bone scaffolds towards load-bearing applications.

2.1.3 Grain Size and Morphology

The mechanical properties of bioceramics are dependent on grain size and morphology. To strengthen the bioceramics or bioglass-bioceramic composites, One can decrease the grain size, increase the grain aspect ratio, make the grains aligned, and change the grain content (crystallization). For example, the bending strength of a dense biosilicate® glass-ceramic (23.8Na$_2$O-23.8CaO-48SiO$_2$-4P$_2$O$_5$ wt%) can be improved from 75 to over 200 MPa by increasing the crystallization from 0 to 40% [31] (Fig. 7). It could be believed that the increase in strength will also work for the porous biosilicate® scaffolds. Shuai, et al. [32] added 30% nano sized HA to macroporous TCP bioceramic scaffolds. The compressive strength increased from 8 to 16 MPa. Feng, et al. [33] developed a macroporous nano HA bioceramic scaffold with a high compressive strength of 18.6 MPa. Feng, et al. [34] also designed and manufactured HA fiber (20%) reinforced CS bioceramic scaffolds (Fig. 8). The compressive strength increased from 3 to 27 MPa.

2.1.4 Second Phase Teinforcement

Second phase reinforcement is an effective strategy for ceramic and glass materials. This method involves incorporating stronger particles/fibers as the second phase to reinforce the matrix. R. Emadi, et al. [35] added nano sized bioglass to increase the mechanical properties of highly macroporous HA bioceramic scaffolds. The

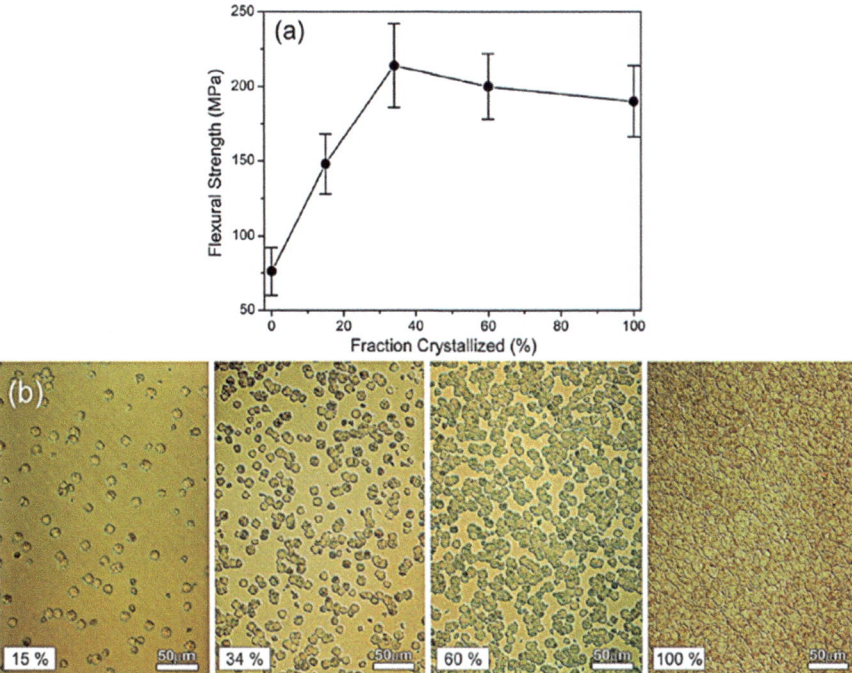

Fig. 7 (**a**) Relationship between bending strength and crystallization fraction of biosilicate glass-ceramic; (**b**) Microstructure of biosilicate glass-ceramic with different crystallization. Reproduced with permission [31] *Copyright © 2016 Elsevier B.V*

porosity was 83% and the pore size was over 800 μm. The compressive strength increased from 0.22 to 1.49 MPa. Sprio, et al. [19] reinforced HA bioceramics with elongated CS grains. The bending strength improved from 42 to 70 MPa. Xu, et al. [20] developed borosilicate glass reinforced 45S5 derived bioglass-ceramics. The compressive strength was increased from 10 to 40 MPa.

2.2 Macrostructure Design: Macropore Shape, Pore size, Macroporosity and Pore Connecting Part Width

2.2.1 Pore Shape

In general, highly macroporous bioceramic/bioglass bone scaffolds have relatively low mechanical properties. However, stronger scaffolds could be fabricated through optimizing the pore shape. Gmeiner, et al. [36] 3D printed a $45.5SiO_2$-$24.25CaO$-$24.25Na_2O$-$6P_2O_5$ bioglass scaffold with hexagon-shape macropores as shown in Fig. 9. The biaxial bending strength achieved 124 MPa. These strong bioglass scaffolds are capable for load-bearing bone reconstruction. Fu, et al. [37] freeze-casted a

Fig. 8 HA fiber reinforced CS bioceramic scaffolds: (**a**) raw CS powders; (**b**) raw HA fibers; (**c**) distribution of CS powders and HA fibers in the composite scaffold; (**d**) and (**f**) macroscopic top view of the scaffold; (**e**) side view of the scaffold. Reproduced with permission [34] *Copyright © 2014 Elsevier Inc.*

13-93 bioglass scaffold with lamellar (elongated and well-aligned) pores (Fig. 10). It had a higher compressive strength (25 MPa) than that of the one with columnar (equiaxial) pores. Fu, et al. [29, 38] also 3D printed a load-bearing 6P53B bioglass scaffold with grid-like (square) macropores, as shown in Fig. 11. The compressive strength reached 136 MPa, comparable to that of cortical bones. Roohani-Esfahani, et al. [39] 3D printed a new family of bioglass-ceramic bone scaffolds

Fig. 9 Stereolithographic 3D printed bioglass scaffold with hexagon-shape macropores: (**a**) macroscopic view; (**b**) Microscopic view. Reproduced with permission [36] *Copyright © 2014 The American Ceramic Society*

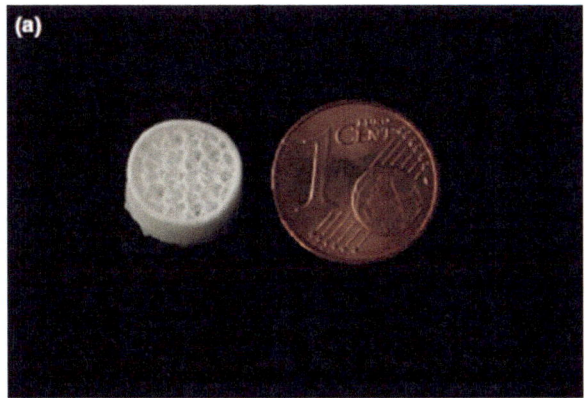

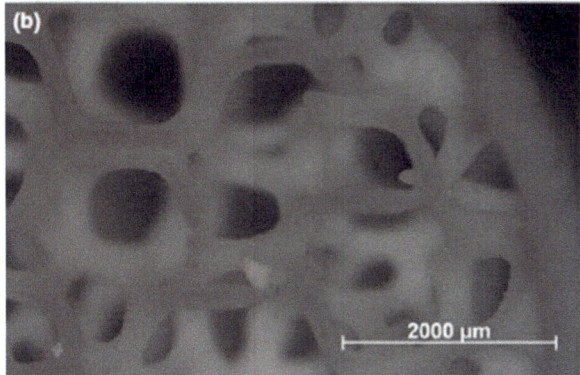

(Sr-Ca$_2$ZnSi$_2$O$_7$-ZnAl$_2$O$_4$-glass phase) with interconnected macropores that had four different shapes: hexagon, curve, rectangle and zigzag (Fig. 12). The hexagon one exhibited the highest mechanical properties: compressive strength of 180 MPa and bending strength of 50 MPa, both for 50% porosity. It was concluded that the mechanical properties of bioceramic/bioglass scaffolds can be highly enhanced by optimizing the macropore shape.

2.2.2 Pore Size and Pore Connecting Part Width

The mechanical properties of bioceramic/bioglass scaffolds are partially dependent on the pore size and pore connecting part width with constant porosity. To get stronger scaffolds, one can decrease the pore size and increase the pore connecting part width. As shown in Fig. 12, the compressive strength and bending strength increased from 90 MPa and 20 MPa to 180 MPa and 50 MPa by decreasing the pore size from 1200 μm to 450 μm [39]. S. Deville, et al. [22] freeze- casted HA scaffolds with controlled porosity and pore connecting part width for load-bearing applications. The compressive strength increased from 65 MPa to 145 MPa, by increasing the pore connecting part width from below 10 μm to 20–40 μm.

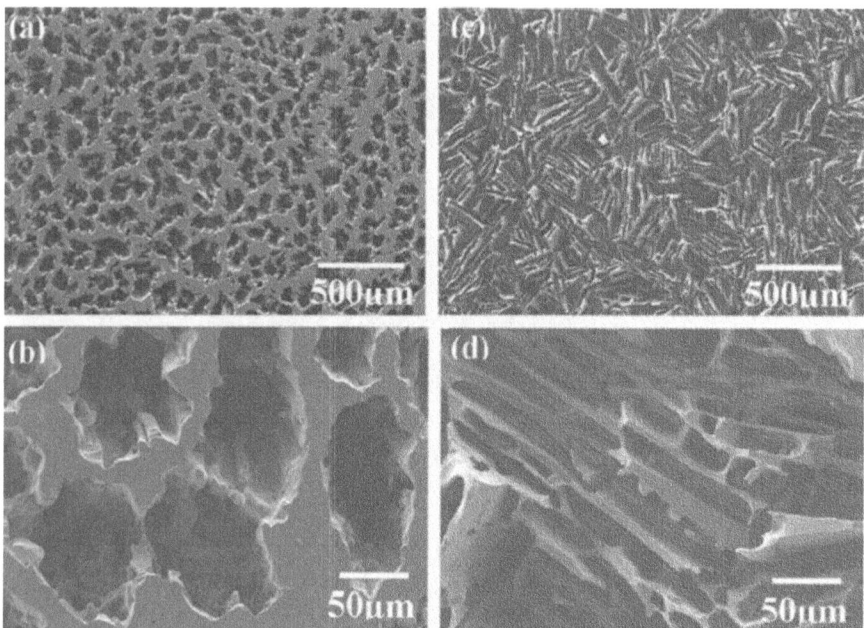

Fig. 10 SEM images showing the cross section structure of 13-93 bioglass scaffolds: (**a**) and (**b**) columnar porous structure; (**c**) and (**d**) lamellar porous structure. Reproduced with permission [37] *Copyright © 2009 Wiley Periodical, Inc.*

2.2.3 Macroporosity

The mechanical properties of bioceramic/bioglass scaffolds could also be improved by decreasing the macroporosity. Zhang, et al. [30] decreased the porosity of HA scaffolds from 70 to 52.5%. The bending strength increased from 9 to 73 MPa. The strength increment was remarkable. Roohani-Esfahani, et al. [39] decreased the porosity of bioglass-ceramic bone scaffolds (Sr-$Ca_2ZnSi_2O_7$-$ZnAl_2O_4$-glass phase) from 70 to 50%. The compressive strength increased from 50–90 MPa to 130–180 MPa. The bending strength increased from 10–20 MPa to 30–50 MPa. Deville, et al. [22] decreased the porosity of HA scaffolds from 56 to 47%. The compressive strength increased from 65 to 145 MPa.

3   Manufacturing Methods

In recent years, some advanced manufacturing methods have been developed to fabricate strong bioceramic/bioglass bone scaffolds for load-bearing applications. 3D printing and freeze casting are able to tailor the macrostructure conveniently for better mechanical properties. The two most popular traditional manufacturing

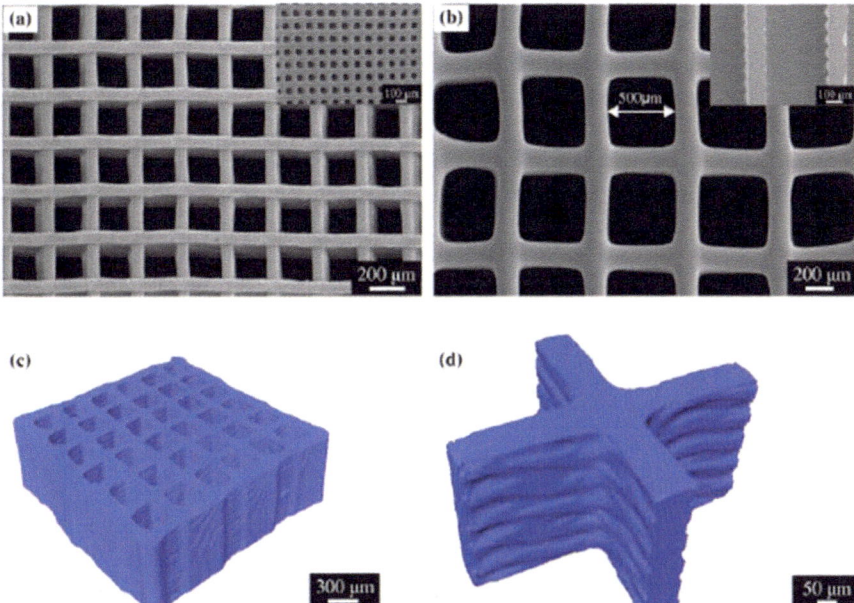

Fig. 11 SEM images showing: (**a**) Surface of as-printed 6P53B biogalss scaffold through a 100 μm nozzle with a spacing of 200 μm, insert displaying the structure after sintering; (**b**) Surface of sintered scaffold with pores of 500 μm. Insert showing cross section; (**c**) Sintered scaffold with regular grids; (**d**) A sectional view of the bonded bioglass filaments. Reproduced with permission [38] *Copyright © 2011 Wiley-VCH Verlag GmbH & Co. KGaA, Weinheim*

methods include slip casting (polymer template burn-out) and thermally bonding particles can also make strong scaffolds by optimizing the processing factors. This part will review the recent advancements of these manufacturing methods and the resultant strong bioceramic/bioglass scaffolds. This review will also comment the advantages and disadvantages of these methods.

3.1 3D printing

3D printing is an advanced rapid manufacturing method that can build strong materials layer-by-layer with fine-tuned structures and customized 3D geometries, as well as high mechanical properties. For bioceramic/bioglass materials, the most popular specific 3D printing techniques include 3D plotting/direct ink writing, robocasting/robotic assisted deposition, stereolithography (SLA), and powder based 3D printing. Other laser related 3D techniques of fused deposition modeling (FDM), laser-assisted bioprinting (LAB), and selective laser sintering (SLS) are mostly used for metals and polymers, as well as composites. These techniques involve fusing the materials and rapid temperature change so that they are not preferable for

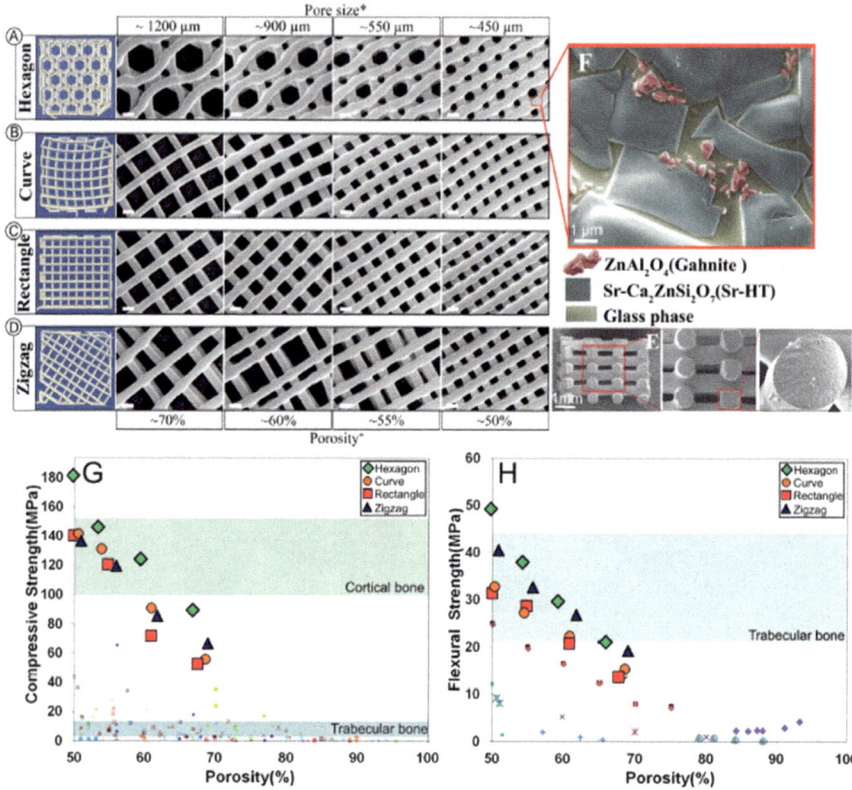

Fig. 12 Direct writing 3D printed bioglass-ceramic bone scaffolds with different pore shapes. (**a**) Hexagonal; (**b**) Curved; (**c**) Rectangular; (**d**) Zigzag; (**e**) SEM images showing the fracture surface of the scaffold, revealing that the filaments are fully dense; (**f**) the scaffolds consisting of three phases of Sr-Ca₂ZnSi₂O₇ grains, ZnAL₂O₄ crystals, and a glassy phase between the grain; (**g**) Compressive strength of scaffolds with distinct pore geometries VS porosity; (**h**) Bending strength VS porosity. Reproduced from Roohani-Esfahani, et al. *Scientific Report 2016* [39]

bioceramics and bioglass that have high melting temperatures and low thermal shock resistance. During the rapid temperature change, bioceramics and bioglass are easy to generate microcracks in that reduce the mechanical properties of the scaffolds. Here this review will only discuss 3D plotting/direct ink writing, robocasting/robotic assisted deposition, stereolithography (SLA), and powder based 3D printing.

3D plotting has been used to manufacture bioceramic/bioglass bone scaffolds including 13-93 bioglass, 6P53B bioglass, 2B6Sr bioglass, 13-93B3 bioglass, strontium doped Ca-Zn-Si-Al-O bioglass-ceramic, HA bioceramic, TCP bioceramic, HA/TCP biphasic bioceramic, CS bioceramic, and Sr₅(PO4)₂SiO₄ bioceramic [25, 26, 28, 29, 36, 38–58]. Through varying the 3D plotting process factors, the porous structures include pore size, porosity, pore shape/morphology, and pore

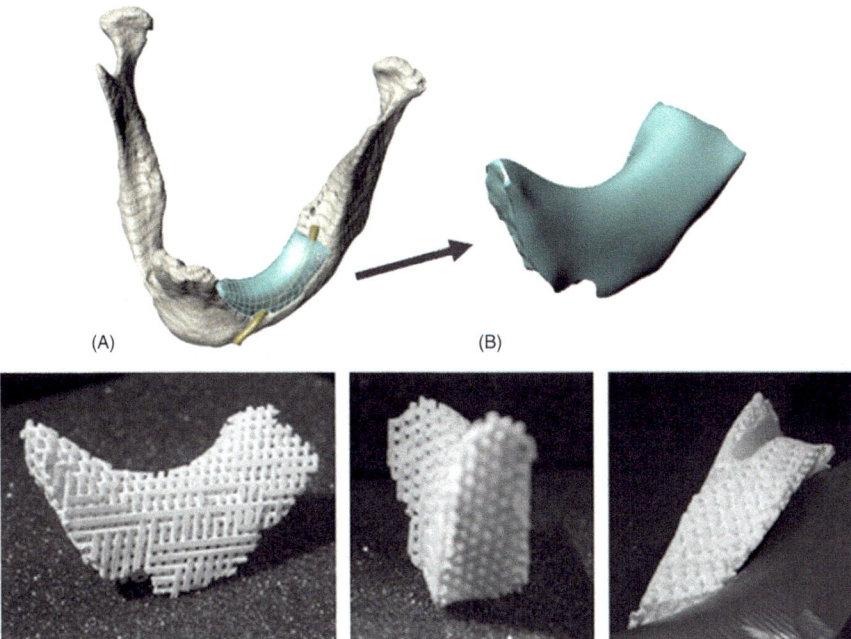

Fig. 13 Customized 3D bioceramic scaffold for mandible reconstruction. (top) 3D model of the scaffold geometry for the mandible defect; (bottom) Various views of the bioceramic scaffold manufactured by direct writing 3D printing. Reproduced with permission [44] *Copyright © 2005 The American Ceramic Society*

connecting part width, can be optimized towards better mechanical properties. Among these studies, some strong bioceramic/bioglass bone scaffolds have been fabricated for load-bearing applications. Cesarano III, et al. [44] 3D plotted a HA bioceramic scaffold with a grid structure and a continuous fully interconnected macroporosity in all directions (Fig. 13). The pore diameter was 270 µm and the pore connecting part had a width of 500 µm. The compressive strength achieved 195 MPa, matching that of cortical bones. This strong scaffold is promising for the repair of load-bearing bone defects (e.g., mandible). Roohani-Esfahani, et al. [39] 3D plotted a new family of bioglass-ceramic bone scaffolds that were composed of strontium doped hardystonite ($Ca_2ZnSi_2O_7$, HT) grains, clusters of submicro gahnite ($ZnAl_2O_4$) crystals, and a glassy phase. As displayed in Fig. 12, the scaffolds have four different porous structures: hexagon, curve, rectangle, and zigzag. The pore diameter ranged from 450 µm to 1200 µm. The porosity was from 50 to 70%. The compressive strength and bending strength achieved were 50–90 MPa and 10–20 MPa respectively for 70% porosity, 130–180 MPa and 30–50 MPa for 50% porosity. The hexagonal structure had the highest mechanical properties. Fu, et al., Jia, et al., and Deliormanli, et al. [28, 29, 38, 45, 59–61] 3D plotted strong bioglass scaffolds of 13-93, 13-93B, 6P53B, and 2BSr. The porosity was around 50%. As

Fig. 14 Bioglass scaffolds with different patterns 3D printed by direct writing. (**a**) Periodic pattern; (**b**) Anatomic pattern with a dense outer layer and a porous inner layer; (**c**) Graded pattern; (**d**) Cross-sections of the scaffolds. Reproduced with permission [29] *Copyright © 2011 Elsevier B.V*

shown in Fig. 14, the scaffolds have a grid porous structure. The compressive strength was 35 MPa–142 MPa that indicated an immediate load-bearing capacity (Fig. 15). Seitz, et al. [41] manufactured strong HA bioceramic scaffolds with powder based 3D printing technique (Fig. 16). The pore diameters in Z and X direction were 447 μm and 569 μm respectively. The pore connecting part width was 330 μm. The compressive strength was 21 MPa. Bian, et al. [58] manufactured a TCP bioceramic scaffold with a SLA 3D printing technique. The porosity was 45%. The pore diameter ranged from 400 to 500 μm. The compressive strength was 23 MPa. By using SLA, Gmeiner, et al. [36] 3D printed a $45.5SiO_2$–$24.25CaO$-$24.25Na_2O$-$6P_2O_5$ bioglass scaffold. The biaxial bending strength achieved 124 MPa. This strong bioglass scaffold is promising for load-bearing bone reconstruction. 3D plotting is the most popular 3D printing method reported in the literature. It can fabricate load-bearing bioceramic/bioglass bone scaffolds. Powder based 3D printing, SLA, and robocasting involve relatively more polymer binders so that the pore connecting part of the printed scaffolds has relatively low density. The mechanical properties are not as strong as the 3D plotted ones. Further research is required to make these 3D printing techniques more powerful towards manufacturing strong scaffolds.

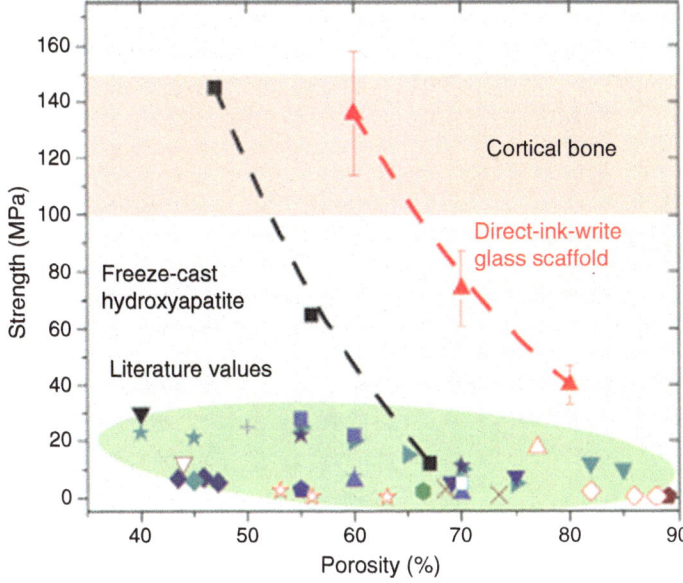

Fig. 15 Compressive strength VS porosity of the bioglass scaffolds manufactured by 3D printing and compared with those fabricated by freeze casting and traditional techniques. Reproduced with permission [29] *Copyright © 2011 Elsevier Ltd.*

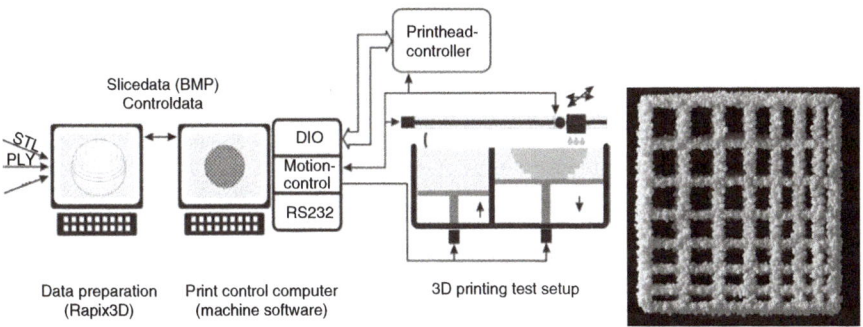

Fig. 16 Bioceramic scaffolds manufactured by powder based 3D printing. (left) The processing; (right) The macroscopic view of the scaffold. Reproduced with permission [41] *Copyright © 2005 Wiley Periodicals, Inc.*

3.2 Freeze Casting

Freeze casting has been drawning increasing attention since it was reported in *Science* (2006) [22]. This powerful manufacturing technique comes from bioinspiration. Some natural materials including coral, nacre, shell, bone, and tooth, are super strong and tough thanks to their well aligned/oriented hierarchical structure from macro to micro/nano scales. Scientists and engineers have been trying very

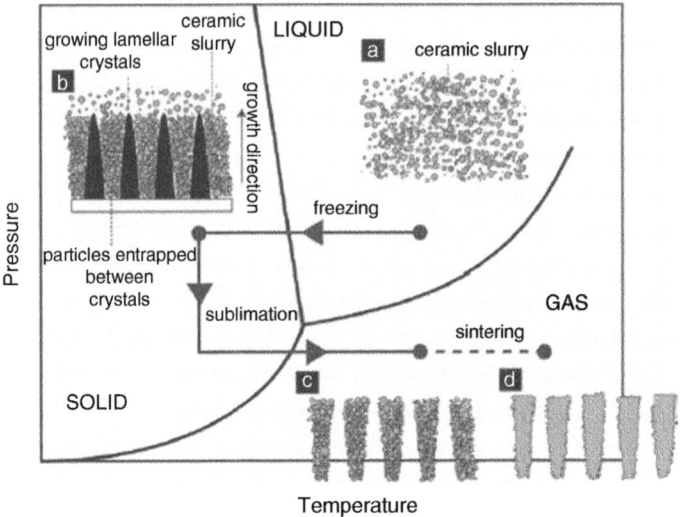

Fig. 17 Freeze casting for manufacturing porous materials. (**a**) The mechanism of freeze casting; (**b**) Compressive strength of freeze-casted HA bioceramic scaffolds compared with those fabricated by traditional techniques. Reproduced with permission [23] *Copyright © 2006 American Association of the Advancement of Science*

hard to transfer the clever structure design from natural materials to synthetic materials. Freeze casting is the ground-breaking method that is able to make bioinspired dense and porous materials with super high mechanical properties for load-bearing applications. As shown in Fig. 17, during directional freezing of ceramic slurry, the ice crystals grow along the temperature decreasing direction and the ceramic particles are pushed and entrapped between the ice dendrites. Afterward, the ice is sublimated by freeze drying that remove the ice through solid-gas phase transition. So, the porous scaffold structure is a negative replica of the ice. By tailoring the freezing kinetics and direction, well-aligned fine structures can be achieved (Figs. 18 and 19). Because the pore wall has a nacre-like micro brick-and-mortar structure, freeze-casted scaffolds behave with super high mechanical properties. As displayed in Fig. 20, the high mechanical properties result from extending and branching of crack propagation that consume more cracking energy. Deville, et al. [22, 23, 62] optimized the freeze casting factors and fabricated strong HA bioceramic scaffolds for repair of load-bearing bone defects. The porosity ranged from 70 to 47%. The compressive strength achieved from 15 to 145 MPa, four times higher than those manufactured by traditional methods. The pore diameter was 100 μm–800 μm, well fulfilling the requirements of bone tissue scaffolds towards transport of nutrients and migration of cells, as well as formation of new bones and vasculature. Fu, et al. [37] manufactured a strong 13-93 bioglass bone scaffold with freeze casting. The scaffold had an oriented porous structure. The compressive strength reached 20 MPa–50 MPa. The pore diameter was around 100 μm. Freeze-casted scaffolds have an immediate load bearing capacity and ideal pore size, but the pore

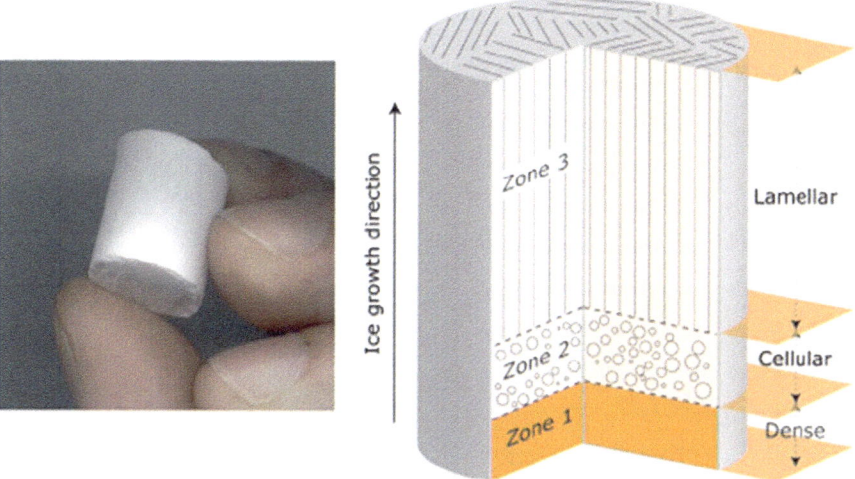

Fig. 18 Zonal structure of freeze-casted bioceramic scaffolds. Reproduced with permission [22] *Copyright © 2006 Elsevier Ltd.*

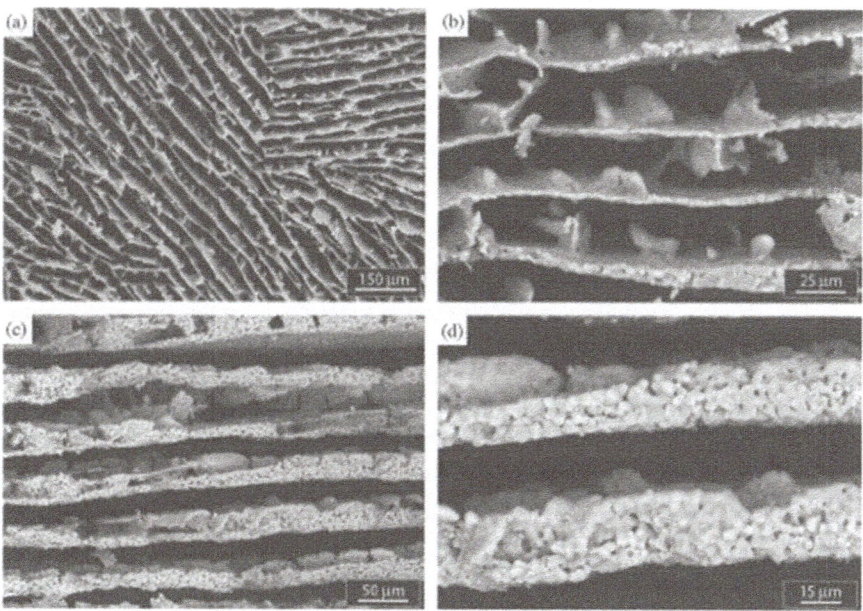

Fig. 19 Typical laminar porous structure of freeze-casted bioceramic scaffolds. Reproduced with permission [22] *Copyright © 2006 Elsevier Ltd.*

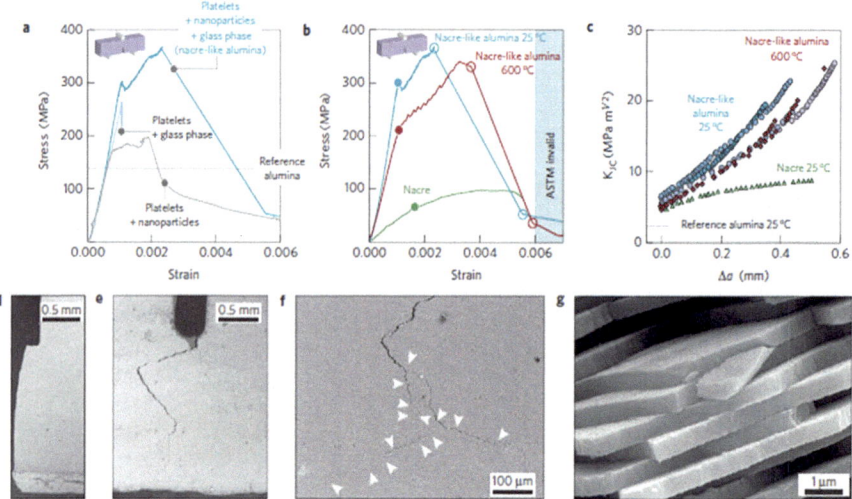

Fig. 20 Toughening mechanism of microcracking and crack bridging for ceramics with well-aligned grains. Reproduced with permission [63] *Copyright © 2014 Macmillan Publisher Ltd.*

interconnectivity is not as high as those fabricated with 3D printing. Another disadvantage is that the freeze-casted scaffolds always have a dense outer layer that requires extra manufacturing to expose the inner porous structure.

3.3   Slip Casting (Polymer Template Burn-Out)

Slip casting is one of the most popular methods that can fabricate highly macroporous scaffolds. It has been used extensively to manufacture porous advanced ceramics. The process involves ceramic slip/slurry preparation, polymer foam casting, squeezing and recasting, drying, and sintering. The scaffold pore size and porosity are mainly dependent on the polymer foam structure and ceramic slurry concentration. Due to the high porosity and high interconnectivity, the slip casting derived scaffolds exhibit low mechanical properties. Lin, et al. [64] slip-casted a bioglass scaffold with a polyurethane sponge as the template. As shown in Fig. 21, the compressive strength ranged from below 1 MPa–6 MPa for 90%–60% porosity. The pore diameter was from 100 μm to 500 μm. Chen, et al. fabricated a highly macroporous bioglass scaffold with slip casting. The porosity was 89%–94%. The bending strength was below 5 MPa [65, 66]. By optimizing the slip casting factors and decrease porosity, stronger bioceramic/bioglass scaffolds can be fabricated. Zhang, et al. [30] decreased the porosity of hydroxyapatite scaffolds from 70 to 52.5% with a double slip casting technique. The bending strength increased from 9 to 73 MPa. The strength increment was remarkable. The porous structure of such strong bioceramic scaffold is shown in Fig. 22. It can be seen that the double slip casted

Fig. 21 The compressive strength versus porosity for slip-casted bioglass scaffolds. Reproduced with permission [64] *Copyright © 2015 Elsevier B.V*

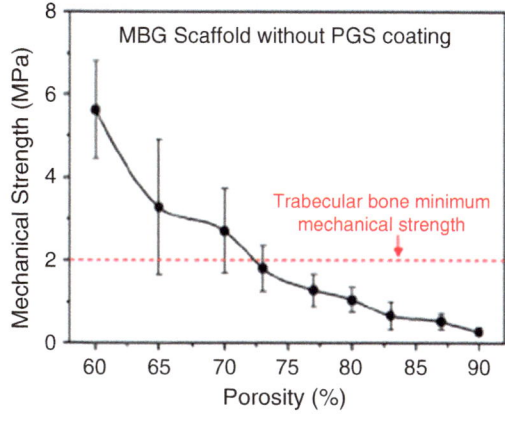

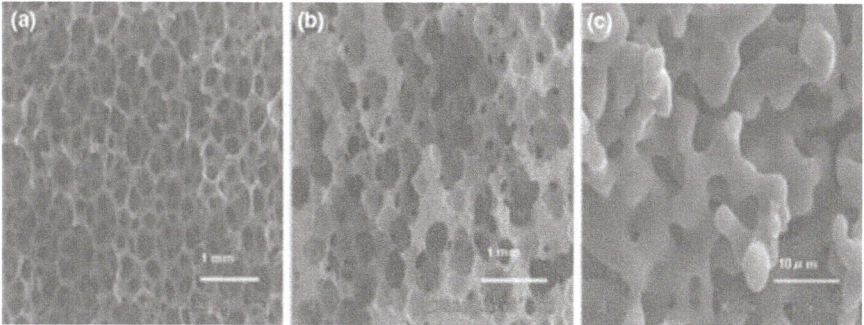

Fig. 22 SEM images showing the porous structure of HA bioceramic scaffolds fabricated by single slip casting and double slip casting. (**a**) Macropores made by single slip casting; (**b**) Macropores made by double slip casting; (**c**) Microporous HA scaffolds obtained by sintering the foam. Reproduced with permission [30] *Copyright © 2011 The American Ceramic Society*

scaffolds had unfavorable pore interconnectivity. So it is still challenging to make load-bearing scaffolds with high interconnectivity by slip casting.

3.4 Thermally Bonding of Particles

Thermally bonding of particles is also popular for manufacturing bioglass bone scaffolds. This porous structure generating method involves glassy phase micro-welding of particles, fibers or spheres at high temperatures. 10 μm–1000 μm interconnected porous structure can be achieved. By tailoring the particle size and shape, various pore diameters and mechanical properties could be achieved. The typical porous structure is shown in Fig. 23. It was reported that the bioglass scaffolds, fabricated by thermally bonding of particles, have a pore size of 50 μm–800 μm, a

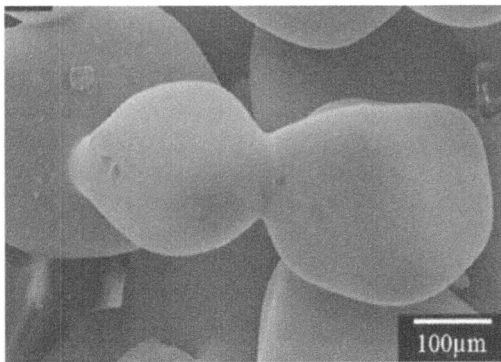

porosity of 40%–80% and a compressive strength of below 50 MPa [59, 60, 67–73]. For example, Fu, et al. [60] manufactured 13-93 bioglass scaffolds by welding the 255–325 μm particles. Scaffolds with porosities of 40–45% and pore diameters of 100–300 μm exhibited a compressive strength of around 22 MPa. Baino, et al. fabricated silicate based bioglass scaffolds by the same method [71]. A high compressive strength of 40 MPa was obtained for the scaffold with a porosity of 35% and a pore diameter of 100–300 μm. Thermally bonding of particles is able to manufacture bioglass bone scaffolds with appropriate porosity, pore size, and mechanical properties. However this method is rarely used to fabricate bioceramic bone scaffolds because the raw materials of dense ceramic particles over 100 μm are not easy to make and melting points of ceramics are significantly higher than glass.

Fig. 24 shows the compressive strength range of bioglass scaffolds fabricated by the four different methods discussed above. As can be seen, 3D printed scaffolds have the highest strength matching that of cortical bones. Freeze-casted scaffolds had the second highest strength. Scaffolds fabricated by thermally bonding particles and slip casting exhibit relatively lower strength. So 3D printing and freeze casting are the most promising methods to manufacture load-bearing bone scaffolds.

4 *In Vitro* Characterization of Load-Bearing Capacity

As mentioned in the introduction section, an ideal bone scaffold for the repair of load-bearing bone defects should possess both excellent biological properties and mechanical properties. It has been well documented that bioceramic/bioglass scaffolds are bioactive, bioresorbable, osteoconductive, and osteoinductive so that they are capable to support osteogenic cell attachment, growth, proliferation, and function *in vitro* [12, 20, 25, 26, 31, 32, 34, 40, 52, 56, 61, 65, 66, 70]. Furthermore numerous *in vivo* studies have demonstrated that bioceramic/bioglass scaffolds promote bone formation and vasculature formation in small animal models (rats and rabbits) [15, 28, 74–78] and large animal models (dogs, sheep, goats, and human)

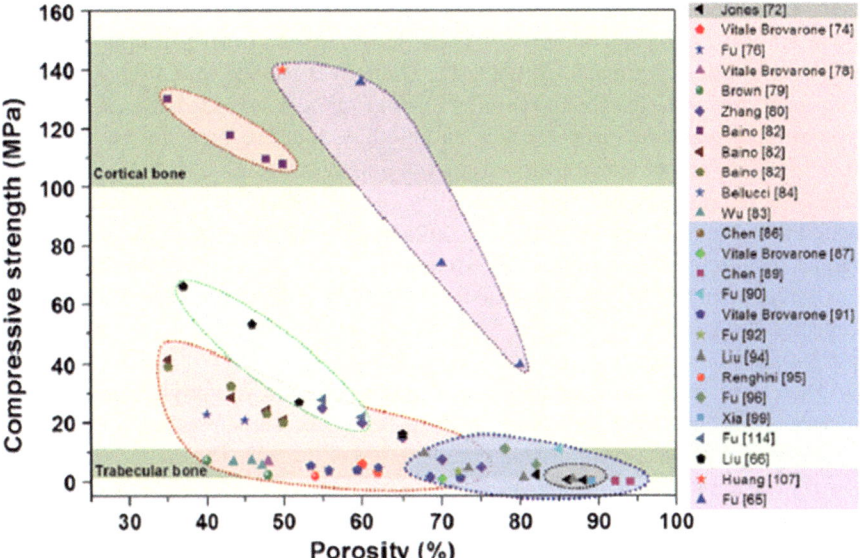

Fig. 24 The compressive strength range of bioglass scaffolds fabricated by the four different manufacturing methods discussed above. Purple area: 3D printing; Green area: freeze casting; Pink area: thermally bonding of particles; Blue area: slip casting. Reproduced with permission [59] *Copyright © 2011 Elsevier B.V*

[14, 31, 74, 79–82]. So they have been clinically used for the reconstruction of non-load-bearing bones. However their main shortcoming of limited mechanical properties makes them far away from load-bearing applications. Load-bearing capacity is especially important for bone scaffolds used to repair critical sized bone defects in load bearing bones such as the skull and the femur or in bones that experience complex deformation during function loading such as the mandible and the spine. The insufficient mechanical property of bioceramic/bioglass bone scaffolds is reported to be one of the most important obstacles for widespread clinical applications in load-bearing bone defect sites [45]. Advanced manufacturing techniques (e.g., 3D printing, freeze casting) have been developed in recent years to fabricate bone scaffolds with load-bearing capacity. A large number of studies reported the processing details, characterized the macro/microstructure, evaluated the *in vitro* biological performance, and assessed the compressive strength and elastic modulus. It should be noted that load-bearing bones experience very complex forces and functions during every day activities. Besides compressive strength and elastic modulus, load-bearing capacity refers to more crucial mechanical properties including bending strength, tensile strength, fracture toughness, weibull modulus (mechanical reliability), and fatigue resistance, which are rarely reported in the literature for bioceramic/bioglass bone scaffolds. This section will review the progress of comprehensive evaluation of load-bearing capacity and highlight the direction of future research on the assessment of mechanical properties of bone scaffolds.

Compressive strength is most commonly used to assess bioceramic/bioglass bone scaffolds because compression is the most relevant *in vivo* loading mode for implanted bone substitutes. Compressive strength is usually tested on cylinder or cubic samples with two flat and parallel loading surfaces. A wide range of compressive strength values for bioceramics and bioglass were reported in the literature. 0.5–195 MPa was obtained for porosities of 35–95% [15, 22–27]. Most of the values are in the range of sponge bones (below 20 MPa). Note that some recent studies stated that strong bone scaffolds were fabricated by advanced manufacturing techniques including 3D printing and freeze casting. Such scaffolds had a high compressive strength matching that of cortical bones indicating their immediate load-bearing capacity. With decreasing the porosity and the pore size, the compressive strength of bone scaffolds could be increased remarkably.

The elastic modulus, also known as Young's modulus, is a physical property of materials that determines the resistance to being deformed elastically (i.e., nonpermanently) under a compressive or tensile force [83, 84]. The elastic modulus is defined as the slope of its stress-strain curve (compression or tensile) within the elastic deformation region [84]. A stiffer material behaves at a higher elastic modulus. Bone scaffolds for load-bearing applications require an appropriate elastic modulus matching that of natural bones. If the elastic modulus of a bone substitute is higher than the bones, "stress shielding" will take place. In this case, the stress will be removed from the bones, which results in the bone density reduction or bone fracture non-healing. This commonly happens when a permanent bone implant is fixed in the body, for example, hip or knee prosthesis. The current bioceramic and bioglass bone scaffolds have a lower modulus than natural bones. For load-bearing applications, internal or external fixation is required. Therefore, there will be no loading or very less loading on the bone scaffolds so that the bone defects could not be regenerated properly because of no stimulus for continued bone remodeling. By Wolff's law, bone in a healthy person or animal will remodel in response to the loads it is placed under, otherwise no satisfactory bone remodeling will occur. So the bone scaffolds should be strong enough to bear loading to some extent for bone remodeling in load-bearing sites. A significant amount of studies reported the elastic modulus of current bioceramic/bioglass scaffolds. It achieved 0.35–35 GPa [43] as summarized in a recent review paper.

In addition to compressive strength and elastic modulus, bending strength is the third commonly evaluated mechanical properties for bone scaffolds. Bending strength of ceramic and glass materials is measured by three-point or four-point bending test. Four-prism shaped specimens with well-polished tensile surfaces are required for this test. The surface polishing is to illuminate the effect of the flaws and defects that can reduce the mechanical properties of flaw sensitive materials. The ASTM approved the testing method only for dense ceramic and glass materials. Bioceramic/bioglass bone scaffolds have a macroporous structure and are much weaker than their dense bulks, so it is difficult to obtain specimens with the required shape and well-polished surfaces (without flaws). Therefore only very limited studies reported bending strength values for bioceramic/bioglass scaffolds. Most of the values are in the range of sponge bones (10–20 MPa) [45]. However a much higher value of 73.6 MPa [30] was achieved by double slip casting. It is still well below those for cortical bones (135–193 MPa) [45]. It should be mentioned that international standards should be developed for testing the bending strength of porous scaffolds.

Tensile strength of ceramic and glass materials is commonly determined on dump-bell-shaped samples held in the grips of a tensile mechanical testing machine. International standards are available for dense bulks according to ASTM. But there are no testing standards for porous scaffolds. It is very complex and difficult to manufacture tensile test specimens, especially the fragile porous ceramic and glass ones. So, the indirect testing of tensile strength has been developed by diametric compression of discs and annuli. A limited number of studies reported tensile strength values of bioceramic bone scaffolds. They are in the range of sponge bones (1–5 MPa) [85–87] and well below those of cortical bones (50–151 MPa) [59]. Due to the compression loading modes for the human bones during most of the everyday activities, the tensile strength of bone scaffolds is not necessary to evaluate towards the load-bearing capacity.

Fracture toughness is a mechanical property that describes the ability of a material to resist fracture. Ductile materials, like metals, have very high fracture toughness and undergo ductile fracture. Brittle materials, like ceramics and glass, have very low fracture toughness and undergo brittle fracture. Bones sit between the ductile materials and brittle materials. Compared with ceramics and glass, bones have significantly higher fracture toughness so that they can bear sudden load at very high stress levels avoiding catastrophic failure. Bone scaffolds implanted at load-bearing sites are subjected to sudden loading from body activities so that it is essential to evaluate their fracture toughness. ASTM describes the standard method for testing the fracture toughness of ceramics and glasses. Beam-shaped specimens containing sharp notches or cracks are placed in the grips under three-point or four-point loading. However it is very difficult to test the fracture toughness of bioceramic/bioglass scaffolds because they are too fragile to machine for obtaining the specimens with required geometry and polished surfaces. Even dense bioceramics and bioglass only have a toughness of below 1 MPa $\cdot$ m$^{1/2}$ much lower than that (2–12 MPa $\cdot$ m$^{1/2}$) of cortical bones [45, 59]. The toughness of porous scaffolds is nearly zero. Due to the intrinsic brittleness, extrinsic toughening is required to make tough bone scaffolds. A related concept of fracture toughness is the work of fracture that expresses the total energy consumed to generate a unit area of fracture surface during complete fracture [88]. Because of the simplicity of the testing, work of fracture is used to assess the fracture toughness. Several studies reported the work of fracture of bioceramic/ bioglass scaffolds [45, 46, 65, 89]. International standards should be developed for testing the work of fracture of porous scaffolds since it may vary due to the difference of specimen geometry, specimen dimension, and testing conditions.

The Weibull modulus is a dimensionless parameter of the Weibull distribution that is used to evaluate the variability in measured strength (compressive, bending, and tensile) of brittle materials and map the reliability (or the probability of failure) at varying stresses [45]. As brittle materials, ceramics and glass are flaw sensitive so that their brittle failure often originates at the physical flaws (weak points) existing in the surface or the body. Therefore the maximum stress that a specimen can be measured to sustain before failure may vary from sample to sample, even under identical testing conditions. If the Weibull modulus is low, the material will have a broadly distributed strength and low mechanical reliability. The porosity has a significant effect on the reliability of bioceramics and bioglass. Some studies have employed the Weibull modulus to evaluate the mechanical reliability of bioceramic/bioglass bone scaffolds [47, 90]. The 3D printed HA and TCP bioceramic scaffolds exhibited a Weibull modulus of 3.2–9.3 based on compression tests [47, 90]. The 3D printed 13-93 bioglass scaffolds had a Weibull modulus of 6.0 and 5.3 [45] based on compression tests and three-point bending tests, respectively. The $CaO-Al_2O_3-P_2O_5$ bioglass-ceramic scaffolds possessed a Weibull modulus of 3–8 based on four-point bending tests. It is crucial and essential to assess the Weibull modulus when evaluating the load-bearing capacity of bioceramic/bioglass bone scaffolds.

Fatigue is the weakening of a material (gradually losing mechanical properties) caused by repetitive and cyclic applied loads. It is deemed to be the progressive and localized structural damage that takes place when a material is subjected to cyclic loading. If the loads are above a certain threshold, micro cracks will begin to generate at the stress concentrating areas such as the surface, persistent slip bands, and grain boundaries/interfaces [91]. Eventually a crack will grow up to a critical size, the crack will propagate suddenly, and the material will fracture. The fatigue resistance can be expressed by fatigue life. ASTM defines fatigue life as the number of stress cycles that a specimen sustains before material failure [92]. For some materials, there is a theoretical value for stress amplitude below which the material will not fail for any number of cycles, called a fatigue limit, endurance limit, or fatigue strength [93]. There are three commonly used methods to measure the fatigue life: the stress-life method, the strain-life method, and the linear-elastic fracture mechanics method [94]. One of the primary causes of bone fracture in humans is repetitive and cyclic loading of bones during daily living [95]. The gradual loss of stiffness and strength throughout cyclic loading resulting from fatigue damage accumulation is hypothesized as the fatigue failure mechanism in bones [96]. It has been demonstrated that both ceramics and glass undergo mechanical degradation under cyclic fatigue loading [97]. Therefore, besides assessment of sudden/catastrophic failure, it is essential to evaluate the fatigue resistance of bioceramic/bioglass scaffolds for load-bearing applications. Only a limited number of investigations reported fatigue behavior of bioceramic/bioglass bone scaffolds. A recent study reported a 3D printed strong bioglass-ceramic scaffold for repair of large-scale and load-bearing bone defects [39]. The scaffold achieved a compressive strength of 110 MPa and a bending strength of 30 MPa. It also behaved with high mechanical reliability. The fatigue life was up to 10^6 cycles at 1–10 MPa compressive cyclic load, and 10^5 cycles at 3–30 MPa compressive cyclic load. Another study evaluated the cyclic fatigue resistance of HA scaffolds using a diametric compression test in Hank's solution for 9×10^5 cycles [86]. The cyclic nature of *in vivo* loading modes makes it essential to assess the fatigue resistance of bone scaffolds for load-bearing applications.

Load-bearing capacity means much more than compressive strength. In addition to compressive strength, it is required to evaluate the elastic modulus, bending strength, fracture toughness, mechanical reliability, and fatigue resistance of bioceramic/bioglass bone scaffolds for load-bearing applications. Current international testing methods are only available for dense ceramic and glass materials. It is urgent to develop relevant mechanic testing methods for porous materials.

5 *In Vivo* Assessment via Load Bearing Bone Defect Model

The ultimate goal of bone tissue engineering is to repair large bone defects *in vivo* with bone scaffolds. Human bones at load-bearing sites undergo super complex loading during everyday activities. Although one can evaluate the mechanical behavior and osteogenesis of bone scaffolds, animal studies are required to assess the biomechanical performance and bone formation/remodeling. Various load-bearing bone defect models have been developed for this purpose, for example, critical sized long bone defect models (femur, tibia, and ulna) [15, 28, 79, 80], spine fusion model [14], and mandible model [98, 99]. The animals used for these models include rat, rabbit, goat, sheep, dog, pig, etc. Small animal studies (rat and rabbit) cost less, but the results may not reflect the real bone formation and bone remodeling provided by the implanted bone scaffolds, because the human body, bone structure, and regenerative ability are significantly different to small animals. Large animal studies cost much more, but the results are more convincing.

Long bone defect models are the most popular for assessment the *in vivo* bone regeneration ability of bone scaffolds under load-bearing conditions. Bi, et al. [15] investigated the repair of a critical sized rat femur defect using strong bioglass scaffolds of 13-93 and 13-93B. The results indicated that under load-bearing conditions, the bioglass scaffolds were able to regenerate bones. After a 12-week post-implantation, the new bone percentage (38%) for bioglass scaffolds was comparable with that (40%) for autografts. Jia, et al. [28] investigated bone regeneration of bioactive glass scaffolds of 13-93 and 2B6Sr in a rabbit femur segmental defect model. After 9 months, the bioglass scaffolds degraded and the femur defect was repaired with new bone formation and remodeling. Lovati, et al. [100] studied the bone formation of HA bioceramic scaffolds in sheep tibia and femur segmental defect models. The results demonstrated that the tibia is characterized by a lower bone regeneration ability compared to the femur.

The spinal fusion model is another popular load-bearing bone defect model. To achieve a stable spine fusion, the bone scaffolds must have good osteogenic, osteoconductive, and osteoinductive abilities; in addition, they should be strong enough to bear loading forces that are applied on the spine during the fusion process. Yuan, et al. [14] evaluated the *in vivo* bone formation ability of HA/TCP bioceramic scaffolds in a dog spinal fusion model. After 12-week post-implantation, the TCP scaffolds produced 20–40% new bones and generated excellent bone integration. The mandible defect model is another promising load-bearing model that can be used to assess the *in vivo* bone regeneration ability of bone scaffolds under load-bearing conditions. Recently Sun, et al. [98] established a critical-size mandibular defect model in growing pigs. It was found that 5 cm is the critical size towards mandible bone self-healing. Shah, et al. [99] developed a load-bearing critical-size composite rabbit mandible bone defect model. It was used to investigate bone regeneration, vascularization, and infection prevention in response to new biomaterial formulations. Cesarano III, et al. investigated the possibility of using strong tissue engineered bioceramic scaffolds to reconstruct the human mandible [44]. Because of the strict regulation, the scaffold did not get approval from FDA for implantation in the human body. However the study demonstrated that the bioceramic scaffold had an immediate load-bearing capacity and an anatomical 3D shape that are required for bone substitutes.

Pioneers have conducted human pilot clinical trials to evaluate the *in vivo* bone regeneration and remodeling performance of bioceramic scaffolds in load-bearing long bone defect models [79]. Four adult patients were involved in this study. 4–7 cm long bone defects in the tibia, ulna, and humerus were used. HA scaffolds were implanted in those defects with internal and external fixation. X-ray and CT results indicated that the scaffolds well integrated with the hosting bones after 5–8 months. The scaffolds were able to sustain loading forces after the fixators were removed. At 2.5 years, the complete bone scaffold integration was identified. After 5 years, the long bones were remodeled to have a tubular structure. Strong bioceramic/bioglass bone scaffolds can be fabricated by advanced manufacturing techniques (3D printing and freeze casting). As the alternative of autografts and allografts, they are promising for the repair of large load-bearing bone defects. To advance the medical translation and commercialization of load-bearing bioceramic/bioglass bone scaffolds, regulations from relevant government sectors should not be too strict. More large animal studies and human preclinical and clinical trials are required.

6   Bioinspiration Design and Future Perspectives

Human bones are strong, tough, reliable, and fatigue resistant so that they are capable to bear the complex loading forces during daily activities. In addition to bone, coral, nacre, bamboo, and animal shell also have an exceptional combination of mechanical properties although their components are quite weak. This is because such strong and tough natural materials have well-aligned hierarchical structure at nano, micro, and macro scales. For example, bone, the composite of 65 wt% carbonated hydroxyapatite (cHA) and 35 wt% collagen, has seven levels in architecture: cHA crystals and mineralized collagen fibrils at a nanometer sale; a fibril array, its corresponding array patterns and osteons at a micrometer scale; and the cortical/concellous bone and the whole bone at a macroscopic scale [45, 101]. The high fracture toughness and mechanical reliability originate from the combination of intrinsic and extrinsic toughening mechanisms. The intrinsic toughening comes from the plastic deformation of area ahead of the crack tip at the scale of under 1 μm, involving the collagen molecule uncoiling, mineralized collagen fibril sliding, and microcracking. The extrinsic toughening comes from crack deflection/twist and crack bridging at the scale of over 1 μm. Through macrostructure design (pore size, pore shape, and porosity), high strength bioceramic/bioglass bone scaffolds can be fabricated by advanced manufacturing techniques including 3D printing and freeze casting. However high fracture toughness, mechanical reliability, and fatigue resistance are also crucial for load-bearing applications. Mechanical reliability and fatigue resistance are dependent on fracture toughness to some extent. The fracture toughness of cortical bones (2–12 MPa · m$^{1/2}$) has an upper range that is much higher than the values for the bioceramics and bioglass (0.5–1 MPa · m$^{1/2}$) [45]. So increasing the fracture toughness would be the primary task to develop load-bearing bone scaffolds. Inorganic materials are intrinsically brittle so that extrinsic toughening is the main method to develop high toughness bioceramic/bioglass scaffolds. Inspired by strong and tough natural materials, scientists and engineers have made great effort to mimic the architecture to develop synthetic materials with high comprehensive mechanical properties. A recent study reported a super tough alumina ceramic with nacre-like micro "brick-and-mortar" structure manufactured by freeze casting [63]. The fracture toughness increased by 10 times, up to 22 MPa · m$^{1/2}$. The low modulus glassy phase between the alumina microplatelets serves as the polymer component in the nacre to extrinsically toughen the alumina ceramic. It is believed that through bioinspiration design, strong, tough, and reliable bioceramic/bioglass bone scaffolds could be developed for load-bearing applications.

References

1. Petite H, Viateau V, Bensaid W, Meunier A, de Pollak C, Bourguignon M, Oudina K, Sedel L, Guillemin G. Tissue-engineered bone regeneration. Nat Biotech. 2000;18(9):959–63.
2. Hollister SJ. Porous scaffold design for tissue engineering. Nat Mater. 2005;4(7):518–24.
3. Place ES, Evans ND, Stevens MM. Complexity in biomaterials for tissue engineering. Nat Mater. 2009;8(6):457–70.
4. Reddi AH. Role of morphogenetic proteins in skeletal tissue engineering and regeneration. Nat Biotech. 1998;16(3):247–52.
5. Langer R, Vacanti JP. Tissue engineering. Science. 1993;260(5110):920.
6. Griffith LG, Naughton G. Tissue engineering--current challenges and expanding opportunities. Science. 2002;295(5557):1009.
7. Bianco P, Robey PG. Stem cells in tissue engineering. Nature. 2001;414(6859):118–21.
8. Perry CR. Bone repair techniques, bone graft, and bone graft substitutes. Clin Orthop Relat Res. 1999;360:71–86.
9. Chu T-MG, Warden SJ, Turner CH, Stewart RL. Segmental bone regeneration using a load bearing biodegradable carrier of bone morphogenetic Protein-2. Biomaterials. 2007;28(3):459–67.
10. Karageorgiou V, Kaplan D. Porosity of 3D biomaterial scaffolds and osteogenesis. Biomaterials. 2005;26(27):5474–91.
11. Shen ZJ, Adolfsson E, Nygren M, Gao L, Kawaoka H, Niihara K. Dense hydroxyapatite–zirconia ceramic composites with high strength for biological applications. Adv Mater. 2001;13(3):214–6.

12. Ahn ES, Gleason NJ, Nakahira A, Ying JY. Nanostructure processing of hydroxyapatite-based bioceramics. Nano Lett. 2001;1(3):149–53.
13. Savaridas T, Wallace RJ, Dawson S, Simpson AHRW. Effect of ibandronate on bending strength and toughness of rodent cortical bone. Bone Joint Res. 2015;4(6):99.
14. Yuan H, Fernandes H, Habibovic P, de Boer J, Barradas AMC, de Ruiter A, Walsh WR, van Blitterswijk CA, de Bruijn JD. Osteoinductive ceramics as a synthetic alternative to autologous bone grafting. Proc Natl Acad Sci. 2010;107(31):13614–9.
15. Bi L, Zobell B, Liu X, Rahaman MN, Bonewald LF. Healing of critical-size segmental defects in rat femora using strong porous bioactive glass scaffolds. Mater Sci Eng C. 2014;42:816–24.
16. Mayr HO, Klehm J, Schwan S, Hube R, Südkamp NP, Niemeyer P, Salzmann G, von Eisenhardt-Rothe R, Heilmann A, Bohner M, Bernstein A. Microporous calcium phosphate ceramics as tissue engineering scaffolds for the repair of osteochondral defects: biomechanical results. Acta Biomater. 2013;9(1):4845–55.
17. Bernstein A, Niemeyer P, Salzmann G, Südkamp NP, Hube R, Klehm J, Menzel M, von Eisenhart-Rothe R, Bohner M, Görz L, Mayr HO. Microporous calcium phosphate ceramics as tissue engineering scaffolds for the repair of osteochondral defects: histological results. Acta Biomater. 2013;9(7):7490–505.
18. Zhang Y, Chen L, Zeng J, Zhou K, Zhang D. Aligned porous barium titanate/hydroxyapatite composites with high piezoelectric coefficients for bone tissue engineering. Mater Sci Eng C. 2014;39:143–9.
19. Sprio S, Tampieri A, Celotti G, Landi E. Development of hydroxyapatite/calcium silicate composites addressed to the design of load-bearing bone scaffolds. J Mech Behav Biomed Mater. 2009;2(2):147–55.
20. Xu S, Yang X, Chen X, Shao H, He Y, Zhang L, Yang G, Gou Z. Effect of borosilicate glass on the mechanical and biodegradation properties of 45S5-derived bioactive glass-ceramics. J Non-Cryst Solids. 2014;405:91–9.
21. Yang J, Hu X, Huang J, Chen K, Huang Z, Liu Y, Fang M, Sun X. Novel porous calcium aluminate/phosphate nanocomposites: in situ synthesis, microstructure and permeability. Nanoscale. 2016;8(6):3599–606.
22. Deville S, Saiz E, Tomsia AP. Freeze casting of hydroxyapatite scaffolds for bone tissue engineering. Biomaterials. 2006;27(32):5480–9.
23. Deville S. Freeze-casting of porous ceramics: a review of current achievements and issues. Adv Eng Mater. 2008;10(3):155–69.
24. Huang TS, Rahaman MN, Doiphode ND, Leu MC, Bal BS, Day DE, Liu X. Porous and strong bioactive glass (13-93) scaffolds fabricated by freeze extrusion technique. Mater Sci Eng C. 2011;31(7):1482–9.
25. Zhu H, Zhai D, Lin C, Zhang Y, Huan Z, Chang J, Wu C. 3D plotting of highly uniform Sr5(PO4)2SiO4 bioceramic scaffolds for bone tissue engineering. J Mater Chem B. 2016;4(37):6200–12.
26. Wu C, Fan W, Zhou Y, Luo Y, Gelinsky M, Chang J, Xiao Y. 3D-printing of highly uniform CaSiO3 ceramic scaffolds: preparation, characterization and in vivo osteogenesis. J Mater Chem. 2012;22(24):12288–95.
27. Xu M, Zhai D, Xia L, Li H, Chen S, Fang B, Chang J, Wu C. Hierarchical bioceramic scaffolds with 3D-plotted macropores and mussel-inspired surface nanolayers for stimulating osteogenesis. Nanoscale. 2016;8(28):13790–803.
28. Jia W, Lau GY, Huang W, Zhang C, Tomsia AP, Fu Q. Bioactive glass for large bone repair. Adv Healthc Mater. 2015;4(18):2842–8.
29. Fu Q, Saiz E, Tomsia AP. Direct ink writing of highly porous and strong glass scaffolds for load-bearing bone defects repair and regeneration. Acta Biomater. 2011;7(10):3547–54.
30. Zhang Y, Kong D, Yokogawa Y, Feng X, Tao Y, Qiu T. Fabrication of porous hydroxyapatite ceramic scaffolds with high flexural strength through the double slip-casting method using fine powders. J Am Ceram Soc. 2012;95(1):147–52.

31. Crovace MC, Souza MT, Chinaglia CR, Peitl O, Zanotto ED. Biosilicate® — a multipurpose, highly bioactive glass-ceramic. In vitro, in vivo and clinical trials. J Non-Cryst Solids. 2016;432(Part A):90–110.
32. Shuai C, Li P, Liu J, Peng S. Optimization of TCP/HAP ratio for better properties of calcium phosphate scaffold via selective laser sintering. Mater Charact. 2013;77:23–31.
33. Feng P, Niu M, Gao C, Peng S, Shuai C. A novel two-step sintering for nano-hydroxyapatite scaffolds for bone tissue engineering. Sci Rep. 2014;4:5599.
34. Feng P, Wei P, Li P, Gao C, Shuai C, Peng S. Calcium silicate ceramic scaffolds toughened with hydroxyapatite whiskers for bone tissue engineering. Mater Charact. 2014;97:47–56.
35. Emadi R, Tavangarian F, Esfahani SIR. Biodegradable and bioactive properties of a novel bone scaffold coated with nanocrystalline bioactive glass for bone tissue engineering. Mater Lett. 2010;64(13):1528–31.
36. Gmeiner R, Mitteramskogler G, Stampfl J, Boccaccini AR. Stereolithographic ceramic manufacturing of high strength bioactive glass. Int J Appl Ceram Technol. 2015;12(1):38–45.
37. Fu Q, Rahaman MN, Bal BS, Brown RF. Preparation and in vitro evaluation of bioactive glass (13-93) scaffolds with oriented microstructures for repair and regeneration of load-bearing bones. J Biomed Mater Res A. 2010;93A(4):1380–90.
38. Fu Q, Saiz E, Tomsia AP. Bioinspired strong and highly porous glass scaffolds. Adv Funct Mater. 2011;21(6):1058–63.
39. Roohani-Esfahani S-I, Newman P, Zreiqat H. Design and fabrication of 3D printed scaffolds with a mechanical strength comparable to cortical bone to repair large bone defects. Sci Rep. 2016;6:19468.
40. Zhang W, Feng C, Yang G, Li G, Ding X, Wang S, Dou Y, Zhang Z, Chang J, Wu C, Jiang X. 3D-printed scaffolds with synergistic effect of hollow-pipe structure and bioactive ions for vascularized bone regeneration. Biomaterials. 2017;135:85–95.
41. Seitz H, Rieder W, Irsen S, Leukers B, Tille C. Three-dimensional printing of porous ceramic scaffolds for bone tissue engineering. J Biomed Mater Res B Appl Biomater. 2005;74B(2):782–8.
42. Miguel C, Marta D, Elke V, Uwe G, Paulo F, Inês P, Barbara G, Henrique A, Eduardo P, Jorge R. Application of a 3D printed customized implant for canine cruciate ligament treatment by tibial tuberosity advancement. Biofabrication. 2014;6(2):025005.
43. Bose S, Vahabzadeh S, Bandyopadhyay A. Bone tissue engineering using 3D printing. Mater Today. 2013;16(12):496–504.
44. Cesarano J, Dellinger JG, Saavedra MP, Gill DD, Jamison RD, Grosser BA, Sinn-Hanlon JM, Goldwasser MS. Customization of load-bearing hydroxyapatite lattice scaffolds. Int J Appl Ceram Technol. 2005;2(3):212–20.
45. Fu Q, Saiz E, Rahaman MN, Tomsia AP. Toward strong and tough glass and ceramic scaffolds for bone repair. Adv Funct Mater. 2013;23(44):5461–76.
46. Bretcanu O, Misra S, Roy I, Renghini C, Fiori F, Boccaccini AR, Salih V. In vitro biocompatibility of 45S5 bioglass®-derived glass–ceramic scaffolds coated with poly(3-hydroxybutyrate). J Tissue Eng Regen Med. 2009;3(2):139–48.
47. Miranda P, Pajares A, Saiz E, Tomsia AP, Guiberteau F. Mechanical properties of calcium phosphate scaffolds fabricated by robocasting. J Biomed Mater Res A. 2008;85A(1):218–27.
48. Franco J, Hunger P, Launey ME, Tomsia AP, Saiz E. Direct write assembly of calcium phosphate scaffolds using a water-based hydrogel. Acta Biomater. 2010;6(1):218–28.
49. Wu C, Luo Y, Cuniberti G, Xiao Y, Gelinsky M. Three-dimensional printing of hierarchical and tough mesoporous bioactive glass scaffolds with a controllable pore architecture, excellent mechanical strength and mineralization ability. Acta Biomater. 2011;7(6):2644–50.
50. Zhang J, Zhao S, Zhu Y, Huang Y, Zhu M, Tao C, Zhang C. Three-dimensional printing of strontium-containing mesoporous bioactive glass scaffolds for bone regeneration. Acta Biomater. 2014;10(5):2269–81.

51. Butscher A, Bohner M, Doebelin N, Hofmann S, Müller R. New depowdering-friendly designs for three-dimensional printing of calcium phosphate bone substitutes. Acta Biomater. 2013;9(11):9149–58.
52. Zocca A, Elsayed H, Bernardo E, Gomes CM, Lopez-Heredia MA, Knabe C, Colombo P, Günster J. 3D-printed silicate porous bioceramics using a non-sacrificial preceramic polymer binder. Biofabrication. 2015;7(2):025008.
53. Seitz H, Deisinger U, Leukers B, Detsch R, Ziegler G. Different calcium phosphate granules for 3-D printing of bone tissue engineering scaffolds. Adv Eng Mater. 2009;11(5):B41–6.
54. Baino F, Vitale-Brovarone C. Three-dimensional glass-derived scaffolds for bone tissue engineering: current trends and forecasts for the future. J Biomed Mater Res A. 2011;97A(4):514–35.
55. Wilson CE, de Bruijn JD, van Blitterswijk CA, Verbout AJ, Dhert WJA. Design and fabrication of standardized hydroxyapatite scaffolds with a defined macro-architecture by rapid prototyping for bone-tissue-engineering research. J Biomed Mater Res A. 2004;68A(1):123–32.
56. Taboas JM, Maddox RD, Krebsbach PH, Hollister SJ. Indirect solid free form fabrication of local and global porous, biomimetic and composite 3D polymer-ceramic scaffolds. Biomaterials. 2003;24(1):181–94.
57. Seol Y-J, Park DY, Park JY, Kim SW, Park SJ, Cho D-W. A new method of fabricating robust freeform 3D ceramic scaffolds for bone tissue regeneration. Biotechnol Bioeng. 2013;110(5):1444–55.
58. Weiguo B, Dichen L, Qin L, Weijie Z, Linzhong Z, Xiang L, Zhongmin J. Design and fabrication of a novel porous implant with pre-set channels based on ceramic stereolithography for vascular implantation. Biofabrication. 2011;3(3):034103.
59. Fu Q, Saiz E, Rahaman MN, Tomsia AP. Bioactive glass scaffolds for bone tissue engineering: state of the art and future perspectives. Mater Sci Eng C. 2011;31(7):1245–56.
60. Fu Q, Rahaman MN, Bal BS, Huang W, Day DE. Preparation and bioactive characteristics of a porous 13-93 glass, and fabrication into the articulating surface of a proximal tibia. J Biomed Mater Res A. 2007;82A(1):222–9.
61. Deliormanlı AM, Rahaman MN. Direct-write assembly of silicate and borate bioactive glass scaffolds for bone repair. J Eur Ceram Soc. 2012;32(14):3637–46.
62. Deville S, Saiz E, Nalla RK, Tomsia AP. Freezing as a path to build complex composites. Science. 2006;311(5760):515.
63. Bouville F, Maire E, Meille S, Van de Moortèle B, Stevenson AJ, Deville S. Strong, tough and stiff bioinspired ceramics from brittle constituents. Nat Mater. 2014;13(5):508–14.
64. Lin D, Yang K, Tang W, Liu Y, Yuan Y, Liu C. A poly(glycerol sebacate)-coated mesoporous bioactive glass scaffold with adjustable mechanical strength, degradation rate, controlled-release and cell behavior for bone tissue engineering. Colloids Surf B: Biointerfaces. 2015;131:1–11.
65. Chen QZ, Thompson ID, Boccaccini AR. 45S5 bioglass®-derived glass–ceramic scaffolds for bone tissue engineering. Biomaterials. 2006;27(11):2414–25.
66. Chen QZ, Efthymiou A, Salih V, Boccaccini AR. Bioglass®-derived glass–ceramic scaffolds: study of cell proliferation and scaffold degradation in vitro. J Biomed Mater Res A. 2008;84A(4):1049–60.
67. Vitale-Brovarone C, Nunzio SD, Bretcanu O, Verné E. Macroporous glass-ceramic materials with bioactive properties. J Mater Sci Mater Med. 2004;15(3):209–17.
68. Vitale-Brovarone C, Verné E, Robiglio L, Martinasso G, Canuto RA, Muzio G. Biocompatible glass–ceramic materials for bone substitution. J Mater Sci Mater Med. 2008;19(1):471–8.
69. Fu Q, Rahaman MN, Sonny Bal B, Brown RF, Day DE. Mechanical and in vitro performance of 13-93 bioactive glass scaffolds prepared by a polymer foam replication technique. Acta Biomater. 2008;4(6):1854–64.
70. Hua Z, Xiao-Jian Y, Jia-Shun L. Preparation and biocompatibility evaluation of apatite/wollastonite-derived porous bioactive glass ceramic scaffolds. Biomed Mater. 2009;4(4):045007.

71. Baino F, Verné E, Vitale-Brovarone C. 3-D high-strength glass–ceramic scaffolds containing fluoroapatite for load-bearing bone portions replacement. Mater Sci Eng C. 2009;29(6):2055–62.
72. Wu S-C, Hsu H-C, Hsiao S-H, Ho W-F. Preparation of porous 45S5 bioglass®-derived glass–ceramic scaffolds by using rice husk as a porogen additive. J Mater Sci Mater Med. 2009;20(6):1229–36.
73. Bellucci D, Cannillo V, Ciardelli G, Gentile P, Sola A. Potassium based bioactive glass for bone tissue engineering. Ceram Int. 2010;36(8):2449–53.
74. Huang Y, Jin X, Zhang X, Sun H, Tu J, Tang T, Chang J, Dai K. In vitro and in vivo evaluation of akermanite bioceramics for bone regeneration. Biomaterials. 2009;30(28):5041–8.
75. Jamshidi Adegani F, Langroudi L, Ardeshirylajimi A, Dinarvand P, Dodel M, Doostmohammadi A, Rahimian A, Zohrabi P, Seyedjafari E, Soleimani M. Coating of electrospun poly(lactic-co-glycolic acid) nanofibers with willemite bioceramic: improvement of bone reconstruction in rat model. Cell Biol Int. 2014;38(11):1271–9.
76. Xin R, Leng Y, Chen J, Zhang Q. A comparative study of calcium phosphate formation on bioceramics in vitro and in vivo. Biomaterials. 2005;26(33):6477–86.
77. Vogel M, Voigt C, Gross UM, Müller-Mai CM. In vivo comparison of bioactive glass particles in rabbits. Biomaterials. 2001;22(4):357–62.
78. Xu S, Lin K, Wang Z, Chang J, Wang L, Lu J, Ning C. Reconstruction of calvarial defect of rabbits using porous calcium silicate bioactive ceramics. Biomaterials. 2008;29(17):2588–96.
79. Maurilio Marcacci EK, Moukhachev V, Lavroukov A, Kutepov S, Quarto R, Mastrogiacomo M, Cancedda R. Stem cells associated with macroporous bioceramics for long bone repair: 6- to 7-year outcome of a pilot clinical study. Tissue Eng. 2007;13(5):947–55.
80. Mastrogiacomo M, Scaglione S, Martinetti R, Dolcini L, Beltrame F, Cancedda R, Quarto R. Role of scaffold internal structure on in vivo bone formation in macroporous calcium phosphate bioceramics. Biomaterials. 2006;27(17):3230–7.
81. Moura J, Teixeira LN, Ravagnani C, Peitl O, Zanotto ED, Beloti MM, Panzeri H, Rosa AL, de Oliveira PT. In vitro osteogenesis on a highly bioactive glass-ceramic (biosilicate®). J Biomed Mater Res A. 2007;82A(3):545–57.
82. Nandi SK, Kundu B, Datta S, De DK, Basu D. The repair of segmental bone defects with porous bioglass: an experimental study in goat. Res Vet Sci. 2009;86(1):162–73.
83. Ferdinand P, Beer ERJ, Dewolf J, Mazurek D. Mechanics of materials. New York: McGraw Hill; 2009.
84. Askeland DR, Phule PP. The science and engineering of materials. 5th ed. Boston, MA: Cengage learning; 2006.
85. Almirall A, Larrecq G, Delgado JA, Martınez S, Planell JA, Ginebra MP. Fabrication of low temperature macroporous hydroxyapatite scaffolds by foaming and hydrolysis of an α-TCP paste. Biomaterials. 2004;25(17):3671–80.
86. Ding S-J, Wang C-W, Chen DC-H, Chang H-C. In vitro degradation behavior of porous calcium phosphates under diametral compression loading. Ceram Int. 2005;31(5):691–6.
87. Charrière E, Lemaitre J, Zysset P. Hydroxyapatite cement scaffolds with controlled macroporosity: fabrication protocol and mechanical properties. Biomaterials. 2003;24(5):809–17.
88. Barinov SM, Sakai M. The work-of-fracture of brittle materials: principle, determination, and applications. J Mater Res. 2011;9(6):1412–25.
89. Peroglio M, Gremillard L, Gauthier C, Chazeau L, Verrier S, Alini M, Chevalier J. Mechanical properties and cytocompatibility of poly(ε-caprolactone)-infiltrated biphasic calcium phosphate scaffolds with bimodal pore distribution. Acta Biomater. 2010;6(11):4369–79.
90. Pernot F, Etienne P, Boschet F, Datas L. Weibull parameters and the tensile strength of porous phosphate glass-ceramics. J Am Ceram Soc. 1999;82(3):641–8.
91. Kim WH, Laird C. Crack nucleation and stage I propagation in high strain fatigue— II. Mechanism. Acta Metall. 1978;26(5):789–99.
92. Ralph I, Stephens AF, Robert R, Stephens HO. Fuchs, metal fatigue in engineering. 2nd ed. Hoboken, NJ: John Wiley & Sons, Inc.; 2001.

93. Bathias. There is no infinite fatigue life in metallic materials. Fatigue & Fracture of Engineering Materials & Structures. 1999;22(7):559–65.
94. Shigley JE, Mitchell LD. Mechanical engineering design. 7th ed. New York: McGraw Hill Higher Education; 2003.
95. Riggs BL, Melton LJ. The worldwide problem of osteoporosis: insights afforded by epidemiology. Bone. 1995;17(5):S505–11.
96. Choi K, Goldstein SA. A comparison of the fatigue behavior of human trabecular and cortical bone tissue. J Biomech. 1992;25(12):1371–81.
97. Ritchie RO, Dauskardt RH. Cyclic fatigue of ceramics. J Ceram Soc Jpn. 1991;99(1154):1047–62.
98. Sun Z, Kennedy KS, Tee BC, Damron JB, Allen MJ. Establishing a critical-size mandibular defect model in growing pigs: characterization of spontaneous healing. J Oral Maxillofac Surg. 2014;72(9):1852–68.
99. Shah SR, Young S, Goldman JL, Jansen JA, Wong ME, Mikos AG. A composite critical-size rabbit mandibular defect for evaluation of craniofacial tissue regeneration. Nat Protocols. 2016;11(10):1989–2009.
100. Lovati AB, Lopa S, Recordati C, Talò G, Turrisi C, Bottagisio M, Losa M, Scanziani E, Moretti M. In vivo bone formation within engineered hydroxyapatite scaffolds in a sheep model. Calcif Tissue Int. 2016;99(2):209–23.
101. Launey ME, Buehler MJ, Ritchie RO. On the mechanistic origins of toughness in bone. Annu Rev Mater Res. 2010;40(1):25–53.

Index

© Springer International Publishing AG, part of Springer Nature 2018
B. Li, T. Webster (eds.), *Orthopedic Biomaterials*,
https://doi.org/10.1007/978-3-319-89542-0

Printed by Printforce, the Netherlands